Emerging Topics in
Computer Vision
and its Applications

Series in Computer Vision - Vol. 1

Emerging Topics in
Computer Vision
and its Applications

Editor

C H Chen

University of Massachusetts Dartmouth, USA

 **World Scientific**

NEW JERSEY · LONDON · SINGAPORE · BEIJING · SHANGHAI · HONG KONG · TAIPEI · CHENNAI

Published by

World Scientific Publishing Co. Pte. Ltd.

5 Toh Tuck Link, Singapore 596224

USA office: 27 Warren Street, Suite 401-402, Hackensack, NJ 07601

UK office: 57 Shelton Street, Covent Garden, London WC2H 9HE

British Library Cataloguing-in-Publication Data
A catalogue record for this book is available from the British Library.

EMERGING TOPICS IN COMPUTER VISION AND ITS APPLICATIONS
Series in Computer Vision — Vol. 1

Copyright © 2012 by World Scientific Publishing Co. Pte. Ltd.

ISBN-13 978-981-4340-99-1
ISBN-10 981-4340-99-5

Printed in Singapore by World Scientific Printers.

PREFACE

From my recollection, computer vision was in early stage of development in the 60's and 70's and there was very limited textbook choice when I first taught computer vision in the mid-70s. Image processing was included as part of computer vision course at that time. The late Prof. A. Rosenfeld was instrumental to the early progress in computer vision and image processing. The influential work on human representation and processing of visual information by David Marr and others have had profound impact later on computer vision. For example texture was found to be very useful for both human and computer vision. There have been rapid progress in both computer vision and image processing in the last forty years. Computer vision and image processing are now two separate courses in most university curriculums. Similar advances were seen in the industry at the same time and computer vision especially is responsible for the advanced automation in the recent years. During the 80's there was much interest in using computer vision in manufacturing and during the 90's the need for robotic vision was greatly increased. The progress in the computer vision applications in fact far exceeds the theoretical advances.

Though currently there are many nice books on computer vision, there is always a need for another one as no single book can capture the major advances in computer vision. The book is much influenced by the well prepared book, Emerging Topics in Computer Vision, edited by Gerard Medioni and Sing Bing, published by Prentice-Hall 2005. The book in particular seeks to provide a balanced coverage of both theory and applications of computer vision, with greater emphasis on modern applications.

The book has three parts: Part 1 is emerging theory and methodology (algorithms) with 8 chapters. Part 2 is emerging applications of computer vision and image processing with 8 chapters, and Part 3 is on modern vision systems and technology with 7 chapters. Some chapters may partially belong to other parts. On the theoretical side, much emphasis is placed on 3-D modeling and analysis and on the application side, the book covers areas like biomedicine, surveillance, human recognition, fish identification, defect inspection, driver assistance, etc. The book's objective is to capture the emerging advances and major results on computer vision so that readers will be able to read in a single and comprehensive volume the modern development, and progress in computer vision.

The first chapter of Part I deals with human post estimation and tracking from video sequence, which has continued to be a challenging problem in computer vision. The second chapter deals with a new approach which reconstructs the high resolution image from its low resolution ones with special reference for face images. The use of graphs and graphical models has been increasingly important in recent years for clustering and recognition. Chapters 3 and 4 present two emerging and powerful approaches with use of the graphs, one being based on a vectorial description of the graph and the other using the time-sliced probabilistic graphical models. Chapter 5 presents a new algorithm to estimate the background for images that are taken by freely moving camera, or images taken from different viewpoints of the same scene. Chapter 6 investigates the relationship between image deformation and their effects on image moments, based on which a registration algorithm is presented that is both computationally efficient and fairly robust to noise. Taking a radical departure from existing approaches, the non-parametric sample-based framework presented in Chapter 7 aims to model image data as a lattice of nonparametric conditional probability distributions estimated via random sampling. The main advantage of the approach is that significantly improved structural preservation at different scales can be achieved for computer vision tasks such as edge detection and image segmentation. Chapter 8 addresses another basic problem which is to determine correspondence of feature-points in a sequence of images Based on the probabilistic framework, the chapter proposes an algorithm for feature-point correspondence that works under relaxed statistical assumptions and demonstrates the effectiveness of the approach by applying it to local flow estimation and tracking.

Chapter 1 of Part II presents a powerful Fourier shape modeling method, called spherical harmonic (SPHARM) method, for processing arbitrarily shaped but simply connected 3D surface data. Among applications of the approach presented is combining shape analysis of brain's substructure with genetic analysis to identify a neurodegeneration biomarker for Alhzheimer's disease. While much of face recognition effort is on faces in visible light, Chapter 2 discusses specifically the eye glass problem in near infrared face recognition. Chapter 3 demonstrates the significant advantage of using phase-only information to detect defect in vision based surface inspection, without prior knowledge of defects. By exploiting some unique advantages of cellular automata (CA), Chapter 4 examines the use of CA in binary and gray scaled image denoising and edge detection. Chapters 5 and 6 present respectively results on computer vision based marine fishery identification and vehicle detection from satellite imagery. One important difficulty in fingerprint classification, as well as other image classification problems, involves extracting representative features that are invariant to translation and rotation. Chapter 7 shows how the Polar Harmonic Transforms has such a desirable property and offers comparable classification performance with the state-of-the art methods using the NIST special fingerprint data base. The final chapter, Chapter 8, of Part II deals with a unique problem in computer vision, which is to detect many-to-many object correspondences across the original multiple images in an unsupervised manner. The method presented can detect and segment identical objects directly from a single image or a handful of images, with the use of object correspondence networks that connect matching objects.

Going back to the 3D world, Chapter 1 of Part III investigates the use of constraints that can help retrieve the 3D metric structure of a scene from two images. In video surveillance, multiple cameras are often used. The problem of how these cameras cooperate with one another such that all existing persons can be followed with an optimal group of cameras is addressed in Chapter 2 by using auction mechanism for camera network. Chapter 3 presents a novel photogeometric framework consisting of a family of methods using off-the-shelf projectors and cameras. The photometric and geometric information is concurrently captured in order to produce a photogeometric modeling system. In Chapter 4, the authors concentrate on image matching which is an essential part of any generic multi-view reconstruction system, with special focus on a particular correspondence growing method called quasi-dense wide baseline matching. The quasi-dense approach can be used to match parts of views of arbitrary and possibly deforming scenes. Chapter 5 investigates a single view traffic sign detection and recognition system with a multi-object, model-based 3D tracker for driver assistance. Chapter 6 examines the problem of extrinsic self-calibration of a multi-camera system. During such a self-calibration the cameras estimate their position and orientation in a common global coordinate system only from the images they record. Robots should be able to localize themselves in order to navigate in the environment, compute a path to a target destination, and recognize that the target destination has been reached. Chapter 7 deals with such robot localization problem by examining issues such as data association, and motion estimation with experimental results using publically available data base.

The very short description of chapters given above is highly incomplete. It is best to let each chapter speaks for itself. The readers can then get a good understanding of the scope, progress and complexity of computer vision problems.

I am grateful to all authors for their important contributions to this book. It is my honor to work with them on this book project. The book is the first of the Series in Computer Vision launched by World Scientific Publishing. The readers may also find some useful chapters in computer vision in the Handbook of Pattern Recognition and Computer Vision, vol. 4, 2010 also published by World Scientific Publishing. We certainly welcome feedbacks from the readers.

May, 2011   C.H. Chen

CONTENTS

Part 1

Emerging Theory and Methodologies in Computer Vision

CHAPTER 1.1

ON THE ESTIMATION OF 3D HUMAN BODY MODELS AND POSE FROM MULTIPLE CAMERAS

Aravind Sundaresan[1] and Rama Chellappa[2]

[1] *Artificial Intelligence Center,*
SRI International, Menlo Park, CA 94025
aravind@ai.sri.com

[2] *Center for Automation Research,*
University of Maryland, College Park, MD 20740

We present a completely automatic algorithm for initializing and tracking the articulated motion of humans using image sequences obtained from multiple cameras. We discuss the challenges in solving this problem and compare our work to some of the state of the art techniques today. We use a detailed articulated human body model composed of sixteen rigid segments that allows both translation and rotation at joints. Voxel data of the subject obtained from the images is segmented into the different articulated chains using Laplacian Eigenmaps. The segmented chains are registered in a subset of the frames using a single-frame registration technique and subsequently used to initialize the pose in the sequence. A temporal registration method is then used to identify the partially segmented or unregistered articulated chains in the remaining frames in the sequence. The tracker uses motion cues such as pixel displacement as well as 2D and 3D shape cues such as silhouettes, motion residues and skeleton curves. The use of complementary cues in the tracking algorithm alleviates the twin problems of drift and convergence to incorrect solutions. The use of multiple cameras also allows us to deal with the problems due to self-occlusion and kinematic singularity. We present tracking results on sequences with different kinds of motion to illustrate the effectiveness of our approach.

1. Introduction

Human pose estimation and tracking, or motion capture, has important applications in a number of fields and is a research area that has made rapid progress in the last decade. Current techniques typically use marker-based techniques, which involve the placement of markers on the body of the subject and capturing the movement of the subject using a set of specialized cameras. The use of *markerless* techniques eliminates the need for the specialized equipment as well as the expertise and time required to place the markers. It can also estimate the pose using anatomically appropriate human body models rather than a set of markers. Motion capture has applications in diverse fields ranging from human motion analysis in clinical studies and sports medicine, to animation in the motion picture and video game industries, and human-computer interaction. In particular, the ability to perform markerless motion capture in real-time has enormous potential in human-computer interaction which has already led to commercial applications in the video game industry such as the Kinect [Microsoft Corporation, 2010].

The typical steps in motion capture [Badler et al., 1993] are (1) model estimation, (2) pose initialization and (3) tracking. Model estimation is the process of estimating the parameters of the human body model such as the shape of the body segments and their articulated structure. Pose initialization refers to the estimation of the pose given a *single* frame.[1] Pose tracking refers to the estimation of a pose in the next frame, given the pose in the current frame. Both (2) and (3) perform pose estimation, but the methods employed are usually different and we list them separately. Different applications have different requirements in terms of the speed of processing and the accuracy of the algorithm. Accordingly pose estimation has been performed with different kinds and number of sensors. This problem has received much attention in the computer vision literature in both the monocular [Ramanan and Forsyth, 2003; Wachter and Nagel, 1999; Sidenbladh et al., 2000; Ju et al., 1996] and multiple camera cases [Chu et al., 2003; Mikić et al., 2003; Mündermann et al., 2007; Sundaresan and Chellappa, 2008; Cheung et al., 2003]. A survey of a number of important pose estimation methods developed in the past decade may be found in [Gavrila, 1999; Moeslund and Granum, 2001; Sigal and Black, 2006]. More recently, a synchronized video and motion capture database and baseline algorithm for human pose estimation has been proposed [Sigal et al., 2010]. Recent developments in 3D sensors such as the Microsoft Kinect [Microsoft Corporation, 2010] and Time-of-flight sensors have provided rich image and depth data, which have led to many algorithms which attain 200 frames per second [Ganapathi et al., 2010; Siddiqui and Medioni., 2010; Shotton et al., 2011]. The availability of multi-modal sensors has also led to methods that fuse the raw data to perform articulated pose estimation [Knoop et al., 2009].

The articulated structure of the human body which is composed of a number of segments, each with its associated shape and pose, makes human pose estimation a challenging task. The complexity of the human body and the range of poses it can assume necessitate the use of a detailed model in order to represent its pose. Body models typically incorporate both the shape of individual body parts and structural aspects such as the articulated connectivity and joint locations of the human body. A common problem faced in image-based methods is that some parts of the body often occlude other parts. It is also difficult to perceive and estimate motion in the

---

[1]By frame, we refer to image(s) obtained at a single time instant; it is one image in the monocular case and a set of images in the multi-camera case.

direction perpendicular to the image plane when using a single camera [Morris and Rehg, 1998], a problem known as "kinematic singularity". Monocular techniques suffer from the above problems of self-occlusion and "kinematic singularities" and multiple or multi-modal cameras are required to estimate pose in a robust and accurate manner. We therefore focus on multi-camera capture techniques in this chapter.

The organization of this chapter is as follows. We present an overview of our algorithm, the challenges faced, and related work in Section 2. We describe our segmentation algorithm that segments the voxel data into different human body parts and estimates the human body model and pose in Section 3. The tracking algorithm which uses multiple cues is briefly described in Section 4. The chapter ends with a summary and concluding remarks in Section 5.

2. Algorithm overview and related work

In this section we briefly describe a number of human body pose estimation algorithms that use different kinds and numbers of sensors and are targeted at different applications. We also list some complete markerless motion capture systems and present an overview of our motion capture system in the context of the current work in human body pose estimation.

2.1. *Related work*

There have been a number of articles presenting a survey of human motion tracking and analysis methods [Gavrila, 1999; Aggarwal and Cai, 1999; Moeslund and Granum, 2001; Sigal and Black, 2006]. Recent depth and 3D sensors provide the ability to extract 3D information from one sensor pair and lead to far more accurate pose estimation algorithms when compared to monocular methods. We focus on some of the important multi-camera and multi-modal techniques followed by a brief discussion of their strengths and limitations. Many of the multi-camera methods were extended from monocular techniques.

Multi-camera methods can also be broadly classified as shape-based and motion-based. Shape-based methods use 2D shape cues such as silhouettes or edges [Kakadiaris and Metaxas, 2000; Delamarre and Faugeras, 1999; Moeslund and Granum, 2000] or 3D shape cues such as voxels [Chu et al., 2003; Mikić et al., 2003; Mündermann et al., 2007; Sundaresan and Chellappa, 2008]. The voxel representation of a person provides cues about the 3D shape of the person and is often used in pose estimation algorithms. Motion-based methods [Yamamoto and Koshikawa, 1991; Yamamoto et al., 1998; Bregler and Malik, 1998] typically use optical flow in the images to perform tracking. The motion-based methods estimate the change in pose and typically assume that the initial pose is available. On the other hand, shape-based methods use absolute cues and can be used to both initialize the pose given a single frame [Chu et al., 2003; Mikić et al., 2003; Mündermann et al., 2007; Sundaresan and Chellappa, 2008], or perform tracking [Delamarre and Faugeras, 1999; Moeslund and Granum, 2000; Sigal et al., 2003, 2004].

Mikić *et al.* [Mikić et al., 2003] and Mündermann *et al.* [Mündermann et al., 2007] perform all the steps in the motion capture using voxel based techniques. They are however limited by the shortcomings of shape-based methods and in the case of [Mündermann et al., 2007], the model is not obtained automatically. Chu *et al.* [Chu et al., 2003] use volume data to acquire and track a human body model and Cheung *et al.* [Cheung et al., 2003] use shapes from silhouette to estimate human body kinematics. However, in [Chu et al., 2003] no tracking is performed, while in [Cheung et al., 2003], the subject is required to articulate one joint at a time in order to initialize the pose. Anguelov *et al.* [Anguelov et al., 2004] describe an algorithm that automatically decomposes an object into approximately rigid parts and obtains their location and underlying articulated structure, given a set of meshes describing the object in different poses. They use an unsupervised non-rigid technique to register the meshes and perform segmentation using the EM algorithm.

The following techniques assume that an initial pose estimate is available and perform tracking using shape and motion cues. [Yamamoto et al., 1998] track human motion using multiple cameras and optical flow. [Bregler and Malik, 1998] also use optical flow and an orthographic camera model. [Gavrila and Davis, 1996] discuss a multi-view approach for 3D model-based tracking of humans in action. They use a generate-and-test algorithm in which they search for poses in a parameter space and match them using a variant of Chamfer matching. [Kakadiaris and Metaxas, 2000] use silhouettes from multiple cameras to estimate 3D motion. [Theobalt et al., 2004] project the texture of the model obtained from silhouette-based methods and refine the pose using the flow field. [Delamarre and Faugeras, 1999] use 3D articulated models for tracking with silhouettes. They use silhouette contours and apply forces to the contours obtained from the projection of the 3D model so that they move towards the silhouette contours obtained from multiple images. [Moeslund and Granum, 2000] perform model-based human motion capture using cues such as depth (obtained from a stereo rig) and the extracted silhouette, while the kinematic constraints are applied in order to restrict the parameter space in terms of impossible poses. [Sigal et al., 2003, 2004] use non-parametric belief propagation to track in a multi-camera set up.

Motion-based trackers suffer from the problem of drift; *i.e.*, they estimate the *change* in pose from frame to frame and as a result the error accumulates over time. On the other hand, shape-based methods rely on absolute cues and do not face the drift problem but it is not possible to extract reliable shape cues in every frame. They typically attempt to minimize an objective function (which measures the error in the pose) and are prone to converge to incorrect local minima. Specifically, background subtraction or voxel reconstruction errors in voxel-based methods result in cases where body segments are missing or adjacent body segments are merged into one. We note that shape cues and motion cues are complementary in nature and it would be beneficial to combine these cues to track pose. We briefly describe our algorithm and discuss how it addresses the above limitations in the following section.

2.2. *Overview of the markerless motion capture system*

We present a detailed articulated model and algorithms for estimating the human body model and initializing and tracking the pose in a completely automatic manner. The architecture of our complete motion capture system is illustrated in Fig. 1. In this section, we briefly describe our human body model followed by an overview of the algorithms for model parameter estimation, pose initialization and pose tracking. The algorithms proposed in this chapter can be used in a number of biomechanical applications, such as gait analysis as well as general human motion analysis.

[Badler et al., 1993] suggest several models to represent human subjects in terms of shape and articulated structure. We find that using modified super-quadrics to represent shapes [Gavrila and Davis, 1996] is reasonably accurate for our purpose, though our approach can accommodate more sophisticated mesh-models. We model the human body as consisting of six articulated chains, namely the trunk, head, two arms and two legs as illustrated in Fig. 2 (c). Our model is based on the underlying skeletal structure of the human body. Each rigid segment is represented by a tapered super-quadric. The model consists of the joint locations and parameters of the tapered super-quadrics describing each rigid segment. The model can be simplified to a skeleton model using just the axis of the super-quadric as illustrated in Fig. 2 (b). The recovery of the human body model is described in detail in [Sundaresan and Chellappa, 2008, 2006a].

We present an algorithm for segmenting volumetric representations (voxels) of the human body by mapping them to Laplacian Eigenspace. We also describe an

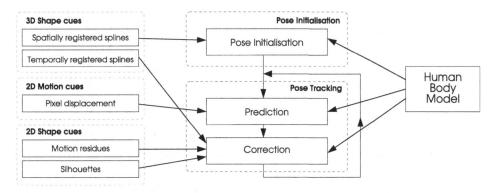

Fig. 1.   The schematic of the motion capture algorithm with the three steps in the dashed boxes on the right and the different cues used on the left.

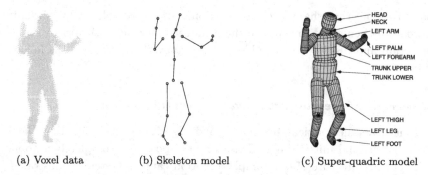

(a) Voxel data	(b) Skeleton model	(c) Super-quadric model

Fig. 2.   Illustration of skeleton and super-quadric model: (a) Voxel data. (b) Segmented voxels. (c) Skeleton model. (d) Corresponding super-quadric model.

application of this algorithm to human body model and pose estimation and provide experimental validation using both synthetic and real voxel data. Some of the key results of the above algorithm are illustrated in Fig. 2. Given a sequence of 3D voxel data of human motion (Fig. 2 (a)), the human body model and pose (Fig. 2 (b)-(c)) are estimated using a sub-set of the frames in the sequence. The human body model consists of rigid segments connected in an articulated tree structure. The pose of each rigid segment is represented in general using a 6-vector (3 degrees of freedom for translation and 3 for rotation). However, in our work we constrain most of the joints to possess only rotational motion (3 degrees of freedom). The full body pose is represented in a parametric form as a stacked vector of the poses of each segment. The concept of segmentation in Laplacian Eigenspace was introduced in [Sundaresan and Chellappa, 2006b] and its application to human body model estimation in [Sundaresan and Chellappa, 2006a]. We describe the theoretical underpinnings of segmentation in Laplacian Eigenspace for human body segmentation in [Sundaresan and Chellappa, 2008].

Our segmentation algorithm can also be viewed as a skeletonization algorithm, that obtains the skeletons of the individual articulated chains similar to [Brostow et al., 2004], which uses voxel data to estimate a novel skeleton representation. Belkin and Niyogi [Belkin and Niyogi, 2003] describe the construction of a representation for data lying in a low dimensional manifold embedded in a high dimensional space and use Laplacian Eigenmaps for dimensionality reduction. While Laplacian Eigenmaps and other manifold methods have been applied to *dimensionality reduction* problems such as classification and face retrieval using Laplacianfaces [He et al., 2005], we map the voxels to a *higher* dimensional space in order to segment the chains. The dimension of this eigenspace depends on the number of chains we wish to segment. Our algorithm is more suited for segmenting articulated objects compared to Isomap [Chu et al., 2003].

Once the pose has been initialized, we present an algorithm for tracking articulated motion of humans using shape and motion cues [Sundaresan and Chellappa, 2009]. We integrate the tracking step with model estimation and pose initialization to build a completely automatic motion capture system, the block diagram of which is illustrated in Fig. 1. We note that the pose initialization algorithm typically works in only a fraction of the frames in a sequence. The failures are typically due both to errors in the processing of the shape cues (*e.g.* voxel reconstruction) and the complexity of the pose itself. A tracking module is therefore essential to complete the motion capture system.

It can be expected that using a single type of image feature leads to a single point of failure in the algorithm and hence it is desirable to use different kinds of shape and motion features or cues. Our algorithm uses both motion cues in the form of pixel displacements as well as 2D and 3D shape cues such as skeleton curves, silhouettes and "motion residues". Thus, our algorithm does not have a single point of failure and is robust. Trackers which use only motion cues suffer from the drift problem due to an accumulation of the tracking error. On the other hand, trackers that use shape cues which are absolute often involve an energy minimization formulation and can converge to the wrong local minima. The motion and shape cues when combined work to alleviate the drift and local minima problem that are manifest when applied separately.

Since we use motion and shape cues in our tracking algorithm, we are able to better deal with cases where the body segments are close to each other such as when the arms are by the side of the body in a typical walking posture. Purely silhouette-based methods, including those that use voxels, experience difficulties in such cases. Indeed, we use a voxel-based algorithm to initialize the pose and initiate the tracking, but the registration algorithm used in the initialization fails in a number of cases where the body segments are too close to each other or when errors in the 2-D silhouette estimation cause holes and gaps in voxel reconstruction. Silhouette or edge-based methods also have problems estimating rotation about the axis of the body segment as it is impossible to detect motion of a sphere or cylinder rotating about their axis by observing only their silhouettes. We also propose a smoothing step that smooths the trunk pose, improving the performance of our tracker.

In our experiments, we use eight to twelve cameras that are placed around the subject as well as data obtained from simulation and 3D scanners. While the tracking algorithm works with fewer than eight cameras, we need at least eight cameras to obtain reasonable voxel reconstruction for the purpose of pose initialization. A visual inspection of the voxel reconstruction obtained using fewer than eight cameras was found to contain "ghost" limbs in a number of frames and was of a poorer quality and unsuitable for pose estimation as was also noted by Mündermann *et al.* [Mündermann et al., 2005]. We note that the prediction module of our tracker requires that the motion between frames be small enough so that pixel displacements can be estimated and the iterative pose estimation algorithm converges. We observe in our experiments that a frame rate of 30 fps suffices for normal human walking motion.

3. Segmentation and body model and pose estimation

The human body can be visualized as six articulated chains connected at joints as illustrated in Fig. 3. The segments labeled b_1, b_2, b_3, b_4, b_5, and b_6 correspond to the trunk, head, left arm, right arm, left leg and right leg respectively. We propose a novel bottom-up method to perform segmentation of the 3D voxel structure into component articulated chains by mapping the voxels to Laplacian Eigenspace (LE) [Sundaresan and Chellappa, 2008]. Having used a bottom-up approach to perform segmentation, we then use a top-down approach using our knowledge of the structure of the human body to register each chain to the human body model (b_i in Fig. 3 (a)) and simultaneously estimate the human body model parameters and pose. The block diagram illustrating the steps in our algorithm is presented in Fig. 4.

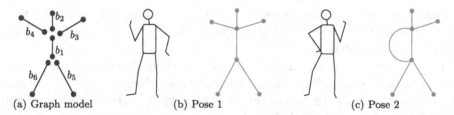

(a) Graph model   (b) Pose 1   (c) Pose 2

Fig. 3.   (a) Human body model comprising of six articulated chains and (b-c) various poses.

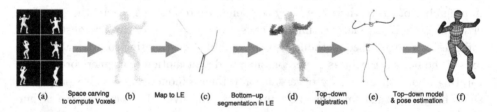

(a) Space carving to compute Voxels   (b) Map to LE   (c) Bottom-up segmentation in LE   (d) Top-down registration   (e) Top-down model & pose estimation   (f)

Fig. 4.   Block diagram describing the steps in the segmentation in Laplacian Eigenspace (LE) to estimate the human body model.

3.1. *Voxel segmentation*

We model the human body as being composed of several articulated chains connected at joints. Each articulated chain may consist of one or more rigid segments connected in a chain and forms a smooth 1D curve in Laplacian Eigenspace (LE) because its length is greater than its thickness. We segment the voxels into these different chains by exploiting the structure and mutual orientation of the 1D curves they form in LE. Since the transformation is based on neighborhood relations between voxels in normal 3D space, it is not much affected by the articulation at

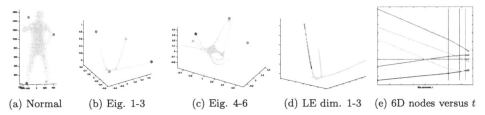

(a) Normal   (b) Eig. 1-3   (c) Eig. 4-6   (d) LE dim. 1-3   (e) 6D nodes versus t

Fig. 5.   Spline initialisation and propagation: The asterisks denote the starting node for the first, second, third and fourth splines in (a)-(c). The propagation of the first spline is illustrated in (d) and (e).

joints. However, at junctions where *three or more* such chains meet, the 1D curves representing different chains diverge in different directions (for *e.g.*, at the neck joint where the head, two arms and trunk meet). We can fit a 1D spline in LE and use the spline fit error at a given node as an indicator of its proximity to a junction. The spline fitting process also enables us to obtain the position of the nodes along their respective 1D curves or articulated chains. All operations described below are performed in LE. We describe the segmentation algorithm using a real example. Figure 5 illustrates the voxel representation of a subject in a pose where there is self-contact between the palm and hip.

We can classify the articulated chains into two types according to whether they are connected at one end (Type 1) or both ends (Type 2) to other chains. In the example in Fig. 5 (a), the two legs, head, and one of the arms are of Type 1, *i.e.*, one end of the chain is free, and the left arm and the trunk are of Type 2, *i.e.*, both ends are attached to other chains. For Type 1 chains, we note that the node at the free end is farthest from other chains. However, for Type 2 chains, the node that is farthest from other chains lies in the middle of the chain. In order to fit a 1D curve on the chain, we begin with a cluster around the starting node and compute the principal axis of the set. To begin with, in the absence of existing splines, we select the node that is farthest from the origin denoted by the asterisk in Fig. 5. The starting node for the second, third and the fourth splines are denoted by asterisks of different shades.

The spline fitting procedure is illustrated using Fig. 5 (d)-(e). Given a set of nodes and the principal axis, we project each node, y_i, onto the principal axis to obtain its site value t_i. The cluster of nodes and the principal axes are plotted in Fig. 5 (d). The nodes which are 6D vectors are plotted against their site parameters, t, in Fig. 5 (e). A 6D spline, f^{EIG}, can be computed to minimize the error given by $\sum_i \|f^{\mathrm{EIG}}(t_i) - y_i\|^2$.[2] The spline is propagated by adding nodes that are closest to the growing end of the spline (for *e.g.*, the dark nodes in Fig. 5) (d). The principal axis used to compute the site value is recomputed locally and an adaptive

---

[2]The spline used is a cubic spline with two continuous derivatives and is computed using the MATLAB spline toolbox function *spap2*.

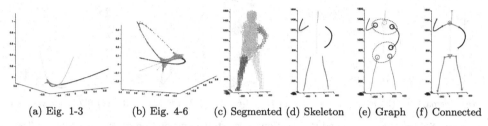

| (a) Eig. 1-3 | (b) Eig. 4-6 | (c) Segmented | (d) Skeleton | (e) Graph | (f) Connected |

Fig. 6. Segmentation and registration in LE: The nodes are segmented in LE (a)-(b). The labels are represented in the original 3D space in (c). The computed skeleton is presented in (d) and the two joints in (e). The correct registration is shown in (f).

thresholding is used to differentiate between high curvature of the 1D spline and the divergence of the spline at a junction. A node is considered an outlier if the spline fit error of that node exceeds a fixed threshold. We note that the number of outliers increases rapidly at a junction because the nodes diverge in widely different directions. The spline propagation is therefore terminated when the number of outliers exceeds a fixed threshold. We show in Fig. 6 (a-c) the successful segmentation of the voxels into different articulated chains although there is contact between the arms and the body.

The spline fitting procedure is stopped when six splines have been discovered. The site value t_i of the nodes in each spline denotes the position of the node along the 1D curve and can be used to compute the curve skeleton in Fig. 6 (d). We compute a 3D smoothing spline with the set of nodes $(t_i, \boldsymbol{v}_i)$ in normal space. The spline $\boldsymbol{f}^{\text{LIMB}}$, seeks to minimize the error given by $\sum_i \left\| \boldsymbol{f}^{\text{LIMB}}(t_i) - \boldsymbol{v}_i \right\|^2$. We thus compute the curve skeleton for each of the splines in normal space. Type 1 chains contain a single spline. Type 2 chains contain two splines which are merged together to form a single spline. We now have a set of splines and construct a graph to describe the connections between the ends of the splines in LE (Fig. 6 (e)). The position of each voxel along the articulated chain is used at various stages of the algorithm.

We use a top-down approach using our knowledge of the human body to perform the registration of the different chains that were segmented using a bottom-up approach. The objective is to identify the segmented chains and resolve possible ambiguities such as those in Fig. 6 (d) so that we can obtain the joint connections shown in Fig. 6 (f). The properties of each spline (length and thickness) and their mutual connectivity are combined probabilistically to compute the most probable labels of each of the limbs. In most cases, the only chain that has non-zero probability of connections at both nodes is the trunk and therefore the number of permutations is greatly reduced. For the example in Fig. 6 (e), the yellow and black chains have equal probability of being identified as the trunk based on the connections alone. The properties of the individual chain help discriminate between the trunk and the arms. The chains are labeled according to the registration with the highest probability. If the probability of the best registration is too low, the frame is discarded as unsuitable for use in model estimation.

3.2. *Body model and pose estimation*

Our objective is to estimate the human body model and pose from the segmented and registered voxels obtained in the previous section. The two sets of parameters of interest are the pose parameters (joint angles) and the body structure (joint locations and super-quadric parameters). In this section, we describe the algorithm to estimate the human body model parameters and the pose. We use a hierarchical approach, beginning with a skeletal model (joint locations and limb lengths, Fig. 7 (b)) and then proceeding to increase the model complexity and refine parameters to obtain a volumetric model (super-quadric parameters, Fig. 7 (c)). The joint locations cannot be reliably estimated from a single frame or pose. We therefore compute the skeleton curve of the subject in a set of key frames where registration is successful. These key frames are spread apart temporally so that a set of distinct poses is obtained. The stature (or height) of the subject is a key parameter that is strongly related to a number of human body model parameters, such as the lengths of long bones in the body [Ozaslan et al., 2003]. Anthropometric studies have been performed on certain demographic groups to study the relationship between stature and the long bones in the body [Choi et al., 1997; Mendonca, 2000]. These studies indicate that we can estimate the lengths of the large bones for an average human subject from the stature. We can construct a skeleton model for the average subject as a function of the stature by scaling the limb lengths and the joint locations by the ratio of the stature of the subject to the stature of the average human. In the first step, we find the optimal stature for the subject using the skeleton model. In the second step, we optimize for the joint locations based on the skeleton model, and in the third step we estimate and optimize for the super-quadric parameters using the full super-quadric model.

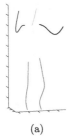   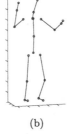   

(a)   (b)   (c)

Fig. 7.   Human body models.

3.3. *Experimental results*

We present results of our experiments on synthetic data obtained from animation models, as well as real data obtained both from 3D laser scans and synchronised video sequences. We illustrate our algorithm on poses with self-contact to illustrate the ability of our algorithm to correctly segment body poses in complex poses. We also present the results of the model estimation algorithm on different sources such as 3D laser scan data as well as voxel reconstructions computed from multi-camera video sequences, and different subjects. These different sources result in voxel data with varying degrees of accuracy.

3.3.1. *Segmentation on video data*

The results of the algorithm on different subjects in both simple and difficult poses are presented in Fig. 8. The voxels in this case were computed using images from multiple cameras. Grey-scale images were captured from twelve cameras. Simple background subtraction was performed to obtain binary foreground silhouettes for each image followed by space carving using the binary silhouettes. We note that the algorithm succeeds in segmenting the voxels into parts based on the 1D structure and the connection between the different body segments. We have successfully performed segmentation and registration in the case of self contact as illustrated in Fig. 8 (a)-(b), which other algorithms (such as [Chu et al., 2003]) do not address. Probabilistic registration allows us to reject improbable registration based on the estimated connections between the segments as well as resolve ambiguities using prior knowledge of the properties of the different segments. The results of the human body model estimation for different subjects (computed from 20 frames) are presented in Fig. 12.

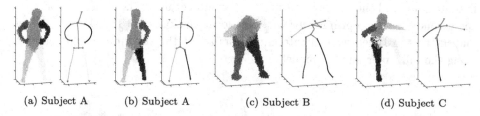

(a) Subject A   (b) Subject A   (c) Subject B   (d) Subject C

Fig. 8.   Segmentation and registration for different subjects and poses.

3.3.2. *HumanEva II dataset*

We show the results of the segmentation and registration algorithm on two sequences from the HumanEva II dataset in Fig. 9. We map the nodes to 5D LE, as the accuracy of voxel reconstruction is low, and we do not gain much by mapping to a higher dimensional space. The algorithm does not find the requisite number of body segments in the majority of the frames principally due to two reasons. The arms are

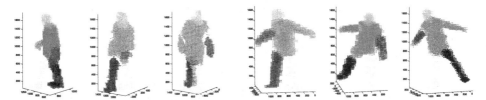

Fig. 9.   Segmented and registered frames from two subjects in the HumanEva II dataset.

too close to the body and obscured in a majority of the cameras and are undetected, or segmented limbs are rejected due to the length of their curve skeleton being too short. The voxel reconstruction algorithm also creates a "ghost limb" as an artifact of the space carving algorithm in certain configurations of the subject with respect to the cameras. It should be noted that both these problems can be solved by the use of more cameras. The problematic frames are rejected automatically. We report results on the Walking (Frames 1-350) and Balancing (800-1222) subsets. A total of 68 frames (around 9% of the total) were segmented and registered.

3.3.3. *Synthetic data set*

We provide results on human body model estimation using a synthetic sequence that has been generated from a known model and known motion sequence. A sample 3D frame and the corresponding voxel data is presented in Fig. 10. The voxel resolution used was 30mm. The human body parameters as well as the pose parameters are known and we can compare the estimated human body model and the motion parameters with the ground truth values. We note that the known body parameters are only the joint locations and not the shape parameters; the 3D animation is a smooth fairly realistic mesh as can be seen in the above figure. The results of the human body model estimation are illustrated in Fig. 10. The sequence had 120 frames, and the six different chains were correctly segmented and registered in 118 of the 120 frames (The registration failed in the remaining two frames and they were automatically discarded). We used 10 equally spaced frames as the key frames in our human body model estimation. The human body model used in this estimation used two rigid segments for the trunk. The human body model used in the other experiments use one rigid segment for the trunk. The pose was computed for all frames using the estimated model. The errors in the joint angles at the important joints are compared in Table 1.

3.3.4. *3D scan data*

The synthetic sequence described in the previous section had limited motion at several joints. We also tested our human body model estimation algorithm on different subjects using laser scan data which provide 3D meshes. Voxels can be

Table 1. Joint angle error for Skeleton and Super-quadric (SQ) optimisation. L and R denote Left and Right sides, while S, E, H, and K denote Shoulder, Elbow, Hip and Knee respectively.

Optim.	Statistic	Trunk	L S	L E	R S	R E	L H	L K	R H	R K
Skeleton	Mean	1.24	8.80	4.20	8.61	5.21	4.09	4.04	3.97	4.82
	Median	1.20	8.51	3.89	8.66	4.98	3.33	3.71	2.68	3.33
SQ	Mean	1.25	7.78	4.25	10.04	5.09	4.18	4.66	3.70	4.96
	Median	1.20	8.13	4.14	9.67	4.97	3.41	4.68	3.22	4.35

Fig. 10. Sample voxel and human body model estimation from synthetic sequence.

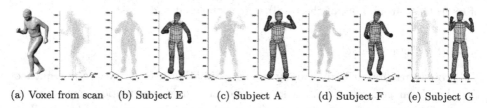

(a) Voxel from scan   (b) Subject E   (c) Subject A   (d) Subject F   (e) Subject G

Fig. 11. Human body model estimation from 3D scan obtained for different subjects.

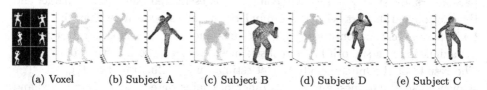

(a) Voxel   (b) Subject A   (c) Subject B   (d) Subject D   (e) Subject C

Fig. 12. Human body model estimation for different subjects from video sequences.

computed from 3D meshes by determining if nodes on a regular 3D grid lie inside the mesh structure. The subject in each case strikes different poses that exercise different joint angles. The subjects are of different heights and build. The voxel was computed from the 3D mesh obtained from the laser scanner. A set of five different poses was used to estimate the human body pose. Each pose is quite different from the other and the 3D scans are quite accurate. As a result, we are able to estimate the human body model parameters from a few frames. The results of the human body model for different subjects is presented in Fig. 11. This experiment illustrates that human body model estimation can be performed using a limited number of frames, provided the poses are varied.

4. Pose tracking using multiple cues

In this section, we describe the pose tracking algorithm. We assume that we are able to perform pose initialization in one or more frames in the sequence. Typically, the pose can be initialized in some of the frames in the sequence but the single-frame registration is unsuccessful in a number of frames and we are left with unregistered skeleton curves. We use a temporal registration scheme by which means we register skeleton curves by exploiting their temporal relations as described in Section 4.1. We then describe the tracking component that tracks the pose in two steps; the prediction step using motion cues, and the correction step using 2D and 3D shape cues in Section 4.2. We use images obtained from eight calibrated cameras. We perform simple background subtraction to obtain foreground silhouettes and space carving to obtain the voxel reconstruction as illustrated in Fig. 13.

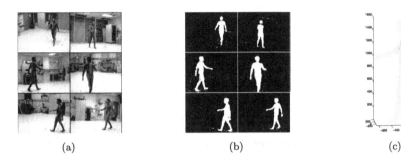

Fig. 13.   Processing images to compute silhouettes and voxels.

4.1. *Pose initialization and temporal registration*

An example of a successfully segmented and registered frame is presented in Fig. 14. However, the single frame registration method does not succeed in all frames due to errors in voxel reconstruction or segmentation, examples of which are presented in Fig. 15. The pose is initialized for a completely registered frame by fitting the skeleton model to skeleton curves in two steps. The trunk is fitted as marked in Fig. 14 (e) the pose of each of the articulated chains are independently estimated as illustrated in Fig. 14 (f). Two examples where registration of skeleton curves to articulated chains in a single frame fails are illustrated in Fig. 15. In one of the examples, the head is missing due to errors in background subtraction. In the other *seven* skeleton curves are discovered instead of six. We therefore use a temporal registration scheme which exploits the proximity of the skeleton curves belonging to the same body segment in temporally adjacent frames. The temporal registration is performed both in forward and reverse directions to check for consistency. Any skeleton curve that is not registered to the same articulated chain in the forward and reverse temporal registration process is not used in the tracking.

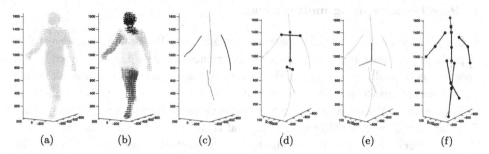

(a)   (b)   (c)   (d)   (e)   (f)

Fig. 14.   Example of registered frame: The various stages from segmentation and registration to pose initialization.

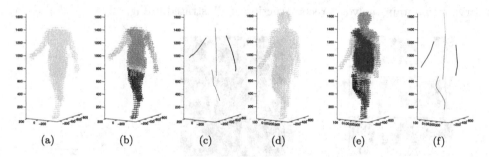

(a)   (b)   (c)   (d)   (e)   (f)

Fig. 15.   Unregistered frame with missing head (a-c) and extra segment (d-f).

4.2. *Pose tracking*

Our tracking algorithm consists of a prediction step and a correction step. In order to estimate the motion of each of the body segments, we first project the body segment onto each image. We call this step pixel-body registration. We then compute the pixel displacement for each body segment in each image using the motion model for a rigid segment. The pixel displacement for a set of bodies in all the images is then stacked in a single matrix equation which we use to estimate the change in 3D pose. Figure 16 (a) illustrates the projection of the body segments onto an image. Different colors denote different body segments. Figure 16 illustrates the pixel displacement computations. Once we have estimated the motion parameters, we can estimate the corresponding "motion residue" which can be considered as the error in the pixel matching for the estimated motion. If the actual motion of a pixel agrees with the estimated motion, then the motion residue for that pixel is zero and otherwise it is some non-zero value. We note that the motion $\psi = 0$ (Fig. 16 (c)) agrees with the motion of the stationary background pixels. However, it does not agree with the motion of the foreground pixels. Figure 16 (d) denotes the estimated pixel displacement for the body segment under consideration. Figure 16 (e) is the motion residue for the estimated ψ. We note that the estimated motion agrees with the actual motion for the foreground pixels (in the mask) but not for the background pixels, *i.e.*, the motion residue for the pixels in the mask is almost zero. Thus, the motion residue provides us with a rough delineation of the location of the body segment, even when the original mask does not exactly match the body segment.

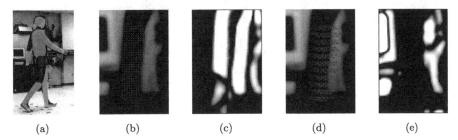

(a)   (b)   (c)   (d)   (e)

Fig. 16. (a) Pixel registration (b) The smoothed image with the foreground mask. (c) Motion residue for $\psi = 0$. (d) The estimated pixel displacement for the mask. (e) Motion residue for estimated ψ that results in the pixel displacement in (c).

We predict the pose at time $t + 1$ given the pose at time t and the pixel displacement computed above. We estimate the pose of the subject in multiple steps, starting at the root of the kinematic chain (trunk) and moving down to the branches (limbs). We note that the motion of the body segment at the end of the chain depends on the motion of all the segments preceding it on that chain and can be substantial. The multi-step prediction described above leads to faster convergence in such cases. The pose can be corrected for all the articulated chains in a given frame that have been registered using 3D shape cues (skeleton curves). The pose parameter search space is centered and bounded around the pose predicted using motion cues. In the absence of 3D shape cues, we use 2D shape cues in the form of silhouettes and motion residues. Thus the algorithm adapts itself to use available spatial cues.

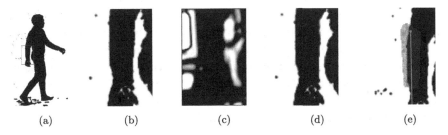

(a)   (b)   (c)   (d)   (e)

Fig. 17. Obtaining unified error image for the forearm: (a) The silhouette at time $t + 1$ (b) A magnified view of the silhouette (c) The motion residue at time t (d) the combined error image (e) error image with the mask corresponding to the segment whose pose we are trying to correct.

We had observed earlier that the motion residue for a given segment provides us with a region that helps us to spatially delineate the segment. We now combine it with the silhouette as illustrated in Fig. 17 to form an error image for that segment.

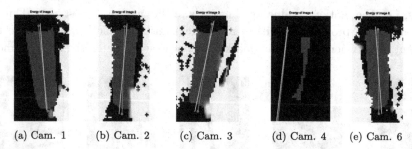

(a) Cam. 1	(b) Cam. 2	(c) Cam. 3	(d) Cam. 4	(e) Cam. 6

Fig. 18. Minimum error configuration: It does not matter if the object is occluded or nearly occluded in some of the images.

The error image is the sum of the silhouette and motion residue and is computed for each camera and each segment along with a mask for the body segment as illustrated in Fig. 17 (e). This error image can be used to compute an objective function in terms of the 3D pose of the segment. Given any 3D pose of the segment we can project the segment onto each image to obtain a mask for the segment (Fig. 17 (e)). The objective function is computed by summing the pixels of the error image that lie in the mask. Our estimate of the 3D pose is the value that minimizes this objective function in *all* the images. The objective function is optimized in a pose parameter space that is centered around the predicted pose using the *lsqnonlin* function in the Matlab non-linear optimization toolbox. We illustrate the results of pose correction for the above example in Fig. 18. The dark line represents the initial position of the axis of the body segment and the light line represents the final position. The final mask location is denoted as a shaded mask and we note that it is well aligned with the silhouette.

It is often beneficial to perform temporal smoothing on the pose vector as it typically improves the performance of the algorithm. We perform a smoothing step that acts on the pose of the *root segment* of the kinematic chain. It is difficult to smooth the entire pose vector due to the articulated constraints between the segments, and we therefore restrict the smoothing to the pose of the trunk segment (root) as it has an impact on the pose of all the body segments. We smooth the pose estimated from the skeleton curves using the smoothing spline function *csaps* in the Matlab Spline Toolbox. The trunk location is interpolated for frames missing the trunk skeleton curve. The translational components of the pose of the trunk for one of the test sequences is presented in Fig. 19 (b)-(c).

4.3. *Experimental results*

We performed tracking on sequences where the subject performs different kinds of motion. The experiments were performed using gray-scale images obtained from eight cameras at a frame rate of 30 frames per second. We present results for two sequences that include the subject walking in a straight line (65 frames, 2 seconds) in

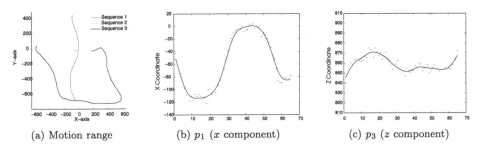

(a) Motion range   (b) p_1 (x component)   (c) p_3 (z component)

Fig. 19. (a) The position of the subject in the world coordinate frame in the three sequences. (b) and (c) denote the raw and smooth translational components represented by dots and lines respectively.

(a) Images from camera 1   (b) Images from camera 3

Fig. 20. Tracking results for sequence 1.

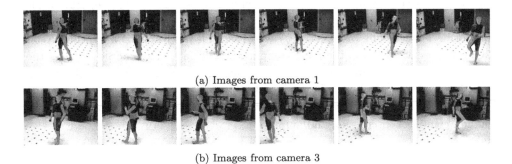

(a) Images from camera 1

(b) Images from camera 3

Fig. 21. Tracking results for sequence 3.

Fig. 20, and walking in a circular path (300 frames, 10 seconds) in Fig. 21. Figure 19 illustrates the motion of the base body in the world coordinate frame in the three sequences. Our experiments show that using only motion cues for tracking causes the pose estimator to drift and lose track eventually, as we are estimating only the *difference* in the pose. This underlines the need for correcting the predicted pose using spatial cues and we observe that the correction step of the algorithm prevents drift in the tracking. We illustrate the results of the tracking algorithm by super-imposing the tracked body model onto the image for two of the eight cameras. The estimated pose of the subject is super-imposed on the images and the success of the tracking algorithm is determined by visual inspection. It is not possible to obtain an objective measure of the pose as the actual pose is not available. The full body pose is successfully tracked in the three sequences as can be observed in the super-imposed video sequences.

5. Summary

We have described a complete motion capture system that includes human body model estimation, pose initialization, and tracking components. We use a descriptive and flexible human body model that allows translation at complex joints such as the shoulder joint. Pose initialization is performed by segmentation of voxels in Laplacian Eigenspace and is particularly suited for extracting the 1D structure and segmenting the different chains when compared to other methods such as Isomap or LLE. The human body model and pose are then simultaneously estimated using the output of the segmentation procedure. We perform temporal registration of partially segmented voxels and use both motion cues and shape cues such as skeleton curves, silhouettes and "motion residues" to perform the tracking. We present results on sequences with different kinds of motion and observe that the several independent cues used in the tracker enable it to perform in a robust manner. The complete motion capture system has been written in Matlab and we note that currently the computational requirements of the system are high primarily due to the number of cameras used in the processing and the inefficiency of the Matlab platform. The tracking process takes several seconds per frame on a Pentium Xeon 2GHz processor. Some of the most computationally intensive modules such as the projection of the human body model onto each of the images can be optimized to greatly speed up the process. We anticipate that our motion capture system can be used in a variety of important applications in biomechanical and clinical analysis, human computer interaction and animation.

References

J.K. Aggarwal and Q. Cai. Human motion analysis: A review. *Computer Vision and Image Understanding*, 73(3):428–440, 1999.

D. Anguelov, D. Koller, H. Pang, P. Srinivasan, and S. Thrun. Recovering articulated object models from 3-D range data. In *Proc. of the Conference on Uncertainty in Artificial Intelligence*, pages 18–26, Banff, Canada, 2004.

N. I. Badler, C. B. Phillips, and B. L. Webber. *Simulating Humans*. Oxford University Press, Oxford, UK, 1993.

Mikhail Belkin and Partha Niyogi. Laplacian eigenmaps for dimensionality reduction and data representation. *Neural Computation*, 15(6):1373–1396, 2003.

C. Bregler and J. Malik. Tracking people with twists and exponential maps. In *Proc. of the IEEE Conference on Computer Vision and Pattern Recognition*, pages 8–15, Santa Barbara, CA, USA, June 1998.

Gabriel Brostow, Irfan Essa, Drew Steedly, and Vivek Kwatra. Novel skeletal representation for articulated creatures. In *Proc. of the European Conference on Computer Vision*, volume 3, pages 66–78, Prague, Czech Republic, May 2004.

K.M. Cheung, S. Baker, and T. Kanade. Shape-from-silhouette of articulated objects and its use for human body kinematics estimation and motion capture. In *Proc. of the IEEE Conference on Computer Vision and Pattern Recognition*, volume 1, pages 77–84, Madison, USA, June 2003.

Byoung Young Choi, Young Moon Chae, In Hyuk Chung, and Ho Suck Kang. Correlation between the postmortem stature and the dried limb-bone lengths of korean adult males. *Yonsei Medical Journal*, 38(2):79–85, 1997.

Chi-Wei Chu, Odest Chadwicke Jenkins, and Maja J. Mataric. Markerless kinematic model and motion capture from volume sequences. In *Proc. of the IEEE Conference on Computer Vision and Pattern Recognition*, volume 2, pages 475–482, Madison, USA, June 2003.

Q. Delamarre and O. Faugeras. 3D articulated models and multi-view tracking with silhouettes. In *Proc. of the International Conference on Computer Vision*, volume 2, pages 716–721, Kerkyra, Corfu, Greece, September 1999.

V. Ganapathi, C. Plagemann, D. Koller, and S. Thrun. Real time motion capture using a single time-of-flight camera. In *CVPR*, 2010.

D. M. Gavrila. The visual analysis of human movement: A survey. *Computer Vision and Image Understanding*, 73(1):82–98, 1999.

D.M. Gavrila and L.S. Davis. 3-D model-based tracking of humans in action: A multi-view approach. In *Proc. of the IEEE Conference on Computer Vision and Pattern Recognition*, pages 73–80, 1996.

Xiaofei He, Shuicheng Yan, Yuxiao Hu, P. Niyogi, and Hong-Jiang Zhang. Face recognition using laplacianfaces. *IEEE Transactions on Pattern Analysis and Machine Intelligence*, 27(3):328–340, 2005.

S. X. Ju, M. J. Black, and Y. Yacoob. Cardboard people: A parameterized model of articulated image motion. In *Proc. of the International Conference on Automatic Face and Gesture Recognition*, pages 38–44, Killington, Vermont, USA, October 1996.

I. A. Kakadiaris and D. Metaxas. Model-based estimation of 3D human motion. *IEEE Transactions on Pattern Analysis and Machine Intelligence*, 22(12):1453–1459, December 2000.

Steffen Knoop, Stefan Vacek, and Rüdiger Dillmann. Fusion of 2d and 3d sensor data for articulated body tracking. *Robot. Auton. Syst.*, 57:321–329, March 2009.

M. C. De Mendonca. Estimation of height from the length of long bones in a portugese adult population. *American Journal of Physical Anthropology*, 112: 39–48, 2000.

Microsoft Corporation. Kinect for Xbox 360, 2010.

Ivana Mikić, Mohan Trivedi, Edward Hunter, and Pamela Cosman. Human body model acquisition and tracking using voxel data. *International Journal of Computer Vision*, 53(3), 2003.

T.B. Moeslund and E. Granum. Multiple cues used in model-based human motion capture. In *Proc. of the International Conference on Face and Gesture Recognition*, pages 362–367, Grenoble, France, March 2000.

T.B. Moeslund and E. Granum. A survey of computer vision-based human motion capture. *Computer Vision and Image Understanding*, 81:231–268, 2001.

D.D. Morris and J. M. Rehg. Singularity analysis for articulated object tracking. In *Proc. of the IEEE Conference on Computer Vision and Pattern Recognition*, pages 289–297, Santa Barbara, CA, USA, June 1998.

Lars Mündermann, Stefano Corazza, Ajit M. Chaudhari, Eugene J. Alexander, and Thomas P. Andriacchi. Most favorable camera configuration for a shape-from-silhouette markerless motion capture system for biomechanical analysis. In *Proc. of SPIE Videometrics*, volume 5665, January 2005.

Lars Mündermann, Stefano Corazza, and Thomas Andriacchi. Accurately measuring human movement using articulated ICP with soft-joint constraints and a repository of articulated models. In *Proc. of the IEEE Conference on Computer Vision and Pattern Recognition*, Minneapolis, MN, USA, June 2007.

Abdi Ozaslan, Haran Tugcu M. Yasar Iscan, Inci Oxaslan, and Sermet Koc. Estimation of stature from body parts. *Forensic Science International*, 132(1):40–45, 2003.

Deva Ramanan and David A. Forsyth. Finding and tracking people from the bottom up. In *Proc. of the IEEE Conference on Computer Vision and Pattern Recognition*, volume 2, pages 467–474, Madison, WI, USA, June 2003.

Jamie Shotton, Andrew Fitzgibbon, Mat Cook, Toby Sharp, Mark Finocchio, Richard Moore, Alex Kipman, and Andrew Blake. Real-time human pose recognition in parts from a single depth image. In *IEEE Conference on Computer Vision and Pattern Recognition*, 2011.

M. Siddiqui and G. Medioni. Human pose estimation from a single view point, real-time range sensor. In *CVCG*, 2010.

Hedvig Sidenbladh, Michael J. Black, and David J. Fleet. Stochastic tracking of 3D human figures using 2D image motion. In *Proc. of the European Conference on Computer Vision*, volume 2, pages 702–718, Dublin, Ireland, June 2000.

L. Sigal, A. Balan, and M. J. Black. Humaneva: Synchronized video and motion capture dataset and baseline algorithm for evaluation of articulated human motion,. *Int. J. Comput. Vis.*, 87(1):4–27, 2010.

Leonid Sigal and Michael Black. Humaneva: Synchronized video and motion capture dataset for evaluation of articulated human motion. Technical Report CS-06-08, Brown University, 2006.

Leonid Sigal, Michael Isard, Benjamin H. Sigelman, and Michael J. Black. Attractive people: Assembling loose-limbed models using non-parametric belief propagation. In *Proc. of the Conference on Neural Information Processing Systems*, pages 1539–1546, Vancouver, Canada, 2003.

Leonid Sigal, Sidharth Bhatia, Stefan Roth, Michael J. Black, and Michael Isard. Tracking loose-limbed people. In *Proc. of the IEEE Conference on Computer*

Vision and Pattern Recognition, volume 1, pages 421–428, Washington, DC, USA, June 2004.

Aravind Sundaresan and Rama Chellappa. Acquisition of articulated human body models using multiple cameras. In *Proc. of the Conference on Articulated Motion and Deformable Objects*, pages 78–89, Port d'Andratx, Mallorca, Spain, July 2006a.

Aravind Sundaresan and Rama Chellappa. Segmentation and probabilistic registration of articulated body model. In *Proc. of the International Conference on Pattern Recognition*, volume 2, pages 92–96, Hong Kong, China, August 2006b.

Aravind Sundaresan and Rama Chellappa. Model driven segmentation and registration of articulating humans in Laplacian Eigenspace. *IEEE Transactions on Pattern Analysis and Machine Intelligence*, 30(10):1771–1785, October 2008.

Aravind Sundaresan and Rama Chellappa. Multi-camera tracking of articulated human motion using shape and motion cues. *IEEE Transactions on Image Processing*, 18(9):2114–2126, September 2009.

Christian Theobalt, Joel Carranza, Marcus A. Magnor, and Hans-Peter Seidel. Combining 3D flow fields with silhouette-based human motion capture for immersive video. *Graphical Models*, 66(6):333–351, 2004.

S. Wachter and H.-H. Nagel. Tracking persons in monocular image sequences. *Computer Vision and Image Understanding*, 74(3):174–192, June 1999.

M. Yamamoto and K. Koshikawa. Human motion analysis based on a robot arm model. In *Proc. of the IEEE Conference on Computer Vision and Pattern Recognition*, pages 664–665, Maui, HI, June 1991.

M. Yamamoto, A. Sato, S. Kawada, T. Kondo, and Y. Osaki. Incremental tracking of human actions from multiple views. In *Proc. of the IEEE Conference on Computer Vision and Pattern Recognition*, pages 2–7, Santa Barbara, CA, USA, June 1998.

CHAPTER 1.2

FACE SUPER-RESOLUTION

Wilman W. W. Zou and Pong C. Yuen

Department of Computer Science, Hong Kong Baptist University,
Kowloon Tong, Hong Kong
{wwzou,pcyuen}@comp.hkbu.edu.hk

1. Introduction

Super-resolution is a technique which reconstructs the high resolution (HR) image from its low resolution (LR) one [van Ouwerkerk (2006)][Park *et al.* (2003)] [Yang and Huang (2010)]. Due to the self-similarity of the face images, a specific super-resolution has been designed to enhance the size of the face region, we call this kind of algorithms as face super-resolution (FSR)(also called that face hallucination).

With the increasing installation of surveillance cameras in public areas, there is an increasing demand of face super-resolution technology for surveillance cameras ranging from a small-scale stand-alone camera applications in banks and supermarkets, to large-scale multiple networked close-circuit television (CCTV) in law enforcement applications in public street. In general, there are two kinds of applications: human visual quality based applications and machine-based applications. Human visual quality based applications need the FSR constructs the images with pictorial information for human interpretation, while the machine-based applications require that the FSR recovers the useful features for automatic machine perception (such as discriminative features for face recognition) [Yang and Huang (2010)].

Comparing with the traditional SR technology [Park *et al.* (2003)], most of the FSR algorithms are machine learning based : Given a set of training high resolution and low resolution face image pairs, denoted as training set $\mathcal{T}$, FSR recovers the missed details of the input LR image I_l and reconstructs the corresponding HR one I_h. Although machine learning based technique can also be used in traditional SR, it is more effective for face images due to the characteristics of human face: face images are self-similar. That means a typical face image consists of two eyes, a mouth and a nose. Face images are located in a face image subspace and are subject to a generic 3D face model. These prior knowledge can be used to compensate the ill-posed SR problem.

This chapter is organized as follows: Section 2 introduces two models to formulate the face super-resolution problem. Section 3 describes the current-state-of-the-art face super-resolution algorithms. In Section 4, the very low resolution (VLR) problem is discussed and some solutions based on face super-resolution technique are given. Finally, Section 5 discusses some open issues of face super-resolution problem.

2. Formulation for Face Super-Resolution

2.1. *Two Models for Face Super-Resolution: The Observation Model and Prior Knowledge Model*

To better understand how face super-resolution works, we introduce two models, namely the observation model and the prior knowledge model, to express how face super-resolution works with the LR images and generates the HR one. Most of the existing super-resolution algorithms are based on these two models.

2.1.1. *Observation Model*

The observation model is used to describe the imaging system generating the low resolution (LR) images.

Denote the LR input image by $I_l(\mathbf{m})$ and the HR image by $I_h(\mathbf{m})$, where $\mathbf{m} = (x, y)$ specifies the pixel location. A generic observation model is used to model the observations as follows,

$$I_i(\mathbf{m}) = E(w_i(\mathbf{m})) * PSF_i(\mathbf{m}) = \int E(w_i(\mathbf{x}))PSF_i(\mathbf{m} - \mathbf{x})d\mathbf{x} \qquad (1)$$

where i can be l or h, E is the intensity function which represents the real image of the face, w_i is the wrapping function which registers the digital pixel to the same face region so that the the observed face image is well alignment, and $PSF_i(\mathbf{m})$ is the point spread function of the digital imaging system, which weights the contribution of the light on the neighborhood of the lighting center $(\mathbf{m})$. In most of the case, the point spread function is not available for the user after the images are acquired. But given some prior knowledge (or assumptions), we can approximate it.

For PSF_h, we assume that HR image has sufficient resolution to represent the intensity of a real face image. Under this assumption, PSF_h can be considered as the Direchlet function $\delta()$ and we have

$$I_h(\mathbf{m}) \approx E(w_h(\mathbf{m})) * \delta(\mathbf{m}) = E(w_h(\mathbf{m})) \qquad (2)$$

For PSF_l, we follow the assumption of Baker and Tanade's method [Baker and Kanade (2002)] that the PSF function is a piece-wise linear function as follows

$$PSF_l(\mathbf{m}) \begin{cases} \dfrac{1}{4s^2}, \|\mathbf{m}\| \leq s \\ 0, \text{ otherwise} \end{cases}$$

If HR and LR images are well aligned, we have $w_h(\mathbf{m}) = w_l(\mathbf{m})$. Combine Eq. (1) and the above equation, given the HR image I_h, the LR observation of I_l is given by

$$I_l(\mathbf{m}) = E(w_l(\mathbf{m})) * PFS(\mathbf{m}) = \int_{\|\mathbf{m}-\mathbf{x}\|<s} E(w_h(\mathbf{x}))PFS_l(\mathbf{m}-\mathbf{x})d\mathbf{x}$$

$$\approx \sum_{\|\mathbf{m}-\mathbf{x}\|<s} \frac{1}{4s^2} I_h(\mathbf{x}) \tag{3}$$

Most of the face super-resolution (FSR) algorithms follow this observation model to justify whether the HR image is "similar" to the LR image. In FSR algorithms, we use data constraint [Liu *et al.* (2007)] to describe this observation model. Reasonably, we assume that the noise in Eq. (3) is subject to Gaussian distribution $\mathcal{N}(0, \Lambda)$ so that the data constraint [Liu *et al.* (2007)] is defined as

$$\varphi_D(I_h) = \|I_l(\mathbf{m}) - \sum_{\|\mathbf{m}-\mathbf{x}\|<s} \frac{1}{4s^2} I_h(\mathbf{x})\|^2_{\Lambda^{-1}} \tag{4}$$

where

$$\|x\|^2_{\Lambda^{-1}} = x^T \Lambda^{-1} x.$$

2.1.2. *Prior Knowledge Model*

Since FSR problem is an ill-posed problem, prior knowledge is required to solve this problem. The prior knowledge is described by the prior knowledge model and is often used to learn informative feature from the training data for reconstructing the HR images. Comparing to the reconstruction-based super-resolution algorithms, machine-learning-based FSR algorithms have larger magnification [Baker and Kanade (2002)] and is more robust to noise [Wang and Tang (2005)]. It is because they made use of more information by prior knowledge model, rather than only using the LR resolution image(s).

Here we use a probabilistic model $P(I_h|\mathcal{T})$ to describe the prior knowledge from the training data $\mathcal{T} = \{(I_l^j, I_h^j), 1 \leq j \leq N\}$. Different algorithms employ different estimators to estimate $P(I_h|\mathcal{T})$, so there is no generic close form solution for the prior knowledge model. We list some assumptions which are useful to use as prior knowledge:

A1 Subspace Assumption The reconstructed HR image (denoted as $\tilde{I}_h$) locates in the same subspace of the training HR images in $\mathcal{T}$. Many of FSR algorithms make this assumption [Wang and Tang (2005)] [Liu *et al.* (2007)][Chakrabarti *et al.* (2007)], and subspace constraints were developed to make use of the prior knowledge of training data.

A2 Homogeneous Assumption The HR image and LR image have the same or similar infrastructure pattern, e.g. they have same weights when projecting HR image and LR image to the image-pairs [Park and Lee (2008)]. The infrastructure pattern can be the distribution of images, such as HR image and LR image share the same distribution after some feature extraction [Li *et al.* (2009)]

A3 Gaussian Noise Assumption The noise follows a Gaussian Distribution. This is always taken into the FSR algorithm to model noise.

2.2. *Two Core Approaches for FSR*

The above two models co-work with each other for super-resolution: the observation model "tells" the FSR algorithm, the reconstructed HR images should corresponding to the observations (LR images), while the prior knowledge model guides the FSR algorithm reconstructing HR images like the training images. We roughly categorize existing FSR algorithms into two categories: MAP-based approach and example-based approach.

2.2.1. *MAP-based Approach*

For MAP-based approach, the FSR processing is implemented by minimizing reconstruction error measured by observation model and at the same time maximizing the prior knowledge likelihood probability. This is done using a Bayesian framework, which trades off between the observation model and the prior knowledge model.

According to Bayesian framework, FSR is implemented by maximizing the conditional probability as follows,

$$\tilde{I}_h = \arg\max_{I_h} P(I_h|I_l, \mathcal{T}) = \arg\max_{I_h} \frac{\overbrace{P(I_l|I_h, \mathcal{T})}^{\text{Observation model}} \overbrace{P(I_h|\mathcal{T})}^{\text{Prior knowledge model}}}{P(I_l|\mathcal{T})}$$

We can see that MAP based approach is formulated as a unconstrained optimization problem. Under the MAP criterion, the core problem of FSR is to model the likelihood for the observation model and model the prior probability for the prior knowledge model. Generally, this approach assumes that the LR image is statistic independent to the training data given the the original HR image and satisfies the assumption of the observation model. $P(I_l|I_h, \mathcal{T})$ is modeled by the Gaussian model, so that maximizing $P(I_l|I_h, \mathcal{T})$ is equivalent to minimizing the observation error (Eq. (5)). For the prior knowledge model $P(I_h|\mathcal{T})$, different algorithms have different methods: Baker and Kanade [Baker and Kanade (2002)] considered the distance between the LR input image patch and the most similar training image patch to estimate $P(I_h)$. Liu *et al.* [Liu *et al.* (2007)] employed non-parameter Markov network to model the HR residual images which is useful to recover the HR image with good visual quality. Subspace method is also employed to restrict the reconstructed HR image locating inside the face subspace, such as PCA subspace [Liu *et al.* (2007)] and KPCA subspace [Chakrabarti *et al.* (2007)].

2.2.2. *Example-based Approach*

For example-based approach, the FSR algorithm is implemented by alternative manner of integrating the observation model and prior knowledge model. Unlike MAP model having a trade off between observation error and the prior knowledge likelihood, example-based approach ensures that the reconstructed HR image satisfies the prior knowledge model, such as A1 and A2. After that, it searches the HR image which has small observation error to the input LR image.

A1 implies that the constructed HR image can be represented by a weighted linear combination of the HR training examples, such as

$$\tilde{I}_h = \sum_i \omega_i I_h^i$$

A2 implies that the weights for HR training examples are the same of the weights for LR training examples. To determine the weights for HR training images, this approach approximates the testing LR image as a weighted linear combination of the LR training examples as follows,

$$\{\omega_i\} - \arg\min \left\| I_l - \sum \omega_i I_l^i \right\|^2$$

The example-based approach can be formulated as a constraint optimization problem as follows:

$$\tilde{I}_h = \arg\min_{I_h} \quad \overbrace{\| I_l - D I_h \|}^{\text{observation model}}$$

$$\text{s.t.} \quad \underbrace{I_h \in span(\{I_h^j, 1 \le j \le N\})}_{\text{prior knowledge model}}$$

The core problem of this approach is to determine the weights. Wang *et al.* [Wang and Tang (2005)] proposed an algorithm using Eigen-transformation while Zhang *et al.* [Zhang and Cham (2008)] performed on DCT domain. Liu *et al.* [Liu *et al.* (2005)] and Jia and Gong [Jia and Gong (2008)] conducted example-base approach on patch tensor space, while Park *et al.* [Park and Lee (2008)] conducted on face texture space and 3D face shape space. Yang *et al.* perform this on to determine the coefficients for sparse coding dictionary [Yang *et al.* (2010)][Yang *et al.* (2008)].

3. The Current-State-of-the-Art

In this section, we introduce five popular FSR algorithms, namely Hallucination Face [Baker and Kanade (2002)], Two Step Framework method [Liu *et al.* (2007)], Eigentransformation based FSR [Wang and Tang (2005)], General FSR [Jia and Gong (2008)] and LPS-GIS FSR [Hu *et al.* (2010)].

3.1. *Hallucinating Face*

Baker and Kanada [Baker and Kanade (2002)] proposed the pioneer FSR algorithm, so-called Hallucinating Face (HF). The core idea of Hallucinating Face is shown in Figure 1. They copy the high resolution details, from the high resolution training images, which is the best match in the corresponding LR training images, as the missing details for the testing input LR image.

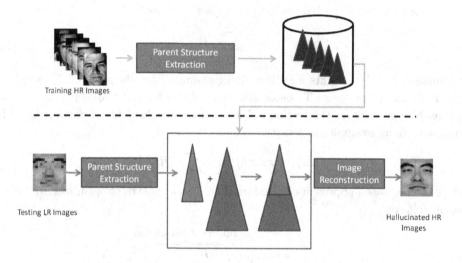

Fig. 1.   The illustration of Face Hallucination. [Baker and Kanade (2002)]

Parent structure vector **PS** is used in their algorithm. This vector consists of five pyramid structures in different feature spaces in multiple resolution. The parent structure vector consists of the first derivation and second derivation of the image and the parent structure vector of training images can be calculated off line.

Given a testing LR image I_l, to recover the HR image I_h, for each pixel **m**, they compare $\mathbf{PS}(I_l)$ to $\{\mathbf{PS}(I_h^j)\}$, and find the most similar training HR image I_h^{jo}, they call this the best match image. The HR part of $\mathbf{PS}(I_h^{jo})$ is fused with $\mathbf{PS}(I_l)$ to construct the parent structure vector of the reconstructed HR image. The HR image is recovered from the this parent structure vector.

This is a straightforward method, but effective method under proper assumptions. This method assumes that if the image has similar pattern in LR image space, it will also have similar pattern in the HR image space. This assumption is often true in many FSR cases. The shortcoming of this method is that the super-resolving process is conducted pixel by pixel, so the reconstructed image may have blocky effect.

3.2. *Two-Step Framework for Face Super-Resolution*

Liu *et al.* [Liu *et al.* (2001)] proposed a novel two-step framework for FSR. They decomposed the HR image into two parts: global face image I_h^g and local face feature image I_h^l. They employed a MAP-based approach (Refer to Section 2.2.1 for details) to reconstruct the I_h^g and I_h^l, and then recover the HR image by $I_h = I_h^l + I_h^g$.

First step: To model the global face image I_h^g, they use Eigenface model (denoting as (W, Λ, μ), where W is the eigenface vectors, Λ is the eigen values and μ is the mean face). They assume that the global face image I_h^g is located inside Eigenface subspace. So they optimize cost function (derived from Eq. (16) in [Liu *et al.* (2001)])

$$f(I_h^g) = \underbrace{\|D(WX+\mu) - I_l\|^2}_{\text{observation constraint}} + \underbrace{\lambda X^T \Lambda^{-1} X}_{\text{prior model constraint}}, X = (W^t I_h^g - \mu)$$

where D is the downsampling operator. The cost function consists of two terms: the first term estimates the error between the global face image I_h^g and input LR face image I_l in LR image space; and the second term models and estimates the prior probability of the global face image I_h^g, after Eigenface model (W, Λ, μ) is determined.

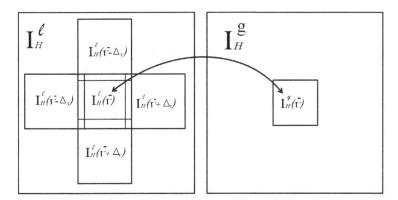

Fig. 2. The patch-based Markov network for high frequency component reconstruction. [Liu *et al.* (2007)]

Second step: The local face feature image I_h^l is modeled by a non-parametric Markov model. They divid the images into overlapped patches and assume that the face image patches are on Markov network, as shown in Figure 2. Following the MAP framework, the I_h^l is inferred from the Markov network by considering the "internal potential function" and "external potential function" as:

$$P(I_h^l | I_h^g) \propto \exp\{E_G^{int}(I_h^l) + E_G^{ext}(I_h^l)\}$$

This process is to compensate the reconstruction error introduced in the first step and keep the reconstructed image away from the blocky effect.

This algorithm takes the A1 and A3 assumptions (Please refer Section 2.1.2 for more details). The experimental results reported in their paper showed that this algorithm can enhance the resolution of facial image with good visual quality. However, this algorithm is computationally expensive due to the optimization problem of non-parametric Markov network.

3.3. *Eigentransformation based Face Super-Resolution*

Hallucinating faces and two step super-resolution are based on MAP framework which need to estimate the likelihood probability and the prior probability, which is often computationally ineffective.

Wang and Tang [Wang and Tang (2003)][Wang and Tang (2005)] proposed an Eigentransformation based FSR system which is fast and effective. Following the framework of example-based super-resolution (Refer Section 2.2.2), they decompose the input LR image as a weighted linear combination of the LR training images, and then replace the LR training images with the HR ones to reconstruct the HR image. The basic idea of their method is given in Figure 3.

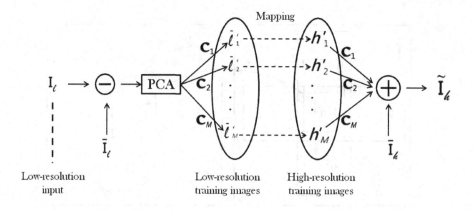

Fig. 3.   Block diagram of using eigentransformation for hallucination. [Wang and Tang (2005)]

The core problem of this algorithm is to calculate the weight for every LR image. They employed the Eigentransformation to calculate the weights. Denote the mean-shift training LR and HR images as $\mathbf{L} = (I_l^1 - \bar{I}_l \; \cdots \; I_l^N - \bar{I}_l)$ and $\mathbf{H} = (I_h^1 - \bar{I}_h \; \cdots \; I_h^N - \bar{I}_h)$, the Eigenface system training from $\mathbf{L}$ as (W, Λ, μ), the weights for LR images can be obtained as

$$\Omega = \mathbf{L}^T W \Lambda^{-\frac{1}{2}}$$

where Ω is vector of N by 1 which contains of the weights for N LR training images. The HR image can be reconstructed by $\tilde{I}_h = \mathbf{H}\Omega + \bar{I}_h$.

In their method, the core assumption made by Wang and Tang is that the structure of LR image space is the same of HR image space, so that the weights for LR images can directly be used to reconstruct the HR image. Eigentransformation based FSR has considerable robustness to white noise due to the advantage of Eigentransformation. However, the assumption may not work in some cases, such as very low resolution problem [Zou and Yuen (2010b)].

3.4. *General Face Super-Resolution*

The problem of face variation were discussed by Jia and Gong [Jia and Gong (2008)]. They proposed a general face super-resolution based on tensor space. This general face super-resolution algorithm reconstructs the HR images with different poses, expressions and lighting conditions from the LR image. They construct a tensor, which consists of images of different resolutions, different pose / illumination / expression, as shown in Figure 4.

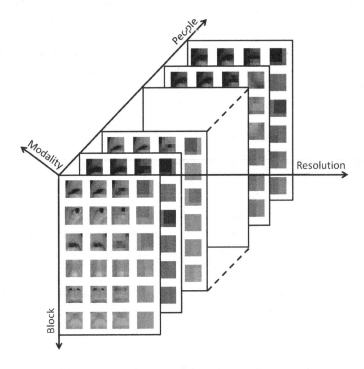

Fig. 4.   Tensor construction illustration using block-wise images of multiple facial expressions at low and high resolution. [Jia and Gong (2008)]

The basic idea of this algorithm is to use the multilinear tensor space (denoted as $\mathcal{A}$) to model the possible appearances of the face, as shown in Figure 4. Tensor

SVD is used to decompose the tensor and represent as

$$\mathcal{A} = \mathcal{Z} \times_1 \mathbf{U}_{idens} \times_2 \mathbf{U}_{modes} \times_3 \mathbf{U}_{resos} \times_4 \mathbf{U}_{blocks} \times_5 \mathbf{U}_{pixels}$$

where $\mathcal{Z}$ is the core tensor, $\mathbf{U}_*$ is a vector which represents different variables such as the identity, mode (different mode means different pose, expression, illumination, etc), resolution, block-position and pixels. The image can be represented as the multi-linear combination of the core tensor by tensor product. If we know the coefficients of identity and mode (e.g. V and s), given a tensor $\mathcal{Z}$, we can recover the image in different resolution by

$$I_r(\mathbf{m}) = \mathcal{Z} \times_1 V \times_2 s \times_3 r \times_4 B(\mathbf{m}) \times_5 P_B(\mathbf{m})$$

where r specifies the resolution of the reconstructed image, $B(\mathbf{m})$ is the block-position of the pixel position $\mathbf{m} = (x, y)$ while $P_B(\mathbf{m})$ is the pixel-position of $\mathbf{m}$ in the block $B(\mathbf{m})$. So the FSR problem is converted to estimated the identity coefficients V and mode coefficients s for HR image. In their method, they employ a MAP model to estimate V and s for LR image. And they assume that the coefficients are the same in HR and LR images, and use the estimated V and s from LR image to product the target HR image.

This method handles the face variation problems, and generates the HR images with different modes from the LR image. However, the linear model may not work well if the non-linear face variation encounter [Zou and Yuen (2010a)]. And the complexity of the appearance tensor (in memory and computational time) is another problem in practical real time applications.

3.5. *LPS-GIS Face Hallucination*

The local pixel structure is also used for FSR. Hu *et al.* [Hu *et al.* (2010)]. proposed to use the local pixel structure of the LR images to predict the global HR super-resolution.

Local pixel structure can be formulated as a set of weights for the neighborhood pixels. Given an image I, the local pixel structure satisfies the following equation,

$$I(\mathbf{m}) \approx \sum_{\|\mathbf{u}\|<\epsilon} w_{\mathbf{u}}(\mathbf{m}) I(\mathbf{m} + \mathbf{u})$$

After working out the local pixel structure $\{w_{\mathbf{u}}(\mathbf{m})\}$, the image can be recovered by minimizing the cost function

$$E_p(I) = \sum_{\mathbf{m}} \left\| I(\mathbf{m}) - \sum_{\|\mathbf{u}\|<\epsilon)} w_{\mathbf{u}}(\mathbf{m}) I(\mathbf{m} + \mathbf{u}) \right\|^2 \tag{5}$$

The super-resolution problem is converted to estimate the local pixel structure of the target HR images. They assumed that if the images are similar in LR image space, they have similar local pixel structures in HR image space. As shown in Figure 5, given a testing LR image, the K most similar training LR images are matched, and then the local pixel structures of K corresponding HR training images are used to calculate the local pixel structure of HR image.

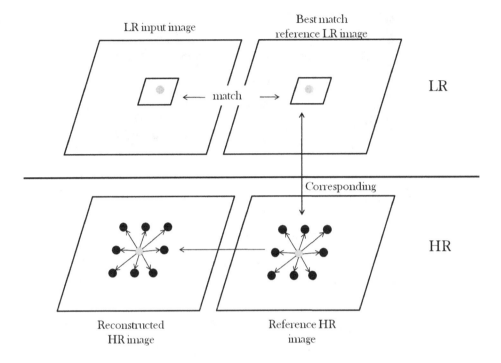

Fig. 5. The local pixel structure is constructed in the LR image space and is used in HR image space to recover the HR image. [Hu *et al.* (2010)]

4. The Very Low Resolution (VLR) Problem

Though many promising FSR algorithms have been proposed, but there still exists some challenge problems, such as very low resolution problem.

Wide-angle cameras are normally used and installed in a way that viewing area is maximized. In turn, the face region in the scene is normally very small. As shown in Figure 6, when the person is not close to the camera, the face region will be very small, e.g. less than 16 x 16 pixels. Working with face image of less than 16 x 16 pixels is called very low resolution face problem [Zou and Yuen (2010b)].

Empirical studies [Lui *et al.* (2009)] showed that minimum face image resolution between 32x32 and 64x64 is required for existing face recognition algorithms. The recognition performance will degrade dramatically, when the testing images is under VLR problem [Lui *et al.* (2009); Hennings-Yeomans *et al.* (2008)]. This is because the VLR face image contains very limited information and many image details have been lost as shown in Figure 6(b). FSR algorithm is a potential approach to solve this problem.

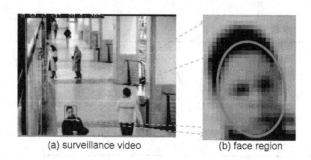

(a) surveillance video   (b) face region

Fig. 6.   A typical frame from a surveillance video (CAVIAR database). The region of the face is in very low resolution (VLR). [Zou and Yuen (2010b)]

4.1. *Limitations of Existing Methods on VLR Face Image*

Many existing super-resolution algorithms can be considered as MAP-based approach or example-based approach.

These two approaches employ the same data constraint

$$\varphi_D(I_h) = \|DI_h - I_l\|^2 \ . \tag{6}$$

MAP-based approach employs the data constraint to model the conditional probability $P(I_l|I_h)$, while example-based methods implicitly use it to determine the weights for the reconstructed HR image.

However, in VLR problem, such a data constraint may not work well because of the limited information carried by the input VLR image due to low-dimension of the VLR image space. A straightforward result is that in VLR image space, the data constraint which makes use of Euclidian norm in LR image space as a metric, cannot be considered as the similarity of two face images in HR image space. Prior knowledge model is employed to alleviate this shortcome, but it is designed for generic face image space. In turn, there is a possibility that reconstructed HR image by existing algorithms may have serious artifacts and/or may not look like the original person. Therefore, making use of data constraint in low resolution pixel domain may not be good under VLR problem. It should be pointed out some existing algorithms do perform the data constraint in other spaces, such as DCT space [Zhang and Cham (2008)] or tensor space [Jia and Gong (2008); Liu *et al.* (2005)] or sparse coding dictionary [Yang *et al.* (2008)], but such spaces are induced from the LR space, and they cannot handle the VLR problem neither.

Theoretically, applying SR technique on the low-resolution (LR) face image, the reconstructed high-resolution (HR) image can be used for face recognition. This approach works well only if the input face image is frontal and captured under good illumination. In order to normalize the face image variations to frontal view before applying the SR method, a two-step approach has been proposed. Li and Lin [Li and Lin (2004)] made use of a view-based model to normalize the pose variations before the SR procedure. Jia and Gong [Jia and Gong (2008)] synthesized the HR images with different poses, lighting conditions and expressions based on tensor space. 3D-model are also adopted for handling the face variations. Mortazavian *et al.* [Mortazavian *et al.* (2009)] employed the 3DMM model to synthesis different pose images. Yu *et al.* [Yu *et al.* (2007)] modeled pose and illumination using bilinear function, and employed pose-illumination-based SR method to reconstruct images. However when the face image resolution is smaller than 16x16 pixels, the results of these normalization algorithms may not be satisfactory.

Moreover, several SR algorithms [Hennings-Yeomans *et al.* (2008); Li *et al.* (2008); Wang and Tang (2003)] have been proposed for recognition purpose, the information of the training data is not fully used, such as class label information which is very important from recognition perspective. This may restrict the performance of the reconstructed HR images in machine-based recognition system.

4.2. *Relationship Learning based Face Super-Resolution*

In VLR problem, the input VLR image space contains very little useful information, so existing data constraint may not be able to estimate the reconstruction error well. A better method is to estimate such error in HR image space. But the challenge is that given a VLR query image, the corresponding HR image is not available. Instead of recovering the HR image directly, Zou and Yuen [Zou and Yuen (2010b)] proposed a new SR algorithm to learn the relationship $\mathcal{R}$ between HR image space and VLR image space. After determining the relationship $\mathcal{R}$, HR image can be reconstructed by $\mathcal{R}$.

Given a set of training HR and VLR image pairs ($\{I_h^i, I_l^i\}_{i=1}^N$), let $\mathcal{R}$ be the relationship between the HR image space and the VLR image space, so that the HR image can be reconstructed from its VLR image and $\mathcal{R}$ as follows,

$$\tilde{I}_h = \mathcal{R}(I_l) + \epsilon \tag{7}$$

where ϵ is the noise. So the reconstruction error in HR image space, $e_h(\tilde{I}_h)$, is estimated by

$$e_h(\tilde{I}_h) = \|\tilde{I}_h - \mathcal{R}(I_l)\|^2 \tag{8}$$

There is no close form solution for the relationship between LR-HR images space. According to Eq. (3), the pixel value of LR image can be calculated by linear combination of the pixels on HR image at the corresponding position. We assume the relationship mapping from LR to HR image space is linear and denote as a matrix R_{rlfsr}. To determine the best R_{rlfsr}, we minimize the mean square error ($\mathcal{E}(R)$),

$$R_{rlfsr} = \arg\min_R \mathcal{E}(R) = \arg\min_R \sum_{i=1}^N \|I_h^i - RI_l^i\|^2 \tag{9}$$

Eq. (9) can be minimized using gradient-descent.

In query stage, given a testing VLR image I_l, the corresponding HR image is recovered as follows,

$$\tilde{I}_h = R_{rlfsr} I_l \tag{10}$$

It can be seen that the reconstruction of HR images is equivalent to applying R_{rlfsr} on the input space. We call this method as relationship learning (RL) based super-resolution.

4.3. *Kernel-based Relationship Learning for Face Super-Resolution*

To overcome the problems in face recognition from video, Zou and Yuen [Zou and Yuen (2010a)] proposed a kernel based face super-resolution algorithm. With the highly complex and nonlinear face image variations, the relationship between LR and HR images will be nonlinear. It is well-known that the nonlinear kernel mapping could transform complex distributed data into high dimensional feature space where the data becomes linear separable and the nonlinear relationship can be learnt.

Because of the nonlinear face variations, the relationship between LR images and HR images may be nonlinear and complicated. There does not exist close-form solution for finding $\mathcal{R}$, so that estimating $\mathcal{R}$ by Eq. (9) is not feasible. Zou and Yuen [Zou and Yuen (2010a)] employed the nonlinear mapping Φ to map the original image to the high dimension feature space. In this feature space, the relationship between HR and LR features can be better modeled by linear approximation. Following Eq. (9), we denote the relationship as R_{kfsr} which can be estimated by

$$R_{kfsr} = \arg\min_R \mathcal{E}(R) = \arg\min_R \sum_{i=1}^{N} \|R\Phi_L(I_l^i) - \Phi_H(I_h^i)\|^2 \tag{11}$$

Let $\mathbf{E}_H = \{e_H^1, e_H^2, \cdots\}$ and $\mathbf{E}_L = \{e_L^1, e_L^2, \cdots\}$ be the orthonormal bases of HR image feature space and LR image feature space. So we have:

$$\min \sum_{i=1}^{N} \|R\Phi_L(I_l^i) - \Phi_H(I_h^i)\|^2$$

$$= \min \sum_{i=1}^{N} \|R \sum_j < \Phi_L(I_l^i), e_L^j > e_L^j - \sum_k < \Phi_H(I_h^i), e_H^k > e_H^k\|^2$$

$$= \min \sum_{i=1}^{N} \|R\mathbf{E}_L f_L^i - \mathbf{E}_H f_H^i\|^2 = \min \sum_{i=1}^{N} \|\mathbf{E}_H^{-1} R\mathbf{E}_L f_L^i - f_H^i\|^2 \tag{12}$$

where f_L^j is the weight (coefficient) of image I_L^j in feature space. Let $R_{\hat{k}fsr} = \mathbf{E}_H^{-1} R\mathbf{E}_L$, to determine R_{kfsr} is equivalent to determine $R_{\hat{k}fsr}$. And we have

$$R_{\hat{k}fsr} = \arg\min_R \sum_{i=1}^{N} \|Rf_L^i - f_H^i\|^2 \tag{13}$$

However, it is computationally expensive to calculate the nonlinear mapping Φ explicitly due to the high dimensionality of the feature space, so it is not feasible to calculate the relationship $R_{\hat{k}fsr}$ by Eq. (13) directly. Kernel trick is used and kernel subspace is employed to represent the features in the kernel feature space. We estimate the reconstruction error on kernel subspace as follows:

$$\mathcal{E}(R) = \sum_{i=1}^{N} \|R\hat{f}_L^i - \hat{f}_H^i\|^2 \tag{14}$$

where $\hat{f}_L^i$ and $\hat{f}_H^i$ are the LR and HR kernel subspace coefficients for representing the image features, respectively. And R_{kfsr} is determined by minimizing Eq. (14).

Given the LR query image, the HR image kernel subspace features can be reconstructed by

$$\hat{f}_H = R_{\hat{k}fsr}\hat{f}_L \tag{15}$$

and $\hat{f}_H$ can be used for recognition directly for any kernel-based face recognition algorithms. Also, if the HR image is required, pre-image learning [Zheng *et al.* (2010)] can be adopted.

4.4. *Discriminative Super-Resolution*

Visual quality is not the only criterion for super-resolution applications, while the discriminability should be considered from machine-based recognition perspective. A discriminative constraint is designed to employ the class label information to learn the proper relationship $\mathcal{R}$, which can reconstruct the HR image with high discriminability.

According to Eq. (10), we get

$$\tilde{I}_h = \sum_{i=1}^{d_L} \mathrm{col}_i(R)I_l(i) \tag{16}$$

where d_L is the dimension of the I_l. $\mathrm{col}_i(R)$ is the i-th column of R, and $I_l(i)$ represent the i-th element of I_l. This implies that the reconstructed HR image locates in a subspace which is spanned by the columns of R, denoted as $\mathbf{S}_R$. According to Eq. (5), R is determined by minimizing the new data constraint and images in $\mathbf{S}_R$ have small reconstruction error. This inspires us to find a better subspace induced by R with other additional constraint(s), so that the reconstructed HR images have more discriminative features, rather than low reconstruction error. In order to further boost the discriminability of the reconstructed HR image, Zou and Yuen [Zou and Yuen (2010b)] proposed to use discriminative constraint integrating with relationship learning based super-resolution, and discriminative super-resolution (DSR) algorithm was proposed.

A natural step of improving the discriminability is to make use of the class label information from the training data. From recognition perspective, they expect the reconstructed HR images should be clustered with the images from the same class, and far away from the images from other classes. Inspired by maximum margin criterion (MMC) [Li et al. (2006)], a discriminative constraint is designed as follows:

$$d(R) = \text{mean}(\{\|I_h^i - RI_l^j\|^2 | \Omega(I_h^i) = \Omega(I_l^j)\})$$
$$- \text{mean}(\{\|I_h^i - RI_l^j\|^2 | \Omega(I_h^i) \neq \Omega(I_l^j)\}) \tag{17}$$

where $\Omega(u)$ is the class label of u. Integrating Eq. (17) with the reconstruction error term, the discriminative super-resolution formula can be written as:

$$R_{dsr} = \arg\min_R \mathcal{E}(R) + \gamma d(R) \tag{18}$$

where γ is a constant. The HR image can be reconstructed after R_{dsr} is determined.

The subspace induced by R_{dsr} is optimized for recognition with respect to Eq. (18). That means the HR images reconstructed by R_{dsr} are located in a subspace where they can be better linear separable. Therefore, the HR image reconstructed by R_{dsr} will contain more discriminability and be better for recognition purpose.

4.5. Experiments

Two experiments are designed to evaluate the performance of the relationship learning based super-resolution methods [Zou and Yuen (2010b)][Zou and Yuen (2010a)], as well as other methods, including Hallucinating Face (HF) [Baker and Kanade (2002)], Eigentransformation based Face super-resolution method (EF) [Wang and Tang (2005)], Kernel prior Face super-resolution (KF) [Chakrabarti et al. (2007)].

4.5.1. Experiments 1: Visual Quality

To estimate the effectiveness of Relationship Learning Based Super-Resolution (RLSR), we apply the relationship mapping of R_{rlfsr} (Eq. (9)) and other methods on CMU-PIE and FRGC databases.

Figures 7(a) and (g) show the input 7x6 query image and original 56x48 HR image. Figures 7(b) - (e) display the results using bicubic interpolation method, HF [Baker and Kanade (2002)] method, EF [Wang and Tang (2005)] method and KF [Chakrabarti et al. (2007)] method. It can be seen that both bicubic interpolation (BC) method and KF method give a relatively blur image and high frequency details cannot be recovered. Both HF method and EF method could recover some high frequent details. However, HF method generates some artifacts which degrade the human visual quality. The visual quality of the reconstructed images from EF method are good. However, when comparing with the original HR image, these HR image does not look like the original HR image. Figure 7(f) shows the results using Zou and Yuen's method [Zou and Yuen (2010b)]. It can be seen that their gives a good visual quality image which also look like the original one. Besides the subjective human visual quality, an objective estimation, namely sum of squared

Table 1. The SSE of different SR methods. [Zou and Yuen (2010b)]

Database	BC	HF [Baker and Kanade (2002)]	EF [Wang and Tang (2005)]	KF [Chakrabarti *et al.* (2007)]	RL [Zou and Yuen (2010b)]
CMU PIE	424.4	475.6	291.9	1143.1	179.9
FRGC 2.0	1259.4	1838.8	1510.1	1707.6	870.5

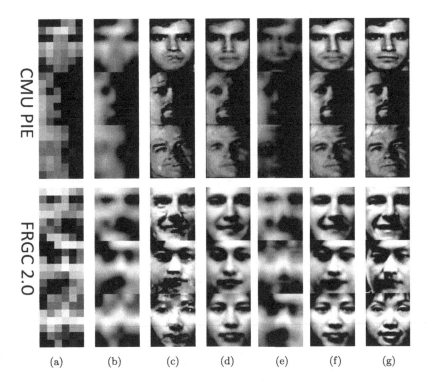

(a)　(b)　(c)　(d)　(e)　(f)　(g)

Fig. 7.　SR results: (a) input VLR images (7 x 6), (b) SR results by Bicubic interpolation, (c) by Hallucinating Face method (HF) [Baker and Kanade (2002)], (d) Eigentransfromation based Face SR method (EF) [Wang and Tang (2005)], (e) KPCA-based Face SR method (KF) [Chakrabarti *et al.* (2007)], (f) RL (R_{rlfsr}) [Zou and Yuen (2010b)], (g) original HR images. The resolution of reconstructed HR images is 56 x 48.

error (SSE), is recorded as shown in Table 1. It can be seen that Zou and Yuen's method [Zou and Yuen (2010b)] method obtained the lowest SSE.

4.5.2. *Experiments 2: Face Recognition*

This experiment is designed to evaluate the performance of the discriminative super-resolution (DSR) algorithm in terms of the recognition results as shown in Figures 8, 9, 10 and 11. We perform the DSR learning with R_{rlfsr} in pixel domain and with R_{kfsr} in kernel feature domain.

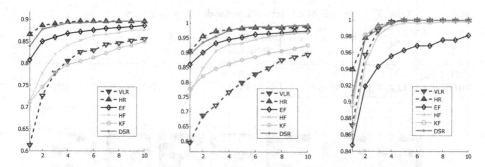

Fig. 8.   Results of DSR with R_{rlfsr} on CMU PIE: Eigenface (left); Kernel PCA (middle); SVM (right). [Zou and Yuen (2010b)]

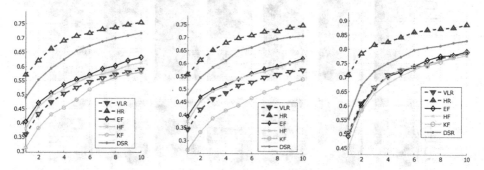

Fig. 9.   Results of DSR with R_{rlfsr} on FRGC: Eigenface (left); Kernel PCA (middle); SVM (right). [Zou and Yuen (2010b)]

The experimental results show that:

- There is a significant drop of recognition accuracy (as high as 30%) for VLR image, comparing with the original HR image, for all recognition engines on both CMU-PIE and FRGC databases.
- The discriminative super-resolution (DSR) method [Zou and Yuen (2010b)] outperforms other SR methods. It implies that the reconstructed HR image using the DSR has high discriminability for recognition purpose.
- The recognition accuracy of Zou and Yuen's [Zou and Yuen (2010b)] method on real surveillance video also have considerable improvement, even though such database is very challenging.

5. Open Issues: The Current Challenges

In the previous sections, we have discussed the popular FSR algorithms and one of the challenging problems: very low resolution problem. Although FSR has been studied for more than ten years, there are some open issues which limit the performance of the practical FSR algorithms.

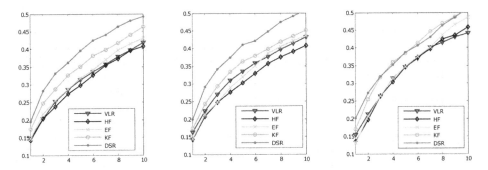

Fig. 10. Results of DSR with R_{rlfsr} on SCface [Grgic *et al.* (2009)]: Eigenface (left); Kernel PCA (middle); SVM (right). [Zou and Yuen (2010b)]

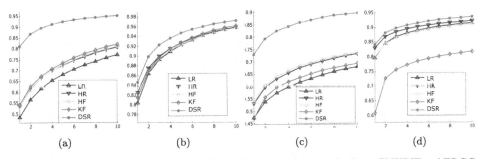

Fig. 11. Results of DSR with R_{kfsr} on different face recognition methods on CMUPIE and FRGC database: (a) KPCA on CMUPIE (b) KDDA on CMUPIE (c) KPCA on FRGC (d) KDDA on FRGC. The results are reported in CMC curves. [Zou and Yuen (2010a)]

As discussed in Section 2.1.1, the image registration is important for the success of the super-resolution algorithms. The performance of the image registration algorithms is bounded in general [Robinson and Milanfar (2004)]. There are many promising algorithms specifically for face images, such as Stasm [Milborrow and Nicolls (2008)], BAM [Liu (2009)]. However, when it comes to the low resolution face, the registration problem becomes challenging.

Another open issue is that the face image variations affect the success of the FSR algorithms. In general, a critical point for the success of FSR is to calculate the similarity of two images (patches). It is well-known that the similarity between face images with variations may not be easily evaluated. There, FSR algorithm may not work well under face variations.

Acknowledgement

This project is partially supported by Science Faculty Research grant of Hong Kong Baptist University and NSFC-GuangdDong research grant U0835005.

References

Baker, S. and Kanade, T. (2002). Limits on super-resolution and how to break them, *IEEE Transactions on Pattern Analysis and Machine Intelligence* **24**, 9, pp. 1167–1183.

Chakrabarti, A., Rajagopalan, A. N. and Chellappa, R. (2007). Super-resolution of face images using kernel pca-based prior, *IEEE Transactions on Multimedia* **9**, 4, pp. 888–892.

Grgic, M., Delac, K. and Grgic, S. (2009). SCface - surveillance cameras face database, *Multimedia Tools and Applications Journal DOI:10.1007/s11042-009-0417-2* .

Hennings-Yeomans, P. H., Baker, S. and Kumar, B. (2008). Simultaneous super-resolution and feature extraction for recognition of low-resolution faces, in *Proceedings of IEEE International Conference on Computer Vision and Pattern Recognition*, pp. 1–8.

Hu, Y., Lam, K., Qiu, G. and Shen, T. (2010). From Local Pixel Structure to Global Image Super-resolution: A New Face Hallucination Framework. *IEEE Transactions on Image Processing* **20**, 2, pp. 433–445.

Jia, K. and Gong, S. (2008). Generalized face super-resolution, *IEEE Transactions on Image Processing* **17**, 6, pp. 873–886.

Li, B., Chang, H., Shan, S. and Chen, X. (2009). Low-Resolution Face Recognition via Coupled Locality Preserving Mappings, *Signal Processing Letters* **17**, 1, pp. 20–23.

Li, B., Chang, H., Shan, S., Chen, X. and Gao, W. (2008). Hallucinating facial images and features, in *Proceedings of IEEE International Conference on Pattern Recognition*, pp. 1–4.

Li, H., Jiang, T. and Zhang, K. (2006). Efficient and robust feature extraction by maximum margin criterion, *IEEE Transactions on Neural Networks* **17**, 1, pp. 157–165.

Li, Y. and Lin, X. (2004). Face hallucination with pose variation, in *Proceedings of IEEE International Conference on Automatic Face and Gesture Recognition*, pp. 723–728.

Liu, C., Shum, H. and Zhang, C. (2001). A two-step approach to hallucinating faces: Global parametric model and local nonparametric model, in *Proceedings of The IEEE International Conference on Computer Vision*.

Liu, C., Shum, H. Y. and Freeman, W. T. (2007). Face hallucination: Theory and practice, *International Journal of Computer Vision* **75**, 1, pp. 115–134.

Liu, W., Lin, D. and Tang, X. (2005). Hallucinating faces: Tensorpatch super-resolution and coupled residue compensation, in *Proceedings of IEEE International Conference on Computer Vision and Pattern Recognition*, Vol. 2, pp. 478 – 484.

Liu, X. (2009). Discriminative face alignment, *IEEE Transactions on Pattern Analysis and Machine Intelligence* **31**, pp. 1941–1954.

Lui, Y. M., Bolme, D., Draper, B. A., Beveridge, J. R., Givens, G. and Phillips, P. J. (2009). A meta-analysis of face recognition covariates, in *Proceedings of International Conference on Biometrics: Theory, Applications and Systems*.

Milborrow, S. and Nicolls, F. (2008). Locating facial features with an extended active shape model, in *LNCS: European Conference on Computer Vision*, Vol. 5305, pp. 504–513.

Mortazavian, P., Kittler, J. and Christmas, W. (2009). A 3-D assisted generative model for facial texture super-resolution, in *Proceedings of IEEE International Conference on Biometrics: Theory, Applications, and Systems*.

Park, J. S. and Lee, S. W. (2008). An example-based face hallucination method for single-frame, low-resolution facial images, *IEEE Transactions on Image Processing* **17**, 10, pp. 1806–1816.

Park, S. C., Park, M. K. and Kang, M. G. (2003). Super-resolution image reconstruction: a technical overview, *IEEE Signal Processing Magazine* **20**, 3, pp. 21–36.

Robinson, D. and Milanfar, P. (2004). Fundamental performance limits in image registration, *IEEE Transactions on Image Processing* **13**, 9, pp. 1185–1199.

van Ouwerkerk, J. (2006). Image super-resolution survey, *Image and Vision Computing* **24**, 10, pp. 1039 – 1052.

Wang, X. and Tang, X. (2003). Face hallucination and recognition, in *LNCS: Advances in Neural Networks*, pp. 486–494.

Wang, X. and Tang, X. (2005). Hallucinating face by eigentransformation, *IEEE Transactions on Systems, Man, and Cybernetics, Part C: Applications and Reviews* **35**, 3, pp. 425–434.

Yang, J. and Huang, T. (2010). Image Super-Resolution: Historical Overview and Future Challenges, *Super-Resolution Imaging. CRC Press* , pp. 1–33.

Yang, J., Tang, H., Ma, Y. and Huang, T. (2008). Face hallucination via sparse coding, in *Proceedings of IEEE International Conference on Image Processing*, pp. 1264–1267.

Yang, J., Wright, J., Huang, T. and Ma, Y. (2010). Image super-resolution via sparse representation, *IEEE Transactions on Image Processing* **19**, 11, pp. 2861–2873.

Yu, J., Bhanu, B., Xu, Y. and Roy-Chowdhury, A. (2007). Super-resolved facial texture under changing pose and illumination, in *Proceedings of IEEE International Conference on Image Processing*.

Zhang, W. and Cham, W. K. (2008). Learning-based face hallucination in dct domain, in *Proceedings of IEEE International Conference on Computer Vision and Pattern Recognition*, pp. 1–8.

Zheng, W., Lai, J. and Yuen, P. (2010). Penalized preimage learning in kernel principal component analysis, *IEEE Transactions on Neural Networks* **21**, 4, pp. 551–570.

Zou, W. and Yuen, P. C. (2010a). Learning the relationship between high and low resolution images in kernel space for face super-resolution, in *Proceedings of International Conference on Pattern Recognition*, Vol. 1, pp. 1152–1155.

Zou, W. and Yuen, P. C. (2010b). Very low resolution face recognition problem, in *Proceedings of IEEE International Conference on Biometrics: Theory, Applications And Systems*.

CHAPTER 1.3

CLASSIFICATION AND CLUSTERING OF VECTOR SPACE EMBEDDED GRAPHS

Kaspar Riesen* and Horst Bunke

Institute of Computer Science and Applied Mathematics, University of Bern, Neubrückstrasse 10, CH-3012 Bern, Switzerland
*riesen@iam.unibe.ch

Due to the ability of graphs to represent properties of entities and binary relations at the same time, a growing interest in graph based object representation can be observed. Yet, graphs are still not the common data structure in pattern recognition and related fields. The reason for this phenomenon is twofold. First, even basic mathematical operations cannot be defined in a standard way for graphs. Second, we observe a significant increase of the complexity of many algorithms when graphs rather than feature vectors are employed. In conclusion, almost none of the standard methods for pattern recognition can be applied to graphs without significant modifications. This chapter is concerned with a novel approach to graph based pattern recognition based on vector space embedding of graphs. The key idea of our embedding framework is to regard dissimilarities of an input graph to some prototypical graphs as a vectorial description of the graph. Our graph embedding framework crucially relies on the computation of graph dissimilarities. In the present chapter the concept of graph edit distance is actually used for this task. Due to the flexibility of graph edit distance, the proposed embedding procedure is particularly able to cope with noisy data and can thus be applied to virtually any kind of graphs. In an experimental evaluation the power and applicability of the proposed graph embedding framework is empirically verified on six graph data sets. The main finding of the experimental evaluation is that the embedding procedure using dissimilarities with subsequent classification or clustering has great potential to outperform traditional approaches in graph based pattern recognition.

1. Introduction

After decades of focusing on independent and identically-distributed representation formalisms, more and more effort is now rendered in various research fields on structured data (e.g. in chemoinformatics[1] or in document analysis[2]). That is, the intensive study of problems where the objects under consideration consist of interrelated entities has emerged rapidly in the last decade. Therefore, a continuously growing interest in graph-based object representation for pattern recognition and machine learning can be observed.[3-5] As a matter of fact, object representation by means of graphs is advantageous compared to vectorial approaches because of two reasons. First, graphs are able to represent not only the values of object properties, i.e. features, but can be used to explicitly model relations that exist between different parts of an object. Second, graphs do not suffer from the constraint of fixed dimensionality. That is, the number of nodes and edges in a graph is not limited a priori and depends on the size and the complexity of the actual object to be modeled. This is in sharp contrast to vectorial representation where the dimensionality of the feature space has to be defined beforehand and all objects are represented by the same number of features.

Yet, one drawback of graphs, when compared to feature vectors, is the significantly increased complexity of many algorithms. Nevertheless, new computer generations, which are now able to more efficiently handle complex data structures, as well as the development of fast approximate algorithms for graph comparison definitively empower researchers to use graphs in their respective problem domains.[3] Yet, another serious limitation in the use of graphs for object classification or clustering arises from the fact that there is little mathematical structure in the domain of graphs. For example, computing the (weighted) sum or the product of a pair of entities, which are elementary operations needed in many standard algorithms in pattern recognition, is not possible in the domain of graphs, or is at least not defined in general for graph structures.

In order to overcome the resulting lack of algorithmic tools applicable to graphs, the traditional approach is to give up the universality of graphs in favor of efficient feature vectors, even in problem domains where graphs would be the method of choice.[6] A promising approach to overcoming the lack of algorithmic tools without losing the power of graphs is offered through graph embedding in vector spaces. Basically, such an embedding of graphs establishes access to the rich repository of algorithmic tools for pattern recognition originally reserved for vectorial data.

A prominent class of graph embedding is based on spectral methods.[7-10] The basic idea is to represent graphs by the eigendecomposition of their structural matrices (i.e. adjacency or Laplacian matrix). The resulting representation exhibits interesting properties for pattern recognition. However, the sensitiveness of the eigendecomposition towards structural errors is a major problem of all spectral methods. That is, spectral methods are not fully able to cope with larger amounts of structural noise, such as missing or spurious nodes or edges. Furthermore, most spectral approaches are only applicable to unlabeled graphs or graphs with severely constrained label alphabets. Only recently, first attempts to overcome this limitations have been reported.[11]

The objective of the present chapter is to define a new class of graph embedding procedures which can be applied to both directed and undirected graphs, as well as to graphs with arbitrary labels on their nodes and/or edges. Furthermore, the novel graph embedding framework is distinguished by its ability to handle structural errors. The novel approach to graph embedding is primarily based on the idea proposed in Ref. 12 where the dissimilarity representation for pattern recognition in conjunction with feature vectors was first introduced. The key idea is to use the dissimilarities of an input graph g to a number of training graphs, termed *prototypes*, as a vectorial description of g. In other words, the dissimilarity representation rather than the original graph representation is used for pattern recognition.

The remainder of this chapter is organized as follows. Next, basic concepts and notation are introduced. Particularly, the concept of graph based representation and the dissimilarity model of graph edit distance are defined. In Sect. 3 the general graph embedding procedure and some basic properties are outlined. Also the relationship of the proposed embedding framework to kernel methods as well as the basic problem of prototype selection (which arises with the proposed procedure) are discussed. In Sect. 3.2.1 several prototype selection strategies are described in detail. Then, in Sect. 4, an experimental evaluation of the embedding framework is carried out. Finally, in Sect. 5, conclusions are drawn.

2. Basic Concepts and Notation

2.1. *Graph Based Pattern Representation*

Definition 1 (Graph). *Let L_V and L_E be a finite or infinite label set for nodes and edges, respectively. A graph g is a four-tuple $g = (V, E, \mu, \nu)$, where V is the finite set of nodes, $E \subseteq V \times V$ is the set of edges, $\mu : V \to L_V$ is the node labeling function, and $\nu : E \to L_E$ is the edge labeling function.*

Definition 1 allows us to handle arbitrarily structured graphs with unconstrained labeling functions. For example, the labels for both nodes and edges can be given by the set of integers $L = \{1, 2, 3, \ldots\}$, the vector space $L = \mathbb{R}^n$, or a set of symbolic labels $L = \{\alpha, \beta, \gamma, \ldots\}$. Given that the nodes and/or edges are labeled, the graphs are referred to as *labeled graphs*. *Unlabeled graphs* are obtained as a special case by assigning the same label ε to all nodes and edges, i.e. $L_V = L_E = \{\varepsilon\}$.

Edges are given by pairs of nodes (u, v), where $u \in V$ denotes the source node and $v \in V$ the target node of a directed edge. The definition given above corresponds to the case of *directed graphs*. Moreover, the class of *undirected graphs* can be modeled by inserting a reverse edge $(v, u) \in E$ with identical label for each edge $(u, v) \in E$.

In the remainder of this chapter, symbol $\mathcal{G}$ will be used to denote the set of all graphs that can be built with labels from L_V and L_E.

2.2. *Graph Matching by Means of Graph Edit Distance*

Various procedures for evaluating proximity, i.e. similarity or dissimilarity, of graphs have been proposed in the literature.[3] The process of evaluating the similarity of two graphs is commonly referred to as *graph matching*. The overall aim of graph matching is to find a correspondence between the nodes and edges of two graphs that satisfies some, more or less, stringent constraints.

Roughly speaking, there are two categories of algorithms in graph matching, viz. *exact graph matching* and *inexact graph matching*. In the former approach, for a matching to be established, it is required that a strict correspondence between the two graphs under consideration, or at least among their subparts, is found.[13] Based on the exact graph matching paradigms of maximum common subgraph and minimum common supergraph, some graph similarity measures have been proposed.[14-16] However, the main restriction of exact graph matching and related similarity measures is the requirement that a significant part of the topology together with the corresponding node and edge labels in two graphs are identical for obtaining a high degree of similarity. In fact, this is not realistic for many real world applications. Especially if the nodes and/or the edges are continuously labeled, the exact matching paradigm is too restrictive.

In order to make graph matching better applicable to real world problems, several error-tolerant, or inexact, graph matching methods have been proposed.[3] In this approach the requirement of strict correspondences between the nodes of the graphs to be matched is substantially relaxed. That is, inexact matching algorithms are endowed with a certain tolerance to errors, enabling them to detect similarities in a more general way than the exact matching approach. For an extensive review of graph matching methods and applications, the reader is referred to Ref. 3.

In this chapter we focus on graph edit distance,[17,18] which is one of the most flexible methods for error-tolerant graph matching. The key idea of graph edit distance is to define the dissimilarity, or distance, of graphs by the minimum amount of distortion that is needed to transform one graph into another. A standard set of distortion operations is given by *insertions*, *deletions*, and *substitutions* of nodes and edges. Other operations, such as *merging* and *splitting* of nodes,[19] can be useful in certain applications but are not considered in this chapter.

Given two graphs, the source graph g and the target graph g', the main idea underlying graph edit distance computation is to delete some nodes and edges from g, relabel (substitute) some of the remaining nodes and edges, and insert some nodes and edges in g', such that g is finally transformed into g'. A sequence of edit operations $e_1, \ldots, e_k$ that transform g into g' is called an *edit path* between g and g'. In Fig. 1 an example of an edit path between two graphs is given. Obviously, for every pair of graphs (g, g'), there exist a number of different edit paths transforming g into g'. Let $\Upsilon(g, g')$ denote the set of all such edit paths. To find the most suitable edit path out of $\Upsilon(g, g')$, one introduces a cost for each edit operation, measuring the strength of the corresponding operation. The idea of such cost functions is to define whether or not an edit operation represents a strong modification of the graph. Hence, between two similar graphs, there should exist an inexpensive edit path, representing low cost operations, while for substantially different graphs an edit path with high costs is needed. Consequently, the *edit distance* of two graphs is defined by the minimum cost edit path between two graphs.

Definition 2 (Graph Edit Distance). *Let* $g = (V, E, \mu, \nu)$ *be the source graph and* $g' = (V', E', \mu', \nu')$ *be the target graph. The graph edit distance between* g *and* g' *is defined by*

$$d(g, g') = \min_{(e_1, \ldots, e_k) \in \Upsilon(g, g')} \sum_{i=1}^{k} c(e_i),$$

where $\Upsilon(g, g')$ *denotes the set of edit paths transforming* g *into* g', *and* c *denotes the edit cost function measuring the strength* $c(e_i)$ *of edit operation* e_i.

Fig. 1. A possible edit path between graph g_1 and g_2 (node labels are represented by different shades of grey).

The computation of the edit distance is usually carried out by means of a tree search algorithm, which explores the space of all possible mappings of the nodes and edges of the first graph to the nodes and edges of the second graph. Due to the fact that the error-tolerant nature of edit distance potentially allows every node of a graph g to be mapped to every node of another graph g', the computational complexity of the edit distance algorithm is exponential in the number of nodes of the involved graphs. This means that the running time and space complexity may be huge even for rather small graphs.

In order to reduce the complexity of graph edit distance computation, *approximate*, or *suboptimal*, matching algorithms can be used instead of exact, or optimal, ones. In contrast to *optimal* error-tolerant graph matching, approximate algorithms do not guarantee to find the global minimum of the matching cost, but only a local one. Usually this approximation is not very far from the global one, but there are no guarantees, of course.[3] The approximate nature of such suboptimal algorithms is typically repaid by polynomial matching time.

In some suboptimal approaches, the basic idea is to perform a local search to solve the graph matching problem, that is, to optimize local criteria instead of global, or optimal ones.[20] In Ref. 21, a linear programming method for computing the edit distance of graphs with unlabeled edges is proposed. The method can be used to derive lower and upper edit distance bounds in polynomial time. Two fast but suboptimal algorithms for graph edit distance computation are proposed in Ref. 22. The authors propose simple variants of an optimal edit distance algorithm that make the computation substantially faster. Another approach to efficient graph edit distance computation has been proposed in Ref. 23. The basic idea is to decompose graphs into sets of subgraphs. These subgraphs consist of a node and its adjacent nodes and edges. The graph matching problem is then reduced to the problem of finding an optimal match between the sets of subgraphs by

means of dynamic programming. In Ref. 24 another efficient algorithm for solving
the problem of graph edit distance computation is introduced. This approach is
somewhat similar to the method described in Ref. 23, i.e. the graph edit distance is
approximated by finding an optimal match between nodes of two graphs and their
local structure. However, in order to find the optimal match, a bipartite matching
procedure is used rather than dynamic programming. Because of its suboptimal
nature, this method does not generally return the optimal edit path, but only an
approximate one. Nonetheless, it has been shown to work well in various graph
classification tasks.[24,25]

3. Graph Embedding by Means of Dissimilarity Representation

3.1. *General Embedding Procedure*

The idea of the proposed graph embedding framework stems from the seminal work
done by Duin and Pekalska[12] where dissimilarities for pattern representation are
used for the first time. Later this method was extended so as to map string representations into vector spaces.[26] In the current work we go one step further and
generalize and substantially extend the methods described in Refs. 12, 26 to the
domain of graphs. The key idea of this approach is to use the distances of an input graph to a number of training graphs, termed *prototype graphs*, as a vectorial
description of the graph.

Assume we have a set of sample graphs, $\mathcal{T} = \{g, \ldots, g_N\}$ from some graph
domain $\mathcal{G}$ and an arbitrary graph dissimilarity measure $d : \mathcal{G} \times \mathcal{G} \to \mathbb{R}$. Note that
$\mathcal{T}$ can be any kind of graph set. However, for the sake of convenience we assume
in the following that $\mathcal{T}$ is a training set of given graphs. After selecting a set of
prototypical graphs $\mathcal{P} \subseteq \mathcal{T}$, we compute the dissimilarity of a given input graph
g to each prototype graph $p_i \in \mathcal{P}$. Note that g can be an element of $\mathcal{T}$ or any
other graph set $\mathcal{S}$. Given n prototypes, i.e. $\mathcal{P} = \{p_1, \ldots, p_n\}$, this procedure leads
to n dissimilarities, $d_1 = d(g, p_1), \ldots, d_n = d(g, p_n)$, which can be arranged in an
n-dimensional vector $(d_1, \ldots, d_n)$.

Definition 3 (Graph Embedding). *Let us assume a graph domain $\mathcal{G}$ is given.*
If $\mathcal{T} = \{g, \ldots, g_N\} \subseteq \mathcal{G}$ is a training set with N graphs and $\mathcal{P} = \{p_1, \ldots, p_n\} \subseteq \mathcal{T}$
is a prototype set with n graphs, the mapping

$$\varphi_n^{\mathcal{P}} : \mathcal{G} \to \mathbb{R}^n$$

is defined as the function

$$\varphi_n^{\mathcal{P}}(g) = (d(g, p_1), \ldots, d(g, p_n)),$$

where $d(g, p_i)$ is any graph dissimilarity measure between graph g and the i-th pro-
totype graph.

Obviously, by means of this definition we obtain a vector space where each axis corresponds to a prototype graph $p_i \in \mathcal{P}$ and the coordinate values of an embedded graph g are the distances of g to the elements in $\mathcal{P}$. In this way we can transform any graph g from the training set $\mathcal{T}$ as well as any other graph set $\mathcal{S}$ (for instance a validation or a test set of a classification problem), into a vector of real numbers. The embedding procedure is illustrated in Fig. 2.

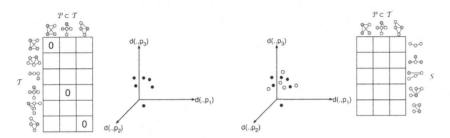

Fig. 2. Based on a prototype set $\mathcal{P} \subseteq \mathcal{T}$ the graphs from the training set $\mathcal{T}$ as well as graphs from some additional data set $\mathcal{S}$ are transformed into a vector of dissimilarities. Black dots represent embedded graphs from $\mathcal{T}$, while white dots are embedded graphs from $\mathcal{S}$. Note that training graphs, which have been selected as prototypes before, have a zero entry in their corresponding vector.

The embedding procedure proposed in this chapter makes use of graph edit distance. Note, however, that any other graph dissimilarity measure can be used as well. Yet, using graph edit distance allows us to deal with a large class of graphs (directed, undirected, unlabeled, node and/or edge labels from any finite or infinite domain). Furthermore, due to the flexibility of graph edit distance, a high degree of robustness against various graph distortions can be expected. Hence, in contrast with other graph embedding techniques (e.g. spectral methods) our approach is characterized by a high degree of flexibility in the graph definition and a high tolerance to structural errors in the graph matching process. Since the computation of graph edit distance is exponential in the number of nodes for general graphs, the complexity of this graph embedding is exponential as well. However, as mentioned in Sect. 2, there exist efficient approximation algorithms for graph edit distance computation with cubic time complexity (e.g. the procedure described in Ref. 24). Consequently, given n predefined prototypes the embedding of one particular graph is established by means of n distance computations with polynomial time.

3.2. *Relation to Kernel Methods*

Dissimilarity embedding is closely related to kernel methods.[27,28] In the kernel approach the patterns are described by means of pairwise kernel functions, and in the dissimilarity approach patterns are described by pairwise dissimilarities. However, there is one fundamental difference between kernels and dissimilarity embeddings. In the former method, the kernel values are interpreted as dot products in some implicit feature space. By means of kernel machines, the pattern recognition task is eventually carried out in this kernel feature space. In the latter approach, the set of dissimilarities is interpreted as a vectorial description of the pattern under consideration. Hence, no implicit feature space, but an explicit dissimilarity space is obtained.

Although conceptually different, the embedding paradigm established by φ_n^P : $\mathcal{G} \to \mathbb{R}^n$ constitutes a foundation for a novel class of graph kernels. One can define a valid graph kernel κ based on the graph embedding $\varphi_n^P : \mathcal{G} \to \mathbb{R}^n$ by computing the standard dot product of two graph maps in the resulting vector space[a]

$$\kappa_{\langle\rangle}(g, g') = \langle \varphi_n^P(g), \varphi_n^P(g') \rangle \quad .$$

Of course, not only the standard dot product can be used but any valid kernel function defined for vectors. For instance an RBF kernel function

$$\kappa_{RBF}(g, g') = \exp\left(-\gamma \|\varphi_n^P(g) - \varphi_n^P(g')\|^2\right)$$

with $\gamma > 0$ can thus be applied to graph maps. We refer to this procedure as *graph embedding kernel* using some specific vector kernel function.

This graph embedding kernel is somewhat similar in spirit to the subgraph kernel proposed in Ref. 30. However, rather than using a potentially infinite set of subgraphs, we restrict our considerations to a finite number of prototypes, i.e. dimensions. Furthermore, rather than counting the number of occurrences of a particular subgraph within some host graph, we employ the edit distance in order to capture graph similarity. Obviously, large values of the kernel function occur when both g and g' are similar to the same prototype graphs. Clearly, whenever both g and g' are similar to the same prototype graphs we may conclude that g and g' are similar to each other. Using the graph edit distance rather than counting the number of occurrences of certain subgraphs allows us to flexibly deal with various kinds of distortions that may affect the underlying graphs.

In a recent book, graph kernels were proposed that directly use graph edit distances.[31] This approach turns an existing dissimilarity measure (e.g. graph edit distance) into a similarity measure by mapping low distance values to high similarity values and vice versa. To this end monotonically decreasing transformations are used. Given the edit distance $d(g, g')$ of two graphs g and g', the similarity kernel is defined, for instance, as $\kappa(g, g') = \exp(-\gamma\, d(g, g'))$. Note that this kernel function is not positive definite in general. However, there is theoretical evidence that using kernel machines in conjunction with indefinite kernels may be reasonable if some conditions are fulfilled.[32] We refer to this method as *similarity kernel*, or *sim* for short, in the following.

Note the fundamental difference between such an approach and our graph embedding procedure. While with the similarity kernel the existing dissimilarity measure is turned into a similarity measure (i.e. a kernel function) and subsequently

---

[a]Note that is approach is very similar to the empirical kernel map described in Ref. 29 where general similarity measures are turned into kernel functions.

plugged into a kernel machine, the dissimilarity embedding of graphs uses the distances to n prototypes as features for a new explicit vectorial description of the underlying object.

3.2.1. *The Problem of Prototype Selection*

The selection of the n prototypes $\mathcal{P} = \{p_1, \ldots, p_n\}$ is a critical issue in graph embedding since not only the prototypes $p_i \in \mathcal{P}$ themselves but also their number n affect the resulting graph mapping $\varphi_n^{\mathcal{P}}(\cdot)$ and thus the performance of the corresponding pattern recognition algorithm. A good selection of n prototypes seems to be crucial to succeed with the algorithm in the embedding vector space. A first and very simple solution might be to use all available training graphs from $\mathcal{T}$ as prototypes. Yet, two severe shortcomings arise with such a plain approach. First, the dimensionality of the resulting vector space is equal to the size N of the training set $\mathcal{T}$. Consequently, if the training set is large, the mapping results in (possibly too) high dimensional feature vectors disabling efficient computations. Second, the presence of similar prototypes as well as outlier graphs in the training set $\mathcal{T}$ is likely. Therefore, redundant and noisy or irrelevant information will be captured in the graph maps which in turn may harm the performance of the algorithms applied subsequently.

The selection of prototypes for graph embedding has been addressed in various papers.[25,33,34] In,[25] for instance, a number of *prototype selection methods* are discussed. These selection strategies use some heuristics based on the underlying dissimilarities in the original graph domain. The basic idea of these approaches is to select prototypes from $\mathcal{T}$ that best reflect the distribution of the training set $\mathcal{T}$ or cover a predefined region of $\mathcal{T}$. The rationale of this procedure is that capturing distances to significant prototypes from $\mathcal{T}$ leads to meaningful dissimilarity vectors. All of the proposed prototype selection methods can be applied class-wise or class-independent. Class-independent selection means that the selection is executed over the whole training set $\mathcal{T}$ to get n prototypes, while in class-wise selection the selection is performed individually for each of the classes occurring in $\mathcal{T}$.

The Border prototype selector (bps), for instance, selects prototypes situated at the border of the training set $\mathcal{T}$.[25] In Fig. 3 (a) and (b) the classindependent and the classwise bps is illustrated, respectively[b].

Another prototype selection strategy is spanning selection (sps). This iterative selection algorithm considers all distances to the prototypes selected before. The first prototype is the set median graph $median(\mathcal{T})$.[35] Each additional prototype is the graph furthest away from the already selected prototype graphs.

The k-Centers prototype selector (k-cps) is a another strategy for prototype selection, which is based on k-medians clustering[36] (cf. Fig. 3 (d)). The prototypes

---

[b]The data underlying these illustrative examples are two-dimensional vectors obtained through multidimensional scaling applied to the original graph edit distances from the Letter data set (this data set is used in the experimental evaluation and is described in detail in Sect. 4).

are given by the set medians of the n disjoint clusters obtained from k-medians clustering of the training set $\mathcal{T}$. In contrast to sps, we note that k-cps avoids selecting prototypes from the border by focusing on graphs that are in the center of densely populated areas (cf. Fig. 3 (c) and (d) for a comparison between sps and k-cps).

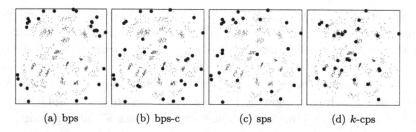

      (a) bps   (b) bps-c   (c) sps   (d) k-cps

Fig. 3. Illustration of the different prototype selectors applied to the training set $\mathcal{T}$ of the Letter data set. The number of prototypes is defined by $n = 30$. The prototypes selected by the respective algorithms are represented by bold dots.

Another solution to the problem of embedding noisy and redundant vectors with too high dimensionality is offered by the following procedure. Rather than selecting the prototypes beforehand, the embedding is carried out first and then the problem of prototype selection is reduced to a feature subset selection problem. That is, we define $\mathcal{P} = \mathcal{T}$ and use all available elements from the training set for graph embedding. Next, a large number of different feature selection strategies can be applied to the resulting large scale vectors eliminating redundancies and noise, finding good features, and reducing the dimensionality. In,[33] for instance, principal component analysis (PCA) and Fisher linear discriminant analysis (LDA) are applied to the vector space embedded graphs. Kernel PCA,[37] rather than traditional PCA, is used for feature transformation in.[34]

4. Experimental Evaluation

In this section we provide results of an experimental evaluation of the proposed embedding procedure. We aim at empirically confirming that the method of prototype based graph embedding and subsequent classification or clustering in real vector spaces is applicable to different graph classification and clustering problems and matches, or even surpasses, the performance of traditional techniques. For graph edit distance computation the suboptimal algorithm introduced in Ref. 24 has been used. This graph edit distance algorithm shows superior performance in time and accuracy compared to other suboptimal algorithms. For prototype selection we focus on three classwise selectors, viz. bps-c, sps-c, and k-cps-c. The classifier used in the embedding space is the support vector machine (SVM).[38] Of course, any other classifier could be used for this purpose as well. However, we feel that the SVM is particularly suitable because of its theoretical advantages and its superior performance that has been empirically confirmed in many practical classification problems.

The k-means algorithm is one of the most popular clustering algorithms in pattern recognition and related areas. In the present chapter this particular algorithm is employed for clustering both the original graph data and the vector space embedded graphs. In the graph domain, the original clustering algorithm is modified in the sense that the underlying distance function is given by the graph edit distance and the mean of a certain cluster is defined as the set median graph. We denote k-means applied to graphs as k-medians and use this system as a reference method for the purpose of comparisons. For clustering of the vector space embedded graphs, a kernelized version of the original k-means algorithm is used.[39]

4.1. *Graph Data Sets*

For evaluation, six data sets from the IAM graph database repository[c] for graph based pattern recognition and machine learning are used.[40] The first graph data set involves graphs that represent distorted letter drawings (Letter). We consider the 15 capital letters of the Roman alphabet that consist of straight lines only (A, E, F, ..., Z). For each class, a prototype line drawing is manually constructed. These prototype drawings are then converted into prototype graphs by representing lines by undirected edges and ending points of lines by nodes. Each node is labeled with a two-dimensional attribute giving its position relative to a reference coordinate system. Edges are unlabeled.

The second graph set (Digit) consists of graphs representing handwritten digits[41] (0, 1, 2, ..., 9). During the recording of the digits, the position of the pen was recorded at constant time intervals. The resulting sequences of (x, y)-coordinates were converted into graphs by inserting nodes in regular intervals between the starting and ending points of a line. Successive nodes are connected by undirected edges. Each node is labeled with a two-dimensional attribute giving its position relative to a reference coordinate system. The edges are attributed with an angle denoting the orientation of the edge with respect to the horizontal direction.

The third graph data set consists of graphs representing fingerprint images out of the four classes *arch*, *left*, *right*, and *whorl* from the Galton-Henry classification system (Fingerprint). The fingerprint database used in our experiments is based on the NIST reference database of fingerprints.[42] In order to obtain graphs from fingerprint images, the relevant regions are binarized and a noise removal and thinning procedure is applied. This results in a skeletonized representation of the extracted regions. Ending points and bifurcation points of the skeletonized regions are represented by nodes. Additional nodes are inserted in regular intervals between ending points and bifurcation points. Finally, undirected edges are inserted to link nodes that are directly connected through a ridge in the skeleton. Each node is labeled

---

[c]`www.iam.unibe.ch/fki/databases/iam-graph-database`

Table 1. Summary of graph data set characteristics, viz. the size of the training (tr), the validation (va) and the test set (te), the number of classes ($|\Omega|$), the label alphabet of both nodes and edges, the average and maximum number of nodes and edges ($\emptyset$/max nodes/edges).

Database	size (tr, va, te)	$|\Omega|$	node labels	edge labels	$\emptyset$/max nodes	$\emptyset$/max edges
Letter	750, 750, 750	15	x, y coordinates	none	4.7/9	4.5/9
Digit	1,000, 500, 2,000	10	x, y coordinates	Angle	11.8/32	13.1/30
Fingerprint	500, 300, 2,000	4	x, y coordinates	Angle	5.4/26	4.4/24
AIDS	250, 250, 1,500	2	Chemical symbol	Valence	15.7/95	16.2/103
Mutagenicity	1,500, 500, 2,337	2	Chemical symbol	Valence	30.3/417	30.8/112
Web	780, 780, 780	20	Word/Frequency	Type	186.1/834	104.6/596

with a two-dimensional attribute giving its position. The edges are attributed with an angle denoting the orientation of the edge with respect to the horizontal direction.

The AIDS data set consists of graphs representing molecular compounds. We construct graphs from the AIDS Antiviral Screen Database of Active Compounds.[43] This data set consists of two classes (*active, inactive*), which represent molecules with activity against HIV or not. The molecules are converted into graphs in a straightforward manner by representing atoms as nodes and the covalent bonds as edges. Nodes are labeled with the number of the corresponding chemical symbol and edges by the valence of the linkage.

In order to convert molecular compounds of the Mutagenicity data set[44] into attributed graphs the same procedure as for the AIDS data set is applied. The Mutagenicity data set is divided into two classes *mutagen* and *nonmutagen*.

In Ref. 2 several methods for creating graphs from web documents are introduced. For the graphs included in this data set (Web), the following method was applied. First, all words occurring in the web document – except for stop words, which contain only little information – are converted into nodes in the resulting web graph. We attribute each node with the corresponding word and its frequency. Next, different sections of the web document are investigated individually. If a word w_i immediately precedes word w_{i+1}, a directed edge from the node corresponding to word w_i to the node corresponding to the word w_{i+1} is inserted in our web graph. The resulting edge is attributed with the corresponding section label. In our experiments we make use of a data set with documents from 20 categories (*Business, Health, Politics, . . .*).

In Table 1 a summary of the graph data set characteristics is given.

4.2. Classification: Reference Systems and Experimental Setup

In contrast with the high representational power of graphs, we observe a lack of general classification algorithms that can be applied in the graph domain. One of the few classifiers directly applicable to arbitrary graphs is the k-nearest-neighbor classifier (k-NN). Given a labeled set of training graphs, an unknown graph is assigned to the class that occurs most frequently among the k nearest graphs (in terms of edit distance) from the training set. The decision boundary of this classifier is a piecewise linear function which makes it very flexible. This classifier in the graph domain will serve us as our first reference system.

The second reference system is the similarity kernel described in Sect. 3.2 (sim) in conjunction with an SVM. Comparing the performance of this similarity kernel function with the performance of the dissimilarity embedding procedure offers us the possibility to understand whether the power of our system is primarily due to the embedding process or to the strength of the kernel classifier.

In the validation phase, the costs of the edit operations are determined first. For all considered graph data sets but the web data, node and edge labels are integer numbers, real numbers, or real vectors. Here the substitution cost of a pair of labels is given by a distance measure (e.g. Euclidean distance), and only the deletion and insertion costs have to be determined. For the sake of symmetry, we assume in all experiments identical costs for node deletions and insertions, and for edge deletions and insertions. Hence only two parameters, i.e. node deletion/insertion cost τ_n, and edge deletion/insertion cost τ_e, need to be validated. For the web data set we followed the cost model used in Ref. 2, i.e. node substitutions are not admissible and the costs of all other edit operations are set to an arbitrary constant. Consequently, no edit cost validation is needed for this data set.

For the second reference system (sim) the same edit distances as for the k-NN classifier are used. Hence, the weighting parameter C of the SVM, which controls whether the maximization of the margin or the minimization of the error is more important, and the meta parameter γ in the kernel function $\kappa(g, g') = \exp(-\gamma \, d(g, g'))$ have to be additionally validated.

For the task of graph embedding in real vector spaces one additional meta parameter has to be validated, namely the number of prototypes n, i.e. the dimensionality of the resulting vector space. In order to determine suitable values of n, each graph set is embedded in a vector space with all of the prototype selectors described in Sect. 3.2.1, varying the dimensionality of the target vector space over a certain interval. With the resulting vector sets, which still consist of a validation, a training, and a test set, an SVM is trained. We make use of an SVM with RBF-kernel where besides the weighting parameter C, the meta parameter γ in the kernel function has to be optimized[d]. In Fig. 4 (a) such an SVM parameter validation on the Fingerprint data set is illustrated.

The SVM optimization is performed on a validation set for every possible dimension of the target space and every prototype selection method. Thus, for each embedding procedure the classification accuracy can be regarded as a function of the dimensionality and the prototype selector. This final optimization is illustrated

---

[d]We make use of the RBF kernel since it is the most widely used kernel and it has been extensively studied in various fields of pattern recognition.[28]

Table 2. Embedding methods vs. reference systems. (①/② Stat. significant improvement over the first/second reference system (k-NN and sim),❷
Stat. significant deterioration compared to the second reference system.)

Data Set	ref. systems		embedding methods		
	k-NN	sim	sps-c	bps-c	k-cps-c
Letter	89.1	92.9	92.3 ①	92.9 ①	92.0 ①
Digit	97.4	98.1	98.6 ①②	98.7 ①②	98.7 ①②
Fingerprint	79.1	82.0	82.0 ①	83.1 ①②	81.7 ①
AIDS	94.9	97.0	98.1 ①②	98.0 ①②	98.1 ①②
Mutagenicity	66.9	68.6	71.6 ①②	70.7 ①	71.8 ①②
Web	80.6	82.9	82.7 ①	80.4 ❷	82.3

on the Fingerprint data set in Fig. 4 (b) where the accuracies for three classwise
prototype selectors (bps-c, sps-c, k-cps-c) and each dimensionality are shown.

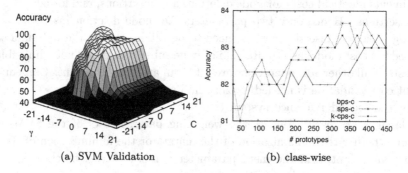

(a) SVM Validation   (b) class-wise

Fig. 4. (a) Validation of the meta parameter tuple (C, γ) for a specific prototype selector and a
certain number of prototypes (the parameter values are on a logarithmic scale to the basis 2). (b)
Validation of three classwise prototype selectors.

By means of this procedure for each prototype selector the number of prototypes
n and the SVM parameters (C, γ) can be determined for every data set on the
validation set. The parameter combination that results in the lowest classification
error is finally applied to the independent test set.

4.3. *Classification: Results and Discussion*

The results obtained on the test set by means of this procedure are reported for
three prototype selectors in Table 2. Let us first compare the classification accuracies achieved by our embedding framework with the first reference system (k-NN).
It clearly turns out that the novel procedure is much more powerful than the traditional k-NN classifier in the graph domain as on all tested applications the embedding based method outperforms the first reference system. Note that 16 out of 17
improvements are statistically significant.

Next we compare the embedding procedure with the similarity kernel (sim). It turns out that the similarity kernel is better on the Letter, Fingerprint, and Web data sets compared to at least one prototype selection method. However, the embedding framework outperforms the similarity kernel ten times on the Digit, Fingerprint, AIDS, and Mutagenicity data sets. Note that nine of the improvements, but only one of the deteriorations are statistically significant. From these findings we can conclude that the power of our novel approach primarily results from the embedding process itself and not from to the strength of the kernel classifier.

For a more detailed analysis of the proposed approach including a more thorough description of the achieved results we refer to Refs. 25, 33, 45.

4.4. *Clustering: Reference System and Experimental Setup*

For k-means clustering to work properly, the number of clusters k has to be defined beforehand (the same accounts for k-medians and kernelized k-means, of course). Since we use the same graph data sets for the clustering evaluation as for the classification experiments, class information is available for all of our graph data sets. For the sake of convenience, we define the number of clusters k to be found as the number of classes occurring in the underlying graph set.

The initialization of k-means is commonly done with a random selection of k objects. However, in the present chapter a deterministic procedure is applied. Let us assume a set of N objects $\mathcal{X} = \{x_1, \ldots, x_N\}$ to be clustered is given. In our case the objects $x_i \in \mathcal{X}$ are either represented by graphs from some graph domain $\mathcal{G}$ or by vector space embedded graphs in some vector space $\mathbb{R}^n$. The set of k initial cluster centers $\mathcal{M}_k = \{m_1, \ldots, m_k\} \subseteq \mathcal{X}$ is constructed by iteratively retrieving the set median from $\mathcal{X}$ ($median(\mathcal{X})$) minus the objects already selected.

In order to compare the kernel k-means clustering algorithm applied to vector space embedded graphs with the conventional k-medians algorithm in the graph domain, we use four different clustering validation indices known as Dunn,[46] C,[47] Rand,[48] and Bipartite index.[49] The Dunn index measures the ratio of the minimum distance of two different clusters and the maximum diameter of a cluster and is considered to be positively-correlated, which means that higher values indicate higher clustering quality. The C index is a ratio where the numerator measures how many pairs of objects of the a nearest neighboring pairs belong to the same cluster. The denominator is a scale factor ensuring that $0 \leq C \leq 1$. The smaller the C index value is, the more frequently do pairs with a small distance belong to the same cluster, i.e. the higher is the clustering quality (negatively-correlated). Rand index measures the consistency of a given clustering according to the ground truth, and therefore higher values indicate better clusterings. Similarly, BP index gives us the maximum possible classification accuracy of the given clustering according to the ground truth. Whereas the two former indices (Dunn and C) do not need any ground truth information, the latter ones (Rand and Bipartite) are defined with respect to the class memberships of the underlying objects.

Note that other clustering validation indices, such as Davies-Bouldin,[50] Calinski-Harabasz,[51] Xie-Beni,[52] $\mathcal{I}$ index,[53] and cluster stability,[54] could be also used. However, we feel that a validation based on the four proposed indices covers the different aspects of cluster quality evaluation quite well, and we leave a more exhaustive analysis involving additional indices to future work.

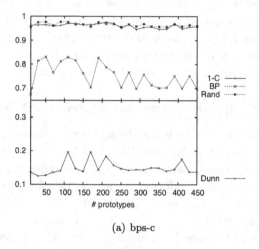

(a) bps-c

Fig. 5.   Validation of clustering indices on the Letter low data set for one prototype selection strategy (bps-c).

Each of our graph sets is divided into three disjoint subsets referred to as training, validation, and test set. The training set is not used for the clustering experiments. Those meta parameters of the clustering algorithm which cannot be directly inferred from the specific application are optimized on the validation set. For k-medians clustering in the original graph domain as well as for dissimilarity based graph embedding, the same parameter values as found for the classification experiments are used. No additional parameter has to be validated for our reference system, the k-medians algorithm. For our novel approach, however, there are three parameters to tune: the prototype selection method, the number of prototypes n (dimensionality of the embedding vector space $\mathbb{R}^n$), and the parameter γ in the RBF kernel. For each of the four validation indices, these three meta parameters are optimized individually on the validation set.

The optimization of the RBF kernel is performed on the validation set for every possible dimension of the embedding space and every prototype selection method. Thus, the four validation indices can be regarded as a function of the dimensionality and the prototype selector. This optimization is illustrated on the Letter data set in Fig. 5 where the scores of all validation indices and each tested dimensionality are shown for one prototype selection method[e]. For each validation index, the best performing prototype selector with optimal parameter combination is applied to the independent test set. These results are reported in the next section.

---

[e]For the sake of convenience, we use $1 - C$ instead of C.

Table 3. Kernel k-means clustering in the embedding space vs. k-medians clustering in the graph domain.

Validation index	wins	ties	losses
Dunn	5	1	0
C	6	0	0
BP	4	0	2
Rand	4	0	2

4.5. *Clustering: Results and Discussion*

The clustering validation indices obtained with the best performing embedding method (kernel k-means clustering in the embedding space) are compared against the index values achieved by the reference method (k-medians clustering in the graph domain) on all data sets. The results are summarized in Table 3.

According to Dunn index, the approach of clustering vector space embedded graphs rather than the original graphs is beneficial on five out of six data sets (i.e. on all data sets but Mutagenicity where both systems achieve the same Dunn index value). Hence, we conclude that the clusters in the embedding space are more compact and better separable than those in the graph domain. Furthermore, on all of the six data sets, the clusterings in the embedding space outperform the original clusterings according to C index. That is, with the novel procedure in the embedding space, pairs of objects with small distances are more frequently in the same cluster than in the original approach. Note that for both indices Dunn and C the class membership is not taken into account but only the size and shape of the clusters.

In case of BP and Rand index we observe four cases where the clustering in the embedding space outperforms the clustering in the graph domain and two data sets (Fingerprint and Web) where the original graph clustering performs better than our novel approach. In general, the clusterings in the embedding space are more accurate and consistent according to the ground truth than the clusterings in the original graph domain.

In order to get a visual impression of the clusterings, the confusion matrices $\mathbf{M} = (m_{ij})_{k \times k}$, where m_{ij} represents the number of elements from class Ω_j occurring in cluster C_i, can be considered. For a confusion matrix $\mathbf{M} = (m_{ij})_{k \times k}$ a $k \times k$ checkerboard plot is generated, where dark squares in the checkerboard represent large values of m_{ij} and bright squares represent small values of m_{ij}. The more elements from class Ω_j occur in cluster C_i, the darker the corresponding square at position (i, j) in the checkerboard plot. For the sake of clarity, the rows of the checkerboard are rearranged such that the squares along the main diagonal are darker than the other squares in the respective row in general. This is achieved by finding a permutation of the rows of the confusion matrix which maximizes the sum of diagonal elements $\sum_{i=1}^{k} m_{ii}$.

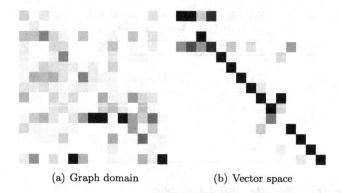

(a) Graph domain   (b) Vector space

Fig. 6. Confusion matrix on the Letter set (clusters vs. classes).

Obviously, the grey scale value of the k squares on the i-th line of these figures visualize how the members of a certain class Ω_i are distributed over the different clusters $(C_1, \ldots, C_k)$. Clearly, the better the clustering according to the real class distribution is, the more dark squares can be observed along the main diagonal of these plots by simultaneously showing only few and rather bright squares in the remaining part of the checkerboard[f].

In Fig. 6 a confusion matrix plot for the Letter data set is given for clusterings in both the graph domain and the embedding vector space. Note that the underlying confusion matrices in the embedding space correspond to clusterings where the BP index is optimal on the validation set. (Similar results have been reported on the other data sets – see Ref. 5.)

These illustrations confirm that by means of the embedding procedure the quality of the clusterings can be improved when compared to the clusterings in the graph domain. Through the embedding process in conjunction with kernel k-means clustering, the number of dark squares along the main diagonal is substantially increased compared to the confusion matrix plots in the graph domain. Simultaneously, the number of grey squares not lying on the main diagonal is decreased. This qualitative comparison well accompanies the quantitative results observed on the same data sets considering BP and Rand indices (which both measure the consistency and accuracy of the clusters found with respect to the underlying classes).

At first glance the meta parameters to be tuned in our novel approach seem to be a drawback compared to the reference system. However, since k-means algorithm is able to find spherical clusters only, these meta parameters establish a powerful possibility to optimize the underlying vector space embedding with respect to a specific validation index.

In case of Dunn's index, for instance, the underlying vector space is optimized such that the resulting clusters are more compact and better separable than the

---

[f]A clustering that perfectly matches the class distribution of the underlying data results in k black squares along the main diagonal while the rest is completely white.

clusters achieved in the original graph domain. In other words, the clusters of the data formed through the embedding process can be much better approximated by non-overlapping ellipses than in the original graph domain. In case of BP or Rand index, by contrast, the embedding and clustering are optimized such that spherical clusters are able to separate the data with a high degree of consistency and accuracy according to the ground truth. Summarizing, the embedding process lends itself to a methodology for adjusting a given data distribution such that the clustering algorithm is able to achieve good results according to a specific validation criterion.

5. Conclusions

For objects given in terms of feature vectors a rich repository of algorithmic tools for classification and clustering has been developed over the last decades. Graphs are a versatile alternative to feature vectors, and are known to be a powerful and flexible representation formalism. The representational power of graphs is due to their ability to represent not only feature values but also relationships among different parts of an object, and their flexibility comes from the fact there are no size or labeling restrictions that constrain the representation of a given object. However, graph based object representation suffers from the fact that there is only little mathematical structure in the graph domain. In contrast to vectors, most of the basic mathematical operations required for classification and clustering do not exist for graphs.

In the present chapter a novel approach to graph embedding, using prototypes and dissimilarities, is proposed. Our embedding procedure explicitly makes use of graph edit distance and can therefore deal with various kinds of graphs (labeled, unlabeled, directed, undirected, etc.). The basic idea of the embedding method is to describe a graph by means of n dissimilarities to a predefined set of graphs termed prototypes. That is, a graph g is mapped explicitly to the n-dimensional real space $\mathbb{R}^n$ by arranging the edit distances of g to all of the n prototypes as a vector. By means of this procedure both statistical classifiers and clustering algorithms can be applied to the resulting graph maps.

The problem of prototype selection is an issue in the proposed graph embedding framework. Three possible solutions are reviewed in the present chapter, viz. heuristic prototype selection,[25] prototype reduction,[45] and feature selection.[33] Three possible algorithms from the first category are described in detail. Note that none of them performs generally the best. That is, the quality of a particular selection strategy depends on the underlying data set. However, in case where no possibility for independent validation exist, one of the classwise prototype selectors could be the method of choice (e.g. sps-c).

In the experimental evaluation, the proposed graph embedding framework is compared to a nearest-neighbor classifier, another kernel-based classifier and

k-medians clustering. In several experiments a high degree of robustness and flexibility of the proposed approach has been empirically verified in this chapter.

References

1. L. Ralaivola, S.J. Swamidass, H Saigo, and P. Baldi. Graph kernels for chemical informatics. *Neural Networks*, 18(8):1093–1110, 2005.
2. A. Schenker, H. Bunke, M. Last, and A. Kandel. *Graph-Theoretic Techniques for Web Content Mining*. World Scientific, 2005.
3. D. Conte, P. Foggia, C. Sansone, and M. Vento. Thirty years of graph matching in pattern recognition. *Int. Journal of Pattern Recognition and Artificial Intelligence*, 18(3):265–298, 2004.
4. D. Cook and L. Holder, editors. *Mining Graph Data*. Wiley-Interscience, 2007.
5. K. Riesen and H. Bunke. *Classification and Clustering of Vector Space Embedded Graphs*. World Scientific, 2010.
6. K. Borgwardt. *Graph Kernels*. PhD thesis, Ludwig-Maximilians-University Munich, 2007.
7. B. Luo, R. Wilson, and E. Hancock. Spectral embedding of graphs. *Pattern Recognition*, 36(10):2213–2223, 2003.
8. T. Caelli and S. Kosinov. Inexact graph matching using eigen-subspace projection clustering. *Int. Journal of Pattern Recognition and Artificial Intelligence*, 18(3):329–355, 2004.
9. R. Wilson and E. Hancock. Levenshtein distance for graph spectral features. In J. Kittler, M. Petrou, and M. Nixon, editors, *Proc. 17th Int. Conference on Pattern Recognition*, volume 2, pages 489–492, 2004.
10. A. Shokoufandeh, D. Macrini, S. Dickinson, K. Siddiqi, and S.W. Zucker. Indexing hierarchical structures using graph spectra. *IEEE Transactions on Pattern Analysis and Machine Intelligence*, 27(7):1125–1140, 2005.
11. W.J. Lee and R. Duin. A labelled graph based multiple classifier system. In J.A. Benediktsson, J. Kittler, and F. Roli, editors, *Proc. 8th Int. Workshop on Multiple Classifier Systems*, LNCS 5519, pages 201–210, 2009.
12. E. Pekalska and R. Duin. *The Dissimilarity Representation for Pattern Recognition: Foundations and Applications*. World Scientific, 2005.
13. J.R. Ullmann. An algorithm for subgraph isomorphism. *Journal of the Association for Computing Machinery*, 23(1):31–42, 1976.
14. H. Bunke and K. Shearer. A graph distance metric based on the maximal common subgraph. *Pattern Recognition Letters*, 19(3):255–259, 1998.
15. W.D. Wallis, P. Shoubridge, M. Kraetzl, and D. Ray. Graph distances using graph union. *Pattern Recognition Letters*, 22(6):701–704, 2001.
16. M.-L. Fernandez and G. Valiente. A graph distance metric combining maximum common subgraph and minimum common supergraph. *Pattern Recognition Letters*, 22(6–7):753–758, 2001.
17. H. Bunke and G. Allermann. Inexact graph matching for structural pattern recognition. *Pattern Recognition Letters*, 1:245–253, 1983.
18. A. Sanfeliu and K.S. Fu. A distance measure between attributed relational graphs for pattern recognition. *IEEE Transactions on Systems, Man, and Cybernetics (Part B)*, 13(3):353–363, 1983.
19. R. Ambauen, S. Fischer, and H. Bunke. Graph edit distance with node splitting and merging and its application to diatom identification. In E. Hancock and M. Vento,

editors, *Proc. 4th Int. Workshop on Graph Based Representations in Pattern Recognition*, LNCS 2726, pages 95–106. Springer, 2003.

20. S. Sorlin and C. Solnon. Reactive tabu search for measuring graph similarity. In L. Brun and M. Vento, editors, *Proc. 5th Int. Workshop on Graph-based Representations in Pattern Recognition*, LNCS 3434, pages 172–182. Springer, 2005.

21. D. Justice and A. Hero. A binary linear programming formulation of the graph edit distance. *IEEE Transactions on Pattern Analysis and Machine Intelligence*, 28(8):1200–1214, 2006.

22. M. Neuhaus, K. Riesen, and H. Bunke. Fast suboptimal algorithms for the computation of graph edit distance. In Dit-Yan Yeung, J.T. Kwok, A. Fred, F. Roli, and D. de Ridder, editors, *Proc. 11.th int. Workshop on Structural and Syntactic Pattern Recognition*, LNCS 4109, pages 163–172. Springer, 2006.

23. M.A. Eshera and K.S. Fu. A similarity measure between attributed relational graphs for image analysis. In *Proc. 7th Int. Conference on Pattern Recognition*, pages 75–77, 1984.

24. K. Riesen and H. Bunke. Approximate graph edit distance computation by means of bipartite graph matching. *Image and Vision Computing*, 27(4):950–959, 2009.

25. K. Riesen and H. Bunke. Graph classification based on vector space embedding. *Int. Journal of Pattern Recognition and Artificial Intelligence*, 23(6):1053–1081, 2009.

26. B. Spillmann, M. Neuhaus, H. Bunke, E. Pekalska, and R. Duin. Transforming strings to vector spaces using prototype selection. In D.-Y. Yeung, J.T. Kwok, A. Fred, F. Roli, and D. de Ridder, editors, *Proc. 11th int. Workshop on Structural and Syntactic Pattern Recognition*, LNCS 4109, pages 287–296. Springer, 2006.

27. B. Schölkopf and A. Smola. *Learning with Kernels*. MIT Press, 2002.

28. J. Shawe-Taylor and N. Cristianini. *Kernel Methods for Pattern Analysis*. Cambridge University Press, 2004.

29. K. Tsuda. Support vector classification with asymmetric kernel function. In M. Verleysen, editor, *Proc. 7th European Symposium on Artifical Neural Networks*, pages 183–188, 1999.

30. J. Ramon and T. Gärtner. Expressivity versus efficiency of graph kernels. In *Proc. First Int. Workshop on Mining Graphs, Trees and Sequences*, pages 65–74, 2003.

31. M. Neuhaus and H. Bunke. *Bridging the Gap Between Graph Edit Distance and Kernel Machines*. World Scientific, 2007.

32. B. Haasdonk. Feature space interpretation of SVMs with indefinite kernels. *IEEE Transactions on Pattern Analysis and Machine Intelligence*, 27(4):482–492, 2005.

33. K. Riesen and H. Bunke. Reducing the dimensionality of dissimilarity space embedding graph kernels. *Engineering Applications of Artificial Intelligence*, 22(1):48–56, 2008.

34. K. Riesen and H. Bunke. Non-linear transformations of vector space embedded graphs. In A. Juan-Ciscar and G. Sanchez-Albaladejo, editors, *Pattern Recognition in Information Systems*, pages 173–186, 2008.

35. X. Jiang, A. Münger, and H. Bunke. On median graphs: Properties, algorithms, and applications. *IEEE Transactions on Pattern Analysis and Machine Intelligence*, 23(10):1144–1151, 2001.

36. L. Kaufman and P. Rousseeuw. *Finding Groups in Data: An Introduction to Cluster Analysis*. John Wiley & Sons, 1990.

37. B. Schölkopf, A. Smola, and K.-R. Müller. Nonlinear component analysis as a kernel eigenvalue problem. *Neural Computation*, 10:1299–1319, 1998.

38. V. Vapnik. *Statistical Learning Theory*. John Wiley, 1998.

39. I.S. Dhillon, Y. Guan, and B. Kulis. Kernel k-means: spectral clustering and normalized cuts. In *Proc. of the 10th ACM SIGKDD Int. conference on Knowledge Discovery and Data Mining*, pages 551–556, 2004.

40. K. Riesen and H. Bunke. IAM graph database repository for graph based pattern recognition and machine learning. In N. da Vitoria Lobo et al., editor, *Proc. of the Int. Workshops on Structural, Syntactic, and Statistical Pattern Recognition*, LNCS 5342, pages 287–297, 2008.

41. E. Alpaydin and F. Alimoglu. *Pen-Based Recognition of Handwritten Digits*. Dept. of Computer Engineering, Bogazici University, 1998.

42. C.I. Watson and C.L. Wilson. *NIST Special Database 4, Fingerprint Database*. National Institute of Standards and Technology, 1992.

43. Development Therapeutics Program DTP. AIDS antiviral screen, 2004. http://dtp.nci.nih.gov/docs/aids/aids_data.html.

44. J. Kazius, R. McGuire, and R. Bursi. Derivation and validation of toxicophores for mutagenicity prediction. *Journal of Medicinal Chemistry*, 48(1):312–320, 2005.

45. K. Riesen and H. Bunke. Dissimilarity based vector space embedding of graphs using prototype reduction schemes. In P. Perner, editor, *Proc. 6th Int. Conference on Machine Learning and Data Mining in Pattern*, LNCS 5632, pages 617–631, 2009.

46. J. Dunn. Well-separated clusters and optimal fuzzy partitions. *Journal of Cybernetics*, 4:95–104, 1974.

47. L. Hubert and J. Schultz. Quadratic assignment as a general data analysis strategy. *British Journal of Mathematical and Statistical Psychology*, 29:190–241, 1976.

48. W. Rand. Objective criteria for the evaluation of clustering methods. *Journal of the American Statistical Association*, 66(336):846–850, 1971.

49. K. Riesen and H. Bunke. Kernel k-means clustering applied to vector space embeddings of graphs. In L. Prevost, S. Marinai, and F. Schwenker, editors, *Proc. 3rd IAPR Workshop Artificial Neural Networks in Pattern Recognition*, LNAI 5064, pages 24–35. Springer, 2008.

50. D.L. Davies and D.W. Bouldin. A cluster separation measure. *IEEE Transactions on Pattern Analysis and Machine Intelligence*, 1(2):224–227, 1979.

51. T. Calinski and J. Harabasz. A dendrite method for cluster analysis. *Communications in statistics*, 3(1):1–27, 1974.

52. X.L. Xie and G. Beni. A validity measure for fuzzy clustering. *IEEE Transactions on Pattern Analysis and Machine Intelligence*, 13:841–847, 1991.

53. U. Malik and S. Bandyopadhyay. Performance evaluation of some clustering algorithms and validity indices. *IEEE Transactions on Pattern Analysis and Machine Intelligence*, 24(12):1650–1654, 2002.

54. L. Kuncheva and D. Vetrov. Evaluation of stability of k-means cluster ensembles with respect to random initialization. *IEEE Transactions on Pattern Analysis and Machine Intelligence*, 28(11):1798–1808, 2006.

CHAPTER 1.4

ACTIVITY MODELING AND RECOGNITION USING PROBABILISTIC GRAPHICAL MODELS

Xiaoyang Wang, Zhi Zeng and Qiang Ji

Rensselaer Polytechnic Institute,
110 Eighth Street, Troy, NY, USA 12180
wangx16@rpi.edu; zengz@rpi.edu; jiq@rpi.edu

In this chapter, we introduce the approaches for visual activity modeling and recognition using the time-sliced probabilistic graphical models, i.e. hidden Markov models (HMM) and dynamic Bayesian networks (DBN). Firstly, we briefly introduce the features used for activity modeling and recognition. Then, three kinds of DBNs, namely, the Generative DBN (GDBN), the Discriminative DBN (DDBN) and the first-order Probabilistic Logics knowledge DBN (PLDBN) are introduced for activity modeling and recognition. The GDBN is a standard generative time-sliced graphical model often used for activity recognition. We also incorporate prior knowledge into the GDBN classification procedure to improve the system performance. The DDBN employs a discriminative parameter learning approach for DBNs so that the criterions during training and testing are consistent. Empirical studies can show the proposed discriminative learning approach outperforms the maximum likelihood or EM algorithm for different activity recognition tasks. The PLDBN provides solutions on learning DBN with domain knowledge for human activity recognition. Different types of domain knowledge, in terms of first order probabilistic logics (FOPLs), are exploited to guide the DBN learning process. The FOPLs are transformed into two types of model priors: structure prior and parameter constraints. The experimental results on PLDBN demonstrate simple logic knowledge can compensate effectively for the shortage of the training data and therefore reduce our dependencies on training data.

1. Introduction

Modeling and recognizing visual activities from videos is becoming one of the most promising applications in the computer vision field. It is fundamental to machine visual surveillance which attracts a growing interest from both academia and industry over the past years. Various graphical, syntactic, and description-based approaches[1] have been introduced for modeling and understanding visual activities. Among those approaches, the time-sliced graphical models, i.e. Hidden Markov Models (HMMs) and Dynamic Bayesian Networks (DBNs), have become the most popular tools. This is not surprising, because graphical models match the nature of activity recognition problem well. These graphical models offer clear Bayesian semantics, and provide expressive representation and propagation of uncertainty over a sequence of video frames.

Existing approaches for visual activity recognition can be mainly classified into two categories: the generative approach and discriminative approach. The generative approaches try to model the joint probability of the activity label and the evidence, and perform activity classification based on the posterior probability with a Bayesian decision scheme. Discriminative approaches, on the other hand, directly model the posterior probability distribution of the class label given the evidence, which avoid the intermediate step to estimate the joint distribution. Common HMMs and DBNs for visual activity recognition are generative models. The usual way of activity recognition with these generative HMMs or DBNs is first to learn one DBN/HMM model for each activity independently through maximum likelihood estimation (MLE), or expectation maximization (EM) if given incomplete training data. And then, the models perform the classification decision through identifying the activity with the highest likelihood given the observations extracted from the raw video, or equivalently, the activity class with the highest posterior probability assuming there is uniform prior over all activities. In this chapter, our generative DBN model for activity recognition is first introduced. This generative DBN model is later combined with prior knowledge using a Bayesian network. The vehicle detection result is an useful prior knowledge that can be used for activity recognition. Based on the output of the vehicle detector, we can compute the new posterior probability given both the evidence and the vehicle detection result. Classification based on this posterior probability performs better than the classification using only the likelihood outputted by the generative DBN model.

The generative approach for HMM or DBN based activity recognition produces a discrepancy between the learning objective and the classification criterion. The models learned by maximum likelihood or EM algorithm for each activity, though capturing the data dependency with the corresponding activity well, may not maximize the classification accuracy between different activities. Our solution to reduce this discrepancy between the learning and classification objective for activity classification is to learn the generative models discriminatively. The learning procedure would maximize the conditional likelihood and ensure a consistent criterion during learning and classification. In this chapter, we will compare our models trained on generative learning and discriminative learning respectively. Experiments on KTH data set[2] showed that, discriminative learning works better than generative learning even for a generative model.

Most of the existing DBN models for activity recognition are learned purely from training data, so when the amount of training data is insufficient, the performance of these models will decrease significantly. One solution alleviating this problem is resorting to various kinds of domain knowledge. First order logic is an expressive language in representing the logic relations in a domain and it is widely applied in many computer vision applications. Its combination with Markov networks, the Markov logic networks (MLN), can deal with rigorous logic reasoning while maintaining the capability of handling uncertainty. In our work, we present a framework to learn the DBN model combining training data with domain knowledge represented with the first order probabilistic logic.

In this chapter, we will first introduce the activity modeling with our DBN models in section 2. The features for activity modeling and recognition will be briefly introduced in this section too. Then, in sections 3, 4 and 5, we will introduce our generative DBN model combined with prior knowledge, our discriminative learning method for the DBN model, and our algorithm of utilizing first order probabilistic logics to guide the DBN learning process respectively. The experiments of these models and methods are discussed in section 6. The chapter ends in section 7 with a summary.

2. Activity Modeling Using DBNs

2.1. *Data Input for Activity Modeling*

The raw data of activities and events are presented in the format of video tracks. A single track and even various examples of the same event can have an arbitrary number of detections. So, in order to enforce a level of consistency for all events, the tracks are partitioned into track intervals during training and testing. A track interval consists of a predefined number of frames and are represented by the corresponding features on each frame. The features used to characterize each event include the position, speed, shape and appearance features.

2.2. *Feature Extraction and Feature Selection*

The features we used for activity recognition includes the position, speed, shape and the appearance features. For feature extraction, we first detect the moving object and extract its silhouette, and then measure the position, speed and shape of the object based on the silhouette. The position feature O_Y is then measured as the distance to a reference point.[3] The speed feature O_V is evaluated as the change of the object center, and the shape feature O_S includes four elements: aspect ratio of the bounding box of the moving object, filling ratio (the area of the object silhouette with respect to the area of the bounding box) and two first-order moments of the silhouette.[4]

The appearance features O_A are selected from a feature pool consisting of the HoG (histogram of oriented gradient) and HoF (histogram of optical flow) features. Here, an Adaboost algorithm is executed on all features in the feature pool, where a decision stump[5] is used as a weak classifier. Since the decision stump chooses the single most discriminative feature that minimizes the overall training error, a count of the selected features results in a ranking of the most discriminative features upon completion of Adaboost. The top 20 most discriminative appearance features O_A are selected on a per event basis and are used as the continuous inputs to the observation nodes of their corresponding model.

In one of our DBN model which incorporates first-order probabilistic logics knowledge,[6] we also used spatio-temporal feature O_{ST} which is the histogram of cuboid types.[7]

2.3. DBN Models

We build three different DBN models for the activity recognition task. The first DBN model, which is called the Generative DBN (GDBN), includes two hidden nodes GM and SA respectively in the first layer. The GM node represents the global motion state, and the SA node represents the appearance and shape state. These two states give a factorized representation of the subject state space. The second layer consists of two measurement nodes OGM and OSA. The OGM node denotes the combination of position feature O_Y and speed feature O_V which are all global motion measurements. The OSA node denotes the appearance features O_A selected from the raw HoG and HoF features using the Adaboost algorithm, and the shape feature O_S. Besides the nodes, there are two types of links in the model: intra-slice links and inter-slice links. The intra-slice links couple different states of the subject to encode their dependencies. And the inter-slice links represent the temporal evolution and capture the dynamic relationships between states at different time. Figure 1 shows the structure of GDBN model.

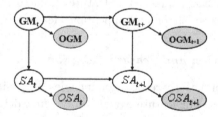

Fig. 1.   The Generative DBN model (GDBN) for activity modeling and recognition.

The second DBN model is called the Discriminative DBN (DDBN), as shown in figure 2. It separates the global motion measurements into *shape* and *motion* for inputs into their own observation nodes, resulting in three hidden nodes per time slice (S, A and V) and corresponding measurement states (OS, OA and OV); but more importantly, it is trained in a discriminative fashion rather than generatively. Typically the model of a particular event is only trained with data from the event of interest using likelihoods; but is tested relative to all events using posterior probabilities. This discrepancy results in non-ideal models for each event given the testing method. To overcome this discrepancy, the DDBN model is trained in a discriminative manner, where each event's model is trained with the data from all events of interest. Promising results are obtained here using the KTH database.

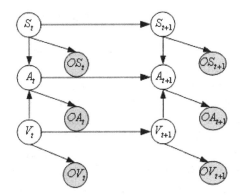

Fig. 2.   The Discriminative DBN model (DDBN) for activity modeling and recognition.

Our third DBN model incorporates the first-order probabilistic logics knowledge (PLDBN). In this PLDBN model, we can decompose the object state space into a set of physical states corresponding to position state Y, shape state S, global speed state V and spatio-temporal state ST. Accordingly, the measurement O consists of four observations: OY, OS, OV and OST. Figure 3 shows an example of our DBN model for activity modeling.

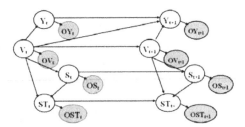

Fig. 3.   The PLDBN model incorporates the first-order probabilistic logics knowledge for activity modeling and recognition.

With the above different modeling strategy and different DBN model structures, we can construct one DBN model for each activity and perform activity recognition through finding the model with the highest likelihood, or finding the model with the highest posterior probability if given prior on the activity classes. Both of them can be evaluated by the forward propagation of dynamic junction tree.[8]

3.   The Generative DBN Model (GDBN) for Activity Recognition

3.1.   *GDBN Model Description*

As shown in figure 1, the GDBN model use the position (O_Y), speed (O_V), shape (O_S) and appearance (O_A) features to recognize different activities. The O_Y and

O_V features are the measurement of the node GM in the GDBN model, and are represented as the observation node OGM in figure 1. Also, the O_S and O_A features are the measurement of node SA in the GDBN model, and are represented as the observation node OSA in the GDBN structure.

The GDBN models the temporal progression of the kinematic and image features for each activity. The collection of trained graphical models consists of all known activities that have enough training data, as well as, one "unknown activity" class; which together form a database of models. An unknown temporal track interval is tested against all models in this database and is assigned the activity id that corresponds to the model with the best fit to the track's feature vector.

The best fit can be the activity with the highest likelihood given corresponding activity model, or the highest posterior probability using additional prior information. Here, we added the vehicle prior information acquired from pre-assumed vehicle detector output. The system performance can be improved using this posterior probability calculated from both the vehicle prior and the GDBN outputted likelihood.

3.2. GDBN Parameter Learning for Activity Recognition

Suppose we have K different kinds of activities to recognize in total. To recognize these K kinds of activities, we need to build K GDBN models. Let the parameters of the K GDBN models to be $\Theta_1, \Theta_2, ..., \Theta_K$ respectively, and the training sequences (extracted features) for activity c is $\mathcal{E}_c$ which contains N_c sequences, i.e. $\mathcal{E}_c = (E_1, E_2, ..., E_{N_c})$. As DBNs usually captures the joint distribution of sequence of variables, it is typically learned by maximizing the log likelihoood of all training sequences. Let the learned parameter for the GDBN model of activity c to be $\hat{\Theta}_c$, then we can write the general learning principle as:

$$\hat{\Theta}_c = \arg \max_{\Theta} P(\mathcal{E}_c | \Theta) = \arg \max_{\Theta} \sum_{E \in \mathcal{E}_c} \log P(E | \Theta) \qquad (1)$$

In the GDBN model, there are two hidden nodes GM and SA that have two and three states respectively. Between the hidden nodes in different time slices, there are three kinds of parameters to learn: the initial state distribution for the GM node, the intra-slice state conditional probability of SA given GM, and the inter-slice state transition probability both from GM_t to GM_{t+1}, and from SA_t to SA_{t+1}. Our observations OGM and OSA are all continuous observations. Given their correspondent hidden states, we suppose the observations of OGM and OSA follow Gaussian distribution. Thus, for each state of the GM and SA nodes, we have a mean vector μ and the covariance matrix Σ to estimate.

For the parameter learning, the data for parameter estimation is incomplete because we do not have direct observations of the hidden state status for each time slice t. As a result, the Expectation Maximization (EM) method for DBN models is adopted to estimate the parameters from the incomplete data. First, the initialization step is taken to give a starting value of the above parameters for the EM procedure. The initial state distribution and the state transition matrics are all initialized randomly. Since the EM algorithm is very sensitive to the initialization of the observation node parameters, the initialization process uses the K-means clustering algorithm to identify the starting mean vector μ and the covariance matrix Σ for each state. After the parameter initialization step, the GDBN parameters are learned by using a maximum of 20 iterations through a DBN EM[9] where a junction tree inferencing engine is used for the expectation step.

With the EM method, we can get the parameters of K models for the K activities to be recognized. Here, we denote the learned K models to be $\hat{\Theta}_1, \hat{\Theta}_2, ..., \hat{\Theta}_K$.

3.3. *Activity Recognition with the Likelihood of GDBN*

The input to the usual activity recognition model is a sequence of video with the track information which includes the starting and ending time of the activity, and the bounding box of the detected objects for this activity. During testing, we extract the same features from the above video and track information. For GDBN inference, suppose the evidence of the testing sequence to be E_T.

We can obtain the likelihood by using the forward propagation of the dynamic junction tree. For the cth model where $c \in [1, K]$, denote the likelihood of the evidence E_T as: $P(E_T|\hat{\Theta}_c)$. The evidence is best fitted to the model with the maximum likelihood. Thus, the classification result $\hat{c}$ can be written as:

$$\hat{c} = \arg \max_{c \in [1,K]} P(E_T|\hat{\Theta}_c) \qquad (2)$$

This classification criterion picks the activity class with the highest likelihood, or equivalently, the activity class with the highest posterior probability assuming there is uniform prior over all activities.

3.4. *Activity Recognition of GDBN Combined with Prior Knowledge*

In the activity recognition problem, there is always some kind of objects that may appear in the activity and would help us identify the activity. For example, for classifying the human activities of *Walking* and *Approaching a Vehicle*, the appearance of a vehicle in the activity bounding box may be an important extra information. But, this information is not thoroughly reflected in our extracted image and kinematic features. On the other hand, for the video sequence itself, it is usually feasible to detect the vehicles from the rest contents in the image.[10] If we take this vehicle detection information as a prior knowledge, and incorporate this prior knowledge into the posterior probability that we will use to classify the input evidence, we can utilize this additional vehicle information systematically.

To describe the causal relationships between the activity class, the physical appearance of a vehicle, the detection of the physical appeared vehicle, and the evidence of the activity sample that we need to classify, we can build up a Bayesian network as shown in figure 4. We call this Bayesian network the Vehicle Prior Bayesian Network (VPBN).

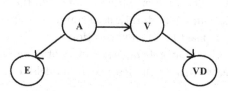

Fig. 4.   The Bayesian network describing the causal relationships when the vehicle detection prior knowledge is incorporated.

In VPBN, the activity node A is the root node of this Bayesian network. It has K discrete values where K stands for the K different kinds of activities to recognize. The node V is the "vehicle" node. Thus, it has two values "1" and "0", where "1" stands for the physical presence of a vehicle in the activity, and "0" stands for no vehicle presence in the activity. For different activities, the probability of the presence of a vehicle is different. The video sequence of the activity *Approaching a Vehicle* would have a much higher probability for the presence a vehicle than the video sequence of the *Walking* activity. And, the link from node A to node V would reflect the probability of the presence (or non-presence) of a vehicle given the value $c \in [1, K]$ of node A.

The node VD stands for the "vehicle detection". In theory, we can have two different vehicle detection results as "Vehicle Detected" and "No Vehicle Detected" for a single video sequence. The link from V to VD would reflect the probability of the vehicle detection when there exist or do not exist a vehicle. We set two values "1" and "0" for the node VD. "1" stands for the "Vehicle Detected; and "0" stands for "No Vehicle Detected". The link from node V to node VD is quantified by $P(VD|V)$, which characterizes the accuracy of the vehicle detector.

The node E is the evidence node which is continuous. The link from node A to node E reflects the probability of evidence E given activity class A. This is the same with the likelihood we estimated in section 3.3.

The VPBN can incorporate the vehicle detection prior knowledge with the existing likelihood provided by GDBN, and they two would provide us an improved posterior probability for classification. Here, we will briefly introduce the learning of VPBN parameters at first, and then discuss the inference with VPBN to get the posterior probability of the activity given the evidence and vehicle detection result, i.e. $P(A|E, VD)$, or the posterior probability $P(A|E)$ when the vehicle detector is not included in the system.

3.4.1. *VPBN Parameter Learning*

In VPBN, we have four sets of parameters to estimate. They are the prior probability of root node A and the three conditional probabilities represented by the three links in the graphical model shown in figure 4. The prior probability of the root node A is assumed to be uniform. And the conditional probability of the node E given A is actually the likelihood outputted by GDBN, as analyzed before. The conditional probability $P(V|A)$ can be learned by estimating the vehicle appearance rate given all the training sequences for every activity. This estimation is straightforward because the data for both two nodes are complete.

The conditional probability $P(VD|V)$ reflects the performance of the applied vehicle detector. $P(VD|V)$ has two independent parameters in its conditional probability table, i.e. $P(VD = 1|V = 0)$ and $P(VD = 1|V = 1)$. In fact, $P(VD = 1|V = 0)$ reflects the false alarm rate (FAR) of the vehicle detector, and $P(VD = 1|V = 1)$ is the positive detection rate (PD). To estimate these two parameters, we can first evaluate the performance of the applied vehicle detector in the data set with the FAR and PD. And then take the evaluated FAR and PD as the learned VPBN parameters. Here, we denote $\alpha = P(VD = 1|V = 1)$ and $\beta = P(VD = 1|V = 0)$ for convenience.

3.4.2. *VPBN Inference with Vehicle Detector*

For the activity classification, we need the posterior probability of the activity given the evidence and the result of a vehicle detection, i.e. $P(A|E, VD)$. Using the VPBN in figure 4, this probability can be calculated as:

$$P(A|E, VD) = \frac{P(A, E, VD)}{P(E, VD)} = \frac{\sum_V P(A)P(E|A)P(V|A)P(VD|V)}{\sum_A \sum_V P(A)P(E|A)P(V|A)P(VD|V)} \quad (3)$$

In section 3.4.1, we have already assumed that the prior probability $P(A)$ is uniform. Considering the nodes V and VD both have only two values, and we have defined $\alpha = P(VD = 1|V = 1)$ and $\beta = P(VD = 1|V = 0)$. Thus, the equation 3 can be:

$$P(A|E, VD = 1) = \frac{P(E|A)[\alpha P(V = 1|A) + \beta P(V = 0|A)]}{\sum_A P(E|A)[\alpha P(V = 1|A) + \beta P(V = 0|A)]}$$

$$P(A|E, VD = 0) =$$
$$\frac{P(E|A)[(1 - \alpha)P(V = 1|A) + (1 - \beta)P(V = 0|A)]}{\sum_A P(E|A)[(1 - \alpha)P(V = 1|A) + (1 - \beta)P(V = 0|A)]} \quad (4)$$

Equation 4 can be used to calculate the conditional probability $P(A|E, VD)$ for all A values. And the calculated probability is in $[0,1]$ space.

With the vehicle detection result, the classification result $\hat{c}$ can be written as:

$$\hat{c} = \arg \max_{c \in [1,K]} P(A = c|E, VD) \quad (5)$$

where $P(A = c|E, VD)$ is calculated in equation 4. This criterion picks the model with the maximum posterior probability $P(A|E, VD)$.

3.4.3. VPBN Inference without Vehicle Detector

When the vehicle detector is not included in the system, we recognize the activity using the posterior probability $P(A|E)$, which is the probability of the activity given the evidence. Calculation of this posterior probability can be unified into the VPBN model, as derived in the following:

$$P(A|E) = \frac{\sum_V \sum_{VD} P(A, E, V, VD)}{\sum_A \sum_V \sum_{VD} P(A, E, V, VD)} \tag{6}$$

The probability $P(A, E, V, VD)$ in equation 6 can be decomposed in chain rules w.r.t the VPBN model as shown in figure 4:

$$\sum_V \sum_{VD} P(A, E, V, VD) = \sum_V \sum_{VD} P(A)P(E|A)P(V|A)P(VD|V)$$

$$= P(A)P(E|A) \sum_V \sum_{VD} \{P(V|A)P(VD|V)\} = P(A)P(E|A)\{\alpha P(V = 1|A) +$$

$$(1 - \alpha)P(V = 1|A) + \beta P(V = 0|A) + (1 - \beta)P(V = 0|A)\}$$

$$= P(A)P(E|A) \tag{7}$$

Substitute equation 7 into equation 6, we have from VPBN model that:

$$P(A|E) = \frac{P(A)P(E|A)}{\sum_A P(A)P(E|A)} \propto P(A)P(E|A) \tag{8}$$

We can pick the activity model with the maximum posterior probability $P(A|E)$ as the classification result.

If we still assume that the prior probability $P(A)$ is uniform, the posterior probability $P(A|E)$ given by VPBN model would be proportional to the likelihood of GDBN. In this case, the activity classification of VPBM without vehicle detector would be the same as activity recognition with the likelihood of GDBN.

4. The Discriminative DBN Model (DDBN) for Activity Recognition

As mentioned above, in classification, our decision is based on comparing the likelihoods or posterior probabilities of different models, so these models are not independently applied. However, the traditional way to learn activity models is to learn the parameters of each DBN model independently through maximum likelihood estimation or EM algorithm. Though ensuring a representative model for each activity, it in general can not guarantee the best performance in classification, because of the discrepancy between the training criterion and test criterion. In this section, we introduce a discriminative learning approach to learn the DBN models for all activity models together, which can guarantee better classification performance on the training set. This Discriminative DBN Model (DDBN) is shown in figure 2.

4.1. *Discriminative Learning Formulation*

The goal of classification is to predict the class label c given the evidence E. Under the Bayesian decision framework, the optimal prediction for data E is the class that maximizes $P(c|E)$. In activity recognition, as we can evaluate the likelihood $P(E|c)$ for each activity c, we can compute the posterior probability based on Bayesian theorem.

$$P(c|E) = \frac{P(E|c)P(c)}{\sum_{c'} P(E|c')P(c')} \tag{9}$$

If we do not have prior on the activities, $p(c|E)$ becomes the normalized likelihood.

As DBN usually captures the joint distribution of sequence of variables, it is typically learned by maximizing the log likelihood of all training sequences.

$$\hat{\Theta} = \arg\max_{\Theta} P(\mathcal{E}|\Theta) = \arg\max_{\Theta} \sum_n \log P(E^n|\Theta)$$

Here $\mathcal{E} = (E^1, E^2, \ldots, E^N)$ denotes all N training sequences.

For the activity recognition problem, with generative learning, we learn the parameters $\hat{\Theta}_c$ of each activity model c independently through maximizing its likelihood (or expected likelihood).

$$\hat{\Theta}_c = \arg\max_{\Theta} P(\mathcal{E}_c|\Theta) = \arg\max_{\Theta} \sum_{E \in \mathcal{E}_c} \log P(E|\Theta)$$

where $\mathcal{E}_c$ denotes the training sequences for activity c. In this way, we can ensure to obtain a representative model for each activity, but it can not guarantee the best performance in classification, since the objective function of the maximum likelihood learning is not consistent with our prediction criterion $P(c|E)$. Hence, a better objective function for learning DBNs for activity recognition would be the conditional log likelihood $CLL(C|\mathcal{E})$.

$$CLL(C|\mathcal{E}) = \sum_{n=1}^{N} \log P(c^n|E^n) = \sum_c CLL(\mathbf{c}|\mathcal{E}_c) = \sum_c \sum_{E \in \mathcal{E}_c} \log P(c|E)$$

where $C = (c^1, c^2, \ldots, c^N)$ denote the activity labels of all N training sequences, $\mathbf{c}$ are the labels of the sequence in $\mathcal{E}_c$ (actually, $\mathbf{c}$ is a vector of c's).

Maximizing the conditional likelihood is not trivial since the CLL objective is non-convex in general. However, we can optimize it locally through gradient search. In this chapter, a Quasi-Newton method called BFGS with the line search under Armijo rule[11] is employed to perform the optimization.

A key step for the optimization is to evaluate the gradient of the CLL. In general, for sample (E, c), the gradient of CLL with respect to model parameter Θ

is

$$\frac{\partial \log P(c|E)}{\partial \Theta} = \frac{\partial \log[P(E|c)P(c)]}{\partial \Theta} - \frac{\partial \log P(E)}{\partial \Theta}$$
$$= \frac{\partial \log P(E|c)}{\partial \Theta} - \frac{\partial \log P(E)}{\partial \Theta} \tag{10}$$

Please note that the second term of equation 10 can be evaluated as the expectation of the first term,

$$\frac{\partial \log P(E)}{\partial \Theta} = E_{P(c|E)}\left[\frac{\partial \log P(E|c)}{\partial \Theta}\right] = \sum_c P(c|E)\frac{\partial \log P(E|c)}{\partial \Theta} \tag{11}$$

Now we consider this gradient with respect to the parameter $\Theta_{c'}$ of a specific activity model c'. With equation 11 and the fact

$$\frac{\partial \log P(E|c)}{\partial \Theta_{c'}} = 0 \quad \text{if } c' \neq c$$

we can get

$$\frac{\partial \log P(c|E)}{\partial \Theta_{c'}} = \begin{cases} (1 - P(c|E))\frac{\partial \log P(E|c)}{\partial \Theta_c} & \text{if } c' = c \\ -P(c'|E)\frac{\partial \log P(E|c')}{\partial \Theta_{c'}} & \text{if } c' \neq c \end{cases}$$

As $P(c|E)$ can be evaluated with equation 9, we mainly focus on computing $\partial \log P(E|c)/\partial \Theta_c$ when evaluating the derivative of the CLL. Please note that $\partial \log P(E|c)/\partial \Theta_c$ is just the derivative of the log likelihood of DBN model c with respect to its own parameters Θ_c.

4.2. Incomplete Data

When the training data are incomplete, or the model has hidden nodes, $P(E|c)$ is not decomposable, so evaluating the derivative of CLL becomes difficult. One natural choice of learning the DBN parameters is the EM algorithm, with the objective function substituted by the CLL. However, in this case, in each maximization step, there is no analytical solution for estimating the parameter and we still need to go through the optimization procedure for the "completed" case. To avoid this double-looped optimization procedure, an efficient way is needed to directly compute the gradient of CLL with incomplete data. We resort this to the existing exact inference algorithms in the hybrid model. In the following parts, we show the derivative $\partial \log P(E)/\partial \theta$ for three parent-child configurations except for the solved DP-DC case.[12]

- DP-CC:
  In the case of discrete parents with continuous child, the derivatives $\partial \log P(E)/\partial \mu_{ij}$ and $\partial \log P(E)/\partial A_{ij}$ can be computed as follows:

$$\frac{\partial \log P(E)}{\partial \mu_{ij}} = -\sum_t P(\pi_{t,i} = j|E)A_{ij}^2 \mu_{ij} + A_{ij}^2 \sum_t E_{p(x_{t,i},\pi_{t,i}=j|E)}\{x_{t,i}\}$$

$$\frac{\partial \log P(E)}{\partial A_{ij}} = -\sum_t P(\pi_{t,i} = j|E)A_{ij}^{-1} \qquad (12)$$

$$-A_{ij}\sum_t E_{p(x_{t,i},\pi_{t,i}=j|E)}\{(x_{t,i} - \mu_{ij})(x_{t,i} - \mu_{ij})^T\}$$

here

$$E_{p(x_{t,i},\pi_{t,i}=j|E)}\{x_{t,i}\} = P(\pi_{t,i} = j|E)E_{p(x_{t,i}|\pi_{t,i}=j,E)}\{x_{t,i}\}$$

$$E_{p(x_{t,i},\pi_{t,i}=j|E)}\{(x_{t,i} - \mu_{ij})(x_{t,i} - \mu_{ij})^T\}$$

$$= P(\pi_{t,i} = j|E)E_{p(x_{t,i}|\pi_{t,i}=j,E)}\{(x_{t,i} - \mu_{ij})(x_{t,i} - \mu_{ij})^T\}$$

where

$$E_{p(x_{t,i}|\pi_{t,i}=j,E)}\{(x_{t,i}-\mu_{ij})(x_{t,i}-\mu_{ij})^T\} = cov[x_{t,i}] + (E[x_{t,i}]-\mu_{ij})(E[x_{t,i}]-\mu_{ij})^T$$

Since $P(\pi_{t,i} = j|E), E[x_{t,i}], cov[x_{t,i}]$ [a] can be obtained through the inference in the hybrid dynamic Bayesian network,[13] we can compute the derivative $\partial \log P(E)/\partial \mu_{ij}$ and $\partial \log P(E)/\partial \Sigma_{ij}$.

- CP-CC:

  With continuous parents and continuous child, we have:

  $$\frac{\partial \log P(E)}{\partial \mu_i} = -A_i^2 \sum_t [\mu_i - E\{x_{t,i}|E\} + W_i E\{\pi_{t,i}|E\}]$$

  $$\frac{\partial \log P(E)}{\partial W_i} = -A_i^2 \sum_t [\mu_i E\{\pi_{t,i}^T|E\} - E\{x_{t,i},\pi_{t,i}^T|E\} + W_i E\{\pi_{t,i}\pi_{t,i}^T|E\}]$$

  $$\frac{\partial \log P(E)}{\partial A_i} = \sum_t [A_i^{-1} - A_i E_{p(x_{t,i},\pi_{t,i}|E)}\{(x_{t,i}-\mu_i-W_i\pi_{t,i})(x_{t,i}-\mu_i-W_i\pi_{t,i})^T\}]$$

  The terms $E_{p(x_{t,i},\pi_{t,i}|E)}\{(x_{t,i} - \mu_i - W_i\pi_{t,i})(x_{t,i} - \mu_i - W_i\pi_{t,i})^T\}$, $E\{x_{t,i}\pi_{t,i}^T|E\}$ and $E(\pi_{t,i}\pi_{t,i}^T)$ can be computed as follows:

  $$E\{x_{t,i}\pi_{t,i}^T|E\} = cov(x_{t,i},\pi_{t,i}) + E(x_{t,i})E(\pi_{t,i})^T$$

  $$E(\pi_{t,i}\pi_{t,i}^T) = cov(\pi_{t,i}) + E(\pi_{t,i})E(\pi_{t,i})^T$$

  $$E_{p(x_{t,i},\pi_{t,i}|E)}\{(x_{t,i} - \mu_i - W_i\pi_{t,i})(x_{t,i} - \mu_i - W_i\pi_{t,i})^T\}$$

  $$= cov(x_{t,i}) + (E(x_{t,i})-\mu_i)(E(x_{t,i})-\mu_i)^T - cov(x_{t,i},\pi_{t,i}) - [E(x_{t,i})-\mu_i]E(\pi_{t,i})^T$$

  $$-W_i E\{\pi_{t,i}(x_{t,i}-\mu_i)^T\} + W_i E(\pi_{t,i}\pi_{t,i}^T)W_i^T$$

  In the above equations, $E(x_{t,i})$, $cov(x_{t,i})$, $E(\pi_{t,i})$ and $cov(\pi_{t,i})$ are the mean and variance of the posterior distribution of $x_{t,i}$ and $\pi_{t,i}$ respectively, $cov(x_{t,i},\pi_{t,i})$ is the covariance of $x_{t,i}$ and $\pi_{t,i}$ with respect to posterior distribution $p(x_{t,i},\pi_{t,i}|E)$, which can all be obtained through the inference of the hybrid Bayesian network.

---

[a] For simplicity, we denote $E_{p(x_{t,i},\pi_{t,i}=j|E)}[x_{t,i}]$ as $E[x_{t,i}]$, $E_{p(x_{t,i},\pi_{t,i}=j|E)}\{(x_{t,i} - E[x_{t,i}])(x_{t,i} - E[x_{t,i}])^T\}$ as $cov[x_{t,i}]$

After obtaining these terms, we can then compute the derivatives $\partial \log P(E)/\partial \mu_i$, $\partial \log P(E)/\partial W_i$ and $\partial \log P(E)/\partial \Sigma_i$.

Up to this point, we can compute the $\partial \log P(E)/\partial \theta$ ($\theta \in \{\mu, \Sigma, W\}$). Further, based on the discussion in section 4.1, we can finally obtain the gradient of the CLL with respect to μ_{ij} and Σ_{ij} of model c.

5. Activity Recognition Using DBN Combined with the First-order Probabilistic Logics Knowledge (PLDBN)

For the activity recognition applications, there often exists some approximate yet generic domain knowledge that governs the physics, kinematics, and dynamics of domain objects. Such knowledge, if exploited, can help regularize the otherwise ill-posed problems. In the PLDBN model, we identify such knowledge in the form of first-order probabilistic logics (FOPLs), and then try to incorporate these knowledge in our activity model.[6]

5.1. *FOPLs for Activity Recognition*

FOPLs is one type of knowledge representation language preserving the expressive power of first-order logic while introducing the probabilistic treatment of uncertainty. Several FOPLs have been proposed,[14] and we keep the formal syntax and semantics defined by Halpern et al.[15]

We use the following alphabet to represent the knowledge for activity recognition:

- Predicates: Is;
- Constants: POS(position), SH(shape), SP(speed), ST(spatio-temporal response), near (NR), far (FA), simple (SI), complex(CO), high(HI), low(LO);
- Connective symbols: $\vee, \wedge, \forall, \neg, |$;
- Variable: t, AS (denotes one of the three constants: POS, SH and SP), s;
- Probability operator: Pr;
- Basic numeric operator: $+, *, =, >$;
- Function: Next;

We describe the domain elements with two sorts of terms: the object term and numeric term. The object term describes the non-numeric basic elements (i.e. "t", "shape", "position", "Next(t)") of the domain, the numeric term describes certain probabilities which are rational numbers in the interval [0 1] (i.e. Pr(Is(position, near, t))). Given these elements, we can interpret the logics of the activity domain with a set of well-formed formula, which, in our case, only consists of the relations between different probabilities. Then, we can further transform the logic formula of the such knowledge into four probabilistic constraints on the conditional probabilities of our activity model in PLDBN.

The first logic is the smoothness logic. It stands for the fact that: the speed of an object at a successive time is more likely to be low if its current speed is low than its current speed is high. In FOPL, it is:

$$Pr[Is(SP, LO, Next(t)) \mid Is(SP, LO, t)] \geq Pr[Is(SP, LO, Next(t)) \mid Is(SP, HI, t)]$$

It can be written in the following probabilistic constraint format:

$$P(V_{t+1} = L \mid V_t = L) \geq P(V_{t+1} = L \mid V_t = H) \tag{13}$$

Here L denotes the low speed state and H denotes the high speed state.

The second logic is the position-motion logic. It stands for the fact that: With a high speed and near position in current frame, an object is more probable to be in far position in next frame than with a low speed and near position in current frame. The FOPL format is:

$$Pr[Is(POS, FR, Next(t)) \mid Is(POS, NR, t)] \wedge Is(SP, HI, t)]$$
$$\geq Pr[Is(POS, FR, Next(t)) \mid Is(POS, NR, t) \wedge Is(SP, LO, t)]$$

Similarly, it can be written in the following probabilistic constraint format.

$$P(Y_{t+1} = F \mid Y_t = N, V_t = H) \geq P(Y_{t+1} = F \mid Y_t = N, V_t = L) \tag{14}$$

Here N denote near position state; and F is far position state.

The third logic is the shape-motion logic. It stands for the fact that: It is more probable for an object to change from simple shape to complex shape with a low speed than with a high speed. We have

$$Pr[\neg Is(SH, CO, Next(t)) \mid Is(SH, SI, t)] \wedge Is(SP, LO, t)]$$
$$\geq Pr[\neg Is(SH, CO, Next(t)) \mid Is(SH, SI, t) \wedge Is(SP, HI, t)]$$

The probabilistic constraint is:

$$P(S_{t+1} = 1 \mid S_t = 0, V_{t+1} = L) \geq P(S_{t+1} = 1 \mid S_t = 0, V_{t+1} = H) \tag{15}$$

Here $S_t = 1$ denotes complex shape and $S_t = 0$ denotes simple shape.

The fourth logic which is the spatio-temporal logic stands for the fact that: An object is more likely to have a high spatio-temporal response if it has a simple shape at current frame and a complex shape at next frame, than if its shape at current frame and next frame are both simple. In FOPL, it is:

$$Pr[\neg Is(ST, HI, Next(t)) \mid Is(SH, SI, t)] \wedge \neg Is(SP, CO, Next(t))]$$
$$\geq Pr[\neg Is(ST, HI, Next(t)) \mid Is(SH, SI, t) \wedge Is(SP, SI, Next(t))]$$

The probabilistic constraint is:

$$P(ST_{t+1} = 1 \mid S_t = 0, S_{t+1} = 1) \geq P(ST_{t+1} = 1 \mid S_t = 0, S_{t+1} = 0) \tag{16}$$

Here $ST_t = 1$ is the high spatio-temporal response and $ST_t = 0$ is low spatio-temporal response.

5.2. *Incorporate the FOPLs with the Activity Model*

Having the four conditional probabilistic constraints mentioned in section 5.1, we can generate two types of model priors. First, the domain knowledge, in terms of qualitative constraints on the model conditional probabilities, can be used as parameter constraints to regularize the parameter learning for the activity model. Secondly, we apply the probability constraints as a soft structure prior, which can allow imperfect specification of the domain knowledge to certain degree. The structure prior, together with the training data, are used to learn the model structure in a Bayesian manner.[6]

We use the log posterior probability (LPP) as the criterion for model structure selection,[16] with the soft structure prior generated from FOPLs. In the case of incomplete training data, a widely adopted approach for DBN model search is the structural EM (SEM) algorithm.[17] One bottleneck of the SEM algorithm is that it requires a large amount of training sequences. Since the data is often limited, but there exists very generic logic knowledge in terms of qualitative constraints about the human activities, we used the constrained structural EM (CSEM) algorithm to learn the model structure combining the training data with these constraints.[6]

With the parameter constraints, the model selection criterion combined with soft structure prior, and the CSEM algorithm for incomplete data, we can then incorporate the domain knowledge in the processing of learning the activity model.

6. Experiments

We trained and tested our GDBN, DDBN and PLDBN models in experiments discussed in this section. The KTH data set[2] is used in the training and testing for all our three models. We compared our performances with the the state-of-art approaches on this dataset, as shown in table 3. We also used the dataset from distant view and the parking lot dataset for GDBN and PLDBN respectively.

6.1. *Generative DBN (GDBN) Model Experiments*

6.1.1. *GDBN Experiments on the Dataset from Distant View*

Firstly, a set of preliminary experiments are performed using the dataset from distant view to train and test the GDBN model without and with the vehicle detection prior knowledge.

This data set from distant view is a set of videos that contains a handful of activities related to vehicles. Thus, we choose this data set to test the concept of incorporating the object detection knowledge with the traditional time-sliced graphical models like HMMs and DBNs. While, videos from this data set are shot far from the activities. The image resolution is relatively low and the contrast between the background and the moving objects is limited. Moreover, we used the computed track both for training and testing. Due to the limited image quality, the quality of the tracker and bounding boxes are also relatively low.

We chose eight activities from this data set to test the GDBN model. The eight activities are *Opening a Trunk* (OAT), *Loading a Vehicle* (LAV), *Entering a Facility* (EAF), *Unloading a Vehicle* (UAV), *Getting into a Vehicle* (GIV), *Getting out of a Vehicle* (GOV), *Exiting a Facility* (XAF), *Closing a Trunk* (CAT). Except for the EAF and XAF, all other six activities are involved with a vehicle. Recognition results on this data set is given in table 1.

Table 1. Comparison of GDBN results without or with the vehicle (Veh.) prior knowledge.

Activity	No Prior	Veh. Prior $\alpha = 0.4$ $\beta = 0.6$	Veh. Prior $\alpha = 0.6$ $\beta = 0.4$	Veh. Prior $\alpha = 0.8$ $\beta = 0.2$	Veh. Prior $\alpha = 1.0$ $\beta = 0$
OAT	.0909	0	.1818	.1818	.1818
LAV	.0500	.0500	.0500	.1000	.1000
EAF	.5000	.3929	.5357	.6429	.6786
UAV	.6087	.5217	.6087	.6087	.6087
GIV	.2000	.2000	.2000	.2000	.2000
GOV	.4118	.4118	.4118	.4706	.4706
XAF	.3421	.2632	.4211	.6053	.6842
CAT	.0667	0	.0667	.1333	.1333
RRNA[a]	.2838	.2299	.3095	.3678	.3822
OARR[b]	.3234	.2635	.3533	.4311	.4551

[a]Recognition Rate Numerical Average over eight activities.
[b]OverAll Recognition Rate.

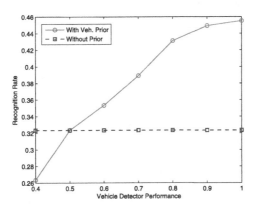

Fig. 5. With or without prior knowledge for GDBN model.

For this data set with limited quality, our GDBN model can generally reach a overall recognition rate of more than 32% for eight activities. And the GDBN model combined with vehicle detection prior knowledge performed better than the tradition GDBN model even if the vehicle detector has only 60 percent of positive detection rate, and as high as 40 percent of false alarm rate. As the vehicle detector performance improved, the GDBN model incorporated with vehicle detection prior knowledge would improve significantly, as shown in figure 5.

6.1.2. GDBN Experiments on KTH Dataset

The KTH dataset[2] is a human activity dataset with 6 basic human activities: *walking*, *jogging*, *running*, *boxing*, *hand waving* and *hand clapping*. Each activity is performed by 25 subjects in four different scenarios: outdoors, outdoors with scale variation, outdoors with different clothes and indoors. So, there are totally 600 video clips in the dataset. In this chapter, we apply our three DBN model discussed in section 2.3, as well as the discriminative learning algorithm, and FOPLs transformed into structural prior and parameter constraints on recognizing these 6 activities.

In KTH data set, we trained the GDBN model with 16 training subjects for each activity, as the way by Yuan et al.[18] and Laptev et al.[19] The recognition rate is given in table 3. The KTH data set consists only single human activities, thus the vehicle detection prior knowledge information is not incorporated in this experiment. We can find that this GDBN performance without prior knowledge is slightly lower than the state of the art performance on the KTH dataset with simple activities. While, the further developed discriminative learning DBN and the PLDBN can be comparable. These results will be discussed in the following.

6.2. Discriminative DBN (DDBN) Experiments on KTH Dataset

6.2.1. Discriminative Learning vs. Generative Learning

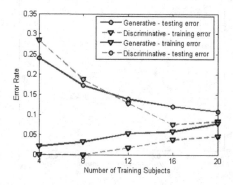

Fig. 6.   Discriminative learning vs. generative learning on KTH dataset.

We first focus on comparing the generative learning approach with the discriminative learning approach with different training size here. Since the basic states of the subject (motion, shape and appearance) are not labeled, the generative learning approach we used is the EM algorithm. For discriminative learning, as our approach can only guarantee a local optimum of the conditional log likelihood, one critical issue is the initialization of the model parameters. In all our experiments, we use the result of the generative learning as the initialization for the discriminative learning.

We can compare the training error of the discriminative learning with generative learning based on the results in figure 6. It can be easily get that the discriminatively learned DBN performs consistently better than generatively learned DBN. When the number of training subjects is small (4 and 8), the discriminatively learned model achieves zero error rate on the training set. This is mainly due to the consistency of the training objective and classification criterion: the conditional likelihood we maximize is closely related to the classification rate on training set.

From figure 6, we can also compare the classification performance of the discriminatively learned model with generatively learned model. When the number of training subjects is large, the discriminatively learned models perform obviously better than the generatively learned models. More specifically, given 16 and 20 subjects for training, the error rates of discriminative learning are 4.5% and 2.5% lower than the generatively learning respectively. However, when the number of training subjects become smaller, discriminative learning suffers more from overfitting than generative learning. We can see that the classification error of the discriminatively learned model is 4.4% higher than generatively learned model given the sequences from 4 subjects for training.

6.2.2. *Comparison with Other Approaches on KTH Dataset*

We also compare our approach with the state-of-art approaches in the KTH dataset. As the results from Yuan et al.[18] and Laptev et al.,[19] we compare our results with 16 training subjects. The comparison is given in table 3. While the classification performance of our DBN model learned with generative learning are about 4% worse than the state-of-art approaches, it can achieve comparable results to the state-of-art approaches if learned discriminatively.

Also, we can see that, our DDBN model performance and the PLDBN model performance are very close when both models are trained with the sequence of 16 subjects. The DDBN here is slightly better. While, PLDBN is very effective especially when we have limited training subjects.

6.3. *PLDBN Model Experiments*

6.3.1. *PLDBN Model Experiments with Parking Lot Dataset*

We first apply our PLDBN algorithm to the problem of recognizing human activities in the parking lot. The data set consists of 108 sequences for 7 activities *walking* (WK), *running* (RN), *leaving car* (LC), *entering car* (EC), *bending down* (BD), *throwing* (TR) and *looking around* (LA). These activities are performed by several people with scale variation, view change and shadow interference. In the experiment, we randomly split the original data set into training set and testing set.

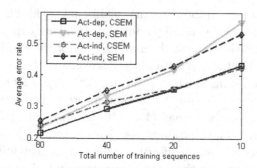

Fig. 7.   Comparison of CSEM and SEM for learning activity-dependent and activity-independent models for PLDBN in parking lot dataset.

Different algorithms are compared using training set with 10, 20, 40, 80 sequences. Each size is tested 10 times and the average recognition error is used for evaluation.

In figure 7, we compare the knowledge-based CSEM with data-based SEM in learning both activity-dependent and activity-independent model structures.

First, we look at the performance of the activity-dependent models learnt with the CSEM algorithm and SEM algorithm. As the number of training sequences decreases, the CSEM algorithm gradually shows its advantage over SEM, which means our knowledge in terms of constraints play more and more important roles on regularizing the structure learning as data size decreases.

From figure 7, we can also find that, with 20 or 10 training sequences, the activity-dependent model obtains comparable results with activity-independent model learnt using CSEM with the same data size, while it performs worse if we learn the structure without constraints. Moreover, the activity-dependent model with CSEM learning (method 1) requires only half training data to obtain comparable result to activity-independent model with SEM learning (method 2) when the data is insufficient. Specifically, with only 10 training sequence, the recognition error of method 1 is 43.2%, while the recognition error of method 2 is 43.0% given 20 training sequence. With 20 training sequence, the recognition error of method 1 is 35.5%; in comparison, the recognition error of method 2 is 35.2% given 40 sequences. Thus, we can see that exploiting the generic logic knowledge in the activity can greatly alleviate the problem of insufficient data.

6.3.2. PLDBN Model Experiments with KTH Dataset

Table 2 compares the knowledge-based CSEM algorithm with the standard SEM algorithm in learning the activity model with different number of training subjects. We can clearly see that, when the number of training subjects is large, CSEM is only marginally better than SEM algorithm. However, when the number of training subjects becomes smaller, the knowledge we exploited gradually play more important role in activity recognition. With the complement of the logic knowledge, the

CSEM algorithm can perform significantly (7.1%) better than the SEM algorithm when the number of training subjects is small.

Table 2.   Comparison of CSEM and SEM with PLDBN on KTH dataset.

# Training Subjects	4	8	12	16	20
SEM	0.760	0.828	0.862	0.880	0.892
CSEM	0.831	0.863	0.904	0.921	0.925

We also compared our approach with the state-of-art approaches on this dataset. As the way discussed before, we use the data from 16 subjects for training. Table 3 shows that we can achieve comparable result to the state-of-art approaches.

Table 3.   Compare our DBN models (GDBN, DDBN and PLDBN) with previous work on KTH dataset.

	Recognition rate
Our method - Generative DBN	88.0%
Our method - Discriminative DBN	92.5%
Our method - PLDBN with SEM	88.0%
Our method - PLDBN with CSEM	92.1%
Yuan et al.[18]	93.3%
Laptev et al.[19]	91.8%

7. Summary

In this chapter, after briefly discussing the input data and the features for activity recognition, we introduced three kinds of Dynamic Bayesian Networks for the activity modeling and recognition problem. The GDBN is a standard generical Dynamic Bayesian Network and it has two hidden nodes reflecting the states of the kinematic feature observations and image feature observations respectively in every time-slice. Other than the traditional classification criterion that choose the model with the highest likelihood, we also introduced our work on combining the likelihood outputted by DBN model with prior knowledge obtained from vehicle detection. Experiments showed that this combination during classification can improve the system performance.

The DDBN is learning the generative DBN models in a discriminative way. It can reduce the discrepancy between the training and testing objective for activity recognition in generative models. Compared to the generative learning approaches, the DDBN approach has a more consistent objective in the training stage with the classification criterion, which can guarantee a better classification performance on the training set. Based on our experiments on the real data from KTH activity data set, we demonstrate the advantage of discriminative learning over generative learning when training data is sufficient.

For the PLDBN model, we focus on exploiting prior knowledge from human activity domain and investigating a constrained structure learning method to learn activity model combining these prior knowledge with training data. The experimental results demonstrate the effectiveness of our knowledge-based learning scheme in reducing the dependence on training data and alleviating the over-fitting problem when data is insufficient. It also shows promise of the activity-dependent structures in improving activity recognition.

References

1. P. Turaga, R. Chellappa, V. Subrahmanian, and O. Udrea, Machine recognition of human activities: A survey, *IEEE Trans. on Circuits and Systems for Video Technology.* **18**(11), 1473–1488, (2008).
2. C. Schuldt, I. Laptev, and B. Caputo. Recognizing human actions: a local svm approach. In *Proceedings of the 17th International Conference on Pattern Recognition*, vol. 3, pp. 32–36, (2004).
3. J. Yamato, J. Ohaya, and K. Ishii, Recognizing human action in time-sequential images using hidden markov model, *CVPR.* (1992).
4. T. Xiang and S. Song, Video behavior profiling for anomly detection, *PAMI.* (2008).
5. W. Iba and P. Langley. Induction of one-level decision trees. In *Proceedings of the Ninth International Conference on Machine Learning*, pp. 233–240, (1992).
6. Z. Zeng and Q. Ji. Knowledge based activity recognition with dynamic bayesian network. In *ECCV 2010*, vol. Part VI, LNCS 6316, pp. 532–546, (2010).
7. P. Dollar, V. Rabaud, G. Cottrell, and S. Belongie. Behavior recognition via sparse spatio-temporal featrues. In *ICCV workshop: VS-PETS*, (2005).
8. Z. Ghahramani, Learning dynamic bayesian networks, *Adaptive Processing of Sequences and Data Structures.* **1387**, (1997).
9. K. Murphy. *Dynamic Bayesian Networks: Representation, Inference and Learning.* PhD thesis, UC Berkeley, Computer Science Division (July, 2002).
10. Z. Sun, G. Bebis, and R. Miller, On-road vehicle detection: A review, *IEEE Transactions on Pattern Analysis and Machine Intelligence.* **28**(5), 694–711, (2006).
11. M. Avriel, *Nonlinear Programming: Analysis and Methods.* (Dover Publishing, 2003).
12. R. Greiner, X. Su, S. Shen, and W. Zhou, Structural extension to logistic regression: discriminative parameter learning of belief net classifiers, *Machine Learning.* **59**, 297–322, (2005).
13. K. Murphy, Inference and learning in hybrid bayesian networks, *Technical report, University of California, Berkeley.* (1998).
14. B. Milch and S. Russell. First-order probabilistic languages: Into the unknown. In *Proceedings of the 16th International Conference on Inductive Logic Programming*, pp. 10–24, (2006).
15. J. Halpern, An analysis of first-order logics of probability, *Artificial Intelligence.* pp. 311–350, (1990).
16. C. Vogler and D. Metaxas, A framework for recognizing the simultaneous aspects of american sign language, *CVIU.* (2001).
17. N. Friedman, The bayesian structural em algorithm, *UAI.* (1998).
18. J. Yuan, Z. Liu, and Y. Wu, Discriminative subvolume search for efficient action detection, *CVPR.* (2009).
19. I. Laptev, M. Marszalek, and C. Schmid, Learning realistic human actions from movies, *UAI.* (2008).

CHAPTER 1.5

BACKGROUND ESTIMATION AND ITS APPLICATIONS

Xida Chen and Yee-Hong Yang

Department of Computing Science, University of Alberta, 4-13 Athabasca Hall,
Edmonton, Alberta, T6G 2E8, Canada
xida@cs.ualberta.ca; yang@cs.ualberta.ca

In this chapter, we present a method to estimate background of a set of related images, which are images that can be taken either from different viewpoints or from the same viewpoint. Only two basic assumptions are required in our methods: the background remains stationary throughout the entire images sequence and background objects are more likely to appear than the transient ones. We propose a novel scanline energy optimization method to select appropriate pixels on each scanline from multiple input images to composite the background. The objective function includes a smoothness measure which integrate image segmentation to represent the similarity of adjacent pixels, and a stationary coefficient that is defined based on the frequency of occurrences of adjacent pixels. We apply linear blending to create a final result after the candidate pixels are selected. The experimental results of our proposed methods are very encouraging. Furthermore, we use the CUDA (Compute Unified Device Architecture) programming language to make full use of the GPU (Graphics Processing Unit) processing power. We implement an efficient graph-based image segmentation algorithm using CUDA for acceleration.

1. Introduction

With the fast advancement of hardware, digital cameras are so popular nowadays that they appear in many different forms from low cost webcams to high end single lens reflex cameras. Despite the proliferation of cameras, their views are still limited by the field of view in the design of the cameras. For people who are not satisfied with images from a single view, they can stitch images from multiple views taken by one camera to form a panorama with a larger field of view. In fact, many cameras come with software that can do image stitching. As well, Brown et al. [3] develop an automatic image stitching system, which creates nice-looking panoramas from multiple view images. The main limitation of this method is that the final image may have undesirable objects in the scene.

For example, when someone is in a room and wants to take images of the beautiful natural scenery outside of the window, the scene may be occluded by the window frames, which will remain in the final panorama when the automatic stitching system is used. Therefore, a method to estimate the background from multiple view images with undesirable foreground objects removed has many practical applications.

In this chapter, we propose a new algorithm to estimate the background for images that are taken by freely moving cameras. A unique feature of the algorithms is to integrate segmentation information into the objective function, the optimization of which gives the estimated background. The algorithms utilize multiple images of a scene and select the appropriate pixels to construct the background.

The proposed algorithms address several problems which normally appear in background estimation.

Figure 1. Two regions from two different images of the same scenes. Although these regions are at the same position, the illumination changes significantly between them.

(1) Illumination difference among input images. When illumination changes significantly among images, the pixel value that corresponds to the same physical point can appear quite differently in different images (Figure 1).
(2) Artifacts when fusing images. Seams could exist in the final panorama due to varying illumination conditions.
(3) Robustness. A robust algorithm should have minimal assumptions of the foreground or the background.

Besides the high quality results, the processing time of an algorithm is also important in real world applications. Therefore, one part of the proposed algorithm is implemented based on the GPU. The GPU is a multithreaded, many-core processor with tremendous computational horsepower and very high internal memory bandwidth. Its highly parallel structure makes it more effective in some forms of computation than the general purpose CPU. In particular, the GPU is a good choice when the same program is executed on many processing elements in parallel. Recently, NVIDIA has developed a programming model for multithreaded processors. CUDA, which is known as a parallel computing architecture, is a C-style programming language that is well suited to expose the parallel capabilities of the GPU to applications.

The organization of this chapter is as follows. Section 2 introduces several related works. Our novel background estimation method is proposed in Section 3. Following that, Section 4 shows the experimental results. Section 5 demonstrates useful applications of background estimation. Then in Section 6, we give more details on the implementation of an image segmentation algorithm on the GPU. Finally we conclude the chapter and introduce future work in Section 7.

2. Related Work

During the last decade, researchers have made significant achievements in the area of image stitching. Algorithms have been proposed to create panorama for scenes without undesirable objects [1] [3] [17] [24] and for scenes with them [8] [10] [25]. As it is commonly known, algorithms in the former category cannot eliminate undesirable objects in the final result. Even for algorithms in the latter category, some of the undesirable objects may still remain in the panorama.

On the other hand, background estimation, as well as background modeling, is a common problem in many areas of computer vision. Researchers have been working on this topic for more than a decade. Indeed, the first algorithm can be traced back 20 years ago to the work of Long and Yang [18] in estimating background for detecting moving objects. Since then, many algorithms have been proposed for background modeling, such as approximated median filtering [19], Kalman filtering [21], a mixture of Gaussians [23], optic flow analysis [13], and mode estimation [20]. Some of these methods have an essential assumption that for every pixel, the background has to appear in more than half of the number of the images. To do a thorough survey in this topic is obviously beyond the space limitation of this chapter. Instead, some of the most related works to ours are discussed.

In computer graphics, Agarwala et al. [1] develop a general and powerful framework for combining a set of images into a single composite image. The framework has been used for a wide variety of applications. For example, to create an image with all the best elements from the input images, to extend the depth of the field, to create panoramic mosaic from multiple images. It can also be applied to background reconstruction. The method for background estimation defines a cost function which includes data penalty and interaction penalty, and is minimized using graph cuts optimization. However, the method cannot always provide good results and requires user interactions for refinement.

Xu et al. [26] propose a simple, yet robust approach for background estimation using loopy belief propagation. The assumption made in this method is that the background objects are stationary in all the input frames and the background is disclosed at least once at each pixel, which is the minimal assumption for background estimation. The method proposes an energy function which is based on visual smoothness only and minimized using loopy belief propagation. This method makes minimal assumption on the input frames and waives the need to tune parameters. It works by composing a visually smooth background image using pixels selected from input frames. However, this approach fails if the background is complex because the energy function has only a smoothness term which measures the smoothness of the background. Therefore, the method cannot always get good results when the background is cluttered.

Recently, Colombari et al. [6] propose a patch-based background initialization technique to construct background in cluttered image sequences. The method first divides the images into overlapping patches, and clusters these patches along the same time-line so that seed patches which are representatives of the largest patches can be detected. Using the seed patches, region growing is applied to generate the background. Since the method assumes that the background is cluttered, visual cuts are more likely to appear in background regions. Hence, the technique is able to extract the background even when some background regions appear only once. However, the assumption that the background is cluttered is not always true in some applications.

Cohen [5] casts the background estimation problem as a labeling problem. The method uses a cost function that includes a data term and a smoothness term. The data term is defined based on color stationariness and motion boundary consistency. The smoothness term is defined in a way to assign a high cost to an area that contains a moving highly textured object. The cost is relatively low for untextured and temporarily still objects. After the cost function is defined, graph cuts optimization is used to minimize the cost. The experimental results shown in Cohen [5] are much better than the results of median filtering.

Similar to the above strategy, Granados et al. [12] propose a novel background estimation method for non-time sequence images. The method replaces the color stationariness penalty with a term that does not require temporal coherence. It adapts the motion boundary penalty and uses a hard constraint for local consistency. Graph cuts is again applied to minimize the energy function. Finally this method applies gradient domain fusion to composite the final background in order to eliminate any visible seams. The scenes presented in the datasets that are used are complex and the results are encouraging. However, this method is much more complicated as compared to Cohen [5].

Comparing to Granados et al. [12], the method in [4] is much simpler while provides similar results. The energy function defined in [4] is also based on MRF formulation. The unique feature of this method is that a predicted term is adapted in the data term. That is, the method first obtains the regions where there are transient objects. These regions are then filled in by using an image inpainting technique. The filled in color is used as predicted color, which is applied in the predicted term. The terms defined in this method is more straightforward than the ones in [12], which makes this method easier to understand.

Motivated by the work of Agarwala et al. [1], we develop a new method to combine information from multiple input images. We propose a cost function and apply a dynamic programming framework to minimize the cost so that the final result is visually smooth. The method incorporates segmentation results in the formulation of the cost function, which improves the quality of our final results. Our method has been incorporated into a depth recovery framework for free viewpoint video rendering, which is only briefly described in [16]. More details and experimental results for this algorithm are provided in this chapter.

3. Proposed Method

The input to our algorithm is a set of related images, which can be taken from the same viewpoint or from different viewpoints with overlapping regions among them. Our goal is to estimate the background. There are two basic assumptions: the background objects are stationary throughout the entire image sequence and they are more likely to appear than the transient ones. These assumptions are commonly used in many background estimation algorithms such as [21] and [12]. The first assumption implies that pixels of the same background object are always in the same position. However, this assumption does not restrict the camera to be stationary. The second assumption provides a constraint that background pixels should appear more frequently in most regions.

Suppose that the input sequence has N input images. Since the images can be taken from different viewpoints, the pixels in every input image need to be transformed to a global coordinate system initially. We estimate the warping function using the method proposed in [3], which is a homography for each input image with respect to the reference image. In the rest of this chapter, we assume that all the images have been projected into a global coordinate system. We denote the color of a pixel in the mth image of the sequence at coordinates (x, y)

as $I_m(x, y)$ where $1 \leq m \leq N$. A pixel in the output background image is denoted as $O(x, y)$. Then the background estimation problem can be formulated as

$$O(x, y) = \sum_{m=1}^{N} \alpha_m(x, y) I_m(x, y) \tag{1}$$

In the above equation, $\alpha_m(x, y)$ is a binary selection function which is specified as

$$\alpha_m(x, y) = \begin{cases} 1 & \text{if } m \text{ is selected} \\ 0 & \text{otherwise} \end{cases} \tag{2}$$

Hence, if $\alpha_m(x, y)$ is set properly for each pixel, then the background is recovered using (1) and (2). The candidate pixels are selected from the input images to composite the background.

We observe that an output image without any foreground object achieves global visual smoothness. In other words, there are significant differences between an image with only the background and an image with both the foreground and the background. The former is visually smoother than the latter one. Therefore, if we have an appropriate cost function to measure smoothness, then the final result with only the background has the minimum cost.

The following is an overview of our method.

(1) Assign a cost to each pair of adjacent pixels according to the proposed cost function (Section 3.1), which includes a smoothness measure (Section 3.2) and a stationary coefficient (Section 3.3).
(2) Apply dynamic programming (DP) to minimize the cost of each scanline (Section 3.4), and determine α_m for each pixel.
(3) Apply linear blending to reduce the seams in the estimated background (Section 3.5).

3.1. *Cost Function*

The proposed cost function is defined as follows.

$$C_{ij}(x, Y) = \left(S_x \left(I_i(x, Y), I_j(x - 1, Y) \right) + \gamma \cdot S_y \left(I_i(x, Y) \right) \right)$$
$$\cdot \rho \left(I_i(x, Y), I_j(x - 1, Y) \right) \tag{3}$$

where $C_{ij}(x, Y)$ is the cost assigned to a pair of adjacent pixels $I_i(x, Y)$ and $I_j(x - 1, Y)$, $1 \leq i, j \leq N$. $S_x \left(I_i(x, Y), I_j(x - 1, Y) \right)$ is the smoothness measure along the x-direction, $S_y \left(I_i(x, Y) \right)$ along the y-direction, γ a coefficient assigned to balance the weight of the smoothness measure along two dimensions which is set to be 1, $\rho\left(I_i(x,Y), I_j(x-1,Y)\right)$ the stationary coefficient.

3.2. *Smoothness Measure*

The smoothness measure is defined to measure how well a pair of adjacent pixels satisfies the visual smoothness constraint. We measure the smoothness in two directions. In this section, we first introduce the smoothness measure along the x-direction and then the y-direction. The definition of our smoothness measure along the x-direction is given as follows.

$$S_x\left(I_i(x,Y), I_j(x-1,Y)\right)$$

$$= \begin{cases} \beta \cdot \sum_{k=r,g,b} \left|I_i^k(x,Y) - I_j^k(x-1,Y)\right| & if\, I_i(x,Y) \in R \text{ and } I_j(x-1,Y) \in R \\ \sum_{k=r,g,b} \left|I_i^k(x,Y) - I_j^k(x-1,Y)\right| & otherwise \end{cases} \quad (4)$$

where $|\cdot|$ denotes the absolute value, R a segment, and r, g, b represent the red, green and blue channels, respectively. It measures the similarity between two neighboring pixels $I_i(x,Y)$ and $I_j(x-1,Y)$ along the same scanline. The parameter β is set to be 0.01 for all the datasets, and the reason of setting it with this particular value is discussed below.

In order to assign a more reasonable cost to each pair of adjacent pixels, the information of image segmentation is integrated into the smoothness measure. Although any segmentation algorithm could be used, we apply a graph-based image segmentation algorithm [9] in this chapter. The interested reader should refer to [9] for more details.

In Equation (4), the penalty is relatively small if two adjacent pixels $I_i(x,Y)$ and $I_j(x-1,Y)$ belong to the same segment R. There are two reasons to take advantage of the segmentation result. First, by construction, a segment is a group of adjacent pixels with similar colors. Second, applying the result of image segmentation can prevent frequent switching among input images when compositing the final background. According to Equation (1), each pixel in the final result is selected from one of the input images appropriately. When image segmentation is applied, pixels in the same segment from the same image will be selected more favorably. Without this bias, adjacent pixels could be selected from different images. Additionally, using segmentation results makes the estimated background less sensitive to illumination change that appears among input images by biasing to select pixels from the same segment. Hence, we set β

to be 0.01 to reflect a lower penalty when two adjacent pixels are in the same segment.

The measure described above encourages the result to be visually smooth along the x-direction. Along the y-direction, we use the information from the previous scanline to measure smoothness. In particular, it is defined as follows.

$$S_y(I_i(x,Y)) = \sum_{k=r,g,b} |I_i^k(x,Y) - I_f^k(x-1,Y)| \tag{5}$$

where i denotes the ith input image. It is defined based on the information from the previous scanline that has been processed, which is the $(Y-1)^{th}$ scanline in Equation (5). For the first scanline of the image, the smoothness term is initialized to be 0 since there is no previous scanline to it. On the $(Y-1)^{th}$ scanline, f denotes the image selected for each x-coordinate, i.e. $\alpha_f(x, Y-1) = 1$. It is noteworthy that f depends on the x-coordinate. In this chapter, however, the dependence is implied to simplify notation. The cost minimization is discussed in Section 3.4.

3.3. *Stationary Coefficient*

We propose a stationary coefficient to satisfy the second assumption, that is, background objects are more likely to appear than the transient ones. The coefficient ρ is defined below.

$$\rho\left(I_i(x,Y), I_j(x-1,Y)\right)$$

$$= \left(1 - \frac{NUM(I_i(x,Y))}{N}\right) \cdot \left(1 - \frac{NUM\left(I_j(x-1,Y)\right)}{N}\right) \tag{6}$$

where N is the number of input images and $1 \leq i,j \leq N$. $NUM(I_i(x,Y))$ denotes the number of pixels from $\{I_1(x,Y), ..., I_i(x,Y), ..., I_N(x,Y)\}$ that are similar to $I_i(x,Y)$. In our experiments, $I_m(x,Y)$ is defined to be similar to $I_i(x,Y)$ if $|I_m(x,Y) - I_i(x,Y)| < 0.1 \cdot I_i(x,Y)$ for all three color channels, where $1 \leq m \leq N$. The stationary coefficient implies that if two adjacent pixels appear frequently, then the corresponding assigned cost is small.

3.4. *Cost Minimization*

Fig. 2 shows how the cost is assigned to two adjacent pixels. To simplify illustrations, only two input images are shown. To extend to more images is straightforward. The figure shows only the Yth scanline. In the figure, I_1

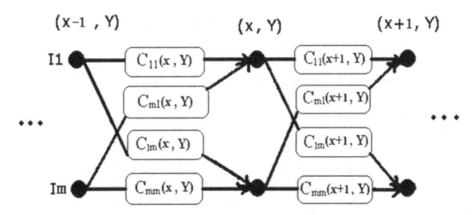

Figure 2. Cost assignment.

represent image #1 and I_m image #m. $C_{1m}(x + 1, Y)$ is the proposed cost for selecting $I_1(x, Y)$ and $I_m(x + 1, Y)$.

After a cost is assigned to every pair of adjacent pixels on a scanline, DP is applied to minimize the aggregated cost for each scanline. The formulation is similar to the one introduced in [7] and is defined as follows.

$$E_Y = \min_m\big(E_Y(L, m)\big) \qquad (7)$$

$$E_Y(x, m) = \begin{cases} 0 & if\ x = 1 \\ \min\ (E_Y(x - 1, 1) + C_{1m}(x, Y), ..., \\ \quad E_Y(x - 1, m) + C_{mm}(x, Y), ...,), ..., \\ \quad E_Y(x - 1, N) + C_{Nm}(x, Y)) & if\ x > 1 \end{cases} \qquad (8)$$

where $1 \le m \le N$. L is the length of a scanline and $1 \le x \le L$. DP is applied to one scanline at a time. In Equation (7), E_Y is the minimum cost of the Yth scanline. In Equation (8), $E_Y(x, m)$ denotes the aggregated cost of the mth row, the xth column on the Yth scanline. The minimum cost is calculated after going through the whole scanline. During backtracking, a pixel at each x-coordinate along the scanline is selected from one of the input images. For each x-coordinate, the selected input image is denoted as f, which is the same as the one used in Equation (5), and the corresponding $\alpha_f(x, Y)$ is set to be 1.

3.5. *Implementation*

In this method, obvious seams could exist in the input image because different regions are selected from different input images. The illumination difference among the input image is the main cause of this artifact. Therefore, blending is required to suppress this problem.

Linear blending is applied to remove the artifact in the result. In order to combine information from the input images, a weighting function $\theta(x, y) = w(x)w(y)$ is assigned to each image, where $w(x)$ varies linearly from 1 at the center of the image to 0 at the edge. Then a weighted sum of the colors is computed by,

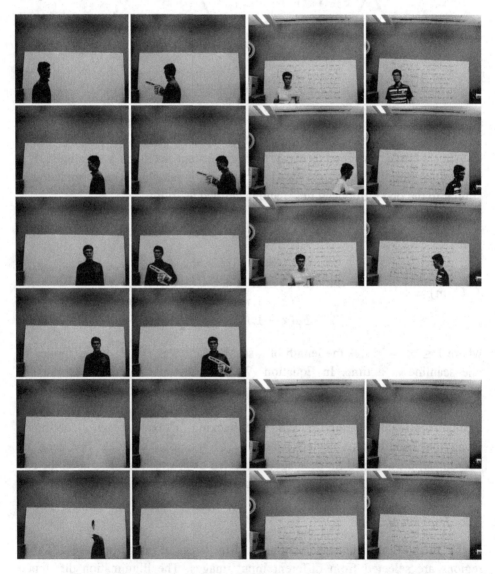

Figure 3. Background estimation of different datasets with various features.

(a) Input images

(b) Estimated background by our method　　　(c) Estimated background by [1]

Figure 4. (a) Four input images from different viewpoints with transient objects. (b) Result by applying our method. (c) Result by applying [1].

$$O^{linear}(x,y) = \frac{\sum_{m=1}^{N} I_m(x,y)\theta_m(x,y)}{\sum_{m=1}^{N} \theta_m(x,y)} \tag{9}$$

where $1 \leq m \leq N$ and $\left|I_m(x,y) - I_f(x,y)\right| < t_H$ for each color channel, $I_f(x,y)$ is introduced in Equation (5) and t_H is specified by the user according to the illumination condition. Typically, t_H is set to be $0.2 \cdot I_f(x,y)$ for all the three color channels. It could be larger when the illumination changes were large among input images.

4. Experimental Results

In our experiments, we generate several datasets and they are divided into different categories. Most of the input images are captured using a digital camera on a tripod. Some datasets generated by other researchers are also used for comparison.

Figure 3 shows some results when we apply our method to four datasets with different features. The results produced by [1] are also provided for comparison, which are shown in the last row. Each column shows an individual datasets and the results. The input images in the first column have a smooth background and smooth foreground. In this dataset, there is a region where the background appears only once. Our method can estimate the background correctly. However,

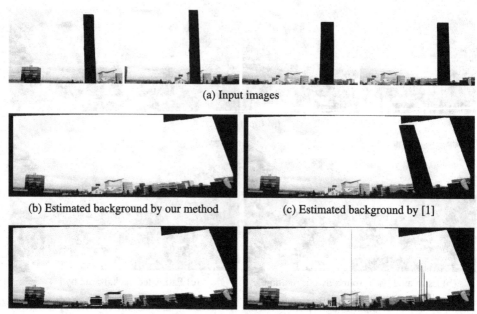

(a) Input images

(b) Estimated background by our method   (c) Estimated background by [1]

(d) Smoothness measure without segmentation (e) Smoothness measure along the y-direction
information   only

Figure 5. (a) Multiple view images with window frames as foreground objects. (b) Estimated
background by our method. (c) Result by applying [1]. (d) Result by our method when using
smoothness measure without segmentation information. (e) Result with smoothness measure along
the y-direction only.

there are error regions in the result using the method in [1]. The background of
the input in the second column is smooth, but the foreground is not. The
background regions in the input images of the third column and the last column
are the same. But one has a smooth foreground and the other has a complex
foreground. In this figure, we demonstrate that our method can produce good
results under various background/foreground conditions.

Figure 4 shows the result when applying our method to four input images
which are taken from different viewpoints, with several transient objects. Both
the background and foreground are complex in the dataset. Figure 4(b) shows our
result. In this dataset, we show that our method can provide correct results even if
the illumination changes significantly among input images. Figure 4(c) indicates
that the result using the background estimation method proposed in [1].

We demonstrate the importance of integrating segmentation information in
Figure 5. There are four input images taken by a moving camera. The window
frame is considered to be foreground object because it is moving relative to the
camera, and the buildings that are far away from the camera are the background

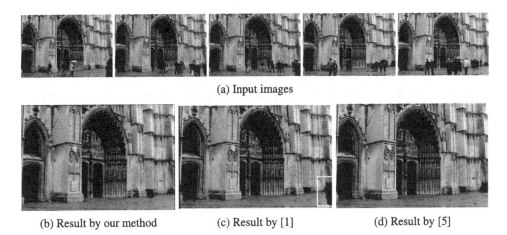

(a) Input images

(b) Result by our method   (c) Result by [1]   (d) Result by [5]

Figure 6. (a) Input images of the *Cathedral* scene. (b) Estimated background by applying our method. (c) Estimated background by applying the method proposed in [1]. (d) Estimated background by applying the method proposed in [5].

(a) A subset of 25 input images

(b) Result by our method   (c) Result by median filtering

Figure 7. Comparison of our method with the median filtering. Both methods are applied to a set of images taken by a traffic surveillance camera.

since their movements relatively to the camera can be ignored. Figure 5(b) is the result by applying our method. For comparison, we give the result produced by [1] as well. This dataset demonstrates the importance of incorporating segmentation information. Figure 5(d) is the result by changing our smoothness measure along the x-direction to be without using any segmentation information. That is, the cost between two neighboring pixels is simply the sum of the absolute difference of their colors. Figure 5(e) shows the result with the smoothness measure along the y-direction only. The red arrows in Figure 5(d) and 5(e) point to the regions that are wrong. Figure 5 shows that our algorithm can create a large field of view with transient foreground objects removed.

Next, we apply our method to the *Cathedral* scene, which is generated by other researchers and originally used in [1]. There are five input images in the sequence, as shown in Figure 6(a). The result by our method is given in Figure 6(b). As comparison, the result by applying the method proposed in [1] is given in Figure 6(c). It is shows that the estimated background by our method is better in the specific region indicated by the yellow rectangle. We also show the result produced by [5] in Figure 6(d) and the red rectangle region shows the error.

5. Applications

We would like to demonstrate that background estimation can be applied to multiple research areas.

5.1. *Traffic Surveillance*

As we know, traffic surveillance is an important practical problem. An important first step is to estimate the background. In order to demonstrate that our method is practical in the real world applications, we apply it to a dataset that is captured by a traffic surveillance camera. The dataset is taken from [5], which has 25 images in total. To conserve space, we only show four of them. Figure 7(b) is our result and Figure 7(c) is the result produced using a median filter. This dataset shows that our method is much better than median filtering.

5.2. *Image-based Rendering*

Our method is first applied to two high-quality video sequences with their depth maps provided by the Microsoft Research Group [27]. The two sequences are provided online and named "Breakdacing" and "Ballet". Both of them can be used in image-based rendering. There are eight different views for each

(c)

(a) Input images   (b) Estimated background

Figure 8. Applying our method to the "Breakdacing" sequence. Top row shows images from one viewpoint and the bottom row is from another viewpoint. (a) The input color images with their corresponding depth maps. (b) Estimated background by our method. (c) Background image used in [15].

sequence, and each sequence contains 100 color images with 100 corresponding depth maps.

Our method is applied to the depth maps and their corresponding color images simultaneously. Figure 8 shows the results when applying our method to the "Breakdacing" sequence. The two left columns are the input images with their corresponding depth maps. The top row is from one viewpoint and the bottom row is from another. The two right columns show the results by applying our method, which are the background for both color images and the depth maps. Similar arrangement of the images is shown in Figure 9 for the "Ballet" sequence. We can see that the estimated background for the depth maps is very smooth, especially in the floor region. The results can be used to improve each individual

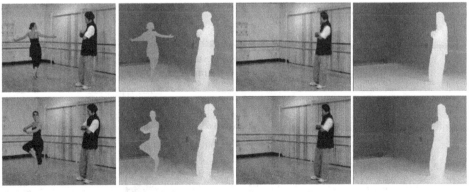

(a) Input images   (b) Estimated background

Figure 9. Applying our method to the "Ballet" sequence. Top row shows images from one viewpoint and the bottom row is from another viewpoint. (a) The input color images with their corresponding depth maps. (b) Estimated background by our method.

input depth map, and therefore the rendering quality can be significantly improved as well. In particular, Figure 8(c) shows the color background used in [15]. Figure 8 clearly shows that our result is much smoother, and has less noise than the one shown in Figure 8(c), especially on the floor. We believe that there could be significant improvements on the results in [15] if our background estimation results were applied.

5.3. *De-fencing*

In our real-world, there are cases where a fence happens to be the foreground of the images. For example, it happens when taking images in the zoo. In this case, the desired background objects will be occluded by the fence. We show that our method can be applied to remove the fence in given images.

In Figure 10, several images with a fence as the foreground object are given as the input images to our method. Figure 10(b-f) show the results with different number of input images. We can see that with the number of input images increases, our method gains more information on the background and hence the fence gradually disappear in the result. Figure 10(g) shows the central part of the result by using the whole image sequence which includes six images. We can see that the fence is completely removed in the central part of that image.

6. GPU Image Segmentation

Because GPUs offer incredible computing resources for both graphics and non-graphics processing, people are focusing on general-purpose computation using

Figure 10. Multiple image de-fencing. (a) Six input images with a fence as foreground object. (b-f) Results with different number of input images. (g) The central part of (f).

the GPU (GPGPU) during the last few years. As a matter of fact, image processing has become a popular topic for acceleration on the GPU as well. The reason is that many image processing methods have sections that consist of a common computation over many pixels. Ahn et al. [2] develop an image processing toolkit on the GPU which contains several techniques such as image segmentation and image enhancement. The segmentation algorithms that implemented are isoperimetric graph partition [11], normalized cut [22] and active contour [14]. However, the method proposed in [9] has not been implemented on the GPU before.

Given an image, the first step of the segmentation algorithm [9] is to transform it into a graph. In the CPU implementation, the weight of each edge in the graph is computed sequentially. This process is parallelized in our GPU implementation. Because the threads on the GPU are executed in parallel, we assign each thread to process one edge in the graph. The only requirement for the weight computation is the intensity information of the adjacent pixels connected by the edge, which is stored in a shared memory. The pseudocode for building a graph is given in Algorithm 1.

Algorithm 1. Building a graph for an image on the GPU

//define the structure of an edge
struct {
    float w;   //the weight of the edge
    int a, b;   //the indices of two adjacent nodes connected by this edge
} edge;
Input: An array containing the color of each pixel in the image
for each thread on the GPU **do**
    assign the indices of two adjacent nodes to edge.a and edge.b
    calculate the weight using the color and assign it to edge.w
end for
synchronize all threads
Output: A group of edges

The second part that is ported to the GPU is the main step to construct the segments. In this step, the algorithm processes each edge sequentially and determines whether or not the two adjacent vertices that are connected by this edge should be merged. In our GPU implementation, we assign one thread to process one edge. However, the result by our implementation can be different from the GPU version. A minimum spanning tree (MST) is maintained for each segment in this algorithm. When an edge is processed, the MST is updated. For example, suppose v_i and v_j are two adjacent pixels connected by an edge e, and

$v_i \in R_i, v_j \in R_j$, where R_i and R_j are two separate segments. To determine whether or not v_i and v_j should be merged, the program uses the information of the MSTs of both R_i and R_j, as well as the size of each segment. We use $P(R_i, R_j)$ to denote a predicate. That is, if $P(R_i, R_j)$ is true, then v_i and v_j should be merged, which means that R_i and R_j should be merged as well. In the CPU version, when e is processed, both segments may have some nodes in them, which means that their size can be large. However, when the threads are running in parallel on the GPU, the segments R_i and R_j may have only one node in each segment, that is, v_i in R_i and v_j in R_j. Since the size and the MST of each segment differ between the GPU and the CPU, different results are produced. We give the pseudocode for this part in Algorithm 2.

Algorithm 2. Segment a graph

Input: A group of edges representing a graph
for each thread on the GPU **do**
    compute $P(R_i, R_j)$ for each edge e connecting two vertices $v_i \in R_i$ and $v_j \in R_j$
    if $P(R_i, R_j) = true$ **then**
      merge R_i and R_j
    else
      do nothing
    end if
end for
synchronize all threads
Output: A group of separate segments

We evaluate our implementation in both **correctness** and **performance**.

Correctness. We use the segmentation results produced by our GPU implementation and redo the experiments of applying segmentation-based

(a) Estimated background with the CPU segmentation results

(b) Estimated background with the GPU segmentation results

Figure 11. Verify the correctness of our GPU implementation by comparing the estimated background with CPU and GPU segmentation results.

method. Figure 11(a) shows the background estimation results by using the CPU segmentation results, and Figure 11(b) is with our GPU image segmentation results. The input images are the same as the one shown in Figure 3(a). The similarity between the images on the left column and the right column demonstrates that our GPU implementation of this algorithm is correct.

Performance. After the correctness is verified, the performance of our GPU implementation is evaluated. We apply our implementation to images of different sizes and compare the running time on the GPU to the CPU. TABLE I shows the comparison of the running time. Figure 12 shows that the curve of running time versus image size for the GPU stays much lower than that of the CPU. This means that the GPU can be used to process much larger size images.

Table I. Comparison between the running time for image segmentation on the CPU and the GPU.

Image Resolution	CPU	GPU
1200*900	30.12s	1.49s
1632*1224	57.49s	2.98s
2308*1732	118.68s	5.57s
3264*2448	247.42s	11.70s

7. Conclusion

We have presented a novel algorithm to estimate background of a set of related images. We demonstrate that our algorithm can be applied to both images taken from different viewpoints and images taken from the same viewpoint. We

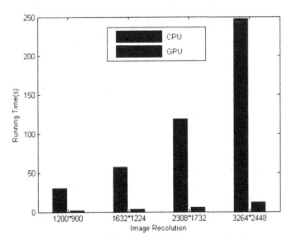

Figure 12. Comparison between the running time of the image segmentation algorithm on the CPU and the GPU.

propose a new segmentation-based cost function and apply a dynamic programming framework to minimize the cost so that the final output image achieves globally visual smoothness. Furthermore, by applying our methods to datasets that are generated by other researchers, we demonstrate that we can obtain comparable results using our proposed methods. In order to shorten the processing time, we implement the graph-based segmentation algorithm on the GPU which is 20 times faster than on the CPU. Notice that there is a threshold t_H when performing linear blending which is defined based on the illumination condition in the input images. A possible research topic in the future is estimate t_H based on input images.

Acknowledgements

The authors would like to thank NSERC, and the University of Alberta for their generous financial support.

References

1. A. Agarwala, M. Dontcheva, M. Agrawala, S. M. Drucker, A. Colburn, B. Curless, D. Salesin, M. F. Cohen. Interactive Digital Photomontage. *In Proc. of ACM SIGGRAPH*, 2004.
2. I. Ahn, M. Lehr, and P. Turner. Image Processing on the GPU. *University of Pennsylvania.* White paper, 2005.
3. M. Brown and D. G. Lowe. Recognising Panorama. *In Proc. of the 9th International Conference on Computer Vision.* Nice, France, 2003.
4. X. Chen, Y. Shen, and Y. H. Yang. Background Estimation using Graph Cuts and Inpainting. *Graphics Interface.* Ottawa, Ontario, 2010.
5. S. Cohen. Background estimation as a labeling problem. *In Proc. of the 10th IEEE International Conference on Computer Vision (ICCV)*, 2005, pp. 1034-1041.
6. A. Colombari, M. Cristani, V. Murino and A. Fusiello. Exemplar-based background model initialization. *In Proc. of the third ACM International Workshop on Video Surveillance & Sensor Networks*, pp.29-36, 2005.
7. T. H. Cormen, C. E. Leiserson, R. L. Rivest and C. Stein. Introduction to Algorithms, Chapter 15.1. *The MIT Press*, 1990.
8. J. Davis. Mosaics of Scenes with Moving Objects. *IEEE Computer Society Conference on Computer Vision and Pattern Recognition.* 1998.
9. P. F. Felzenszwalb and D. P. Huttenlocher. Efficient Graph-Based *Image Segmentation. International Journal of Computer Vision.* 2004.
10. N. Gracias, A. Gleason, S. Negahdaripour, and M. Mahoor. Fast Image Blending using Watershed and Graph Cuts. *In Proc. of British Machine Vision Conference.* 2006.
11. L. Grady and E. Schwartz. Isoperimetric Graph Partitioning for Image Segmentation. *IEEE Transactions on Pattern Analysis and Machine Intelligence*, 2006.
12. M. Granados, H. Seidel, H. Lensch. Background estimation from non-time sequence images. *In Graphics Interface*, 2008, pp. 33-40.

13. D. Gutchess, M. Trajkovic, E. Cohen-Solal, D. Lyons, A. K. Jain. A background model initialization algorithm for video surveillance. *In Proc. of the International Conference on Computer Vision (ICCV)*, 2001, pp. 733-740.

14. M. Kass, A. Witkin, and T. Terzopoulos. Snakes: Active Contour Models. *International Journal of Computer Vision*, 1987.

15. E. S. Larsen, P. Mordohai, M. Pollefeys and H. Fuchs. Temporally Consistent Reconstruction from Multiple Video Streams Using Enhanced Belief Propagation. *In Proc. of the International Conference on Computer Vision*. Rio de Janeiro, Brazil, 2007.

16. C. Lei, X. Chen, and Y. H. Yang. A New Multiview Spacetime-Consistent Depth Recovery Framework for Free Viewpoint Video Rendering. *In Proc. of International Conference on Computer Vision. Kyoto*, Japan, 2009.

17. A. Levin, A. Zomet, S. Peleg, and Y. Weiss. Seamless Image Stitching in the Gradient Domain. *8th European Conference on Computer Vision (ECCV 2004)*, 2004.

18. W. Long, Y.-H. Yang. Stationary background generation: An alternative to the difference of two images. *Pattern Recognition*, 23, 1990, 1351-1359.

19. N. Mcfarlane, C. Schofield. Segmentation and tracking of piglets in images. *Machine Vision and Applications*, 8(3), 1995, 187-193.

20. M. Piccardi, T. Jan. Mean-shift background image modeling. *In Proc. of the International Conference on Image Processing (ICIP)*, 2004, pp. 3399-3402.

21. C. Ridder, O. Munkelt, H. Kirchner. Adaptive background estimation and foreground detection using kalman filtering. *In Proc. of the International Conference on Recent Advances in Mechatronics (IJRAM'95)*, 1995, pp. 193-199.

22. J. Shi and J. Malik. Normalized Cuts and Image Segmentation. *IEEE Transactions on Pattern Analysis and Machine Intelligence*, 2000.

23. C. Stauffer, W. E. L. Grimson. Adaptive background mixture models for real-time tracking. *In Proc. of the 1999 Conference on Computer Vision and Pattern Recognition (CVPR)*, 1999, pp. 2246-2252.

24. R. Szeliski and H.-Y. Shum. Creating Full View Panoramic Image Mosaic and Texture-mapped Models. *Computer Graphics (Proc. of SIGGRAPH 2007)*. Los Angeles, 2007.

25. M. Uyttendaele, A. Eden, and R. Szeliski. Eliminating Ghosting and Exposure Artifacts in Image Mosaics. *IEEE Computer Society Conference on Computer Vision and Pattern Recognition*. Kauai, Hawaii, 2001.

26. X. Xu, T. S. Huang. A loopy belief propagation approach for robust background estimation. *In IEEE Computer Society Conference on Computer Vision and Pattern Recognition (CVPR)*, 2008.

27. C. L. Zitnick, S. B. Kang, M. Uyttendaele, S. Winder, and R.Szeliski. High-quality video view interpolation using a layered representation. *ACM Transactions on Graphics*, 23(3):600-608, 2004.

CHAPTER 1.6

NONRIGID IMAGE REGISTRATION USING MOMENTS

Wei Liu and Eraldo Ribeiro*

Computer Vision and Bio-Inspired Computing Laboratory,
Florida Institute of Technology,
Melbourne, FL 32901, U.S.A.
*eribeiro@cs.fit.edu

Image moments have been widely used for designing shape descriptors that are invariant to rigid transformations. This chapter focuses on the use of image moments to achieve incremental nonrigid registration of images. Here, we address the problem of estimating non-rigid deformation fields based on image moment variations. We propose a non-rigid registration method that is able to recover the deformation field between two images without solving the correspondence problem. This is achieved by representing deformation-induced variations in image moments as a system of quadratic functions. Our method uses polynomials to both parameterize the deformation field and to define image moments. The method was tested on MPEG-7 shapes and cardiac MRI sequences.

1. Introduction to image registration

Image registration is a fundamental problem in computer vision and image analysis with many applications in shape retrieval,[1] object recognition,[2] and medical image analysis.[3] Roughly speaking, the goal of registration is to align images that have been deformed by some geometric transformation. In this chapter, we study registration methods based on a classic integral image transform known as image moments.

Image moments have been widely used in computer vision to achieve geometric invariance for applications such as shape retrieval and recognition.[4] Obtaining invariance to more complex deformations (e.g., affine deformation[5]), and the use of other integral image representations (e.g., wavelet[6] and Fourier transform[7]) have been goals of moment-based approaches to image analysis. For image registration, rather than focusing on the moment's invariance properties, one can use moments as a measurement of the deformation between pairs of images. In this chapter, we investigate the relationship between image deformations and their effect on image moments. Based on this relationship, we are able to design a registration algorithm that is both computationally efficient and fairly robust to noise. More importantly, this study reveals an avenue for future research on image registration

---

*Corresponding author.

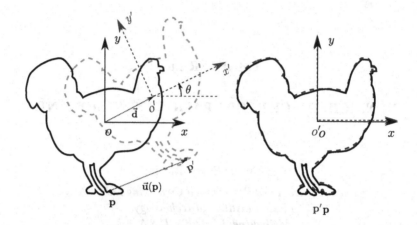

Fig. 1. Registering images undergoing rigid transform. Given two images (left), the goal of image registration is to align them (right), by calculating a coordinate transform represented using parametric models (e.g., in the case of rigid transform, the parameters are rotation angle θ and the shifting of coordinate origin $\vec{\mathbf{d}}$). Alternatively, the coordinate transform can also be represented as pixel-wise displacements (deformation) $\vec{\mathbf{u}}(\mathbf{p})$.

using integral image transforms which can lead to the development of registration methods that work in transformed domain (e.g., compressed image data).

The chapter is organized as follows. First, we introduce the image registration problem, followed by a review of image moments and their application to shape retrieval. Secondly, we elaborate on the relationship between changes of image moments and image deformations. Based on this relationship, we show how to incorporate image moments into registration methods. Finally, we demonstrate some of these methods on both synthetic and real-world images, and conclude the chapter by providing directions for future study.

1.1. *Problem statement and main components*

Suppose that we want to register the two shapes shown in Figure 1. The registration task essentially deforms the *source image* or *template image* (dashed-line) to be aligned with the *target image* (solid-line). Three subproblems are involved in solving this task:

How to represent the geometric transformation between the images. Generally, image deformation can be represented using a coordinate mapping[a] $\mathbf{x}' = \mathbf{T}(\mathbf{x})$, where $\mathbf{x}$ and $\mathbf{x}'$ are the coordinates in the original and deformed images, respectively. Alternatively, the geometric transformation can be modeled by a pixel-wise displacement $\mathbf{x}' = \mathbf{x} + \mathbf{u}(\mathbf{x})$, where $\mathbf{u}(\mathbf{x})$ is a translation vector of an image pixel at $\mathbf{x}$, and the vector field $\mathbf{u}(\mathbf{x})$ is often called the *deformation field*. $\mathbf{T}(\mathbf{x})$ and $\mathbf{u}(\mathbf{x})$ are often represented by parametric functions of some parameter $\lambda \in \Lambda$, where Λ is the set of all possible parameter values.

---

[a]In this chapter, we use bold-font letters to indicate coordinate vectors, e.g., $\mathbf{x} = (x \ y)^{\top} \in \mathbb{R}^2$. In later discussion, we also use parenthetical superscripts to index the components of $\mathbf{x}$, such as $\mathbf{x} = (x^{(1)}, x^{(2)})^{\top}$.

How to measure similarities between deformed images. Given the large number of ways images can be aligned, a registration algorithm should be able to quantify the quality of a given alignment, i.e., decide that one transformation is "better" than another. Given an image as a 2-D function $f(\mathbf{x})$, we need to define a similarity measure $\mathbf{E}(f'(\mathbf{x}'), f(\mathbf{x})) \in \mathbb{R}$ between the deformed source image $f'(\mathbf{x}')$ and a target image $f(\mathbf{x})$.

How to optimize the image alignment. Given the similarity measure $\mathbf{E}(f'(\mathbf{x}'), f(\mathbf{x}))$, the registration problem is then converted to one of maximizing the similarity measure, by searching for the "best" estimate of the parameters $\hat{\lambda}$ as follows:

$$\hat{\lambda} = \arg \max_{\lambda \in \Lambda} \mathbf{E}(I'(\mathbf{x}'), I(\mathbf{x})). \tag{1}$$

Using (1), image registration is finally formulated as a numerical optimization problem that can be solved computationally, often allowing for the use of readily available off-the-shelf optimization algorithms. Next, we focus ourselves on the first two subproblems, and start introducing our choices of deformation model and similarity measures.

1.2. *Image deformation model*

There exist a number of parametrized models for image deformations. A deformation is called a *rigid* transformation when the size and shape of transformed objects remain unchanged, i.e., it consists of a rotation and a translation. This transformation is given by:

$$\mathbf{x}' = \underbrace{\begin{pmatrix} \cos\theta & \sin\theta \\ -\sin\theta & \cos\theta \end{pmatrix}}_{\mathbf{R}(\theta)} \mathbf{x} - \begin{pmatrix} d_x \\ d_y \end{pmatrix}. \tag{2}$$

Here, θ is the rotation angle, $\mathbf{R}(\theta)$ is the rotation matrix, and $\mathbf{d} = (d_x, d_y)^\mathsf{T}$ is the displacement vector from the coordinate origin (see Figure 1). A more general transformation model can be described by generalizing the rotation matrix in Equation 2 to obtain:

$$\mathbf{x}' = \begin{pmatrix} a_1 & a_2 \\ b_1 & b_2 \end{pmatrix} \mathbf{x} + \begin{pmatrix} a_3 \\ b_3 \end{pmatrix}. \tag{3}$$

The transformation in (3) represents a larger group of so-called *affine transformations*. In addition to rotation and translation, an affine transformation contains scaling, reflection, and shearing. Figure 2 shows examples of different types of image deformations.

Furthermore, real-world object deformation often include elastic motion (e.g., stretching and squeezing[8]) and articulated motions (e.g., hand gesture and human motion[9]). These transformations require more sophisticated mathematical models to represent them. In this chapter, we focus ourselves on elastic deformations. These transformations are typical in applications such as medical image analysis[3] and shape-matching.[10]

Elastic deformations usually cannot be properly represented by the affine model defined in (3). Instead, such deformations are usually modeled by nonlinear elastic transformations including Free-Form Deformation (FFD)[11] model that uses B-splines, Thin-Plate Splines

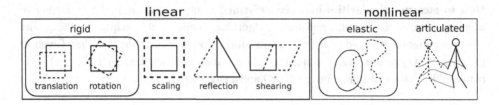

Fig. 2.   Types of image deformation: rigid, affine, elastic, and articulated motion.

(TPS)[12] based on radial basis functions, and polynomial deformation models.[13]   In this chapter, we adopt the polynomial deformation model.   Some reasons for this choice of model are as follows. First, polynomial deformation models can be combined with image moments in a way that leads to an efficient and robust registration method. Secondly, polynomial models naturally extend the affine model in (3), by including higher-order monomials (nonlinear terms). For nonlinear deformations, it is common practice to model the deformation field $\mathbf{u}(\mathbf{x})$ instead of the transformation function $\mathbf{T}(\mathbf{x})$. The polynomial representation of a 2-D deformation field $\mathbf{u} = (u(x,y), v(x,y))^\mathsf{T}$ is given by:

$$u(x,y) = \sum_{s=0}^{N}\sum_{t=0}^{N} a_{s,t}\, x^s y^t \quad \text{and} \quad v(x,y) = \sum_{s=0}^{N}\sum_{t=0}^{N} b_{s,t}\, x^s y^t. \tag{4}$$

which are linear combinations of monomials, and can be written as:

$$\mathbf{u} = \sum_{i}^{M} \mathbf{a}_i \gamma_i = \mathbf{A}^\mathsf{T}\Gamma, \tag{5}$$

where $\mathbf{A}$ is a coefficient matrix with columns $\mathbf{a}_i$, and $\Gamma = (\gamma_1, \ldots, \gamma_M)^\mathsf{T}$ is a vector of basis functions, with $\gamma_i = x^s y^t$ for some $s, t \geq 0$. The polynomial deformation model in (4) contains the affine model as a special case. The compact representation in (5) simplifies our later discussion, and facilitates future generalization to other deformation models using different basis functions. In fact, popular deformation models such as FFD and TPS can be written as linear combinations of basis functions. With the deformation representation at hand, we proceed in the next section by defining a similarity measure that will guide the optimization method to find the "best" deformation parameters given images as input.

1.3. *Similarity measures*

We commence by introducing a simple and classic similarity measure that is based on the assumption that image intensity does not change during object deformation.[14]   Formally, this *intensity-constancy* assumption can be written as follows:

$$f(\mathbf{x}) = f'(\mathbf{x}'), \quad \mathbf{x}' = \mathbf{x} + \mathbf{u}(\mathbf{x}), \quad \mathbf{x} \in \Omega, \tag{6}$$

where Ω is the image domain. Based on this assumption, we can measure the similarity between two images, $f(\mathbf{x})$ and $f'(\mathbf{x}')$, using the sum-of-squared-distances (SSD):

$$E_{ssd} = -\int_{\Omega} \left(f(\mathbf{x}) - f'(\mathbf{x}') \right)^2 d\mathbf{x}, \tag{7}$$

with $\mathbf{x}' = \mathbf{T}(\mathbf{x})$. When $f(\mathbf{x})$ and $f'(\mathbf{x}')$ are in perfect alignment, E_{ssd} is at its maximum (i.e., $E_{ssd} = 0$). Otherwise, $E_{ssd} < 0$. This classic similarity measure has been extended in many ways for different applications. For a detailed survey of existing similarity measures, please refer to.[15] Here, we discuss our extension to E_{ssd}. Our extension starts by modifying the intensity-constancy assumption in (6), and multiplying both sides of that equation with a kernel (test) function ϕ_l, and integrating as follows:

$$\int_{\Omega} f(\mathbf{x})\phi_l(\mathbf{x})d\mathbf{x} = \int_{\Omega} f'(\mathbf{x}')\phi_l(\mathbf{x}')d\mathbf{x}', \quad l = 1, \dots, N. \tag{8}$$

We can see that (6) is a *stronger* condition as it leads to (8), but not vice versa. For example, consider the two binary shapes. If we set $\phi_l = 1$, then its integrations for $f(\mathbf{x})$ and $f'(\mathbf{x}')$ are the shapes' areas. However, shapes of similar areas are not necessarily similar. An image-similarity measure can be obtained using the weak-form constraint in (8), by minimizing the sum-of-squared-differences of integral transforms, i.e.,

$$E_{weak} = -\sum_{l=1}^{N} \left(\langle f, \phi_l \rangle - \langle f', \phi_l \rangle \right)^2. \tag{9}$$

Here, $\langle f, \phi_l \rangle$ and $\langle f', \phi_l \rangle$ denote the integrals of $f(\mathbf{x})$ and $f'(\mathbf{x}')$, respectively, with test (kernel) functions ϕ_l. In principle, the weak form in (8) holds for arbitrary test functions ϕ_l. Here, there are two points to consider: (a) Why would we want to use a weaker condition for image registration rather than using the stronger intensity-constancy constraint? (b) Is E_{weak} enough to ensure image similarity?

On the first issue, the intensity-constancy assumption (strong form) may not always hold in practice due to image noise or illumination changes. The strong form in (6) essentially states that individual image pixels in the source image should be similar to the corresponding ones in the target image, disregarding the fact that individual image intensities may become corrupted making them unreliable for image registration. In contrast, the test functions $\phi_l(\mathbf{x}')$ in our weak form may act like lenses to "filter" out irrelevant information of compared images. For example, we may choose $\phi_l(\mathbf{x}')$ to be high-frequency wavelet functions to filter out low-frequency image information, and make the registration method robust to global illumination changes.[6] In other words, by choosing specific test functions, the weak form allows us to gauge image similarities based on higher-level image properties rather than relying on low-level pixel intensities.

On the suitability of the weak-form constraint, its usage is sufficient to enforce image similarity, given a sufficient number of test functions. In the extreme case, if we choose the test functions to be Delta sampling functions $\phi_l(\mathbf{x}) = \delta(\mathbf{x})$, then the weak-form E_{weak} becomes equivalent to E_{ssd}. The test functions also implicitly carry coarse-scale image features. These coarse-scale information helps the registration method to "look at the big picture", allowing for faster alignment by improving the algorithm's convergence speed.

In this chapter, our derivation uses a specific type of test function: monomials such as $\phi_l(x, y) = x^s y^t$ with $s, t \geq 0$. For this choice of test functions, the integral transforms in (8) become equivalent to image moments.[4] Before we explore the registration method further, let us take a look first at both the definition and classic applications of image moments.

2. Image moments and shape description

By using monomials as test functions, their integral with an image $f(\mathbf{x})$ coincides with the definition of image moments which is given by:

$$\langle f, \phi_l \rangle = \langle f, x^p y^q \rangle = M_{p,q} = \iint_{\Omega} x^p y^q f(x, y) \, dx \, dy, \qquad (10)$$

for some $p, q \geq 0$, and $M_{p,q}$ is called image moment of order (p, q). The quantity in (10) is also called *raw moment*, and can be modified to form the *central moments*:

$$\mu_{p,q} = \iint_{\Omega} (x - \overline{x})^p (y - \overline{y})^q f(x, y) \, dx \, dy, \qquad (11)$$

where $\overline{x} = M_{1,0}/M_{0,0}$ and $\overline{y} = M_{0,1}/M_{0,0}$ are the coordinates of the *centroid*.

Image moments have been previously used as part of shape descriptors that are invariant (moment invariants) to rigid transforms. In these cases, the 2-D shapes are treated as piecewise-constant images, i.e., $f(\mathbf{x}) = 1$ for $\mathbf{x}$ inside the shape, and, $f(\mathbf{x}) = 0$ otherwise. Image moments are generally *not* invariant to rigid transformations by themselves. For example, $M_{0,0} = \mu_{0,0}$ coincides with a shape's area, and remains constant under rotations, but is not invariant to image scaling. To address this issue, Hu[4] has shown that certain combinations of image moments remain constant regarding to image scaling, rotation, and translation. Hu's work inspired a large number of applications in recognizing patterns undergoing rigid transformations. Later, a number of works extended Hu-invariants using complex-valued[16] or other high-dimensional moments.[17]

Most works on moment invariants aim at compensating for the changes of image moments under rigid transforms, and at calculating measurements that are invariant to image deformation. Here, we ask a question from a different viewpoint, in contrast with the classic work: do changes in image moments tell us something about the deformation? For example, changes in the shape's area ($M_{0,0}$) certainly tell us something about possible scaling of an image. Can we determine the relationship between between changes in image moments and incremental deformations in images, and use it for image registration? It turns out there is a principled way to do that. Furthermore, this relationship can be simplified in two different but related ways that facilitate efficient image registration. One way leads to a recent work by Domokos and Kato[2] on affine shape registration, and the other leads to our work on nonlinear image registration.[13] In the following sections, we study this relationship, and explain how to exploit it for efficient and robust image registration.

3. Registration based on weak-form and image moments

3.1. *Variation of integral transforms*

Our goal in this section is to study how image deformation affects image moments. We will begin with the integral transform of the source image as given by the right-hand side of Equation 8 (i.e., $\langle f', \phi_l \rangle$). By substituting the transformation $\mathbf{x}' = \mathbf{T}(\mathbf{x})$ into the integral transform $\langle f', \phi_l \rangle$, and making the appropriate change of variables, we obtain:[4]

$$\int_\Omega f'(\mathbf{x}')\phi_l(\mathbf{x}')d\mathbf{x}' = \int_\Omega f(\mathbf{x}) \underbrace{|\mathbf{J_T}| \phi_l(\mathbf{T}(\mathbf{x}))}_{\text{deformed test function}} d\mathbf{x}. \tag{12}$$

Here, $|\mathbf{J_T}|$ is the determinant of the Jacobian matrix given by:

$$\mathbf{J_T} = \begin{pmatrix} \frac{\partial \mathbf{T}_x}{\partial x} & \frac{\partial \mathbf{T}_x}{\partial y} \\ \frac{\partial \mathbf{T}_y}{\partial x} & \frac{\partial \mathbf{T}_y}{\partial y} \end{pmatrix}. \tag{13}$$

where $\mathbf{T}_x$ and $\mathbf{T}_y$ are the two components of transformed coordinates $\mathbf{T}(\mathbf{x})$. Under the change of coordinates, the role of the Jacobian determinant in (12) is to measure the infinitesimal area scaling effect of $\mathbf{T}$.[18] Figure 3 shows the deformation of an infinitesimal rectangle area $dx\,dy$ on the image. The rectangle is stretched into a parallelogram. The area of this parallelogram can be calculated from the cross-product of vectors along its two sides. These two vectors can be calculated from the partial derivatives of $\mathbf{T}$ given by:

$$\vec{\mathbf{v}}_1 = \left(\frac{\partial \mathbf{T}_x}{\partial x}\, dx,\ \frac{\partial \mathbf{T}_y}{\partial x}\, dx \right)^\mathsf{T} \quad \text{and} \quad \vec{\mathbf{v}}_2 = \left(\frac{\partial \mathbf{T}_x}{\partial x}\, dy,\ \frac{\partial \mathbf{T}_y}{\partial x}\, dy \right)^\mathsf{T}. \tag{14}$$

According to the Jacobian matrix in (13), the area of this parallelogram equals to $\|\vec{\mathbf{v}}_1 \times \vec{\mathbf{v}}_1\| = |\mathbf{J}|\, dx\, dy$.

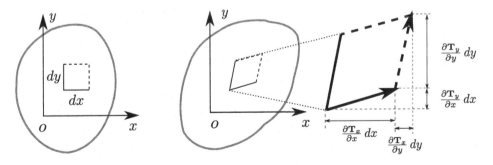

Fig. 3. Deformation of an infinitesimal rectangle $dx\,dy$. The rectangle is deformed into a parallelogram, and its area is related to $dx\,dy$ through the Jacobian matrix.

Equation 12 shows that integral transforms of a deformed image can be equivalently obtained using the original image and a deformed test function. However, this equation is not suitable for image registration. On one hand, the Jacobian matrix of an arbitrary deformation $\mathbf{T}$ may not have a simple analytic form to facilitate computations. On the other hand, deforming an arbitrary test function (i.e., $\phi_l(\mathbf{T}(\mathbf{x}))$) can be also complicated. To simplify Equation 12, and make it useful for image registration, we make assumptions about two components: (1) the image deformation $\mathbf{T}$; (2) the class of test functions ϕ_l.

First, we can restrict $\mathbf{T}$ to be an affine transformation. Then, from its definition in (3), we can easily show that its Jacobian matrix has constant determinant, $|\mathbf{J_T}| = a_1 b_2 - b_1 a_2$. This greatly simplifies the Jacobian matrix calculation. Indeed, intuitively, an affine transformation changes shape's area uniformly throughout the image domain, and $|\mathbf{J_T}|$ can be calculated by dividing the total areas of the source and target shapes. If we make a further assumption, and use monomials as our test functions, then Equation 12 can be simplified into a system of polynomial equations. The reason for this is that affine transformations of a monomial test function are polynomials, i.e., linear combination of monomials. To verify this assertion, it suffices to show the affine transformation of a monomial $\phi_l(\mathbf{x}) = x^s y^t$, for some $s, t \geq 0$. Substituting the affine transformation defined in (5) into $\phi_l(\mathbf{T}(\mathbf{x}))$, we obtain:

$$\phi_l(\mathbf{T}(\mathbf{x})) = (a_1 x + a_2 y + a_3)^s (b_1 x + b_2 y + b_3)^t = \sum_{i=0}^{s} \sum_{j=0}^{t} c_{i,j} x^i y^j. \tag{15}$$

Here, $c_{i,j} = \sum_{m+n=i} a_1^m b_1^n \sum_{m+n=j} a_2^m b_2^n$. Substituting (15) into the general variation of integral transforms in Equation 12, and moving the constant Jacobian matrix determinant $|\mathbf{J_T}|$ out of the integration, we obtain:

$$\underbrace{\langle f'(\mathbf{x}'), x^s y^t \rangle}_{M'_{s,t}} = |\mathbf{J_T}| \sum_{i=0}^{s} \sum_{j=0}^{t} c_{i,j} \underbrace{\langle f(\mathbf{x}), x^i y^j \rangle}_{M_{i,j}}. \tag{16}$$

Interestingly, we have obtained an equation of image moments. Given the deformed image $f'(\mathbf{x}')$, we can calculate its moments $M'_{s,t}$. For each (s, t), we obtain an equation as in (16), where the unknowns $c_{i,j}$ are polynomials of the affine transformation parameters $a_i, b_i, i = 1, 2, 3$, and the coefficients $M_{i,j}$ can be calculated from the target image $f(\mathbf{x})$. After obtaining a number of these equations, an over-determined polynomial system can be solved. This is the general idea of Domokos and Kato's method[2] for planar shape registration, and we have introduced it as one of the two registration methods based on image moments. Next, we describe a second registration method that uses image moments.

Domokos and Kato[2] use affine deformations, which are restrictive to many practical applications. Our proposed method relaxes this limitation and handle nonlinear image deformations.[13] To begin, we do not assume affine deformation. Instead, we assume the image deformation is relatively *small*, and is *smooth*. Intuitively, this is often the case for many real-world elastic objects that deform *incrementally* between image frames. The exact meaning of a deformation being small and smooth will become clear as we further develop our method. We will later choose a specific deformation model (i.e., a polynomial model), but our discussion still starts with general image deformations, to reveal as much fundamental observations as possible. In the following section, we will first show how this assumption simplifies the calculation of the Jacobian matrix determinant $|\mathbf{J_T}|$.

3.2. *Incremental deformation*

For incremental nonrigid motions, it is convinient to represent the transformation as $\mathbf{x}' = \mathbf{T}(\mathbf{x}) = \mathbf{x} + \mathbf{u}(\mathbf{x})$, where $\mathbf{u}(\mathbf{x})$ is the pixel-wise displacement (the incremental part). Please refer to Figure 1 for an example. Let us now take a look at what the Jacobian matrix under this representation. For clarify, we study the 2-D case, and write $\mathbf{u}(\mathbf{x})$ using two components as $\mathbf{u}(\mathbf{x}) = (u(x,y), v(x,y))^{\mathsf{T}}$. The transformation is then written as $\mathbf{T}(\mathbf{x}) = (x + u(x,y), y + v(x,y))^{\mathsf{T}}$, and the Jacobian matrix is given as follows:

$$\mathbf{J_T} = \begin{pmatrix} 1 + \frac{\partial u}{\partial x} & \frac{\partial u}{\partial y} \\ \frac{\partial v}{\partial x} & 1 + \frac{\partial v}{\partial y} \end{pmatrix}, \tag{17}$$

which in turn has a determinant given by:

$$|\mathbf{J_T}| = 1 + \frac{\partial u}{\partial x} + \frac{\partial v}{\partial y} + \underbrace{\frac{\partial u}{\partial x}\frac{\partial v}{\partial y} - \frac{\partial u}{\partial y}\frac{\partial v}{\partial x}}_{\text{higher-order term}}. \tag{18}$$

The higher-order term in (18) complicates the computation of $|\mathbf{J_T}|$. It is common in approximation theory to ignore the higher-order terms as they are usually much smaller numerically compared to the result. However, to justify that the higher-order terms in our case are indeed negligible, we need to make sure that the partial derivatives $\frac{\partial u}{\partial x}$, $\frac{\partial v}{\partial y}$, $\frac{\partial u}{\partial y}$, and $\frac{\partial v}{\partial x}$ are comparably small. To achieve this, we assume the partial derivatives of $\mathbf{u}(\mathbf{x})$ to be bounded by a small number, i.e., $\left|\frac{\partial \mathbf{u}(\mathbf{x})}{\mathbf{x}}\right| \leq 1 - \epsilon$, for some $\epsilon > 0$ that controls the approximation accuracy. Intuitively, it means that the deformation field $\mathbf{u}(\mathbf{x})$ is differentiable and does not fluctuate significantly at any point. This is precisely what we mean by saying that the image deformation is "smooth"[b]. Now, $|\mathbf{J_T}|$ can be approximated by:

$$|\mathbf{J_T}| \approx 1 + \frac{\partial u}{\partial x} + \frac{\partial v}{\partial y} = 1 + \text{div}(\mathbf{u}), \tag{19}$$

where $\text{div}(\mathbf{u})$ is the divergence of the deformation field is given by:

$$\text{div}(\mathbf{u}) = \sum_j \frac{\partial u^{(j)}}{\partial x^{(j)}} = \sum_j \sum_i a_i^{(j)} \frac{\partial \gamma_i}{\partial x^{(j)}}. \tag{20}$$

To avoid confusion with indices of polynomials, we use both lower-case letters and letters with parenthetical superscripts to indicate the components of a vector. Like the Jacobian matrix, the divergence of a deformation field also has an intuitive geometric interpretation. Both the determinant of Jacobian matrix and divergence are related to changes in the area of infinitesimal rectangle elements. While the determinant of a Jacobian matrix indicates the ratio of area changes, the divergence approximates the *amount* of area changes.

Let us remind ourselves that all our assumptions and derivations aim at calculating variations of the integral transform in Equation 15, and we have greatly simplified the

---

[b]An alternative induction of our method exists without making this assumption. It is based on variational calculus and is beyond the scope of this chapter.

calculation of the determinant of a Jacobian matrix. The second difficulty still to overcome is to calculate the deformed kernel function $\phi_l\left(\mathbf{T}(\mathbf{x})\right)$. Since we have assumed that the deformation field $\mathbf{u}$ is small, we can approximate $\phi_l\left(\mathbf{T}(\mathbf{x})\right)$ using Taylor's series. Also, it is useful now to assume that the deformation field $\mathbf{u}(\mathbf{x})$ is represented as a linear combination of basis functions as in (5). Then, we obtain the approximated $\phi_l\left(\mathbf{T}(\mathbf{x})\right)$ as follows:

$$\phi_l\left(\mathbf{x}+\mathbf{u}\right) = \phi_l(\mathbf{x}) + \mathbf{u}^{\mathsf{T}}\frac{\partial\phi_l}{\partial\mathbf{x}} = \sum_j\sum_i a_i^{(j)}\frac{\partial\phi_l}{\partial x^{(j)}}\gamma_i. \tag{21}$$

With the linearized versions of the Jacobian determinant and the deformed kernel function at hand, we can then obtain an expression for the variation of a general integral transform with respect to an incremental deformation. This expression is found by substituting (19) and (21) into (12), and ignoring higher-order terms, to obtain:

$$
\begin{aligned}
\langle f', \phi_l\rangle &\approx \int_\Omega \left(\phi_l(\mathbf{x}) + \frac{\partial\phi_l}{\partial\mathbf{x}}\mathbf{u}\right) f(\mathbf{x})\left(1 + \operatorname{div}(\mathbf{u})\right) d\mathbf{x} \\
&\approx \underbrace{\int_\Omega \phi_l(\mathbf{x})f(\mathbf{x})d\mathbf{x}}_{\text{original integral}} + \underbrace{\int_\Omega \frac{\partial\phi_l}{\partial\mathbf{x}}\mathbf{u}f(\mathbf{x})d\mathbf{x}}_{\text{shape variation}} + \underbrace{\int_\Omega \phi_l(\mathbf{x})f(\mathbf{x})\operatorname{div}(\mathbf{u})d\mathbf{x}}_{\text{area variation}} \\
&= \langle f, \phi_l\rangle + \sum_j\sum_i a_i^{(j)}\int_\Omega \frac{\partial\phi_l}{\partial x^{(j)}}\gamma_i f(\mathbf{x})d\mathbf{x} + \sum_j\sum_i a_i^{(j)}\int_\Omega \phi_l(\mathbf{x})\frac{\partial\gamma_i}{\partial x^{(j)}}f(\mathbf{x})d\mathbf{x} \\
&= \langle f, \phi_l\rangle + \sum_j\sum_i a_i^{(j)}\left(\langle f, \frac{\partial\phi_l}{\partial x^{(j)}}\gamma_i\rangle + \langle f, \phi_l\frac{\partial\gamma_i}{\partial x^{(j)}}\rangle\right).
\end{aligned}
\tag{22}
$$

In Equation 22, we have approximated the relationship between variations of integral transforms $\langle f, \phi_l\rangle$ and incremental image deformations $\mathbf{u}(\mathbf{x})$. It is worth pointing out that the second step in (22) has three components, corresponding to the original integral, the variation caused by shape deformation, and the variation caused by the area changes resulted from the deformation's stretching and squeezing effect. More importantly, this relationship is expressed as a linear function in the deformation parameters $a_i^{(j)}$. In principle, we may stack the linear equations obtained for each test function into an overdetermined system, and solve this system for $a_i^{(j)}$.

However, we can further simplify Equation 22 by choosing both the test functions ϕ_l and the deformation bases γ_i to be monomials, and convert (22) into a linear system of image moments. This technique brings two advantages. First, image moments enhance the robustness of our method. Secondly, image moments can be recursively calculated, bringing down the computational cost in calculating the integral transforms. In the following section, we study how to combine polynomial deformation model and monomial test functions into this general linear equation.

3.3. *Efficient computation using polynomials*

From Equation 22, the incremental variation of the integral transform w.r.t the deformation parameters, i.e., the gradient of $\langle f, \phi_l \rangle$ is given by:

$$\frac{\partial \langle f, \phi_l \rangle}{\partial a_i^{(j)}} = \langle f, \frac{\partial \phi_l}{\partial x^{(j)}} \gamma_i \rangle + \langle f, \phi_l \frac{\partial \gamma_i}{\partial x^{(j)}} \rangle = \langle f, \frac{\partial \phi_l \gamma_i}{\partial x^{(j)}} \rangle. \tag{23}$$

It turns out that Equation 23 can be efficiently calculated when ϕ_l and γ_i are both monomials (i.e., $x^s y^t$ with $s, t \in \mathbb{Z}$, $s \geq 0$, and $t \geq 0$). Since monomials are closed under differentiation and multiplication, the terms $\partial \phi_l \gamma_i / \partial x^{(j)}$ are also monomials. Thus, in our method, we use the monomials $x^s y^t$ for both the test function and the deformation basis functions, so the integral transforms are still equivalent to classical image moments, but we have extended the affine deformation model to a polynomial one. Substituting $\phi_l = x^p y^q$ and $\gamma_i = x^s y^t$ into (23), we have:

$$\frac{\partial M_{p,q}}{\partial a_{s,t}} = \frac{\partial \langle f, \phi_l \rangle}{\partial a_{s,t}} = \langle f, \frac{\partial x^{p+s} y^{q+t}}{\partial x} \rangle = (p+s) M_{p+s-1,q+t}, \tag{24}$$

and

$$\frac{\partial M_{p,q}}{\partial b_{s,t}} = \frac{\partial \langle f, \phi_l \rangle}{\partial a_{s,t}} = \langle f, \frac{\partial x^{p+s} y^{q+t}}{\partial y} \rangle = (q+t) M_{p+s,q+t-1}. \tag{25}$$

We can then substitute (24), (25), and (10) into the linearized integral-transform variation in Equation 22, to obtain:

$$M'_{p,q} \approx M_{p,q} + \sum_{s=0}^{N} \sum_{t=0}^{N} (a_{s,t}(p+s) M_{s+p-1,t+q}) + \sum_{s=0}^{N} \sum_{t=0}^{N} (b_{s,t}(q+t) M_{s+p,t+q-1}). \tag{26}$$

In other words, the moments of the deformed image can be approximated as a linear combination of the original image moments. Equation 26 coincides with our previous derivation based on approximation theory.[13]

3.4. *Image registration using the weak form*

Given the above ingredients, we can now optimize the quantity E_{weak} in Equation 9 to obtain an estimate of the deformation field $\mathbf{u}$. Formally, we want to find the deformation field that maximizes the following functional:

$$\mathbf{u} = \arg \max_{\mathbf{u}} E_{weak} \left(f'(\mathbf{x}'), f(\mathbf{x}) \right) = \arg \max_{\mathbf{u}} - \sum_{l=1}^{N} (\langle f, \phi_l \rangle - \langle f', \phi_l \rangle)^2. \tag{27}$$

We can solve (27) in two different ways. First, as Equation 22 defines a linear constraint for each test function, we can stack these constraints to form an over-determined linear system. Then, we adopt a fixed-point iteration method in which we solve that linear system for each iteration step. After obtaining the incremental deformation parameter $\Delta \mathbf{a}$, we can warp the template image according to the updated deformation parameter. In each iteration, the similarity between the deformed template image and the target image increases until convergence. Algorithm 1 summarizes the process using fixed-point iterations. Note that, instead of using the image from the previous iteration, we obtain a deformed template image from the original input for each iteration of the algorithm. This procedure helps avoid propagating numerical errors due to repetitive image interpolations.

Algorithm 1. Moment-based image registration using fixed-point iteration.

> **Input**: Template I_s, target image I_t, deformation basis γ_i, and test functions ϕ_i.
> **Output**: Deformation parameters **a**.

1: Initialize $\mathbf{a}^0 \leftarrow 0$
2: Initialize currently deformed template image $I_s^0 = I_s$
3: Calculate image moments of the target image $\langle I_t, \phi_i \rangle$
4: **repeat**
5:     Calculate image moments of the deformed template image $\langle I_s^k, \phi_i \rangle$
6:     Solve the linear system given by stacking Equation 22
7:     Obtain current incremental deformation parameters $\Delta \mathbf{a}$
8:     Update $\mathbf{a}^{k+1} = \mathbf{a}^k + \Delta \mathbf{a}$ and warp $I_s^{k+1} \leftarrow I_s^0 \left(\mathbf{x} + \mathbf{u}(\mathbf{a}^{k+1}) \right)$
9: **until** convergence

Alternatively, we can simply use a gradient-descent algorithm to minimize E_{weak}. Gradient-descent algorithms starts from an initial solution, and use the gradient at current solution to guide the search for a better solution. The required gradient calculations in our method are performed in the transform domain as shown by Equation 23. There are many off-the-shelf gradient-descent methods available, so we simply choose one and plug into our algorithm. We summarize the gradient-descent process in Algorithm 2.

Algorithm 2. Moment-based image registration using gradient-descent

> **Input**: Template I_s, target image I_t, deformation basis γ_i, and kernel functions ϕ_i.
> **Output**: Deformation parameters **a**.

1: Initialize $\mathbf{a}^0 \leftarrow 0$
2: Initialize $I_s^0 \leftarrow I_s$
3: Calculate moments of the target image $\langle I_t, \phi_i \rangle$
4: **repeat**
5:     Calculate moments of the source image $\langle I_s^k, \phi_i \rangle$
6:     Calculate Jacobian matrix J from $\langle I_s, \phi_i \rangle$ and $\langle I_t^k, \phi_i \rangle$ (Equation 23)
7:     Update $\mathbf{a}^k$ by gradient-descent using J
8:     Warp $I_s^{k+1} \leftarrow I_s^k(\mathbf{x} + \mathbf{u}(\mathbf{a}^k))$
9: **until** convergence

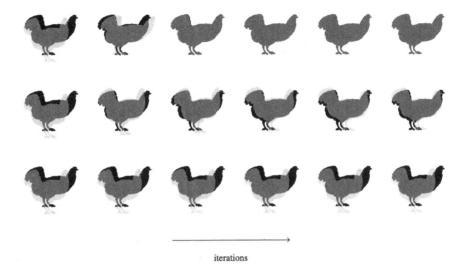

iterations

Fig. 4.   Shape registration convergence. Row 1: our method using fixed-point iteration. Row 2: our method using gradient-descent. Row 3: Lucas-Kanade's method. Template images (gray) and target images (black).

We now would like to compare our method with Domokos and Kato's method introduced in Section 3.1. First, our method is based on a polynomial deformation model in contrast to Domokos and Kato's affine model. Thus, ours is able to handle more general image deformations. Secondly, our method register images *incrementally*, iteratively warping a template image towards the target images by solving a system of linear equations, while Domokos and Kato's calculates the affine deformation in a single step by solving a system of polynomial equations. There are pros and cons with Domokos and Kato's single-step approach. On one hand, their method is highly efficient as solving a polynomial system can be very fast. On the other hand, the single-step approach are sensitive to image occlusions and noise. In the following section, we compare the two methods using both synthetic and real-world images. Additionally, we compare our method with the classic Lucas-Kanade optical-flow approach that is based on the strong-form intensity-constancy constraint.

4. Evaluation

In this section, we perform evaluations on a number of images. We used binary images from a shape dataset as well as gray-scale medical images. For the shape dataset, we compared our method with the classic Lucas-Kanade method[14] and Domokos and Kato's method.[2] We then show an application of our method to images obtained from a microscope. The source code of our method and some of the test data used in the experiments are available on-line for download [c].

---

[c] http://www.cs.fit.edu/~eribeiro/icpr2010moments/

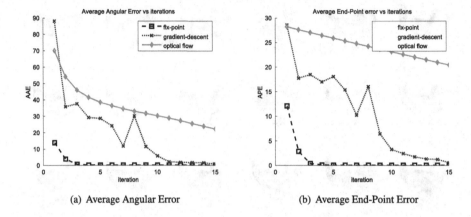

(a) Average Angular Error   (b) Average End-Point Error

Fig. 5. Algorithms' convergence. a) Average Angular-Error. b) Average End-Point Error. Our fixed-point iteration method converged very fast (in less than five iterations). While our gradient-descent implementation converged with fluctuations, it showed faster convergence than the classic Lucas-Kanade optical-flow method.

Our method has two parameters: the number of basis functions M in our deformation model (Equation 5), and the number of test functions N in our weak form (Equation 9). N must be larger than the number of unknowns ($2 \times M$) so that the linear system of moments (Equation 26) is over-determined. However, if N is too large, higher-order moments may introduce numerical inaccuracy. In general, it is difficult to select an optimal value for N. In our experiments, we chose N empirically. For the affine deformation experiments, we used deformation basis $\Gamma = (1, x, y)^{\mathsf{T}}$ (i.e., $M = 3$) and we set $N = 12$. For the second-order polynomials with $\Gamma = (1, x, y, x^2, xy, y^2)^{\mathsf{T}}$, we set $N = 20$.

4.1. Affine shape registration

First, we evaluated our method using the MPEG-7 shape dataset.[19] The shapes were distorted by random affine transformations to create synthetic target shapes, and then the original shapes were registered to the target shapes using both our method and the classic Lucas-Kanade method.[14] We implemented the Lucas-Kanade method using the same polynomial deformation model adopted by our method, so that the only difference between the two implementations is that our method used the integral transform weak-form constraint, while Lucas-Kanade's used the strong-form (i.e., the intensity-constancy constraint). Figure 4 shows qualitative results comparing registrations obtained from two implementations of our method (i.e., fixed-point and gradient-descent) and with the Lucas-Kanade method. In this case, both implementations of our method converged faster.

For a quantitative comparison, we need to set up the criteria for comparing the registration results, i.e., a final deformation field $\mathbf{u}$, against the ground-truth deformation field $\mathbf{u}'(\mathbf{x})$. Since $\mathbf{u}(\mathbf{x})$ is controlled by its parameters, in principle we can measure the registration error by calculating Euclidean distance between the ground-truth and output parameters as $e = \|\mathbf{a} - \mathbf{a}'\|$. However, this error metric is biased to certain components of $\mathbf{a}$ with large magnitude. As a result, we borrowed two concepts, the Average Angular Error (AAE) and the Average End-Point Error (APE), from optical-flow methods to evaluate our deformation fields. Here, we treat a deformation field as a flow-field. AAE measures the vectors' directional error, and APE measures the relative magnitude (length) error. Formally, AAE and APE are the average of pixel-wise error metrics defined as follows:

$$e_{\text{ang}}(\mathbf{x}) = \arccos\left(\mathbf{u}(\mathbf{x}) \cdot \mathbf{u}'(\mathbf{x})\right), \quad \text{and} \quad e_{\text{len}}(\mathbf{x}) = \frac{|\mathbf{u}(\mathbf{x}) - \mathbf{u}'(\mathbf{x})|}{max(|\mathbf{u}(\mathbf{x})|, |\mathbf{u}'(\mathbf{x})|)}. \quad (28)$$

Then, we performed a three-way comparison between the two implementations of our method and the Lucas-Kanade method. All these methods are of iterative nature, so we calculated both AAE and APE of the estimated deformation fields at each iteration. Figure 5 shows the quantitative evaluation of their convergence. Here, the fixed-point implementation converged in less than five iterations, while the gradient-descent implementation fluctuated. Even with the fluctuations, our gradient-descent implementation still converged faster than the classic Lucas-Kanade optical-flow method. Approximately, the optical-flow method reduced the registration error at a constant speed, but our fixed-point iteration achieved the same reduction at an exponential speed. This result supports our conjecture that the test functions in our weak-form formulation carry large-scale ("big picture") image information, which is similar to the idea used in hierachical Lucas-Kanade method,[20] and allows the algorithm to converge faster. Our experiments also showed that the fixed-point method is much more robust and converges faster than the gradient-descent method. Moreover, gradient-descent search requires many evaluations of the Jacobian matrix. In the following experiments, we will report results produced by the fixed-point method.

In addition to achieving faster convergence, our method was also robust to image occlusion and noise. Figure 6 shows examples of registering occluded and noisy shapes. For these images, we added significant amount of salt-and-pepper noise with probability of occurrence of 0.3. Salt-and-pepper noise randomly flips image intensities, and can be challenging for optical-flow methods since it violates the pixel-wise intensity-constancy assumption. Figure 7 illustrates the convergence of our method in comparison with the Lucas-Kanade method. Lucas-Kanade's converged slower for noisy images, while our method converged in five iterations with a sharp reduction in registration error. We believe that our method's good performance is due to the use of moments and the weak-form registration. Image moments are calculated using the entire image, making it robust to noise.

In the last set of experiments using the affine model, we compared our algorithm with Domokos and Kato's method[d].[2] As introduced earlier, Domokos and Kato's method is an efficient one-step registration algorithm. Here, we focus on the comparison of robustness and registration accuracy. For this, we selected six shapes from the MPEG-7 dataset (chicken-20, beetle-6, cattle-4, bell-1, and camel-4). For each shape, we added salt-and-pepper noise with increasing probability of occurrence. At each noise level, the shapes were

---

[d]source code available for download at `http://www.inf.u-szeged.hu/~kato/`

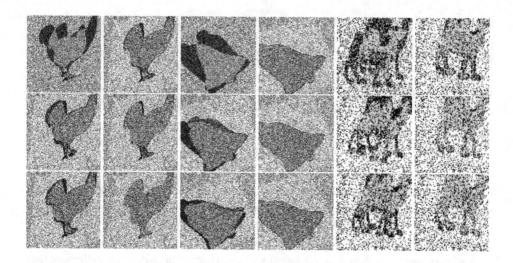

Fig. 6. Registration convergence for noisy and occluded images. The source image (gray) is iteratively warped to the target image (black), shown as a sequence from top to bottom. For each shape: first row shows Lucas-Kanade's method, and second row is our method. Noise: salt-and-pepper noise with probability of occurrence of 0.2. Deformed shapes contain some occlusion, and have low resolution (the last rows).

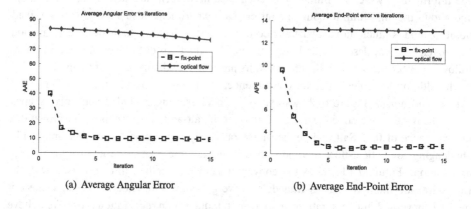

(a) Average Angular Error   (b) Average End-Point Error

Fig. 7. Convergence of shape registration algorithms. a) Average Angular-Error. b) Average End-Point Error. For high noise levels, the optical-flow method converges slow. The convergence of our method is less affected.

distorted using 10 randomly generated affine transformations, and the images were subsequently registered using both methods. Figure 8(a) shows the average APE as a function of noise level. For all noise levels, our method performed consistently better than Domokos and Kato's, with more than 30 percent of reduction in the average registration error. Additionally, our method had a much lower APE variance, especially when the noise level exceeded 0.3 occurrence probability. We found that the large variance in their method's accuracy was partially due to sensitivity to image occlusions. Figure 8(b) shows examples where Domokos and Kato's method failed while our method produced good alignment. It seems that at higher noise levels, very small amounts of occlusion might distort their results significantly. It is also worth pointing out that, like many existing shape-registration methods, both ours and Domokos-Kato method may become trapped in local minima.

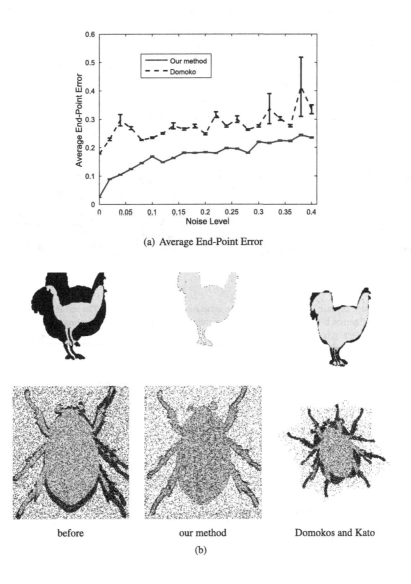

(a) Average End-Point Error

before   our method   Domokos and Kato

(b)

Fig. 8.   Comparison with Domokos-Kato's method. a) Average End-point Error (APE) as a function of noise level. Our method not only produced more accurate result, but also had lower registration-error variance. b) Occluded shapes. Domokos and Kato's was affected by occlusions, but our method could still register the shapes.

4.2. Nonlinear image deformation

In this set of experiments, we tested our method on a set of gray-level images undergoing nonlinear image deformations. Here, we performed registration of noisy medical images. Domokos and Kato's method cannot handle nonlinear deformation, so we left it out, and compared our method to the Lucas-Kanade method only. In one set of experiments, we synthesized nonlinear deformations of MRI brain images using randomly generated second-order polynomials, i.e., the deformation basis functions were $\Gamma = \left(1, x, y, x^2, xy, y^2\right)^{\mathsf{T}}$. In addition, the images were contaminated with significant amounts of additive salt-and-pepper noise with probability of occurrence of 0.2. Figure 9 shows a registration sequence using our method and Lucas-Kanade's. Figure 10 shows the APE of our method and Lucas-Kanade method as a function of iteration number. Under nonlinear deformation, the convergence time of our method doubled to around ten iterations. However, our method still converged exponentially, and was much faster than the optical-flow method.

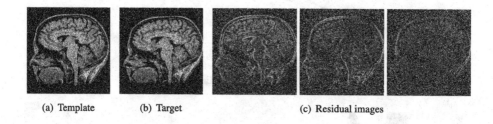

   (a) Template   (b) Target   (c) Residual images

Fig. 9.   Registration of brain MRI images. a-b) The template and target images. c) The residual images calculated as the absolute difference between the deformed template and the target images. The images undergo nonlinear deformation. Our method is able to align gray-scale MRI brain image despite significant noise levels.

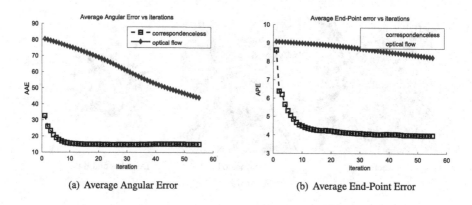

      (a) Average Angular Error   (b) Average End-Point Error

Fig. 10.   Registration convergence under nonlinear deformation. a) Average Angular-Error. b) Average End-Point Error. Under nonlinear deformation, both our method and the optical-flow method converged slower than in the linear deformation case. However, our method still was significantly faster than the optical-flow method.

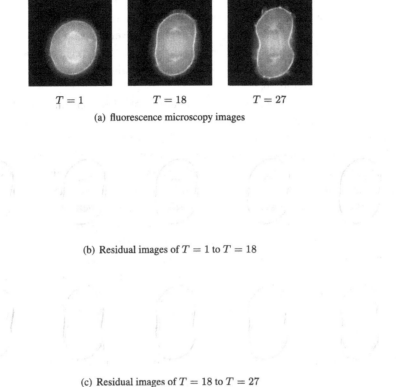

(a) fluorescence microscopy images

(b) Residual images of $T = 1$ to $T = 18$

(c) Residual images of $T = 18$ to $T = 27$

Fig. 11. Registration of deforming cells using our method. a) Three frames of a cell undergoing nonlinear deformation under a fluorescent microscope. b-c) Inter-frame registration residuals shown in dark regions.

4.3. *Application to microscopy image registration*

Finally, we performed a set of experiments to estimate nonlinear cell deformation. Figure 11(a) shows three frames from a sequence of fluorescence microscopy images of a living cell. Figure 11(b) and Figure 11(c) show the registration process. Similar to the previous example, we used a nonlinear deformation model to represent nonrigid cell motion. Since ground-truth was not available for the cell images, we show the registration quality by overlaying the source and target images. Notice that, given the global nature of our polynomial deformation model, the registration quality of our method is slightly worse than methods based on local deformation models such as B-Splines.[8] However, local deformation models require computationally intensive optimization.

5. Conclusion and further readings

In this chapter, we have studied two recent registration methods using image moments, specifically, Domokos and Kato's method for affine registration,[2] and our incremental method for nonlinear registration.[13] We introduced the basic framework for solving image-registration problem, and then discussed about image moments, including its definition and its application in object recognition. By studying the relationship between the changes of image moments and the image deformations, we are able to register images using image moments, in a robust and efficient way. Finally, we evaluated these methods based on both synthetic and real-world images.

It is worth pointing out that there are a number of other registration methods using image moments. Recently, Flusser et al.[21] proposed the use of *implicit moment invariants* for shape recognition. The method essentially aligns images by matching their moments. The underlying idea of Flusser's work is similar to both Domokos and Kato's and our methods, but approaches the problem from a classic invariance perspective, and uses a general minimization scheme without exploiting efficient calculation of the Jacobian matrix. The work in[21] also extends previous moment-based registration methods[22–25] that are restricted to affine transformations. Please refer to[26] for a survey of image moments used for pattern recognition, and related works on medical image registration.[27,28]

Nevertheless, moment-based registration methods have some limitations. For instance, calculation of image moments is computationally extensive. Although this computational cost is partially offset by faster convergence of moment-based methods, improved computational efficiency can be obtained if we change the monomial test functions to orthogonal basis functions such as wavelets[6] and trigonometric harmonics.[7] In this way, the integral transforms in our method correspond to well-known wavelet and Fourier transforms. Both of these transforms have been used previously for affine image registration, but few works address the problem of nonlinear image registration.

Secondly, our deformation field is limited to low-order polynomials due to the global support and numeric sensitivity of higher-order polynomials. It would be interesting to extend the polynomial model to local deformation models such as splines,[11] thin-plate splines,[12] and meshless models.[8] Our weak-form formulation of the registration problem has provided a general framework for the aforementioned extensions.

Finally, both Domokos and Kato's method and ours belong to an emerging group of so-called correspondence-less registration methods,[29] that do not explicitly rely on the correspondence between image pixels or features. Instead, image similarity is indirectly measured in the domain of integral transform. Integral transforms can be used as compressed images representations, they may facilitate efficient group and data-driven registration, where a large number of images may be compares *simultaneously*. Finally, since integral transforms are also frequently used in content-adaptive compression methods,[30] content-adaptive registration may also be an interesting topic for future investigation.

References

1. H. Chen and B. Bhanu. Global-to-local non-rigid shape registration. In *Intnl. Conf. on Pattern Recognition, Vol 4*, pp. 57–60, Los Alamitos, CA, USA, (2006). IEEE Computer Society.
2. C. Domokos and Z. Kato, Parametric estimation of affine deformations of planar shapes, *Pattern Recognition*. **43**(3), 569–578, (2010). ISSN 0031-3203.

3. J. Maintz and M. Viergever, A survey of medical image registration, *Medical Image Analysis.* **2** (1), 1–36, (1998).

4. M. K. Hu, Visual pattern recognition by moment invariants, *IEEE Transactions on Information Theory.* **IT-8**, 179–187 (February, 1962).

5. K. Ito, T. Aoki, E. Kosuge, R. Kawamata, and I. Kashima. Medical image registration using Phase-Only Correlation for distorted dental radiographs. In *IEEE International Conference on Pattern Recognition*, pp. 1–4, (2008).

6. Y. Bentoutou, N. Taleb, K. Kpalma, and J. Ronsin, An automatic image registration for applications in remote sensing, *IEEE Trans. on Geo. and Remote Sens.* **43**(9), 2127, (2005).

7. E. De Castro and C. Morandi, Registration of translated and rotated images using finite Fourier transforms., *IEEE Trans. on Pattern Anal. and Machine Intell.* **9**(5), 700–703, (1987).

8. W. Liu and E. Ribeiro. A meshless method for variational nonrigid 2-d shape registration. In *International Symposium on Visual Computing*, pp. 262–272, (2010).

9. R. Filipovych and E. Ribeiro, Robust sequence alignment for actor-object interaction recognition: Discovering actor-object states, *Comp. Vision and Image Understanding.* **115**(2), 177–193, (2011). ISSN 1077-3142.

10. X. Huang, N. Paragios, and D. N. Metaxas, Shape registration in implicit spaces using information theory and free form deformations, *IEEE Trans. Pattern Anal. Mach. Intell.* **28**, 1303–1318 (August, 2006).

11. D. Rueckert, L. Sonoda, C. Hayes, D. Hill, M. Leach, and D. Hawkes, Nonrigid registration using free-form deformations: application to breast MR images, *IEEE Transaction on Medical Imaging.* **18**(8), 712–721, (1999).

12. G. Rohde, A. Aldroubi, and B. Dawant, The adaptive bases algorithm for intensity-based nonrigid image registration, *IEEE Trans. on Med. Imag.* **22**(11), 1470–1479, (2003).

13. W. Liu and E. Ribeiro. Estimating nonrigid shape deformation using moments. In *IEEE International Conference on Pattern Recognition*, pp. 185–188. IEEE, (2010).

14. B. Lucas and T. Kanade. An iterative image registration technique with an application to stereo vision. In *International joint conference on artificial intelligence*, vol. 3, pp. 674–679, (1981).

15. B. Zitova and J. Flusser, Image registration methods: a survey, *Image and vision computing.* **21** (11), 977–1000, (2003).

16. M. R. Teague, Image analysis via the general theory of moments, *Journal of the Optical Society of America (1917-1983).* **70**, 920–930 (August, 1980).

17. M. Schlemmer, M. Heringer, F. Morr, I. Hotz, M. Hering-Bertram, C. Garth, W. Kollmann, B. Hamann, and H. Hagen, Moment invariants for the analysis of 2d flow fields, *IEEE Transactions on Visualization and Computer Graphics.* **13**(6), 1743–1750, (2007).

18. W. Rudin, *Real and complex analysis.* (Tata McGraw-Hill, 2006).

19. N. Thakoor, J. Gao, and S. Jung, Hidden markov model-based weighted likelihood discriminant for 2-D shape classification, *IEEE Trans. on Image Proc.* **16**(11), 2707–2719, (2007).

20. L. Lucchese, G. Doretto, and G. Cortelazzo, A frequency domain technique for range data registration, *IEEE Trans. on Pattern Anal. and Machine Intell.* pp. 1468–1484, (2002).

21. J. Flusser, J. Kautsky, and F. Šroubek, Implicit Moment Invariants, *International journal of computer vision.* **86**(1), 72–86, (2010). ISSN 0920-5691.

22. J. Flusser and T. Suk, A moment-based approach to registration of images with affine geometric distortion, *IEEE Trans. on Geo. and Remote Sens.* **32**(2), 382–387, (2002). ISSN 0196-2892.

23. C. Shah, Y. Sheng, and L. Smith, Automated image registration based on pseudoinvariant metrics of dynamic land-surface features, *Geoscience and Remote Sensing, IEEE Transactions on.* **46** (11), 3908–3916 (Nov., 2008). ISSN 0196-2892. doi: 10.1109/TGRS.2008.2000636.

24. J. Sato and N. Hollinghurst, Image registration using multi-scale texture moments, *Image and Vision Computing.* **13**(5), 496–513, (1995). ISSN 0262-8856.

25. X. Dai and S. Khorram, A feature-based image registration algorithm using improved chain-code representation combined with invariant moments, *Geoscience and Remote Sensing, IEEE Transactions on.* **37**(5), 2351–2362 (Sept., 1999). ISSN 0196-2892. doi: 10.1109/36.789634.
26. J. Flusser, T. Suk, and B. Zitov, *Moments and moment invariants in pattern recognition.* (Wiley Online Library, 2009). ISBN 0470699876.
27. T. Makela, P. Clarysse, O. Sipila, N. Pauna, Q. Pham, T. Katila, and I. Magnin, A review of cardiac image registration methods, *IEEE Trans. on Med. Imag.* **21**(9), 1011–1021, (2003).
28. D. Shen and C. Davatzikos, HAMMER: hierarchical attribute matching mechanism for elastic registration, *IEEE Transaction on Medical Imaging.* **21**(11), 1421–1439, (2003).
29. J. Pokrass, A. M. Bronstein, and M. M. Bronstein. A correspondence-less approach to matching of deformable shapes. In *Scale Space and Variational Methods*, pp. 262–272, (2011).
30. S. Chang, B. Yu, and M. Vetterli, Adaptive wavelet thresholding for image denoising and compression, *IEEE Trans. on Image Processing.* **9**(9), 1532–1546, (2002).

CHAPTER 1.7

NONPARAMETRIC SAMPLE-BASED METHODS FOR IMAGE UNDERSTANDING

Alexander Wong*, Akshaya Mishra, David A. Clausi and Paul W. Fieguth

Department of Systems Design Engineering, University of Waterloo,
200 University Ave. West, Waterloo, ON, Canada N2L 3G1
* a28wong@engmail.uwaterloo.ca; * alex.s.wong@gmail.com

This chapter presents a nonparametric framework for image understanding based on models constructed via conditional sampling. Rather than modeling imaging data as a lattice of fixed values (one intensity per pixel), or as a set of parametric distributions (such as a Gaussian distribution), we explore the modeling of imaging data as a lattice of nonparametric conditional probability distributions estimated via random sampling. Such a framework facilitates the development of robust stochastic techniques for tackling computer vision challenges. We will illustrate the application of this framework using computer vision examples of denoising, edge detection, and segmentation.

1. Introduction

The physical world in which we live in is a highly complex environment, composed of a vast diversity of structures, phenomena, and events that have different semantic characteristics depending on the observed scale.

A wonderful example of this is the human body, which is vastly complex in its construction and can have very different functional interpretation when studied at different scales. At the meter scale, the body is conceptually meaningful to interpret based on external form and build. However, at the centimeter scale it is difficult to interpret overall form and build, and it might be more appropriate to discuss individual organs. At the millimeter scale one would start asserting semantic meaning into the heterogeneous tissue structure within an organ, while at the micrometer scale we might find only meaning in the individual cells within the tissue structure.

Applying this metaphor to an image of the physical world, it may be important to understand that various image structures are only meaningful when studied and investigated at certain scales. On the basis of this motivation we have developed multi-scale methods and models for describing, understanding, and analyzing images in a meaningful manner, thus allowing us to tackle important computer vision challenges when dealing with images having complex structural content.

A particularly powerful approach for multi-scale image modeling and understanding is scale space theory [1], where the modeling and study of image structures at different scales within an image are facilitated via the representation of the underlying image content as a single-parameter family of images, parameterized by *scale*, where there is a monotonic decrease in the fine scale structure as scale increases. Scale space theory has proven to be an extremely powerful tool for a wide variety of image processing and computer vision applications where the context of scale is important, such as in denoising [5, 7, 8], edge detection [2, 4, 5], color enhancement [6], segmentation [3, 9, 13, 18], and blur estimation [10, 11]. In traditional scale space approaches, the image data is modeled as a lattice of fixed values, where a single intensity value acts as the representation at a particular pixel at each scale. Unfortunately, such approaches produce multi-scale models with poor image structural localization and high smoothing at larger scales. Therefore, alternative multi-scale approaches for image modeling and understanding that avoids these problems is highly desired.

In this chapter, we represent a non-parametric sample-based framework for multi-scale image understanding and analysis. Taking a radical departure from existing approaches, this non-parametric sample-based framework aims to model data as a lattice of nonparametric conditional probability distributions estimated via random sampling. Such a stochastic representation of imaging data at different scales facilitates the use of robust stochastic techniques to tackle difficult image processing and computer vision problems.

The chapter is organized as follows. The theory behind linear and non-linear scale space approaches for image modeling and understanding is provided in Section 2. A non-parametric sample-based framework is introduced in Section 3. The application of the proposed framework to denoising, edge detection, and segmentation is described in Section 4.

2. Existing multi-scale approaches for image understanding

Given the benefits of scale space for multi-scale modeling and understanding of images, several approaches have been proposed. These approaches are deterministic in nature and can typically be divided into two groups:

1. linear approaches, and
2. nonlinear approaches,

discussed below.

2.1. *Linear approaches*

Linear scale space approaches model an image as a family of sub-images, derived via a set of linear scale space operators. The most common linear approach is the Gaussian approach introduced by Witkin [1] and Koenderink and Van Doorn [12] where, given an image $f(\underline{x})$, the multi-scale representation at scale t is given by the convolution

$$L_t(\underline{x}) = f(\underline{x}) * w_t \tag{1}$$

where w_t is the scale-related Gaussian weighting function

$$w_t(\underline{a}) = \prod_{i=1}^{n} \frac{1}{\sqrt{2\pi t}} \exp\left[-\frac{a_i^2}{2t}\right]. \tag{2}$$

The linear approach to scale space analysis asserts the following [12]:

Causality: Structures at coarser scales must exist at finer scales, and no new structures can be introduced going from finer scale to coarser scales.

Homogeneity and Isotropy: The linear scale space operator must be space and scale invariant.

In particular, the causality assertion ensures the monotonic reduction of fine scale structures as scale increases, thus allowing for clear separation between structures at different scales. It was later shown by Perona and Malik that the isotropy assertion is not a necessity for scale space analysis [4].

2.1.1. *Advantages and Disadvantages*

The main advantage of the linear approach to multi-scale modeling is its computational and theoretical simplicity, allowing the properties and characteristics of such linear approaches to have been very well studied [1, 2].

However, there are disadvantages of the linear approaches, as is illustrated by the multi-scale modeling of the house image shown in Fig. 1, which leads to three observations of fundamental limitations of linear scale space approaches:

- First, the structures at coarser scales are poorly localized, which means that such structures are not sharp and do not coincide with their true locations in the original image.
- Second, a large portion of the corner information that exists in the original image no longer exist at coarser scales. As a result, linear approaches do not facilitate the extraction of meaningful structural features such as corners and edges at coarse scales, which limits image processing and computer vision tasks where such information is useful or necessary. In particularly, the structural degradation at coarser scales can be attributed to the independent nature of the scale space operator, which does not take into account the underlying image content, resulting in the mixing of intensities from different structures, leading to inter-region smoothing and poor structural localization.
- Third, multi-scale modeling using linear approaches, as illustrated in the second row of Fig. 1, are sensitive to high levels of noise at fine scales, which can be an issue for computer vision tasks such as edge detection and image segmentation.

This is illustrated by comparing the edge map for the original and noise cases, where the edge map for the noise case virtually has no structure. This sensitivity to high levels of noise is a result of the insufficiency of local information being used by linear approaches to provide robust models under such situations.

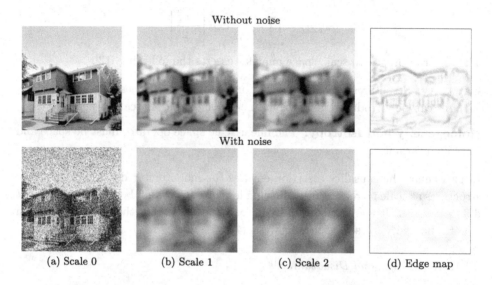

Fig. 1. Multi-scale representations of an image without noise (top) and with noise (bottom) contamination using linear approaches. The columns represent three scales and the corresponding edge map across the scales.

2.2. *Nonlinear approaches*

To address the obvious limitations of linear scale space approaches, a variety of nonlinear approaches have been introduced [4, 5, 7, 8]. The basic concept behind nonlinear scale space based approaches is, as before, the modeling of an image as a family of derived images, but now based on *nonlinear* scale space operators. Since many of the limitations of linear approaches are due to the isotropic nature of the linear operators, resulting in poor structural localization and inter-region smoothing, nonlinear scale space approaches replace the isotropic operators with nonlinear anisotropic operators that can potentially better preserve large scale structures at coarse scales.

The generalized diffusion equation

$$\frac{\partial L}{\partial t} = c_t(\underline{x})\nabla^2 L + \nabla \cdot (c\nabla L), \tag{3}$$

acts as the basis for existing nonlinear approaches, where $c_t(\underline{x})$ and ∇ denote the diffusion coefficient and the gradient, respectively. The most popular nonlinear scale

Without noise

With noise

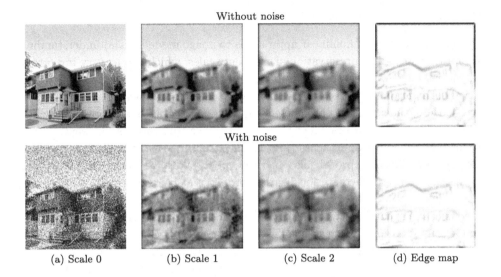

(a) Scale 0   (b) Scale 1   (c) Scale 2   (d) Edge map

Fig. 2. Multi-scale representations of the same image in Fig. 1, but now using the nonlinear approach introduced by Perona and Malik [4]. While better than the linear approaches in Fig. 1, the structures at the coarser scales when using the nonlinear approach also becomes poorly localized and degraded, which is particularly evident in the edge map.

space approach was first introduced by Perona and Malik [4], which introduces two important assertions to scale space based multi-scale modeling [4]:

Immediate Localization: Structures at each scale should be sharp and coincide with the true locations of the structures in the original image.

Piecewise Smoothing: Intra-region smoothing is preferred over inter-region smoothing.

In an attempt to enforce these properties in the scale space, a non-negative diffusion coefficient $c_t(\underline{x})$ was introduced by Perona and Malik into Eq. (3), where the coefficient is a function h of the gradient magnitude,

$$c_t(\underline{x}) = h\left(\|\nabla L_t(\underline{x})\|\right) \qquad (4)$$

which serves to enforce the above two properties in the following manners. First, the diffusion coefficient $c_t(\underline{x})$ limits diffusion when $\nabla L_t(\underline{x})$ is large, thus better preserving structural detail and hence better enforcing the immediate localization property. Second, $c_t(\underline{x})$ encourages intra-region smoothing by encouraging diffusion in directions of lower gradient, hence better enforcing the piecewise smoothing property. A discrete implementation of the nonlinear approach introduced by Perona and Malik for the 2D case can be defined as as

$$L_t(x,y) = L_{t-1}(x,y) + \lambda\left(\sum_i\sum_j c_t(x+i,y+j)\nabla L_{t-1}(x+i,y+j)\right), \qquad (5)$$

where the iteration is initialized, as $L_0(x,y) = f(x,y)$.

2.2.1. Advantages and disadvantages

The main advantage of nonlinear approaches to image modeling and understanding is improved structural preservation at different scales, which can be very important for various image processing and computer vision tasks such as edge detection and image segmentation. This improvement over linear approaches is illustrated in Fig. 2. Here, the structures at coarser scales are noticeably better preserved when using nonlinear approaches when compared to linear approaches. However, while better than the linear approaches (Fig. 1), the structures at the coarser scales when using the nonlinear approach also becomes poorly localized and degraded, thus not satisfying the immediate localization property at coarser scales. As with the case of linear approaches, the structural delocalization and degradation exhibited at coarser scales can be attributed to the reliance of nonlinear scale space operators on local information, which is insufficient for preserving large scale structures at coarser scales. Furthermore, as with the case of linear approaches, such nonlinear approaches exhibit high sensitivity to noise at finer scales (see the noise case in Fig. 2), resulting in poor multi-scale models under such situations.

Given that the limitations of both linear and nonlinear deterministic approaches relate to the fact that local information is insufficient for structural preservation and modeling at coarse scales, the exploration of alternative approaches that better utilize global information to better preserve structures at coarse scales and improve noise robustness is desired.

3. Non-parametric sample-based approaches

Given that the issues associated with existing scale space approaches are associated with the limitations of using only local information, one is motivated to devise multi-scale modeling approaches that better utilize relevant global information from the entire image. To facilitate for improved multi-scale image modeling and understanding without significant overhead from utilizing global information, Mishra et al. [14] proposed a non-parametric sample-based approach, which we will generalize here for the purpose of multi-scale modeling and understanding.

Let X denote a set of sites in a discrete lattice $\mathcal{L}$ where the image is defined, with $\underline{x} \in X$ denoting a site in $\mathcal{L}$. Let the image be defined as $f = \{f(\underline{x}) | \underline{x} \in X\}$, and the corresponding gradient magnitude defined as $G_i = \{G_i(\underline{x}) | \underline{x} \in X\}$. Furthermore, let us define the relationship between adjacent scales in the multi-scale model as

$$L_{t-1}(\underline{x}) = L_t(\underline{x}) + C_t(\underline{x}), \tag{6}$$

where $L_t = \{L_t(\underline{x}) | \underline{x} \in X\}$ is the representation at scale t, $C_t = \{C_t(\underline{x}) | \underline{x} \in X\}$ is the residual inter-scale structure at scale t, both being random fields on X, and $L_0(\underline{x}) = f(\underline{x})$ is the base representation. As such, the multi-scale modeling process is essentially a recursive inverse problem

$$\hat{L}_t\left(\underline{x}\right) = \arg_{\hat{L}_t} \min \left\{ E\left(\left(\hat{L}_t\left(\underline{x}\right) - L_t\left(\underline{x}\right)\right)^2 | L_{t-1}\left(\underline{x}\right)\right)\right\}, \tag{7}$$

with the Bayesian least-squares solution given by [15]

$$\hat{L}_t\left(\underline{x}\right) = \int L_t\left(\underline{x}\right) p\left(L_t\left(\underline{x}\right) | L_{t-1}\left(\underline{x}\right)\right) dL_t\left(\underline{x}\right). \tag{8}$$

Looking at Eq. (8), it can be observed that the multi-scale model produced by the sample-based approach introduced by Mishra et al. is comprised of the base representation $L_0(\underline{x})$, and a series of conditional probability distributions $p\left(L_1\left(\underline{x}\right) | L_0\left(\underline{x}\right)\right), p\left(L_2\left(\underline{x}\right) | L_1\left(\underline{x}\right)\right), \ldots, p\left(L_{N-1}\left(\underline{x}\right) | L_{N-2}\left(\underline{x}\right)\right)$, where N is the number of scales. To obtain these conditional probability distributions, a quasi-random density estimation approach is employed where only highly relevant samples from across $L_{t-1}\left(\underline{x}\right)$ are utilized. By employing such an approach, the limitations of linear and nonlinear deterministic approaches associated with utilizing only local information are alleviated.

3.1. *Estimation of conditional probability distributions*

To estimate the conditional probability distributions at each scale, a quasi-random density estimation strategy is employed. First, a set of m samples is drawn from a Sobol quasi-random sequence [16] with respect to $\underline{x}$. This initial use of a Sobel quasi-random sequence allows low discrepancy samples to be drawn for use in estimating $p\left(L_t(\underline{x}) | L_{t-1}(\underline{x})\right)$. To determine samples with high relevancy, a Gaussian mixture model is constructed based on the distribution $p\left(L_{t-1}(\underline{x})\right)$ created from the samples drawn from the Sobol sequence. Given this Gaussian mixture model, the final sample set Ω used to estimate $p\left(L_t(\underline{x}) | L_{t-1}(\underline{x})\right)$ can be defined as all samples that fall within the Gaussian distribution to which $L_{t-1}(\underline{x})$ belongs within the Gaussian mixture model. An example illustrating this sample selection process is found in Fig. 3.

Finally, given Ω, the estimated conditional probability distribution $\hat{p}\left(L_t(\underline{x}) | L_{t-1}(\underline{x})\right)$ is given by

$$\hat{p}\left(L_t(\underline{x}) | L_{t-1}(\underline{x})\right) = \frac{p^*\left(L_t(\underline{x}) | L_{t-1}(\underline{x})\right)}{\int\limits_0^1 p^*\left(L_t(\underline{x}) | L_{t-1}(\underline{x})\right) dL_t(\underline{x})}, \tag{9}$$

where the distribution p^* quantifies sample relevancy:

$$p^*\left(L_t(\underline{x}) | L_{t-1}(\underline{x})\right) =$$

$$\frac{1}{\sqrt{2\pi}\sigma_{L_t}} \sum_{k=\Omega} f_T\left(k\right) \exp\left(-\frac{1}{2}\left(\frac{L_t - L_{t-1}(\underline{x}_k)}{\sigma_{L_{t-1}}}\right)^2\right), \tag{10}$$

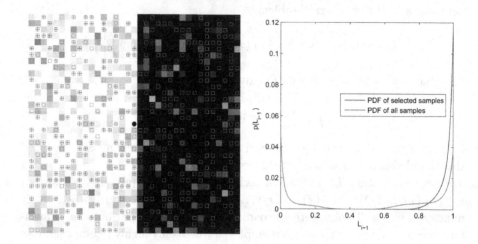

Fig. 3. A synthetic example showing the results of samples used to estimate the conditional probability distributions. The black circle represents the site under consideration, the squares represent the samples drawn from the Sobol sequence, and the crosses represent the final samples used to estimate the conditional probability distributions. The corresponding distribution $p(L_{i-1}(x))$ is shown at the bottom.

where $f_T(k)$ is the total assessment measure of sample relevancy. In designing $f_T(k)$ to accurately assess the relevancy of a particular sample, a number of different factors can be put into consideration. Relevant samples should have image intensities that are close to that at site $\underline{x}$ to avoid inter-region smoothing and poor structural localization. Relevant samples should also be similar image gradient magnitudes to better preserve the underlying structural details, Furthermore, spatial closeness is also an important indicator of sample relevancy. Therefore, based on these factors, the total assessment measure $f_T(k)$ can be formulated as the product of individual assessment measures for the different factors:

$$f_T(k) = f_I(k)\, f_G(k)\, f_V(k), \tag{11}$$

where $f_I(k)$, $f_G(k)$, and $f_V(k)$ assess sample relevancy based on image intensity, image gradient magnitude, and spatial distance respectively:

$$f_I(k) = \exp\left(-\frac{1}{\rho_{I_{t-1}}}(L_{t-1}(\underline{x}) - L_{t-1}(\underline{x}_k))^2\right), \tag{12}$$

$$f_G(k) = \exp\left(-\frac{1}{\rho_{G_{t-1}}}(G_{t-1}(\underline{x}) - G_{t-1}(\underline{x}_k))^2\right), \tag{13}$$

and

$$f_V(k) = \exp\left(-\frac{1}{\rho_{V_{t-1}}}\|(\underline{x}) - (\underline{x}_k)\|_2\right), \tag{14}$$

and ρ_{I_t}, ρ_{G_t} and ρ_{V_i} denote regularization constants for scale t, with ρ_{I_t}, ρ_{G_t} defined as the median of local standard deviations over a sliding window:

$$\rho_{I_t} = median(\sigma_{I_t}(j,k)), \qquad (15)$$

$$\rho_{G_t} = median(\sigma_{G_t}(j,k)). \qquad (16)$$

An example of this conditional probability distribution estimation is shown in Fig. 4.

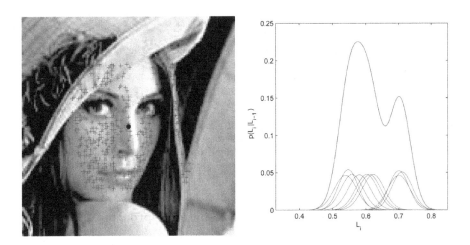

Fig. 4. An example of the conditional probability distribution estimation for a site in an image. The black circle represents the site s, while the crosses represent the samples. As seen in the estimate of $p\left(L_t(\underline{x})|L_{t-1}(\underline{x})\right)$ in the bottom figure, each sample contributes to the overall estimate based on their corresponding attributes.

3.1.1. *Benefits of nonparametric approach*

The main advantage of the nonparametric sample-based approach is that significantly improved structural preservation at different scales can be achieved, which can be very important for various image processing and computer vision tasks such as edge detection and image segmentation. This improvement over linear (Fig. 1) and nonlinear (Fig. 2) deterministic approaches is illustrated in Fig. 5. Structures at coarser scales are significantly better preserved when compared to linear and nonlinear deterministic approaches. From the edge map, the multi-scale model constructed using the nonparametric sample-based approach provides structures at the coarser scales that are well localized and sharp, thus better satisfying the immediate localization property at coarser scales. Furthermore, such a nonparametric sample-based approach exhibits greater noise robustness at all scales, resulting in more accurate multi-scale models under such situations.

Without noise

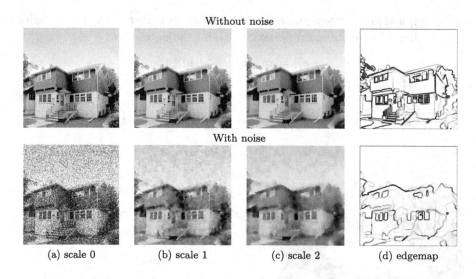

With noise

    (a) scale 0   (b) scale 1   (c) scale 2   (d) edgemap

Fig. 5.   Multi-scale representations of an image of a house without (top) and with (bottom) noise contamination using the nonparametric sample-based approach. The structures at coarser scales are significantly better preserved when compared to linear (Fig. 1) and nonlinear (Fig. 2) deterministic approaches, with structures at the coarser scales that are well localized and sharp even under noisy situations.

4. Applications of nonparametric sample-based approach

Such a nonparametric sample-based approach (which we will refer to as NPS) should be explored in image processing and computer vision tasks to illustrate its effectiveness. First, let us explore the application of this non-parametric sample-based framework for the purpose of image denoising and edge detection. Examples of its potential for image denoising and edge detection is illustrated in Figs. 6, 7, 8 and 9 where the image was contaminated by additive Gaussian noise with standard deviations of $\sigma = \{10\%, 40\%\}$ of the dynamic range of the images.

Given the structural localization and noise robustness gained from such a nonparametric sample-based approach, another potential application in computer vision is for the purpose of image segmentation. To illustrate this, image segmentation is performed on a series of images contaminated by high levels of Gaussian noise using a level set based active contour [17] method and Otsu binary thresholding [19] method. The integrated scale space and level set approach based image segmentation result is presented in Fig. 11. The segmentation result of Otsu binary threholding method is presented in Fig. 12.

5. Future directions

In summary, a nonparametric framework for image understanding based on models constructed via conditional sampling was explored in this chapter. By modeling

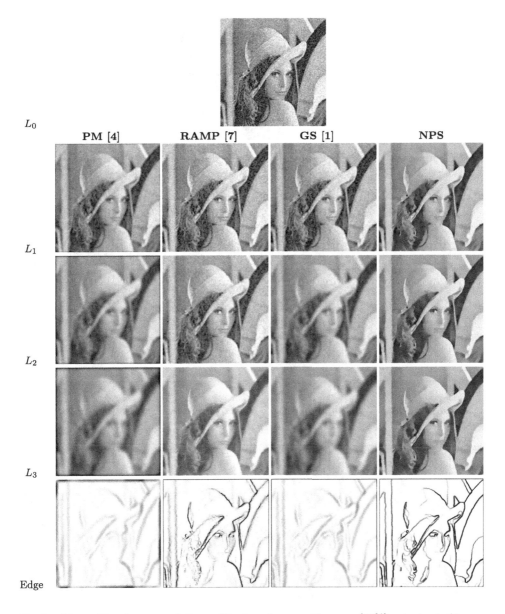

Fig. 6. The multi-scale representations of the Lena image, with a $\sigma = \{10\%\}$ noise level. Observe the significant absence of blurring in the representations produced using the nonparametric sample-based approach.

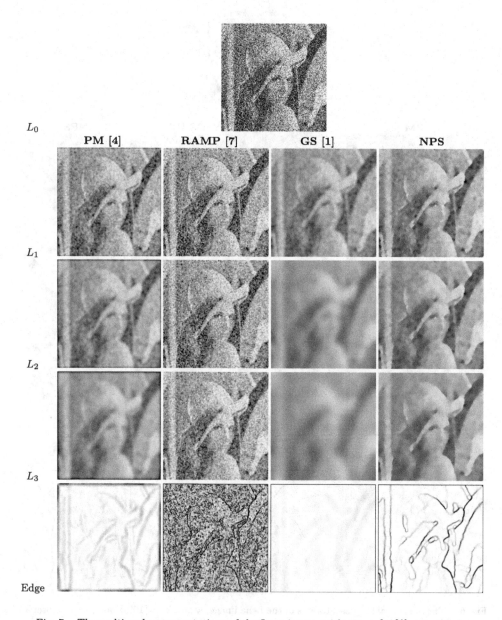

Fig. 7.   The multi-scale representations of the Lena image, with a $\sigma = \{40\%\}$ noise level.

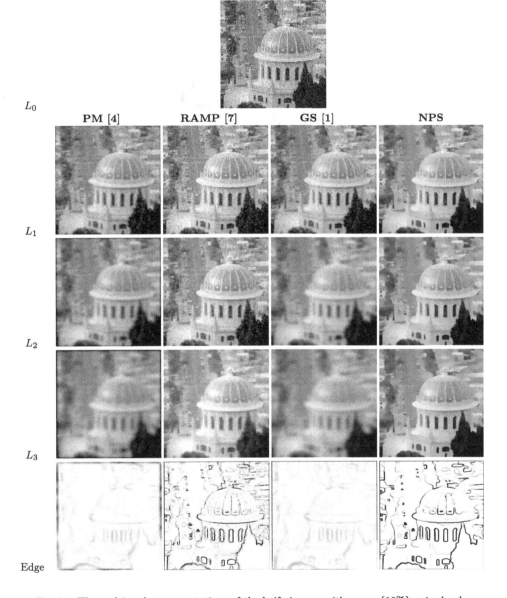

Fig. 8. The multi-scale representations of the haifa image, with a $\sigma = \{10\%\}$ noise level.

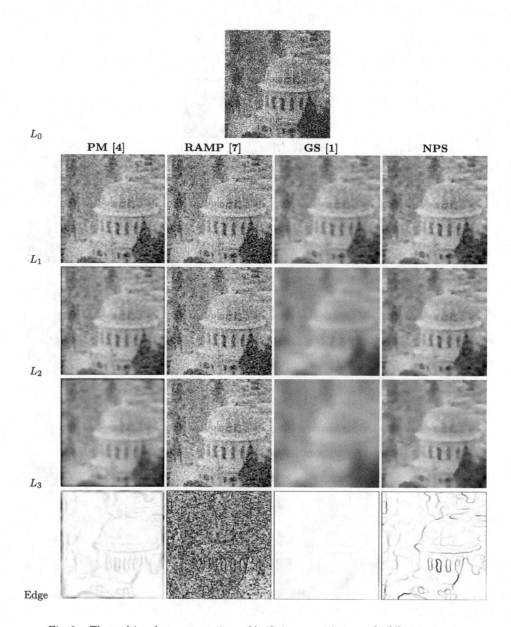

Fig. 9.   The multi-scale representations of haifa image, with a $\sigma = \{40\%\}$ noise level.

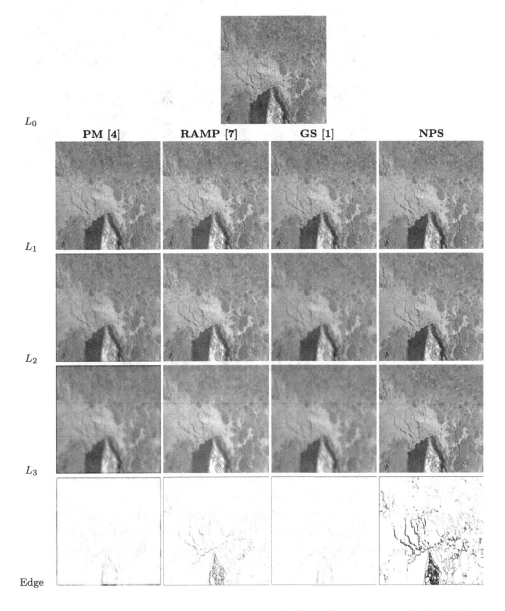

Fig. 10. The multi-scale representations of the image a RADARSAT sea ice image (RADARSAT-1 imagery 2004 CSA, All Rights Reserved).

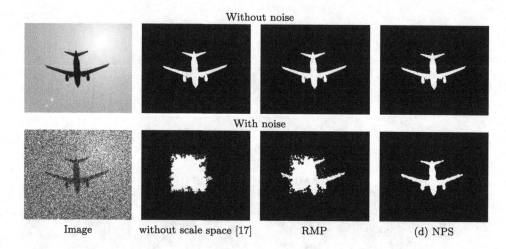

Fig. 11.   Illustration of the effectiveness of the proposed scale space in image segmentation. Proposed approach helps the active contour based [17] segmentation method to find the object of interest in the presence of high noise.

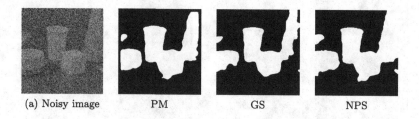

Fig. 12.   Illustration of the effectiveness of the proposed scale space in image segmentation. Scale space approaches helps the Otsu binary threshold based segmentation method to find the object of interest in the presence of high noise.

images as a lattice of nonparametric conditional probability distributions estimated via random sampling, robust techniques for tackling image processing and computer vision challenges can be devised, as illustrated by examples shown such as denoising, edge detection, and segmentation. A possible extension to this nonparametric sample-based framework that is worth exploring is the possibility of incorporating texture characteristics as one of the objective functions used to judge sample relevancy, which can be particularly beneficial for tasks such as texture segmentation and denoising.

References

[1] A. Witkin, "Scale-scale filtering," Proc. 7th International Joint Conference on Artificial Intelligence, pp. 1019-1022, 1983.

[2] T. Lindberg, "Edge detection and ridge detection with automatic scale selection," Proc. IEEE Computer Society Conference on Computer Vision and Pattern Recognition, pp. 465-470, 1996.

[3] A. Mishra, A. Wong, K. Bizheva, and D.A. Clausi, "Intra-retinal layer segmentation in optical coherence tomography images", Optics Express, Vol. 17, No. 26, pp. 23719-23728, 2009.

[4] P. Perona and J. Malik, "Scale-space and edge detection using anisotropic diffusion," IEEE Transactions on Pattern Analysis and Machine Intelligence, Vol. 12, No. 7, 629-639, 1990.

[5] J. Yu, Y. Wang, and Y. Shen, "Noise reduction and edge detection via kernel anisotropic diffusion," Pattern Recognition Lettersm, Vol. 29, pp. 1496-1503, 2008.

[6] D. Jobson, Z. Rahman, and G. Woodell, "A multiscale retinex for bridging the gap between color images and the human observation of scenes", IEEE Transactions on Image Processing, Vol. 6, No. 7, pp. 965-976, 1997.

[7] G. Gilboa, N. Sochen, and Y. Zeevi, "Image enhancement and denoising by complex diffusion process," IEEE Transactions on Pattern Analysis and Machine Intelligence, Vol. 25, pp. 1020-1036, 2004.

[8] G. Gilboa, "Nonlinear scale space with spatially varying stopping time," IEEE Transactions on Pattern Analysis and Machine Intelligence. Vol. 30, No. 12, pp. 2175-2187, 2008.

[9] A. Petrovic, O. Divorra, and P. Vandergheynst, "Multiresolution segmentation of natural images: from linear to nonlinear scale space representations," IEEE Transactions on Image Processing, Vol. 13, No. 8, pp. 1104-1114, 2004.

[10] J. Elder and S. Zucker, "Scale space localization, blur, and contour-based image coding," Proc. IEEE Computer Society Conference on Computer Vision and Pattern Recognition, pp. 27-34, 1996.

[11] J. Elder and S. Zucker, "Local scale control for edge detection and blur estimation," Proc. IEEE Transactions on Pattern Analysis and Machine Intelligence, Vol. 20, No. 7, pp. 699-716, 1998.

[12] J. Koenderink and A. Van Doorn, A., "The structure of images," Biological Cybernetics, pp. 363-370, 1984.

[13] A. Wong, A. Mishra, J. Yates, D.A. Clausi, P. Fieguth, and J. Callaghan, "Intervertebral disc segmentation and volumetric reconstruction from peripheral quantitative computed tomography imaging," IEEE Transactions on Biomedical Engineering, Vol. 56, No. 11, pp. 2748-2751, 2009.

[14] A. Mishra, A. Wong, D.A. Clausi, and P. Fieguth, "Quasi-random nonlinear scale," Pattern Recognition Letters, Vol. 31, No. 13, pp. 1850-1859, 2010.

[15] P. Fieguth, "Statistical Image Processing and Multidimensional Modeling", Springer, 2010.

[16] I. Sobol, "Uniformly Distributed Sequences with an Additional Uniform Property," USSR Computational Mathematics and Mathematical Physics, Vol. 16, pp. 236-242, 1997.

[17] T. F. Chan, L. A. Vese, "Active contours without edges", IEEE Transactions on In Image Processing, Vol. 10, No. 2, pp. 266-277, 2001.

[18] A. Mishra, P.W. Fieguth and D.A. Clausi, "Decoupled Active Contour (DAC) for Boundary Detection", *IEEE Transactions on Pattern Analysis and Machine Intelligence*, Vol. 33, No. 2, pp. 310-324, 2011.

[19] N. Otsu, "A threshold selection method from gray-level histograms", IEEE Trans. Sys., Man., Cyber., Vol. 9, No. 1, pp. 6266, 1979.

CHAPTER 1.8

A PROBABILISTIC FORMULATION FOR THE CORRESPONDENCE PROBLEM

Ron Tal[1] and Minas E. Spetsakis[2]

Centre for Vision Research, York University, Toronto, Ont. Canada
[1] *rontal@cse.yorku.ca*
[2] *minas@cse.yorku.ca*

The problem of determining correspondence of feature-points in a sequence of images has received considerable attention from the computer vision community and an effective solution will be a boon in many applications. Unfortunately, the problem itself is ill-posed and is further complicated by many sources of uncertainty in the image. In this chapter we review the evolution of proposed solutions from the literature and argue that a general probabilistic formulation is desired. The generality of the probabilistic framework which we introduce is further indicated by rederiving one of the classical solutions to the problem, as a special case. Based on our probabilistic framework, we propose an algorithm for feature-point correspondence that works under relaxed statistical assumptions. In order to accurately model the joint probability of feature appearance and position we introduce a Monte-Carlo technique that uses an extended EM algorithm to approximate the distribution using a mixture of Gaussians. We demonstrate the effectiveness of this approach by applying it to local flow estimation and tracking.

1. Introduction

The problem of determining feature-point correspondence in multiple images and its variants, like optical flow and contour matching are at the core of many computer vision problems such as motion segmentation, tracking, stereo, structure from motion, etc. Generally speaking the problem of matching in all its forms is ill posed and a great variety of assumptions are employed in all the classical solutions [Barron *et al.* (1994)]. First, it is necessary to review the problem and discuss how approaches to solving it have evolved over the years.

1.1. *Feature-point correspondence*

Generally speaking, the goal of feature-point correspondence is to determine the new location of some known feature based on its appearance. Since in order to determine correspondence the feature must maintain many of its physical characteristics, early formulations relied on the brightness constancy assumption. More specifically, the observed feature appearance remains identical, subject to changes in position only

$$I(x + u, y + v, t + 1) - I(x, y, t) = 0 \qquad (1)$$

where u and v are horizontal and vertical image motion parameters, respectively. If we assume that inter-frame image motion is small, we can derive a linear approximation of (1) by taking the first order Taylor Series

$$I_x u + I_y v + I_t = 0, \qquad (2)$$

where I_x and I_y are the image derivatives in the corresponding direction in space and I_t is the image derivative in time. As we have one linear equation and two unknowns our problem is ill-posed. This is a direct mathematical consequence of the aperture problem. To get around this, it is common to incorporate the smoothness assumption, which states that neighboring pixels exhibit very similar inter-frame displacement. The Horn and Schunk algorithm [Horn and Schunk (1981)] is one of the most popular early techniques and it applied by formulating the brightness and smoothness constraints at every pixel by introducing a regularization term

$$H = I_x u + I_y v + I_t + \lambda(u_x^2 + v_x^2 + u_y^2 + v_y^2). \qquad (3)$$

Eqn. (3) then solves for the motion parameters by minimizing the sum of H over a small neighborhood (such as the one depicted in Fig. 1) centered around the feature-point

$$[u, v]^T = \arg\min_{u,v}(\sum_{i=1}^{N} H) \qquad (4)$$

A slightly different contemporary approach is the Lucas and Kanade algorithm [Lucas and Kanade (1981)], which rose from obscurity after a comprehensive study by Barron et al [Barron et al. (1994)] has indicated that it outperforms other approaches and it became a benchmark for later motion estimation techniques. They apply the smoothness constraint by assuming that the variables (i.e. the motion parameters) are identical for all pixels in a small neighborhood giving us an overdetermined system of equations

$$I_{x,1}\delta x + I_{y,1} u = -I_{t,1}$$
$$\vdots \qquad (5)$$
$$I_{x,N}\delta x + I_{y,N} v = -I_{t,N}$$

which can be expressed in vector form as

$$\sum_{i=1}^{N} \begin{bmatrix} I_{x,i} \\ I_{y,i} \end{bmatrix}^T \begin{bmatrix} u \\ v \end{bmatrix} = -\sum_{i=1}^{N} I_{t,i} \qquad (6)$$

which we can reformulate as a least of squares problem, giving rise to their well-known solution

$$\sum_{i=1}^{N} \begin{bmatrix} I_{x,i}^2 & I_{x,i}I_{y,i} \\ I_{x,i}I_{y,i} & I_{y,i}^2 \end{bmatrix} \begin{bmatrix} u \\ v \end{bmatrix} = -\sum_{i=1}^{N} \begin{bmatrix} I_{x,i}I_{t,i} \\ I_{y,i}I_{t,i} \end{bmatrix}. \tag{7}$$

This approach is remarkably accurate when the assumptions are met. Motion has to be smooth, brightness has to remain constant and since it makes use of a least-squares estimator, noise has to be normally distributed. For a more detailed comparison of traditional approaches, we refer the reader to the work of Barron *et al.*

However, one can envision many situations where these assumptions are violated. For once, motion can be large, rendering a local Taylor approximation inaccurate. Additionally, changes of up to 15% in brightness are relatively common and are practically undiscernible to the human eye while rendering the entire formulation useless. Other sources of noise include both camera and electronic(due to fluctuations of photon and electron arrival), specular distortion and digitization.

(a) Feature position and its asso-   (b) Feature position and its associated appearance at frame 1   ciated appearance at frame 2

Fig. 1.   Association of feature position and appearance across multiple frames. The feature-point is represented by its position $\vec{x} = [x, y]^T$ and by its corresponding appearance $\vec{I}(\vec{x})$, which is the neighborhood of pixels centered on $\vec{x}$.

1.2. *Towards more general assumptions*

To address these issues, several methods have been used to relax one or more of these assumptions. At the core of most such methods is a novel formulation that specifically address, often heuristically, a new paradigm regarding one assumption or another.

One way to address violations of the brightness constancy assumption is to explicitly incorporate it in the set of unknowns. The Cornelius and Kanade algorithm

[Cornelius and Kanade (1986)] extends the Lucas and Kanade algorithm by explicitly incorporating additive brightness changes to (1)

$$I(x + u, y + v, t + \delta t) - I(x, y, t) = \delta I. \tag{8}$$

The small motion assumption is applied to (8), providing a local linear constraint on motion and brightness change parameters

$$I_x u + I_y v + I_t - \delta I = 0. \tag{9}$$

This gives rise to an over-determined set of linear equations that has similar form to (7).

Negahdaripur [Negahdaripour (1995)] has proposed a new definition of the feature-point correspondence constraint model that incorporates both geometric and radiometric cues from the image that addresses violation of the brightness constancy assumption and problems caused by specular reflections. His algorithm had a similar formulation to the one used by Lucas and Kanade with additional unknowns corresponding to affine brightness change model

$$I(x + u, y + v, t + \delta t) - m(x, y, t)I(x, y, t) - I(x, y, t) - c(x, y, t) = 0, \tag{10}$$

where $m(x, y, t)$ is a multiplicative factor of brightness change, and $c(x, y, t)$ is an additive factor of brightness change. These equations are then solved using a least-squares formulation

$$\sum_{i=1}^{N} \begin{bmatrix} I_{x,i}^2 & I_{x,i}I_{y,i} & -I_{x,i}I_i & -I_{x,i} \\ I_{x,i}I_{y,i} & I_{y,i}^2 & -I_{y,i}I_i & -I_{y,i} \\ -I_{x,i}I_i & -I_{y,i}I_i & I_i^2 & I_i \\ -I_{x,i} & -I_{y,i} & I_i & 1 \end{bmatrix} \begin{bmatrix} \delta x \\ \delta y \\ \delta m \\ \delta c \end{bmatrix} = -\sum_{i=1}^{N} \begin{bmatrix} I_{x,i}I_{t,i} \\ I_{y,i}I_{t,i} \\ -I_{t,i}I_i \\ -I_{t,i} \end{bmatrix}. \tag{11}$$

This approach is more powerful than its predecessors, working under a variety of lighting change conditions. However, it does not address problems that stem from non-Gaussian sources of noise and large inter-frame displacement.

Perhaps the most popular techniques for handling both large and small inter-frame displacements is to apply a more traditional approach hierarchically using Gaussian pyramids of the image [Bergen et al. (1992)].

1.3. Non-Gaussian sources of uncertainty

Black and Anandan [Black and Anandan (1993)] use robust statistics and formulate the problem of determining flow using a robust M-Estimator [Meer et al. (1991)] in order to reduce the effect of outliers on accuracy. M-Estimators were developed independently, and later shown to be a special case of EM estimators [McLachlan and Krishnan (2007)].

Jepson and Black [Jepson and Black (2004)] have adapted the EM clustering algorithm for probabilistic mixture models in order to detect multiple motions and thus, handle motion boundaries. The inovation in this approach from a statistician's perspective is that the EM is not simply a best fit of a Gaussian Mixture Model (GMM), but also assigns physical meaning to each component in the mixture.

Other methods handle non-Gaussianity by using particle filtering techniques [Black and Fleet (2000); Isard and Blake (1998)]. This approach works well for problems with low dimensionality. In [Wills *et al.* (2003)] a RANSAC based algorithm is used for determining multiple motion parameters for the purpose of motion segmentation with the presence of large inter-frame motion.

1.4. *Towards a general probabilistic framework*

While the summary of previous approaches is by no means exhaustive, a few trends are clear. In early works, simple statistical and computational techniques were used to arrive at an estimate for a solution to the correspondence problem. The limited efficacy of these techniques to handle elaborate sources of uncertainty meant that their application is most effective when strict assumptions are used. Later work followed advances in research on statistical estimation. More specifically, there is a strong trend to weaken the assumptions regarding sources of uncertainty. In this work we equate the generality of an approach with weakening the underlying assumptions.

Previously, we have argued for further improvements on the correspondence problem a general probabilistic formulation is desired [Tal and Spetsakis (2010)]. We start by developing a probabilistic framework that is general enough to incorporate an arbitrary model of uncertainty. We show that by incorporating the assumptions used by Lucas and Kanade (namely small motion, brightness constancy, smoothness and Gaussian sources of uncertainty) into our general probabilistic framework, we can re-derive their equations.

To keep the probability that governs the relationship between feature position and its appearance general and meaningful we model it as a GMM. We also employ a Monte-Carlo technique [Press *et al.* (1992)] to collect samples of changes to feature-point appearance due to motion and other sources of uncertainty. The Expectation Maximization (EM) clustering algorithm [Dempster *et al.* (1997)] is then used to fit a GMM on the samples.

2. Our Formulation

Usually, strong assumptions regarding possible changes to feature appearance are made first, and a formulation that relies on these assumptions follows. Our approach is to first develop a general probabilistic framework and only then incorporate restrictive assumptions later. This way, the algorithm's specifics can be adjusted based on the application, visual conditions and available computing power. In this section we present the probabilistic formulation. In the following section we present an algorithm that stems from the probabilistic formulation.

2.1. The probabilistic framework

Given a point $\vec{x}$ on the image plane, we define its *appearance* $\vec{I}(\vec{x})$ to be a vector that comprises all the intensities of the pixels in a small neighborhood around point $\vec{x}$

$$\vec{I}(\vec{x}) = \begin{bmatrix} I\left[\vec{x} + \vec{d_1}\right] \\ \vdots \\ I\left[\vec{x} + \vec{d_k}\right] \end{bmatrix} \tag{12}$$

where I is the image as measured by the camera and $\vec{d}_{i=1..k}$ are the positions of the pixels from the neighborhood relative to $\vec{x}$. If two points $\vec{x}_a$ and $\vec{x}_b$ are projections of the same world object on images I_a and I_b respectively, the appearance of their corresponding neighborhoods will change due various sources of noise such that

$$\vec{I_b}(\vec{x}_b) = \vec{I_a}(\vec{x}_a) + \eta \tag{13}$$

where η is a noise vector whose behavior is dictated by the pdf $p(\eta \mid \gamma)$, and γ is a set of parameters that define the noise model. In the simplest case the noise model parameters are just a mean and a variance, but in our case will incorporate camera noise, changing illumination, sub-pixel image jitter and one can easily add other noise components such as affine deformations etc. The conditional probability density of $\vec{I_b}(\vec{x}_b)$ can be written as

$$p(\vec{I_b}(\vec{x}_b) \,|\vec{x}_a, I_a, \gamma). \tag{14}$$

This distribution can be determined given the noise model and can then be used, as we show later, to compute

$$p(\vec{x}_a | \vec{I_b}(\vec{x}_b), I_a, \vec{x}_b, \gamma) \tag{15}$$

from which the position $\vec{x}_a$ in image I_a that corresponds to the neighborhood $\vec{I_b}(\vec{x}_b)$ can be estimated.

The pdf in Eqn. (15) expresses the probabilistic model of $\vec{x}_a$ given all that is normally available in a correspondence problem. In such situations we are given neighborhood $\vec{I_b}(\vec{x}_b)$ and we try to find the best match in an image I_a. The position of the match, $\vec{x}_a$ can depend on $\vec{I_b}(\vec{x}_b)$ and I_a alone but in some cases the pdf $p(\vec{x}_a \mid \vec{x}_b)$ is available as a prior and we may take advantage of it. Eqn. (15) from Bayes rule becomes

$$p(\vec{x}_a | \vec{I_b}(\vec{x}_b), I_a, \vec{x}_b, \gamma) = \frac{p(\vec{I_b}(\vec{x}_b) \mid \vec{x}_a, I_a, \vec{x}_b, \gamma) p(\vec{x}_a \mid I_a, \vec{x}_b, \gamma)}{p(\vec{I_b}(\vec{x}_b) \mid I_a, \vec{x}_b, \gamma)}. \tag{16}$$

In Eqn. (16) the denominator $p(\vec{I_b}(\vec{x}_b) \mid I_a, \vec{x}_b, \gamma)$ is constant with respect to the maximization variable $\vec{x}_a$ and can be ignored. The numerator can be used in the form that appears in Eqn. (16), or can be seen as the joint probability of position and appearance

$$p(\vec{I_b}(\vec{x}_b) \mid \vec{x}_a, I_a, \vec{x}_b, \gamma) p(\vec{x}_a \mid I_a, \vec{x}_b, \gamma) = p(\vec{x}_a, \vec{I_b}(\vec{x}_b) \mid I_a, \vec{x}_b, \gamma). \tag{17}$$

Now, given the appearance vector $\vec{\mathsf{I}}_b(\vec{x}_b)$, we can estimate the corresponding position in image I_a by maximizing the joint probability of image appearance

$$\hat{\vec{x}}_a = \arg\max_{\vec{x}_a} \left(p(\vec{x}_a, \vec{\mathsf{I}}_b(\vec{x}_b) \mid I_a, \vec{x}_b, \gamma) \right). \tag{18}$$

This solution can form the basis for algorithms for feature point matching, tracking and optical flow as follows. Using an initial feature neighborhood $\vec{\mathsf{I}}_n$ centered on a feature point $\vec{x}_n$ and a noise and motion model, we first compute a distribution that relates a possible position of the feature given changes in appearance in between frames $p(\vec{x}_{n+1}, \vec{\mathsf{I}}_n(\vec{x}_n) \mid I_{n+1}, \vec{x}_n, \gamma)$. Given a vector $\vec{\mathsf{I}}_{n+1}$ representing a neighborhood in a different image, we can determine its most likely position by maximizing

$$\vec{x}_{n+1} = \arg\max_{\vec{x}_{n+1}} \left(p(\vec{x}_{n+1}, \vec{\mathsf{I}}_n(\vec{x}_n) \mid I_{n+1}, \vec{x}_n, \gamma) \right) \tag{19}$$

This expression provides us with a powerful framework for the estimation of correspondence of two points in an image sequence under various conditions of noise and motion.

2.2. *Reduction to known solution by incorporating restrictive assumptions*

To demonstrate the generality and power of this formulation we derive from it the Lucas and Kanade algorithm by introducing its underlying assumptions into the probabilistic framework. The Lucas and Kanade algorithm is perhaps the most successful of the early differential methods for computing the correspondence of points in two images. While it remained obscure for over a decade, a comparative study by Barron, Fleet and Beauchemin [Barron *et al.* (1994)] showed that it is far superior to the Horn and Schunk algorithm [Horn and Schunk (1981)] and replaced it as the default benchmark. The algorithm is derived around three core assumptions: brightness constancy, such that feature appearance does not change due to motion; small motion, such that a first-order Taylor series can be used for approximation; and smoothness of flow, which is used to turn an ill-posed problem into an over-determined least-of-squares problem by assuming that motion is identical for every pixel within a small neighborhood. The solution is in the form

$$\vec{\mathbf{u}}_n = \mathbf{M}'^{-1}\vec{b} \tag{20}$$

where

$$\mathbf{M}' = \sum_{i=1}^{N} \begin{bmatrix} I_{x,i}^2 & I_{x,i}I_{y,i} \\ I_{x,i}I_{y,i} & I_{y,i}^2 \end{bmatrix}, \tag{21}$$

$\vec{\mathbf{u}}_n$ is the vector of motion parameters $[u, v]^T$ at frame n,

$$\vec{b} = -\sum_{i=1}^{N} \begin{bmatrix} I_{x,i}I_{t,i} \\ I_{y,i}I_{t,i} \end{bmatrix}, \tag{22}$$

I_x and I_y are the directional derivatives of the image, I_t is the time derivative of the image and the summation is over the small neighborhood around a feature. In reality, M' may be singular and thus non-invertible. In order to have stable behavior, in most practical implementations of the algorithm Eqn. (20) is modified to include a stabilization constant

$$\vec{u}_n = (M' + \varepsilon 1)^{-1} \vec{b} \tag{23}$$

where ε is usually a small number.

We show later in this section how we can relax these assumptions, along with several other ones, to obtain a more general solution by introducing our own noise model. The joint probability of feature position and its appearance is implicitly assumed Gaussian in the original paper [Lucas and Kanade (1981)], *i.e.*,

$$p(\vec{x}_{n+1}, \vec{I}_n \mid I_{n+1}, \vec{x}_n, \gamma) = G(\vec{x}_{n+1}, \vec{I}_n; \vec{\mu}_{\vec{x}, \vec{I}}, C_{\vec{x}, \vec{I}}) \tag{24}$$

thus, Eqn. (18) can find the value for $\vec{x}_{n+1}$ by minimizing the Mahalanobis distance

$$M_{xI} = \begin{bmatrix} \vec{I}_n - \vec{\mu}_I \\ \vec{x}_{n+1} - \vec{\mu}_{\vec{x}} \end{bmatrix}^T C_{\vec{x}, \vec{I}}^{-1} \begin{bmatrix} \vec{I}_n - \vec{\mu}_I \\ \vec{x}_{n+1} - \vec{\mu}_{\vec{x}} \end{bmatrix} \tag{25}$$

where

$$\vec{\mu}_I = \begin{bmatrix} I_{n+1}\left[\vec{\mu}_{\vec{x}} + \vec{d}_1\right] \\ \vdots \\ I_{n+1}\left[\vec{\mu}_{\vec{x}} + \vec{d}_k\right] \end{bmatrix} = \vec{I}_{n+1}. \tag{26}$$

Unless we have a prior for $\vec{x}_{n+1}$, $\vec{\mu}_{\vec{x}}$ is simply $\vec{x}_n$, vector $\vec{I}_n$ is the image data in the neighborhood of the feature-point at time n and vector $\vec{I}_{n+1}$ is the image data in the neighborhood of the same feature at time $n+1$. The joint mean of $\vec{x}_{n+1}$ and $\vec{I}_n$ is

$$\vec{\mu}_{\vec{x}, \vec{I}} = \begin{bmatrix} \vec{\mu}_I \\ \vec{\mu}_{\vec{x}} \end{bmatrix}. \tag{27}$$

The joint covariance matrix of $\vec{x}_{n+1}$ and $\vec{I}_n$

$$C_{\vec{x}, \vec{I}} = E\left\{ \begin{bmatrix} \vec{I}_n - \vec{I}_{n+1} \\ \vec{x}_{n+1} - \vec{x}_n \end{bmatrix} \begin{bmatrix} \vec{I}_n - \vec{I}_{n+1} \\ \vec{x}_{n+1} - \vec{x}_n \end{bmatrix}^T \right\} \tag{28}$$

can be partitioned in the following manner

$$C_{\vec{x}, \vec{I}} = \begin{bmatrix} C_{II} & C_{Ix}^T \\ C_{xI} & C_{xx} \end{bmatrix} \tag{29}$$

where

$$\mathbf{C}_{II} = E\left\{\left(\vec{\mathbf{I}}_n - \vec{\mathbf{I}}_{n+1}\right)\left(\vec{\mathbf{I}}_n - \vec{\mathbf{I}}_{n+1}\right)^T\right\}$$

$$\mathbf{C}_{xx} = E\left\{\left(\vec{x}_{n+1} - \vec{x}_n\right)\left(\vec{x}_{n+1} - \vec{x}_n\right)^T\right\}$$

$$\mathbf{C}_{xI} = E\left\{\left(\vec{x}_{n+1} - \vec{x}_n\right)\left(\vec{\mathbf{I}}_n - \vec{\mathbf{I}}_{n+1}\right)^T\right\}.$$

Thus the matrix can be efficiently inverted using the method of inversion by partitioning [Press *et al.* (1992)]

$$\mathbf{C}_{\vec{x},\vec{\mathbf{I}}}^{-1} = \left[\begin{array}{c:c} \mathbf{S}_{II} & \mathbf{S}_{xI}^T \\ \hdashline \mathbf{S}_{xI} & \mathbf{S}_{xx} \end{array}\right] \tag{30}$$

where

$$\mathbf{S}_{II} = \left(\mathbf{C}_{II} - \mathbf{C}_{Ix}\mathbf{C}_{xx}^{-1}\mathbf{C}_{xI}\right)^{-1} \tag{31}$$

$$\mathbf{S}_{xI} = -\left(\mathbf{C}_{xx}^{-1}\mathbf{C}_{Ix}\right)\left(\mathbf{C}_{II} - \mathbf{C}_{Ix}\mathbf{C}_{xx}^{-1}\mathbf{C}_{xI}\right)^{-1} \tag{32}$$

$$\mathbf{S}_{xx} = \left(\mathbf{C}_{xx} - \mathbf{C}_{xI}\mathbf{C}_{II}^{-1}\mathbf{C}_{Ix}\right)^{-1} \tag{33}$$

Using (30) M_{xI} becomes

$$\begin{aligned} M_{xI} &= \left(\vec{\mathbf{I}}_n - \vec{\mathbf{I}}_{n+1}\right)^T \mathbf{S}_{II}\left(\vec{\mathbf{I}}_n - \vec{\mathbf{I}}_{n+1}\right) \\ &+ \left(\vec{x}_{n+1} - \vec{x}_n\right)^T \mathbf{S}_{xI}\left(\vec{\mathbf{I}}_n - \vec{\mathbf{I}}_{n+1}\right) \\ &+ \left(\vec{\mathbf{I}}_n - \vec{\mathbf{I}}_{n+1}\right)^T \mathbf{S}_{xI}^T\left(\vec{x}_{n+1} - \vec{x}_n\right) \\ &+ \left(\vec{\mathbf{I}}_n - \vec{\mathbf{I}}_{n+1}\right)^T \mathbf{S}_{xx}\left(\vec{x}_{n+1} - \vec{x}_n\right) \end{aligned} \tag{34}$$

We can find the value for $\vec{x}_{n+1}$ that minimizes M_{xI} by taking partial derivative with respect to $\vec{x}_{n+1}$ and equating it to zero

$$\frac{\partial M_{xI}}{\partial \vec{x}_{n+1}} = 2\left(\vec{\mathbf{I}}_n - \vec{\mathbf{I}}_{n+1}\right)\mathbf{S}_{xI}^T + 2\left(\vec{x}_{n+1} - \vec{x}_n\right)^T \mathbf{S}_{xx} = 0 \tag{35}$$

from where we get

$$\vec{x}_{n+1} = \vec{x}_n - \mathbf{S}_{xI}^T \mathbf{S}_{xx}^{-1}\left(\vec{\mathbf{I}}_n - \vec{\mathbf{I}}_{n+1}\right) \tag{36}$$

By substituting the inversion by partitioning equations we arrive at the solution

$$\vec{x}_{n+1} = \vec{x}_n + \mathbf{C}_{xI}\mathbf{C}_{II}^{-1}\left(\vec{\mathbf{I}}_n - \vec{\mathbf{I}}_{n+1}\right). \tag{37}$$

The noise in the image measurements in [Lucas and Kanade (1981)] can be modeled by

$$\vec{\mathbf{I}}_n(\vec{x}_n) = \vec{\mathbf{I}}_{n+1}(\vec{x}_{n+1}) + \eta_n \tag{38}$$

where η_n is implicitly assumed to be an independent, identically distributed (i.i.d) additive noise. Using the Lucas and Kanade assumptions of small motion with no prior regarding direction, the variance of $\vec{x}_{n+1}$ is simply $\mathbf{C}_{xx} = \sigma_{xx}^2 \mathbb{1}$ and we can approximate $\vec{\mathrm{I}}_n$ using a first order Taylor series

$$\vec{\mathrm{I}}_n(\vec{x}_{n+1}) = \vec{\mathrm{I}}_n(\vec{x}_n) + \nabla \vec{\mathrm{I}}_n^T (\vec{x}_{n+1} - \vec{x}_n) \tag{39}$$

where

$$\nabla \vec{\mathrm{I}}_n = \begin{bmatrix} I_{n,x}\left[\vec{x}_n + \vec{d}_1\right] I_{n,y}\left[\vec{x}_n + \vec{d}_1\right] \\ \vdots \qquad\qquad \vdots \\ I_{n,x}\left[\vec{x}_n + \vec{d}_k\right] I_{n,y}\left[\vec{x}_n + \vec{d}_k\right] \end{bmatrix}. \tag{40}$$

Now, we can apply Eqns. (39) and (38) to Eqns. (28) and (29)

$$\mathbf{C}_{xI} = \sigma_{xx}^2 \nabla \vec{\mathrm{I}}_n^T \tag{41}$$

and

$$\mathbf{C}_{II} = \sigma_{xx}^2 \nabla \vec{\mathrm{I}}_n \nabla \vec{\mathrm{I}}_n^T + \sigma_{\eta\eta}^2 \mathbb{1} \tag{42}$$

We invert $\mathbf{C}_{II}$ by applying the Woodbury identity [Press *et al.* (1992)] to Eqn. (42), and from Eqns. (37), (41) and (42)

$$\vec{x}_{n+1} = \vec{x}_n + (\frac{\sigma_{\eta\eta}^2}{\sigma_{xx}^2}\mathbb{1} + \mathrm{M}')^{-1}\nabla \vec{\mathrm{I}}_n^T \left(\vec{\mathrm{I}}_n - \vec{\mathrm{I}}_{n+1}\right) \tag{43}$$

where

$$\mathrm{M}' = \nabla \vec{\mathrm{I}}_n^T \nabla \vec{\mathrm{I}}_n = \sum_{i=1}^{N} \begin{bmatrix} I_{x,i}^2 & I_{x,i}I_{y,i} \\ I_{x,i}I_{y,i} & I_{y,i}^2 \end{bmatrix} \tag{44}$$

which is the matrix used for the solution proposed by Lucas and Kanade. If we set $\frac{\sigma_{xx}^2}{\sigma_{\eta\eta}^2} = \varepsilon$, then $\mathrm{M}' + \varepsilon\mathbb{1} = \mathrm{M}$ and we can rewrite Eqn. (43) as

$$\vec{x}_{n+1} = \vec{x}_n + \mathrm{M}^{-1}\nabla \vec{\mathrm{I}}_n^T \left(\vec{\mathrm{I}}_n - \vec{\mathrm{I}}_{n+1}\right), \tag{45}$$

and after rearranging (45), we get

$$\vec{x}_{n+1} - \vec{x}_n = \mathrm{M}^{-1}\nabla \vec{\mathrm{I}}_n^T \left(\vec{\mathrm{I}}_n - \vec{\mathrm{I}}_{n+1}\right). \tag{46}$$

By applying the nomenclature used by Lucas and Kanade, $\vec{x}_{n+1} - \vec{x}_n = \vec{\mathbf{u}}_n$ is the motion vector associated with frame n and $\vec{\mathrm{I}}_{n+1} - \vec{\mathrm{I}}_n = I_t$ the time derivative of the image. We now make the observation that

$$-\nabla \vec{\mathrm{I}}_n^T \cdot I_t = -\sum_{i=1}^{N} \begin{bmatrix} I_{x,i}I_{t,i} \\ I_{y,i}I_{t,i} \end{bmatrix} = \vec{b} \tag{47}$$

We can now rewrite Eqn. (46)

$$\vec{\mathbf{u}}_n = M^{-1}\vec{b},$$ (48)

which has the same form as Eqn. (20). An interesting observation to make is that when deriving the Lucas and Kanade formulation by explicitly considering the mathematical implications of their assumptions, the use of a stabilization constant is supported by theory and not by practice only.

Now that we have rederived the solution proposed by Lucas and Kanade using our probabilistic framework, we can relax the Lucas and Kanade assumptions that were incorporated into our framework. In the following sections a more general noise model will be developed, taking into account random fluctuations of light intensity and affine deformation.

2.3. *Modeling probabilities using a mixture of Gaussians*

In order to relax the implicit Gaussian assumption, we use a Gaussian Mixture Model (GMM) to attain a better approximation of any arbitrary joint pdf of image appearance and position. The pdf from (24) can thus be reformulated as:

$$p(\vec{x}_{n+1}, \vec{I}_n \mid I_{n+1}, \vec{x}_a, \gamma) = \sum_{i=1}^{N} \pi_j G(\vec{x}, \vec{I}; {}^j\vec{\mu}_{\vec{x},\vec{I}}, {}^j\mathbf{C}_{\vec{x},\vec{I}})$$ (49)

where π_i is the mixture prior, ${}^j\vec{\mu}_{\vec{x},\vec{I}}$ is the mean and ${}^j\mathbf{C}_{\vec{x},\vec{I}}$ the covariance matrix of the jth component. There are two distinct problems that we must solve: fitting and maximization. The method that we use to fit a joint pdf for position and image appearance is described later in this section. In the previous section we have shown how we could find the maximum likelihood value for $\vec{x}$ given $\vec{I}$ simply by maximizing the log-likelihood, when using the assumptions employed in the Lucas and Kanade formulation. However, when representing the pdf with a GMM, this cannot be directly done as we cannot derive a simple closed form solution for $\vec{x}$. Instead, we first determine the mixture component k that is the most likely cause of our data $\vec{I}$ by minimizing the Mahalanobis distance as determined in Eqn. (25) with parameters

$$^k\mu = \left[\begin{array}{c} ^k\vec{\mu}_I \\ \hline ^k\vec{\mu}_x, \end{array}\right]$$ (50)

$$^k\mathbf{C}_{\vec{x},\vec{I}} = \left[\begin{array}{c|c} ^k\mathbf{C}_{II} & ^k\mathbf{C}_{Ix}^T \\ \hline ^k\mathbf{C}_{xI} & ^k\mathbf{C}_{xx} \end{array}\right]$$ (51)

and $\vec{x}$ is then calculated as in (37):

$$\vec{x} = {}^k\vec{\mu}_x + {}^k\mathbf{C}_{xI}\,{}^k\mathbf{C}_{II}^{-1}\left(\vec{I} - {}^k\vec{\mu}_I\right).$$ (52)

Ideally, the larger the number of Gaussian components used, the better the approximation of the pdf. However, there are a few practical considerations to keep in mind. A large number of components will require a larger number of samples, reducing efficiency. In our tests, we have found that there is no advantage of using more than eight mixture components.

3. The Algorithm

3.1. *General description*

In the previous section we have outline our theoretical formulation for the feature-point correspondence problem. Our framework is general enough to accommodate a large variety of statistical assumptions. In this section we outline how our framework can form the basis to a more general algorithm that estimates a solution to the correspondence problem, when weaker assumptions are incorporated.

If we have a good guess at the joint pdf of feature-point appearance and its position than the solution can be easily derived from the development presented in the previous section. However, often it is not readily available. In our method we use a Monte-Carlo technique, where observations are synthesized using a noise model and a GMM is fitted using the EM algorithm. This method is intended for points that are good for tracking like the ones obtained with the KLT feature detector [Shi and Tomasi (1994)].

3.2. *Noise model*

The noise model should reflect possible changes to image appearance between two frames. We use a similar noise model to the one utilized by Wong and Spetsakis [Wong and Spetsakis (2004)]. Besides inter-frame motion and independent, identically distributed camera noise that were considered in our previous example, we can expand our noise model to contain a wider range of changes and deformations such as random fluctuations of illumination and pixelwise jitter. We model the image appearance at a given time as a function of the appearance at the previous frame:

$$
\begin{aligned}
\vec{I}_{n+1}[\vec{x}_b] \;=\; & \vec{I}_n[\vec{x}_a + \vec{u}_n] + \vec{\eta}_n \\
& + Diag(\vec{I}_x[\vec{x}_a + \vec{u}_n])\vec{\varpi}_n \\
& + Diag(\vec{I}_y[\vec{x}_a + \vec{u}_n])\vec{\epsilon}_n \\
& + \vec{I}_n[\vec{x}_a + \vec{u}_n]\beta + \vec{1}\alpha
\end{aligned}
\tag{53}
$$

where $Diag(\vec{I}_x)$ is a diagonal matrix consisting of the elements of $\vec{I}_x$ and similarly for $\vec{I}_y$. The random variables are: $\vec{\eta}_n$, a random vector representing i.i.d camera noise, $\vec{\varpi}_n$ and $\vec{\epsilon}_n$, also i.i.d random vectors representing small horizontal and vertical pixelwise motion(jitter) within a neighborhood (like leaves fluttering in the wind), β, a random scalar that reflects multiplicative change of illumination with respect to the original image and α, a random scalar reflecting additive change of illumination as proposed by Negahdaripour [Negahdaripour (1998)] that it is normally sufficient to model brightness change as affine. The inter-frame motion of the feature is represented by the flow vector $\vec{u}_n$, which is the unknown random vector that reflects our expectations of feature motion between frames.

3.3. *Generative model for probability distributions*

Now that we have introduced a realistic uncertainty model and an efficient way to determine the most likely position of a feature point, we need a way to fit a pdf that reflects our model. We approximate the GMM using a Monte-Carlo approach: we first take random samples of neighborhoods that reflect changes of feature appearance due to both displacement and noise (as described by the noise model, introduced in the previous section) and then fitting a GMM using the EM clustering algorithm.

The random samples used by the Monte-Carlo method should reflect our assumptions regarding noise and displacement. For each feature that we want to track we take N random samples where the formula for the ith sample can be written as

$$
\begin{aligned}
P_i = \; & \vec{I}_n[\vec{x}_a + \vec{u}_n] + \vec{\eta}_n \\
& + Diag(\vec{I}_x[\vec{x}_a + \vec{u}_n])\vec{\varpi}_n \\
& + Diag(\vec{I}_y[\vec{x}_a + \vec{u}_n])\vec{\epsilon}_n \\
& + \vec{I}_n[\vec{x}_a + \vec{u}_n]\beta + \vec{1}\alpha
\end{aligned}
\tag{54}
$$

where P_i is a possible appearance of the neighborhood around the feature in the subsequent frame and $\vec{u}_n$, $\vec{\eta}_n$, $\vec{\varpi}_n$, $\vec{\epsilon}_n$, β, and α are instances of the random variables discussed above. The random variables presented during the discussion of our noise model are used to generate a random observation for each sample. Their distribution parameters can be modified according to a specific application or when certain knowledge about the scene is known or computed from empirical data [Wong and Spetsakis (2004)]. For example, in a stereo problem the random motion $\vec{u}$ can be distributed according to the epipolar constraint. In our tracking examples presented in Section 4.2, no prior information regarding the motion of vehicles was assumed, thus $\vec{v}$ was isotropically distributed about the initial feature position with $\sigma = 7\,pix$. The other random variables are generated in accordance of their corresponding parameters in the noise model. For example, ϖ and ϵ are i.i.d vectors generated according to a normal distribution with $\sigma = 0.25$ to represent small subpixel jitter. $\vec{\eta}$ represents camera noise and is thus an i.i.d normal vector with $\sigma = 2\,greylevels$.

The samples that we generate according to the noise model provide us with statistical behavior of feature appearance. After generating a sufficient number of sample points, we find the GMM that best fits the samples using the EM algorithm, where each pixel position relative to the center of every neighborhood sample is a separate dimension. In our tests we have used 5-by-5 neighborhoods, giving us 27 dimensions in total, 25 for appearance and 2 for position.

4. Results

We demonstrate the applicability of our framework for determining feature-point correspondence. In order to conduct a quantitative evaluation of our method, we used a standard and popular database of image sequences with ground-truth motion [Baker *et al.* (2007)] and compared our results to those attained using a hierarchical implementation of the Horn and Schunk algorithm [Horn and Schunk (1981)] and to the Black and Anandan [Black and Anandan (1996)] method. In a second set of experiments we used a traffic video sequence captured using a regular digital camera, we demonstrate how our method can be utilized for tracking.

4.1. *Quantitative comparison*

In the first series of experiments we compared the performance of two well known and popular algorithms; hierarchical Horn and Schunk (HS) and Black and Anandan (BA) versus the algorithm presented in the previous section (TS). We run the experiment on a set of trackable features from the ground-truthed database, repeatedly, limiting our scope to points where motion is relatively small (below 5 pixels). In addition to the inter-frame variation available in images in the database, we introduced synthetic additive camera noise, as well as global change of illumination that ranges from 0% to 10%. In all the experiments we used the same noise model parameters for our algorithm. Obviously, the most important parameter for such an experiment is β (multiplicative change of illumination) and was assigned a zero mean and a standard deviation of 5%. Additionally, our method assumes isotropic motion with standard deviation of 3.5 around its initial position, as a prior. All the methods examined in this study had similar performance when change of illumination is small. However, when change of illumination is increased beyond 6% the other two methods diverged whereas the effect on our method was minor (see Fig. 2). The other two methods performed slightly better at low intensity changes, this is to be expected since our method assumes change of illumination with a standard deviation of 5%, whereas the actual change is almost zero.

4.2. *Qualitative evaluation using real-world data*

In order to provide a practical demonstration of our method, we have captured a short sequence of traffic video using a simple point-and-shoot digital camera. Using this sequence, we have have experimented with tracking various vehicles in the scene. Along with vehicle motion and global motion caused by vigorously shaking the camera , the sequence contains typical CCD noise as well as automatic adjustment of the shutter. One such experiment is shown in Fig. 3. We begin by selecting a feature point and we then iteratively compute its likely position in the subsequent frame. In the simple example provided in the figure the initial position of the track is given in Fig. 3(a) and its position after 10 frames is given in Fig. 3(b).

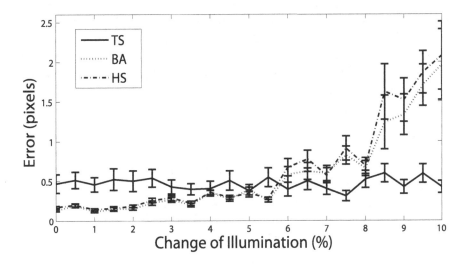

Fig. 2. A comparison of the three methods for feature-point correspondence, under varying degrees of global illumination change.

(a) A selected feature at frame 19   (b) The feature point as traced after 10 frames

Fig. 3. Tracking of a single feature point using real-world data. Our algorithm successfully keeps up with a moving vehicle. The tracked feature point has moved 38 pixels over the course of the sequence. Motion varies between 0.1 and 13 pixels in between frames. The tracker drifted about 5 pixels away from the original target, which is consistent with the results shown in Fig. 2.

Overall, we have ran our algorithm on 15 feature-points corresponding to various vehicles in the scene. On average, the tracked objects were displaced 29.7 pixels over a 10 frame sequence. For all but one of the features considered, we were able to track the vehicle throughout, with average loss of 0.35 pixels per frame, a figure which is consistent with the baseline error reported in Fig. 2. As we did not provide our algorithm with a motion prior, small motion will be assigned higher likelihood and thus causing a drift in the direction opposite to the dominant motion of the target. For the remaining feature for which are method failed to track consistently over the sequence, we believe it is due to the similarity of appearance between the tracked feature and a tree it occludes midway.

As the immediate neighborhood of a single feature point is insufficient for robust tracking of an object in space, we propose to construct a tracking algorithm that utilizes our method as the basis for estimating correspondence. In such a system, an object can be modeled as a cluster of feature points under an affine motion model. The object's state can be used to determine good motion priors for efficient correspondence estimation.

5. Conclusion

In this chapter, we have presented a general probabilistic formulation to the correspondence problem. Based on our probabilistic formulation, we developed an algorithm that uses the probabilistic framework to model uncertainty regarding changes in image appearance. We have further shown that when our probabilistic framework is used in conjunction with the assumptions inherent to the Lucas and Kanade algorithm, we can reduce our solution to theirs.

Methods developed around such strong assumptions can not be easily generalized to handle additional sources of noise without losing limiting their applicability. Our formulation is flexible in the sense that a great variety of noise models can be easily incorporated. Since a closed form of the joint probability density function of feature position and its appearance cannot be derived for all noise models, we approximate it using a generative approach. We use a Monte-Carlo technique to generate random samples of possible appearance of image data under our noise model and fit a GMM using the EM clustering algorithm.

Using synthetic image sequences that simulate relaxed assumptions, we have demonstrated that our method performs better than two of the most popular algorithms in the literature under relaxed statistical assumptions. Our comparative study included hierarchical implementations of Horn and Schunk algorithm and Black and Anandan robust flow. We have demonstrated the effectiveness of our algorithm for feature-point tracking using sequences of freeway traffic. Our algorithm has performed very well when tracking good features at the absence of occlusions.

References

Baker, S., Scharstein, D., Lewis, J. P., Roth, S., Black, M. J. and Szeliski, R. (2007). A database and evaluation methodology for optical flow, *In Proc. Eleventh IEEE International Conference on Computer Vision.*

Barron, J. L., Fleet, D. J. and Beauchemin, S. S. (1994). Performance of optical flow techniques, *International Journal of Computer Vision* **12**, 1, pp. 43–77.

Bergen, J. R., Anandan, P., Hanna, K. J. and Hingorani, R. (1992). Hierarchical model-based motion estimation, *Proceedings of the Second European Conference on Computer Vision*, pp. 237–252.

Black, M. J. and Anandan, P. (1993). Framework for the robust estimation of optical flow, *Fourth International Conference On Computer Vision*, pp. 231–236.

Black, M. J. and Anandan, P. (1996). The robust estimation of multiple motions: Parametric piecewise-smooth flow fields, *Computer Vision and Image Understanding* **63**, 1, pp. 75–104.

Black, M. J. and Fleet, D. J. (2000). Probabilistic detection and tracking of motion boundaries, *International Journal of Computer Vision* **38**, 3, pp. 231–245.

Cornelius, N. and Kanade, T. (1986). Adapting optical-flow to measure object motion in reflectance and x-ray image sequences, in *Proc. of the ACM SIGGRAPH/SIGART interdisciplinary workshop on Motion: representation and perception* (Elsevier North-Holland, Inc., New York, NY, USA), pp. 145–153.

Dempster, A. P., Laird, N. M. and Rubin, D. B. (1997). Maximum likelihood from incomplete data via the em algorithm, *Journal of the Royal Statistical Society Series B* **39**, 1, pp. 1–38.

Horn, B. K. and Schunk, B. G. (1981). Determining optical flow, *Artificial Intelligence* **17**.

Isard, M. and Blake, A. (1998). Condensationconditional density propagation for visual tracking, *International Journal of Computer Vision* **29**, pp. 5–28.

Jepson, A. and Black, M. J. (2004). Mixture models for optical flow computation, *Proceedings of IEEE Conference on Computer Vision and Pattern Recognition*, pp. 760–761.

Lucas, B. and Kanade, T. (1981). An iterative image registration technique with an application in stereo vision, *DARPA IU Workshop*, pp. 121–130.

McLachlan, G. J. and Krishnan, T. (2007). *The EM Algorithm and Extensions* (John Wiley & Sons Inc.).

Meer, P., Mintz, D., Rosenfeld, A. and Kim, D. Y. (1991). Robust regression methods for computer vision: A review, *International Journal of Computer Vision* **6**, pp. 59–70.

Negahdaripour, S. (1995). Revised representation of optical flow for dynamic scene analysis, Proc. ISCV, Coral Gables, Fla.

Negahdaripour, S. (1998). Revised definition of optical flow: Integration of radiometric and geometric cues for dynamic scene analysis, *IEEE Transactions on Pattern Analysis and Machine Intelligence* **20**, 9, pp. 961–979.

Press, W. H., Teukolsky, S. A., Vetterling, W. T. and Flannery, B. P. (1992). *Numerical Recipes in C: The Art of Scientific Computing*, 2nd edn. (Cambridge University Press).

Shi, J. and Tomasi, C. (1994). Good features to track, *Proceedings of IEEE Conference on Computer Vision and Pattern Recognition*, pp. 293–600.

Tal, R. and Spetsakis, M. E. (2010). Probabilistic framework for feature-point matching, in *Proceedings of the 2010 Canadian Conference on Computer and Robot Vision* (IEEE Computer Society), pp. 1–8.

Wills, J., Agarwal, S. and Belongie, S. (2003). What went where, *Proceedings of IEEE Conference on Computer Vision and Pattern Recognition*, pp. 37–44.

Wong, K. Y. and Spetsakis, M. E. (2004). Motion segmentation by em clustering of good features, *Conference on Computer Vision and Pattern Recognition Workshops*, pp. 166–166.

Part 2

Emerging Applications of Computer Vision

CHAPTER 2.1

FOURIER METHODS FOR 3D SURFACE MODELING AND ANALYSIS

Li Shen*, Sungeun Kim, Jing Wan, John D. West and Andrew J. Saykin

Department of Radiology and Imaging Sciences,
Indiana University School of Medicine,
950 W Walnut St, R2 E124, Indianapolis, IN 46202
{shenli,sk31,wanjing,jdwest,asaykin}@iupui.edu

Shape analysis is a fundamental topic in computer vision and graphics, and plays increasingly prominent roles in a wide range of practical applications in engineering, biology, physics, chemistry, geosciences, medicine, and entertainment. This chapter presents the most recent development of a principled and effective Fourier method, the spherical harmonic (SPHARM) description, for modeling and analyzing 3D surface data. SPHARM is a highly promising method for modeling arbitrarily shaped but simply connected 3D objects, where protrusions and intrusions can be effectively handled. A standard SPHARM processing pipeline includes three major steps: spherical parameterization, SPHARM expansion and SPHARM registration. The original and enhanced methods used in each step are presented, and the existing software packages are discussed. Several example applications are given to demonstrate the effectiveness of SPHARM as a promising method applicable to many areas such as computer graphics, medical imaging, bioinformatics, and other related geometric modeling and processing fields.

1. Introduction

Shape analysis is a fundamental topic in computer vision and graphics, and plays increasingly prominent roles in a wide range of practical applications in engineering, biology, physics, chemistry, geosciences, medicine, and entertainment. Capacity to acquire high throughput imaging and geometric data quickly and inexpensively has grown dramatically in recent years. These data offer new opportunities to study shape in numerous applications. Traditional geometric morphometrics[1-7] has been focused on studying landmarks, which are often very coarse representations of object geometry, and so cannot handle complex shape configurations. This chapter addresses the most recent development of principled and effective Fourier methods and tools for modeling and analyzing surface shape data.

---

*Corresponding author. Phone: +1 317 278 0498. E-mail: shenli@iupui.edu.

Fourier-based techniques were first studied to describe two-dimensional (2D) shapes. For example, a closed 2D contour could be modeled by a one-parameter function $r(\theta)$, that maps the distance from a specified origin to each point on the contour as a function of the angle in the polar coordinate system, and this radial function $r(\theta)$ could then be expressed in terms of a Fourier series (e.g., elliptical Fourier methods[8]). This approach could also be expanded to model more complex, non-star-shaped contours by employing two parametric functions $x(\theta)$ and $y(\theta)$.[9,10]

Spherical harmonics are the extension of Fourier techniques to three dimensions and are particularly well suited to modeling 3D surface data. Spherical harmonics were first used as a type of parametric surface representation for radial or stellar surfaces $r(\theta, \phi)$ by Ballard et al.[11,12] An extended method, called SPHARM, was proposed by Brechbühler et al.[13] to model more general shapes, where three functions of θ and ϕ were used to represent a surface. These spherical harmonic methods have recently received a lot attention, and have been studied and applied to applications in various fields including computer vision,[11,13] graphics,[14–18] medical image analysis,[19–22] bioinformatics,[23–26] and evolutionary biology.[10,27,28]

In this chapter, we focus on describing the SPHARM framework for 3D surface modeling and analysis. SPHARM is a powerful Fourier shape modeling method for processing arbitrarily shaped but simply connected 3D objects. As a highly promising method, SPHARM has been widely used in several domains including medical imaging. However, its primary use has been focused on modeling small or moderately-sized surfaces that are relatively smooth, due to challenges related to its applicability, robustness and scalability. This chapter presents not only the original SPHARM method but also several recent developments that address these issues and show that the use of SPHARM can expand into broader areas.

1.1. *SPHARM Description*

The SPHARM method was proposed by Brechbühler et al.[13] to model arbitrarily shaped but simply connected 3D objects. It is essentially a Fourier transform technique that defines a 3D surface using three spherical functions and transforms them into three sets of Fourier coefficients in the frequency domain. Three steps are often involved in a typical SPHARM processing pipeline: (1) spherical parameterization, (2) SPHARM expansion, and (3) SPHARM registration.

Spherical parameterization creates a continuous and uniform mapping from the object surface to the surface of a unit sphere, and its result is a bijective mapping between each point $\mathbf{v}$ on a surface and a pair of spherical coordinates θ and ϕ:

$$\mathbf{v}(\theta, \phi) = (x(\theta, \phi), y(\theta, \phi), z(\theta, \phi))^T.$$

Fig. 1(a) shows a hippocampal surface extracted from a magnetic resonance imaging (MRI) scan and Fig. 1(f) shows its spherical parameterization. This parameterization is an area preserving mapping computed using Brechbühler's method.[13]

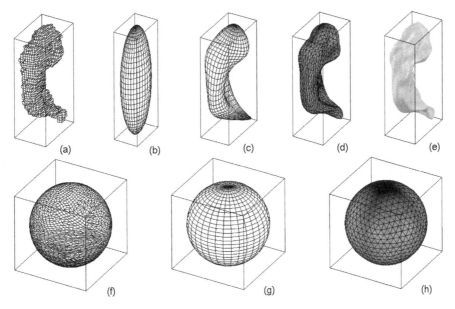

Fig. 1. Sample SPHARM reconstruction: (a) A hippocampal surface, (b-e) its SPHARM reconstructions using coefficients up to degrees 1, 5, 10 and 15, respectively, (f) its spherical parameterization, (g) a regular mesh grid on the sphere that is used for reconstructions shown in (b) and (c), (h) an icosahedral subdivision at level 3 that is used for reconstructions shown in (d) and (e).

SPHARM expansion expands the object surface into a complete set of spherical harmonic basis functions Y_l^m, where Y_l^m denotes the spherical harmonic of degree l and order m and it is essentially a Fourier basis function defined on the sphere. The expansion takes the form:

$$\mathbf{v}(\theta, \phi) = \sum_{l=0}^{\infty} \sum_{m=-l}^{l} \mathbf{c}_l^m Y_l^m(\theta, \phi), \tag{1}$$

where $\mathbf{c}_l^m = (c_{xl}^m, c_{yl}^m, c_{zl}^m)^T$. The Fourier coefficients $\mathbf{c}_l^m$ up to a user-desired degree can be estimated by solving a linear system. The object surface can be reconstructed using these coefficients, and using more coefficients leads to a more detailed reconstruction. A sample reconstruction case is shown in Fig. 1: The original model is shown in (a) and its SPHARM reconstructions using coefficients up to degrees 1, 5, 10 and 15 are shown in (b-e), respectively. Different spherical sampling schemes can be used to reconstruct the object. Reconstructions shown in (b) and (c) are created using a regular mesh grid on the sphere (g), and reconstructions shown in (d) and (e) are created using a level 3 icosahedral subdivision (h).

SPHARM registration creates a shape descriptor (i.e., excluding translation, rotation, and/or scaling) from a normalized set of SPHARM coefficients, which are comparable across objects. While parameterization and expansion steps are necessary for modeling an individual shape, this registration step is optional and mainly required for pair-wise or group analysis of 3D models.

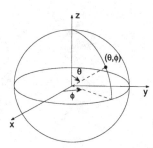

Fig. 2.   Notational convention for spherical coordinates: for the point (θ, ϕ) on the unit sphere, θ is taken as the polar (colatitudinal) coordinate, and ϕ as the azimuthal (longitudinal) coordinate.

This chapter is organized as follows. Sections 2-4 describes methods used in spherical parameterization, SPHARM expansion, and SPHARM registration, respectively. Section 5 discusses the existing SPHARM software tools and presents example applications to demonstrate the effectiveness of SPHARM. Section 6 concludes the chapter.

2. Spherical Parameterization

In order to model a surface using SPHARM, we first need to create a continuous and uniform mapping from the object surface to the surface of a unit sphere so that each vertex on the object surface can be assigned a pair of spherical coordinates (θ, ϕ). This process is called *spherical parameterization*. To match the definition of spherical harmonics,[29] the following convention for spherical coordinates (θ, ϕ) is used in this chapter: θ is taken as the polar (colatitudinal) coordinate with $\theta \in [0, \pi]$, and ϕ as the azimuthal (longitudinal) coordinate with $\phi \in [0, 2\pi)$ (Fig. 2). Thus, the north pole has $\theta = 0$ and the south pole has $\theta = \pi$.

A satisfactory mapping often requires a minimization of some types of distortions. Three typical distortions are length, angle and area distortions.[30] A mapping is isometric (length preserving) if and only if it is conformal (angle preserving)[31,32] and equiareal (area preserving).[13,33,34] Thus, an isometric mapping is an ideal mapping without any length, angle, or area distortion. However, isometric mappings only exist in very special cases, for example, the mapping of a cylinder onto a plane that transforms cylindrical coordinates into Cartesian coordinates. For SPHARM surface modeling, an equiareal mapping is more attractive than a conformal mapping because we want to treat each area unit on object surface equally by assigning the same amount of parameter space to it. However, an uncontrolled equiareal mapping often contains excessive angle or length distortions that are not desired. Therefore, an ideal spherical mapping for SPHARM should be equiareal with minimized or controlled angle and length distortions. Below, we briefly review two parameterization methods employed by two existing SPHARM software packages,[35,36] one designed for voxel surfaces and the other for triangular surfaces.

2.1. *Voxel Surfaces*

We first describe the spherical mapping approach proposed by Brechbühler *et al.* in the original SPHARM paper.[13] This approach aims to create an equal area mapping while minimizing angle distortions. It is applicable only to a voxel surface (e.g., the surface of a binary volume, see Fig. 1(a)), since it exploits the uniform quadrilateral structure of a square surface mesh. The approach has been implemented in SPHARM-PDM,[35] and consists of two steps: (1) initialization, and (2) optimization.

In the first step, an initial parameterization is constructed by creating a harmonic map from the object surface to the parameter surface. For colatitude θ, two poles are selected in the surface mesh by finding the two vertices with the maximum and minimum z coordinate in object space. Then, a Laplace equation (Eq. (2)) with Dirichlet conditions (Eq. (3) and Eq. (4)) is solved for colatitude θ:

$$\nabla^2\theta = 0 \qquad \text{(except at the poles)} \tag{2}$$

$$\theta_{north} = 0 \tag{3}$$

$$\theta_{south} = \pi \tag{4}$$

Since our case is discrete, we can approximate Eq. (2) by assuming that each vertex's colatitude (except at the poles') equals the average of its neighbours' colatitudes. Thus, after assigning $\theta_{north} = 0$ to the north pole and $\theta_{south} = \pi$ to the south pole, we can formulate and solve a system of linear equations for colatitude θ by considering all the vertices. For longitude ϕ, the same approach can be employed except that longitude is a cyclic parameter. To overcome this problem, a "date line" is introduced. When crossing the date line, longitude is incremented or decremented by 2π, depending on the crossing direction. After slightly modifying the linear system according to the date line, the solution for longitude ϕ can also be achieved.

In the second step, the initial parameterization is refined to obtain an area preserving mapping by solving a constrained optimization problem, where a few constraints were established for preserving area and mesh topology and an objective function is formed for minimizing angle distortion. An iterative procedure is used to perform two operations alternately: (1) satisfying the constraints using the Newton-Raphson method,[37] and (2) optimizing the objective function using a conjugate gradient method.[37] More details are available in Brechbühler *et al.*[13,38]

The parameterization result is a bijective mapping between each vertex $\mathbf{v} = (x, y, z)^T$ on a surface and a pair of spherical coordinates (θ, ϕ). We use $\mathbf{v}(\theta, \phi)$ to denote such a mapping, meaning that, according to the mapping, $\mathbf{v}$ is parameterized with the spherical coordinates (θ, ϕ). Taking into consideration the x, y, and z coordinates of $\mathbf{v}$ in object space, the mapping can be represented as follows:

$$\mathbf{v}(\theta, \phi) = (x(\theta, \phi), y(\theta, \phi), z(\theta, \phi))^T. \tag{5}$$

The key idea in the surface parameterization step is to achieve a homogeneous distribution of parameter space so that the surface correspondence across subjects can be obtained in a later step. Note that equal area parameterization implies such a surface correspondence that the corresponding parts of two surfaces by design occupy the same amount of surface area.

2.2. General Triangular Surfaces

The traditional method[13] described above aims to create an equal area mapping as well as minimize angle distortions. This is very effective in analyzing small and moderately-sized structures extracted from volumetric images (e.g., MRI, CT). However, its scalability is limited due to the expensive optimization procedure and, in addition, it is applicable only to voxel surfaces. CALD[17] is a newer method that extends the traditional method and can be applied to general triangular meshes. CALD has been implemented and available in the SPHARM-MAT package.[36]

We first define a few relevant concepts. Let $M = \{t_i\}$ be a triangle mesh and let Ψ be a continuous invertible map which maps M to another mesh $\Psi(M) = \{\Psi(t_i)\}$. $A(\cdot)$ is used to denote the area of a triangle or a mesh. The area distortion cost (ADC) C_a with respect to Ψ is defined as follows. For each triangle $t_i \in M$,

$$C_a(t_i, \Psi) = \frac{A(\Psi(t_i))}{A(t_i)}.$$

This measures the local ADC of a single triangle. For each mesh vertex v in M,

$$C_a(v, \Psi) = \frac{\sum_{t_i \in M_v} A(\Psi(t_i))}{\sum_{t_i \in M_v} A(t_i)},$$

where M_v is the set of triangle incident upon v. This measures the local ADC around a single vertex. For the whole parametric mesh M,

$$C_a(M, \Psi) = \frac{\sum_{t_i \in M} max(C_a(t_i, \Psi), \frac{1}{C_a(t_i, \Psi)}) A(\Psi(t_i))}{A(\Psi(M))}. \tag{6}$$

This measures the overall ADC for the whole mesh. By taking

$$max\left(C_a(t_i, \Psi), \frac{1}{C_a(t_i, \Psi)}\right)$$

as the ADC contribution from each triangle, we treat contraction and expansion equally, and so always have $C_a(M, \Psi) \geq 1$. The worst ADC is defined as follows

$$C_a^W(M, \Psi) = max\left\{max\left(C_a(t_i, \Psi), \frac{1}{C_a(t_i, \Psi)}\right) | t_i \in M\right\}. \tag{7}$$

Let M be the object surface mesh and $\Psi(M)$ be the current parameterization. For each vertex v in M, we use (θ_v, ϕ_v) to denote its parameterization on the unit sphere. The distribution of ADCs for parameterization $\Psi(M)$ can then be formulated as a spherical function $F_\Psi(\theta, \phi)$ such that for each v in M, $F_\Psi(\theta_v, \phi_v) = C_a(v, \Psi)$. We call $F_\Psi(\theta, \phi)$ the *area scaling ratio function* (ASRF in short) of Ψ. For an equal area mapping Ψ, its ASRF should be 1 for any value of (θ, ϕ).

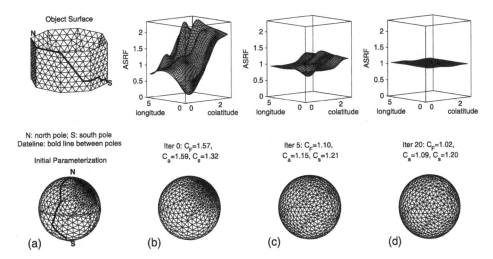

Fig. 3.   Sample smoothing procedure: (a) the top plot shows the object surface, and the bottom plot shows the initial parameterization; (b–d) the top plot shows the ASRF after iteration 0, 5, or 20, and the bottom plot shows the corresponding parameterization. C_F (average distortion of ASRF), C_a (overall ADC), and C_s (overall LDC) are also provided for each parameterization.

To measure length distortion, the stretch concept by Sander et al.[39] is adopted. They considered the case of mapping from a planar domain to 3D surface and at any point in the planar domain, two singular values of the 3×2 Jacobian matrix were computed to represent the largest and smallest length distortions when a vector in the 2D domain was mapped to the 3D surface. In our case, the length distortion cost (LDC) C_s with respect to a given mesh mapping Ψ from M to $\Psi(M)$, is defined as follows. Given a mesh mapping Ψ from M to $\Psi(M)$,

$$C_s(M, \Psi) = \sqrt{\frac{\sum_{t_i \in M} (\Gamma(t_i)^2 + \frac{1}{\gamma(t_i)^2}) A(\Psi(t_i))}{2A(\Psi(M))}}, \text{and} \tag{8}$$

$$C_s^W(M, \Psi) = max \left\{ max \left(\Gamma(t_i), \frac{1}{\gamma(t_i)} \right) | t_i \in M \right\}, \tag{9}$$

where $\Gamma(t_i)$ and $\gamma(t_i)$ are the largest and smallest length distortions for a triangle t_i. While $C_s(M, \Psi)$ measures the average length distortion cost (LDC) for the whole mesh M, $C_s^W(M, \Psi)$ defines the worst LDC. The largest and smallest length distortions are directly computed from the length of three corresponding sides between t_i and $\Psi(t_i)$. Again, contraction and expansion are equally treated in both definitions.

The CALD algorithm is designed to achieve the goal of minimizing the ADC $C_a(M, \Psi)$ while controlling the LDC $C_s(M, \Psi)$. It starts from an initial parameterization and performs local and global smoothing methods alternately until a solution is approached. Mesh smoothing[40,41] can be described as a procedure that relocates mesh vertices to improve the mesh quality without changing mesh topology. Note that smoothing operates only on the parameter mesh (e.g., bottom row in Fig. 3) and does not affect the object mesh (e.g., top panel in Fig. 3(a)) and its shape.

The CALD framework contains three key components. The *initial parameterization* step is an extension of Brechbühler's method[13] for triangle meshes. A *local smoothing* step and a *global smoothing* step are developed to improve the quality of the parameterization iteratively. The local smoothing step aims to minimize the ADC at a local submesh by solving a linear system and also to control its worst LDC at the same time. The global smoothing step calculates the distribution of ADCs (captured by ASRF) for all the mesh vertices and tries to equalize them over the whole sphere. The overall algorithm combines the local and global methods together, and performs each method alternately until a solution is achieved. Algorithmic details are available in Shen et al.[17] Fig. 3 shows a sample smoothing procedure, where the ASRF gets closer and closer to 1 over the iterative procedure.

Besides CALD, conformal mapping has also been used for spherical parameterization in some SPHARM studies.[16] It has solid mathematical foundation but tends to introduce large area distortion that may not be ideal for establishing surface correspondence between models in statistical shape analysis. Two recent studies[18,42] use progressive meshes and stretch metrics to minimize vector length distortion and their parameterization results look very promising for handling large scale graphics models. Since spherical parameterization has been extensively studied, people can often find a method that fits their needs.

3. SPHARM Expansion

For convenience, we focus our discussion on expanding a single spherical function:

$$f(\theta, \phi) = \sum_{l=0}^{\infty} \sum_{m=-l}^{l} a_l^m Y_l^m(\theta, \phi), \tag{10}$$

For SPHARM, we can apply the same method to expand $x(\theta, \phi)$, $y(\theta, \phi)$, and $z(\theta, \phi)$ separately. Given a function $f(\theta, \phi)$ and a user-specified maximum degree L_{max}, our task is to extract coefficients a_l^m in Eq. (10) for $l \leq L_{max}$ and $|m| \leq l$.

There are two types of approaches for computing a_l^m: one uses numerical integration;[24,43] the other formulates a linear system and solves it using least square fitting (LSF).[13] The naive numerical integration method is inefficient and so not applicable to large models. Healy et al.[43] proposed a fast algorithm to accelerate the integration procedure using a divide-and-conquer strategy; and several studies[14,16,18] employed this method. However, to use Healy's method, a preprocessing step is required to remesh the model using a regular spherical mesh grid (e.g., Fig. 1(g)). This step is inconvenient and also it tends to introduce additional remeshing errors. In contrast, the LSF method[13] is very easy to implement and works directly on original object meshes even if they are irregular. The LSF method seems to be a more popular method in SPHARM studies, especially in brain imaging. Below, we briefly describe the LSF approach.

The input data for spherical harmonic expansion contain a spherical function $f(\theta, \phi)$ and a user-specified maximum degree L_{max}. The spherical function is described by a set of spherical samples (θ_i, ϕ_i) and their function values $f_i = f(\theta_i, \phi_i)$, for $1 \leq i \leq n$. According to Eq. (10), we can formulate a linear system as follows.

$$
\begin{pmatrix} y_{1,1} & y_{1,2} & y_{1,3} & \cdots & y_{1,k} \\ y_{2,1} & y_{2,2} & y_{2,3} & \cdots & y_{2,k} \\ \vdots & \vdots & \vdots & & \vdots \\ y_{n,1} & y_{n,2} & y_{n,3} & \cdots & y_{n,k} \end{pmatrix} \begin{pmatrix} x_1 \\ x_2 \\ x_3 \\ \vdots \\ x_k \end{pmatrix} = \begin{pmatrix} f_1 \\ f_2 \\ \vdots \\ f_n \end{pmatrix}
$$

where $y_{i,j} = Y_l^m(\theta_i, \phi_i)$, $j = l^2 + l + m + 1$, and $k = (L_{max} + 1)^2$. Here an indexing scheme is used to assign a unique index $j = l^2 + l + m + 1$ to every pair (l, m). Least square fitting is used to solve the above system for $(x_1, x_2, ..., x_k)^T$, since $n \neq k$ in almost all the cases. Because each $x_j \equiv \hat{a}_l^m$ is an estimate of the original coefficient a_l^m for $j = l^2 + l + m + 1$, we can reconstruct the original function as follows:

$$
\hat{f}(\theta, \phi) = \sum_{l=0}^{L_{max}} \sum_{m=-l}^{l} \hat{a}_l^m Y_l^m(\theta, \phi) \approx f(\theta, \phi). \tag{11}
$$

The more degrees one uses, the more accurate the reconstruction $\hat{f}(\theta, \phi)$ is.

The bottleneck of the LSF method is its limited scalability, because most methods for solving large linear systems[44,45] are either designed for sparse or symmetric matrices or not easy to implement. This might be the reason that SPHARM has been mostly used for modeling small or moderately sized 3D surfaces. Recently an iterative residual fitting (IRF) method[19,46] has been proposed to overcome this limitation and facilitates the opportunity of creating large scale SPHARM models.

The basic idea behind the IRF method is simple and follows the properties of spherical harmonic transform. *First*, these harmonics form a coarse-to-fine hierarchy. If we just use a few low degree harmonics to expand a spherical function $f(\theta, \phi)$, we get a low-pass filtered reconstruction. If we use more degrees, more details are included in the reconstruction. The IRF method takes advantage of this coarse-to-fine hierarchy. It starts from a low degree reconstruction and then iteratively adds more details into our model by involving higher degree harmonics. *Second*, spherical harmonics form an orthonormal basis and geometric information is stored in different frequency channels. Thus, if we first extract information from low frequency channels, the residual (*i.e.*, $f(\theta, \phi)-$ its reconstruction) will exactly contain information in high frequency channels. To add in more details to our model, we can simply use a few higher degree harmonics to fit the residual. Algorithmic details about IRF are available in Shen and Chung.[19,46]

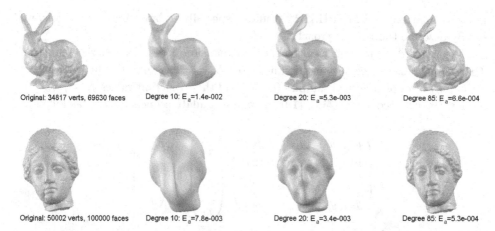

Fig. 4. Original models and their SPHARM reconstructions of degrees 10, 20 and 85. Reconstruction error E_a is the root square measure distance between original and reconstructed models.

The IRF method breaks a large linear system into several small linear systems and thus SPHARM modeling can be easily done at a large scale on standard workstations with average configuration using a standard linear solver. Note that a degree l SPHARM model involves $(l+1)^2 \times 3$ complex coefficients. The experiments in Shen et al.[47] show that a degree 85 SPHARM model described by 22,188 complex coefficients can reasonably capture surface details of an original model with 100,002 vertices and 200,000 faces that is described by 900,006 variables in total. Shown in Fig. 4 are a few SPHARM models of large scale 3D surfaces.

4. SPHARM Registration

Shape registration finds correspondences among shapes and is the basis for performing shape comparison and analysis. The traditional method[13] uses the first order ellipsoid (FOE) for SPHARM registration, and works only if this ellipsoid is a real ellipsoid and may not work well in many other cases. SHREC[48] is a newer SPHARM registration method that minimizes the mean square distance between corresponding surface parts and works for general cases. Both FOE and SHREC are designed for relatively smooth surfaces without landmarks. However, landmarks often contain important prior knowledge about the objects. For example, they tend to have critical biological or anatomical meanings in biological or biomedical applications and should not be ignored in the registration procedure. A novel STPS-based (spherical thin plate spline based) SPHARM alignment method[47] has recently been proposed and it can register SPHARM models together under pre-existing landmark constraints as well as control parametric area and length distortions. We discuss these three methods in this section.

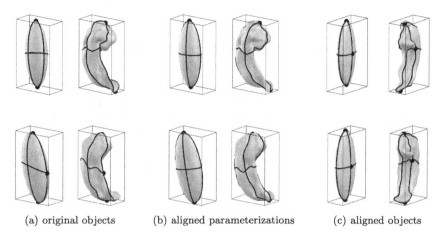

| (a) original objects | (b) aligned parameterizations | (c) aligned objects |

Fig. 5. SPHARM registration using first order ellipsoids (FOEs). Each row shows one sample hippocampus. Each of (a-c) shows the FOE on left and degree 15 reconstruction on right. Parameterization is indicated by the lines on the surface, including equator $\theta = \pi/2$ and four longitudinal lines $\phi = -\pi/2, 0, \pi/2, \pi$. The north and south poles and the point $(\pi/2, 0)$ are shown as dots.

4.1. FOE Alignment

SPHARM registration is a common operation in shape analysis. It creates a normalized set of SPHARM coefficients, which are comparable across objects, to form a shape descriptor (*i.e.*, excluding translation, rotation, and scaling). *Scaling invariance* can be achieved by adjusting the coefficients so that the object volume is normalized. Ignoring the degree 0 coefficient results in *translation invariance*. By design, the degree one reconstruction is an ellipsoid for any SPHARM model (Fig. 1(b)). We call it the *first order ellipsoid* (FOE). *Rotation invariance* can be achieved by aligning the FOE.[13]

Fig. 5 demonstrates the registration of SPHARM models by aligning the FOEs. Each row shows the processing of one sample hippocampus. Each of (a-c) shows the FOE on left and degree 15 reconstruction on right. In (a), the original pose and parameterization are shown. Note that the correspondence between two SPHARM models is implied by the underlying parameterization: two points with the same parameter pair (θ, ϕ) on two surfaces are defined to be a corresponding pair. Thus, in (b), the FOE is used to align the parameterization in the parameter space and establish the surface correspondence: although the object pose stays the same, the parameter net on each FOE is rotated to a canonical position such that the north pole is at one end of the longest main axis, and the crossing point of the zero meridian and the equator is at one end of the shortest main axis. In (c), the FOE is used to adjust the object pose in the object space: the FOE is rotated to make its main axes coincide with the coordinate axes, putting the shortest axis along x and longest along z. Now we can see that these two hippocampi are aligned to a canonical position in both parameter space and object space. Algorithmic details about this method are available in Brechbühler et al.[13,38]

4.2. *SHREC Alignment*

Clearly, the FOE method registers SPHARM objects in both parameter space and object space. However, it works only if the FOE is a real ellipsoid but not an ellipsoid of revolution or a sphere.[48] In the latter case, we cannot find a unique set of main axes to align things together. To overcome this limitation, we can register two models together by minimizing their root mean squared distance (RMSD). The RMSD can be calculated directly from SPHARM coefficients. Let S_1 and S_2 be two SPHARM surfaces, where their SPHARM coefficients are formed by $\mathbf{c}_{1,l}^m$ and $\mathbf{c}_{2,l}^m$, respectively, for $0 \leq l \leq L_{max}$ and $-l \leq m \leq l$. The RMSD between S_1 and S_2 can be calculated as follows:[21]

$$\text{RMSD} = \sqrt{\frac{1}{4\pi} \sum_{l=0}^{L_{max}} \sum_{m=-l}^{l} \|\mathbf{c}_{1,l}^m - \mathbf{c}_{2,l}^m\|^2}. \tag{12}$$

Inspired by the iterative closest point (ICP) strategy,[49] the key idea of the SHREC algorithm (*SP HARM REgistration with ICP*)[48] is to first establish an initial alignment and then perform the following two steps alternately until RMSD is minimized: (1) optimizing surface correspondence, and (2) minimizing the distance between the corresponding surface parts. We assume that the initial alignment makes two models roughly aligned in both object space and parameter space. Thus, to optimize the surface correspondence, we can rotate the parameterization of one SPHARM model to best match the other's, since the underlying parameterization defines the correspondence between different SPHARM surfaces. A naive solution for rotating the parameterization of a SPHARM model is to recalculate the SPHARM coefficients using the rotated parameterization. However, this requires to solve three linear systems and is time-consuming. To accelerate the process, we use a rotational property in the harmonic theory and rotate SPHARM coefficients without recalculating the SPHARM expansion.

Let $\mathbf{v}(\theta, \phi) = \sum_{l=0}^{\infty} \sum_{m=-l}^{l} \mathbf{c}_l^m Y_l^m(\theta, \phi)$ be a SPHARM parametric surface. After rotating the parameter net on the surface in Euler angles $(\alpha\beta\gamma)$, the new coefficients $c_l^m(\alpha\beta\gamma)$ can be calculated as follows [24,50,51]

$$c_l^m(\alpha\beta\gamma) = \sum_{n=-l}^{l} D_{mn}^l(\alpha\beta\gamma) \, c_l^n \tag{13}$$

where

$$D_{mn}^l(\alpha\beta\gamma) = e^{-i\gamma n} d_{mn}^l(\beta) e^{-i\alpha m},$$

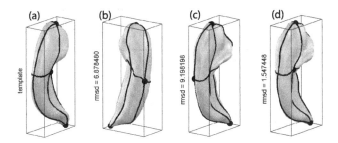

Fig. 6. Sample result of SHREC: (a) template, (b) object, (c) after ICP, (d) after SHREC.

$$d_{mn}^l(\beta) = \sum_{t=max(0,\,n-m)}^{min(l+n,\,l-m)} (-1)^t$$
$$\times \frac{\sqrt{(l+n)!(l-n)!(l+m)!(l-m)!}}{(l+n-t)!(l-m-t)!(t+m-n)!t!}$$
$$\times \left(cos\frac{\beta}{2}\right)^{(2l+n-m-2t)} \left(sin\frac{\beta}{2}\right)^{(2t+m-n)}.$$

We employ a sampling-based strategy that fixes one parameterization and rotates the other to optimize the surface correspondence by minimizing the RMSD defined in Eq. (12). The rotation space can be sampled nearly uniformly using icosahedron subdivisions. This assigns rotation angles to β and γ. Let n be the number of icosahedral samples. The expansion coefficients are then rotated through $(0\beta\gamma)$ and then by $\sqrt{n}$ equal steps in α using Eq. (13), evaluating the RMSD at each orientation. The result is the best orientation that minimizes the RMSD. A hierarchical sampling approach can also be used to achieve accurate registration efficiently.[48]

The SHREC strategy starts from an initial alignment. One approach for getting the initial alignment is as follows: (1) create surface samples for SPHARM models, (2) use ICP to align the surface samples between models, and (3) use the parameterization rotation method described above to align the models in the parameter space. For any object whose FOE is a real ellipsoid, we can also use the FOE method to get the initial alignment. Thus, in the initial registration, SPHARM objects are roughly aligned in both object space and parameter space.

To optimize the initial alignment, we can alternately improve the alignment in the object space and the alignment in the parameter space. The alignment in the parameter space can be improved by the parameterization rotation method described earlier. To improve the alignment in the object space, given an roughly established surface correspondence, we can simply create corresponding surface samples between SPHARM models and use a quaternion-based algorithm to align two corresponding point sets together in a least squares sense.[49] For convenience, we call this quaternion-based algorithm as CPS (*i.e.*, aligning corresponding point sets). Algorithm 1 summarizes this iterative procedure and presents our SHREC method.

Algorithm 1 SHREC: SPHARM registration with ICP

1: Set up initial alignment
2: **repeat**
3:    Run CPS to align models in the object space
4:    Align models in the parameter space
5: **until** No more changes in the parameter space

Figure 6 shows a sample run of Algorithm 1 on registering two hippocampi. In (a), the template is shown. In (b), the original orientation and parameterization of the individual is shown. In (c), the result after running ICP is shown. We can see that, in the object space, the individual is roughly aligned to the template; however, the lines show that their parameter nets are not aligned at all. In (d), it shows the result after registering the parameter net, and now we can see that the individual and the template are aligned pretty well in both object space and parameter space. The RMSD between the individual and the template is shown in each of (b-d).

4.3. *Landmark-guided Alignment*

Both FOE and SHREC are designed for relatively smooth surfaces without landmarks. However, landmarks often contain important prior knowledge about the objects. They tend to have critical biological or anatomical meanings in biological or biomedical applications and should not be ignored in the registration procedure. Here we present a STPS-based (spherical thin plate spline based) SPHARM alignment method[47] that can register SPHARM models together under pre-existing landmark constraints as well as control parametric area and length distortions.

Spherical thin plate spline (STPS)[52] is an extension of 2D thin plate spline to a spherical domain and has been used to deform spherical parametric domains[53,54] (Fig. 7). STPS on a spherical domain S^2 minimizes a bending energy $J(u)$, subject to $u(P_i) = z_i, i = 1, 2, \cdots, n$, where $P_i \in S^2$ and z_i is the fixed displacement at P_i. This bending energy is formulated as

$$J(u) = \int_0^{2\pi} \int_0^{\pi} (\Delta u(\theta, \phi))^2 \sin \phi d\theta d\phi, \tag{14}$$

where $\theta \in [0, \pi]$ is latitude, $\phi \in [0, 2\pi]$ is longitude, and Δ is the Laplace-Beltrami operator. The solution on the sphere is given by

$$u_n(P) = \sum_{i=1}^{n} c_i K(P, P_i) + d \tag{15}$$

where $\mathbf{c}$ and d are determined by

$$\mathbf{K}_n \mathbf{c} + d\mathbf{T} = \mathbf{z}, \mathbf{T}'\mathbf{c} = 0,$$

(a) Template 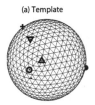 (b) Individual 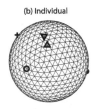 (c) STPS–aligned Individual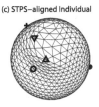

Fig. 7. STPS result for registering an individual (b) to the template (a): the landmarks on the STPS-aligned individual (c) are perfectly aligned with those on the template (a).

$\mathbf{K}_n$ is the $n \times n$ matrix with i, jth entry $K(P_i, P_j)$, $\mathbf{T} = (1, ..., 1)'$, and $\mathbf{z} = (z_1, ..., z_n)'$. $K(X, Y)$ between two arbitrary points $X, Y \in S^2$ is defined as follows:

$$K(X, Y) = \frac{1}{4\pi} \int_0^1 (\log h)(1 - \frac{1}{h})(\frac{1}{\sqrt{1 - 2ha + h^2}} - 1)dh$$

where $a = \cos(\gamma(X, Y))$ and $\gamma(X, Y)$ is the angle between X and Y. However, this is not in a computable closed form expression. Instead, thin plate pseudo-spline on the sphere $R(X, Y)$, defined in Wahba,[52] is used for landmark-guided registration.

For convenience, we assume that objects are already aligned to one another in the object space by some alignment methods such as ICP[49] and their SPHARM descriptions are known (pre-calculated). STPS is employed to match the landmark positions of an individual object in the spherical parametric domain to the corresponding landmark positions of the template object and subsequently transform the underlying parametric mesh of the individual object (e.g., Fig. 7). This method distorts the parametric mesh of the individual object and can introduce additional errors to some extent to the reconstructed shape of the individual object. Therefore, mesh distortion cost functions are employed and these distortion costs are used as selection criteria for finding best rotation angles in the proposed algorithm.

Distortion of the parametric mesh by STPS is measured by calculating the area distortion cost (ADC) and the length distortion cost (LDC). The overall and worst costs for the whole parametric mesh are measured to evaluate the performance of the STPS-based algorithms. The average and worst ADCs as well as the average and worst LDCs are measured by the corresponding cost functions, defined by Eqs. 6, 7, 8, and 9, respectively. These measures can be used to calculate the minimum overall distortion cost in a variety of ways.[47]

A basic alignment method (B-STPS) directly applys the STPS algorithm to two sets of landmarks with known correspondence between the sets, defined in a parametric mesh. However, this naive approach can severely distort the parametric mesh of an individual object and result in a distorted reconstruction without re-sampling the parametric mesh, especially when one or more landmarks are located near the north or south pole.[47]

To avoid large distortion introduced by STPS, a sampling-based strategy can be employed that rotates the individual's landmarks on the sphere using Euler angles $(\alpha\beta\gamma)$ in order to find the best oriented landmarks for applying STPS. The rotation

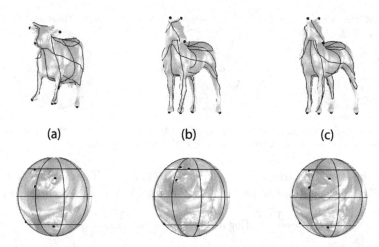

(a)   (b)   (c)

Fig. 8.   Sample results of STPS registration. The first column shows a template model. The second and the third columns exhibit an individual model before and after STPS-registration respectively. A quadrilateral mesh and a bump map on sphere indicate the correspondence between each object and its parameterization. Landmarks are shown as black dots on the surface.

space can be sampled nearly uniformly using icosahedral subdivisions (see Fig. 1(h) for an icosahedral subdivision mesh at level 3). This assigns rotation angles to α and β. However, it is noticed that the rotation along the z axis does not affect the distortion level of an STPS result. Therefore, γ is always set to zero.

To reduce computation time, a hierarchical sampling scheme proposed in Shen et al.[48] can be adopted to sample the rotation space instead of using all the icosahedral samples (H-STPS): For each sampling point, the hierarchical approach moves it to the north pole, and the entire parametric mesh is rotated accordingly. Then, STPS is applied to each rotated parametric mesh and the best K rotation angles are selected for minimizing the over all distortion costs. This process is repeated for a higher level of sampling mesh until a certain criterion is satisfied.

While running H-STPS, it is observed that template's landmarks, if located near the poles, could also result in large distortion of the individual's parametric mesh. This observation suggests that both template's and individual's landmarks should be adjusted to be as far away from the poles as possible to avoid large STPS distortions. A simple approach is to rotate not only individual's but also template's parametric mesh before B-STPS or H-STPS is applied to the objects. After the application of the B-STPS or H-STPS method, the distorted mesh is rotated accordingly so that the individual result matches the original template instead of a rotated template. In general, H-STPS and rotational B-STPS outperform B-STPS, and rotational H-STPS performs the best in terms of minimizing mesh distortion cost. Further details are available in Shen et al.[47]

Fig. 8 show the rotational H-STPS result of aligning a horse model to a cow model: the top row shows the objects and the bottom row shows their spherical parameterization. Bump maps are used to visualize the correspondence between the object and its parameterization. Landmarks and a coarse mesh grid are also shown on each surface. The first column shows the template model (i.e., cow). The second and the third columns exhibit the individual model (i.e., horse) before and after SPHARM registration respectively. Comparing the original individual with the template, you will notice that their landmarks are aligned only in the object space but not in the spherical parameter space. However, comparing the registered individual with the template, you can see that their landmarks are aligned not only in the object space but also in the parameter space.

5. SPHARM Tools and Applications

5.1. *Software Tools*

SPHARM-PDM[35,55] and SPHARM-MAT[36] are two open-source SPHARM software tools, and are both available at www.nitrc.org. SPHARM-PDM implements the traditional method proposed in the original SPHARM paper[13] and is applicable only to voxel surfaces. It employs the FOE alignment[13] and Procrustes alignment[56] for surface registration. The point-based models computed by SPHARM-PDM can be combined with a statistical tool shapeAnalysisMANCOVA[57] to perform quantitative morphological analysis. SPHARM-PDM is implemented in C/C++.

SPHARM-MAT (SPHARM Modeling and Analysis Toolkit) is designed as a synergistic effort to SPHARM-PDM. SPHARM-MAT has a user friendly interface to call SPHARM-PDM components. In addition, SPHARM-MAT implements the CALD method for spherical parameterization and is applicable to general triangular surfaces. SPHARM-MAT also implements SHREC for efficient SPHARM alignment in more general cases. SPHARM-MAT is a Matlab-based tool with user friendly interface, providing an alternative for users to choose the platform they favor and giving opportunity for tool comparison and cross-validation.

For statistical shape analyses, a Matlab-based surface analysis tool SurfStat[58,59] is highly suggested to be used together with SPHARM-MAT. SurfStat can perform statistical analysis of univariate and multivariate surface and volumetric data using linear mixed effects models and random field theory.

5.2. *Applications*

We first demonstrate a SPHARM application in *imaging genetics*, an emerging trans-disciplinary area where genetic effects are studied using imaging as phenotypes. As an underdeveloped area, combining shape analysis of brain's substructure with genetic analysis has strong potential to identify a neurodegeneration biomarker for many diseases, such as Alzheimer's disease (AD) and mild cognitive impairment (MCI). For example, in Wan et al.,[60] genetic effects of candidate single nucleotide polymorphism (SNPs) were examined on hippocampal volume and shape using an

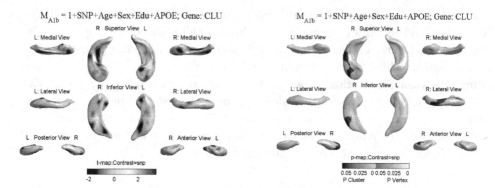

Fig. 9. Genetic effect of rs9314349-CLU (39 AA versus 58 AG and 16 GG) on hippocampal surface. Left panel shows the map of t statistics, and right panel shows the map of corrected P values (only $p < 0.05$ shown). All the significant regional changes are associated with positive t values, indicating local surface expansion in AG/GG group compared with AA group.

existing MCI cohort.[61] In the study, each pair of hippocampal surfaces were modeled using SPHARM coefficients up to degree 15 and registered to an atlas (average of hippocampal surfaces for all healthy controls) using SHREC method. Surface signals were then extracted as the deformation along the surface normal direction of the atlas. General linear model was performed using the SurfStat tool[58,59] to detect association between hippocampal surface signals and candidate SNPs in three different genetic models (additive, dominant, recessive). Age, gender, education and intracranial volume (ICV) were used as covariates in all analyses, with and without controlling for APOE E4 status, respectively. In the analyses, six out of fourteen SNPs showed significant effects (corrected $p \leq 0.05$) of regional shape changes. Shape analysis confirmed the same dominant or recessive model discovered by volume analysis, and provided detailed surface mapping of localized genetic effects (Fig. 9). This study demonstrated that genetic analysis of quantitative shape features as phenotypic data had considerable potential for examining disease mechanisms from a novel perspective.

After seeing an application in imaging genetics, let us review a few other possible applications of the SPHARM method. The most intuitive application is to describe 3D models. In particular, the IRF method (see Section 3) enables large-scale SPHARM expansion that can reasonably capture fine surface details for complicated 3D surfaces (e.g., Fig. 4) and allows for many useful applications. SPHARM description of large-scale 3D models has a few advantages over the traditional description of triangulated surfaces. First, it is a more compact representation than a triangulated surface in many cases. This compactness property can be used for geometric compression.

Second, one can operate not only in the spatial domain but also in the frequency domain. Taking a lower order SPHARM reconstruction can naturally achieve the goal of surface smoothing and filtering. See Fig. 4 for a few samples. This property

Fig. 10. Morphing using STPS-registered SPHARM models (bottom row) exhibits more intuitive and natural morphing sequences than that using the original SPHARM models (top row).

has been used in a couple of prior studies[14,18] for surface smoothing and filtering. It can also be used for level of details representation and transferring.

Third, the level of details applications can be done not only via the frequency domain but also via the spatial domain. A SPHARM reconstruction is essentially a remeshed original model. We can use different spherical sampling schemes with different sampling resolutions for SPHARM reconstruction.[47] Objects can be reconstructed using regular spherical mesh grids (Fig. 1(g)) and icosahedral subdivisions (Fig. 1(h)) at different sampling resolutions.

In addition, SPHARM descriptor of large-scale 3D models with the improved alignment by landmark-guided registration method can subsequently contribute to the feasibility of performing pair-wise processing or group analysis of these 3D models. For example, such a SPHARM registered result (e.g., Fig. 8) can help improve a morphing sequence between two 3D models. Shown in Fig. 10 are two morphing sequences: (1) the top row is between a template (cow in Fig. 8(a)) and an original individual (horse in Fig. 8(b)), and (2) the bottom row is between the same template and an STPS-registered individual (horse in Fig. 8(c)). Morphing the template to the registered individual achieves a much better effect than to the original individual, since the intermediate shape of morphing to the registered individual looks more natural than those of morphing to the original individual. Besides morphing, there are many other SPHARM applications requiring models being registered with landmark guidance, such as surface-based morphometry in biomedical imaging[19–22,25] and morphological analysis in evolutionary biology.[10,27,28]

6. Conclusions

SPHARM is a highly promising Fourier method for modeling arbitrarily shaped but simply connected 3D objects, where protrusions and intrusions can be effectively handled. Thanks to its underlying spherical parameterization, SPHARM is suitable for many surface manipulation and analysis applications, including texture mapping, morphing, remeshing, compression, statistical modeling, and surface-based morphometry, because of the following reasons: (1) processing is much easier on the spherical domain than on an irregular mesh; (2) the spherical domain is continuous, allowing for more flexible processing on it than on a regular mesh; and (3) processing can be done not only in the spatial domain but also in the frequency domain. This chapter have discussed traditional and enhanced methods used for SPHARM surface modeling and analysis as well as relevant software tools and example applications. These methods and tools greatly enable the potential of applying the SPHARM method to a broad range of areas such as computer graphics, medical imaging, CAD/CAM, bioinformatics, and other related geometric modeling and processing fields.

Acknowledgement

This work was supported in part by NIBIB R03 EB008674/EB008674S1, CTSI CBR/CTR (based on NIH UL1RR025761), NIA R01 AG19771, NIA 1RC-2AG036535, NIA P30 AG10133, NCI R01 CA101318, NIH U54 EB005149, NIAAA U01 AA017123, NIAAA U01 AA014809, Foundation for the NIH, and Indiana Economic Development Corporation (IEDC #87884). Graphics models were downloaded from Hugues Hoppe's website[42] at Microsoft Research.

References

1. F. L. Bookstein, *Morphometric tools for landmark data : geometry and biology.* (Cambridge University Press, Cambridge [England] ; New York, 1991).
2. F. J. Rohlf and L. Marcus, A revolution in morphometrics, *Trends in Ecology and Evolution.* **8**, 128–132, (1993).
3. L. F. Marcus, M. Corti, et al., *Advances in morphometrics.* (Plenum, New York, 1996).
4. C. G. Small, *The statistical theory of shape.* (Springer, New York, 1996).
5. I. L. Dryden and K. V. Mardia, *Statistical shape analysis.* (John Wiley and Sons, New York, 1998).
6. C. Klingenberg, M. Barluenga, et al., Shape analysis of symmetric structures: quantifying variation among individuals and asymmetry, *Evolution.* **56**, 1909–20, (2002).
7. C. Klingenberg and L. Monteiro, Distances and directions in multidimensional shape spaces: implications for morphometric applications, *Sys. Biology.* **54**, 678–88, (2005).
8. F. J. Rohlf and J. W. Archie, A comparison of fourier methods for the description of wing shape in mosquitoes (diptera: Culicidae), *Sys. Zoology.* **33**(3), 302–317, (1984).
9. F. P. Kuhl and C. R. Giardina, Elliptic fourier features of a closed contour, *Computer Graphics and Image Processing.* **18**, 236–258, (1982).
10. L. Shen, H. Farid, and M. A. McPeek, Modeling three-dimensional morphological structures using spherical harmonics, *Evolution.* **63**(4), 1003–16, (2009).
11. D. H. Ballard and C. M. Brown, *Computer Vision.* (Prentice-Hall, N.J., 1982).
12. R. Schudy and D. Ballard. Towards an anatomical model of heart motion as seen in 4-D cardiac ultrasound data. In *Com. App. in Rad. & Ana. of Rad. Im.*, (1979).
13. C. Brechbühler, G. Gerig, and O. Kubler, Parametrization of closed surfaces for 3D shape description, *Computer Vision and Image Und.* **61**(2), 154–170, (1995).
14. T. Bulow, Spherical diffusion for 3D surface smoothing, *IEEE Trans. on Pattern Analysis and Machine Intelligence.* **26**(12), 1650–1654, (2004).

15. T. Funkhouser, P. Min, et al., A search engine for 3d models, *ACM Transactions on Graphics.* **22**(1), 83–105, (2003).
16. X. Gu, Y. Wang, et al., Genus zero surface conformal mapping and its application to brain surface mapping, *IEEE Trans. on Med. Im.* **23**(8), 949–958, (2004).
17. L. Shen and F. Makedon, Spherical mapping for processing of 3d closed surfaces, *Image and Vision Computing.* **24**(7), 743–761, (2006).
18. K. Zhou, H. Bao, and J. Shi, 3D surface filtering using spherical harmonics, *CAD.* **36** (4), 363–375, (2004).
19. M. K. Chung, K. M. Dalton, L. Shen, et al., Weighted fourier series representation and its application to quantifying the amount of gray matter, *IEEE Trans. on Med. Im.* **26**(4), 566–81, (2007).
20. G. Gerig and M. Styner. Shape versus size: Improved understanding of the morphology of brain structures. In *MICCAI'2001, LNCS 2208*, pp. 24–32, (2001).
21. G. Gerig, M. Styner, et al. Shape analysis of brain ventricles using SPHARM. In *IEEE MMBIA*, pp. 171–178, (2001).
22. L. Shen, J. Ford, et al., A surface-based approach for classification of 3D neuroanatomic structures, *Intelligent Data Analysis.* **8**(6), 519–542, (2004).
23. W. Cai, X. Shao, and B. Maigret, Protein-ligand recognition using spherical harmonic molecular surfaces: towards a fast and efficient filter for large virtual throughput screening, *J Mol Graph Model.* **20**(4), 313–328, (2002).
24. D. Ritchie and G. Kemp, Fast computation, rotation, and comparison of low resolution spherical harmonic molecular surfaces, *J. Comp. Chem.* **20**, 383–395, (1999)
25. L. Shen, A. J. Saykin, et al. Morphometric analysis of hippocampal shape in mild cognitive impairment: An imaging genetics study. In *IEEE 7th Int. Symp. on Bioinformatics and Bioengineering*, pp. 211–217, Boston, MA, (2007).
26. R. J. Morris, R. J. Najmanovich, A. Kahraman, and J. M. Thornton, Real spherical harmonic expansion coefficients as 3d shape descriptors for protein binding pocket and ligand comparisons, *Bioinformatics.* **21**(10), 2347–2355, (2005).
27. M. McPeek, L. Shen, et al., The tempo and mode of 3-dimensional morphological evolution in male reproductive structures, *American Naturalist.* **171**(5), E158–78, (2008).
28. M. McPeek, L. Shen, et al., The correlated evolution of 3-dimensional reproductive structures between male and female damselflies, *Evolution.* **63**(1), 73–83, (2009).
29. E. W. Weisstein. "Spherical Harmonic" from mathworld–a wolfram web resource: http://mathworld.wolfram.com/sphericalharmonic.html.
30. M. S. Floater and K. Hormann. Surface parameterization: a tutorial and survey. In *Multiresolution in Geometric Modelling.* Springer, (2004).
31. X. Gu and S. T. Yau, Computing conformal structures of surfaces, *Communications in Information and Systems.* **2**, 121–146, (2002).
32. S. Haker, S. Angenent, et al. Conformal surface parameterization for texture mapping, *IEEE Trans. on Vis. and Comp. Graph.* **6**(2), 181–189, (2000). ISSN 1077-2626.
33. M. Quicken, C. Brechbühler, et al. Parameterization of closed surfaces for parametric surface description. In *Comp. Vis. and Pat. Recog.*, vol. 1, pp. 354–360, (2000).
34. A. Sheffer, C. Gotsman, and N. Dyn, Robust spherical parametrization of triangular meshes, *Computing.* **72**(1–2), 185–193, (2004).
35. SPHARM-PDM: http://www.nitrc.org/projects/spharm-pdm/.
36. SPHARM-MAT: http://imaging.indyrad.iupui.edu/projects/spharm/.
37. W. H. Press, *Numerical Recipes in C : the Art of Scientific Computing.* (Cambridge University Press, Cambridge [Cambridgeshire]; New York, 1992), 2nd edition.
38. C. Brechbühler. *Description and Analysis of 3-D Shapes by Parametrization of Closed Surfaces.* PhD thesis, IKT/BIWI, ETH, Zurich, (1995).

39. P. V. Sander, J. Snyder, S. J. Gortler, and H. Hoppe. Texture mapping progressive mesh. In *SIGGRAPH 2001, Computer Graphics Proceedings*, pp. 409–416, (2001).
40. S. Canann, J. Tristano, and M. Staten. An approach to combined laplacian and optimization-based smoothing for triangular, quadrilateral, and quad-dominant meshes. In *Proc. of the 7-th International Meshing Roundtable*, pp. 479–494, (1998).
41. L. Freitag, On combining laplacian and optimization-based mesh smoothing techniques, *AMD Trends in Unstructured Mesh Generation, ASME.* **220**, 37–43, (1997).
42. E. Praun and H. Hoppe, Spherical parametrization and remeshing, *ACM Transactions on Graphics.* **22**(3), 340–349 (July, 2003). ISSN 0730-0301.
43. D. Healy, D. Rockmore, et al., FFTs for the 2-sphere - improvements and variations, *The Journal of Fourier Analysis and Applications.* **9**(4), 341–385, (2003).
44. R. Barrett, M. Berry, et al., *Templates for the Solution of Linear Systems: Building Blocks for Iterative Methods, 2nd Edition.* (SIAM, Philadelphia, PA, 1994).
45. R. Freund, G. Golub, and N. Nachtigal. Iterative solution of linear systems. In ed. A. Iserles, *Acta Numerica 1992*, pp. 57–100. Cambridge University Press, (1992).
46. L. Shen and M. K. Chung. Large-scale modeling of parametric surfaces using spherical harmonics. In *IEEE 3DPVT 2006*, UNC, Chapel Hill, (2006).
47. L. Shen, S. Kim, and A. J. Saykin, Fourier method for large scale surface modeling and registration, *Comput Graph.* **33**(3), 299–311, (2009).
48. L. Shen, H. Huang, et al. Efficient registration of 3D SPHARM surfaces. In *4th Canadian Conf. on Computer and Robot Vision*, pp. 81–88, Montreal, QC., (2007).
49. P. J. Besl and N. D. McKay, A method for registration of 3-D shapes, *IEEE Trans. on PAMI.* **14**(2), 239–256, (1992).
50. G. Burel and H. Hennocq, Determination of the orientation of 3D objects using spherical harmonics, *Graphical Models and Image Processing.* **57**(5), 400–408, (1995).
51. H. Huang, L. Shen, et al., Surface alignment of 3D spherical harmonic models: application to cardiac MRI analysis, *MICCAI 2005.* **8**(Pt 1), 67–74, (2005).
52. G. Wahba, Spline interpolation and smoothing on the sphere, *SIAM Journal of Scientific and Statistical Computing.* **2**, 5–16, (1981).
53. G. Zou, J. Hua, and M. Dong. Integrative information visualization of multimodality neuroimaging data. In *Pac. Conf. on Comp. Graph. and Appl.*, pp. 473–476, (2007).
54. G. Zou, J. Hua, and O. Muzik. Non-rigid surface registration using spherical thin-plate splines. In *MICCAI 2007*, pp. 367–374, (2007).
55. M. Styner, I. Oguz, S. Xu, C. Brechbhler, D. Pantazis, and G. Gerig. Framework for the statistical shape analysis of brain structures using SPHARM-PDM. In *In Insight Journal, Special Edition on the Open Science Workshop at MICCAI*, (2006).
56. F. L. Bookstein, Shape and the information in medical images: A decade of the morphometric synthesis, *Comp. Vision and Image Under.* **66**(2), 97–118 (05, 1997).
57. B. Paniagua, M. Styner, M. Macenko, et al., Local shape analysis using mancova, *The Insight Journal* (07. 2009).
58. K. J. Worsley. Surfstat: http://www.math.mcgill.ca/keith/surfstat.
59. K. J. Worsley, M. Andermann, M. Koulis, et al., Detecting changes in non-isotropic images, *Human Brain Mapping.* **8**, 98–101, (1999).
60. J. Wan, S. Kim, L. Shen, et al. Association analysis of candidate snps on hippocampal volume and shape in mild cognitive impairment and older adults with cognitive complaints. In *ICAD2010:International Conference on Alzheimer's Disease*, (2010).
61. A. J. Saykin, H. A. Wishart, et al., Older adults with cognitive complaints show brain atrophy similar to that of amnestic mci, *Neurology.* **67**(5), 834–42, (2006).

CHAPTER 2.2

NEAR INFRARED FACE RECOGNITION AND EYEGLASSES

Stan Z. Li and Dong Yi

Center for Biometrics Security Research &
National Laboratory of Pattern Recognition,
Institute of Automation, Chinese Academy of Sciences,
95 Zhongguancun Donglu, Beijing, 100190, P. R. China
szli, dyi@cbsr.ia.ac.cn

Near infrared (NIR) face recognition has emerged as a successful technology for overcoming illumination changes in face recognition. With years of development, the NIR face recognition technology been in practical use with success, and products have appeared in the market. However, there is room to improve. Dealing with eyeglasses is one direction. Particular to the NIR technique is the specular reflections caused by active NIR illumination.

In this chapter, we introduce NIR face recognition as an emerging technology. We then present a method for dealing with problems caused by eyeglasses, that based on a sparse feature representation learned in a space of local features. Advantages are demonstrated by experiments with a large NIR face database. The results, in comparison with other occlusion robust or synthesis based methods, are analyzed.

1. Introduction

Face recognition performance is subject to factors including changes in illumination, pose and expression [Zhao *et al.* (2003)]. Among these extrinsic factors, the influence of illumination is the most common and important one in practice, especially in cooperative-user applications. Much effort has been made to model and correct illumination changes on faces in visible light (VIS) images.

Near infrared (NIR) face recognition [Li *et al.* (2007)] has emerged as an effective technology for overcoming illumination changes in face recognition. Active frontal lights is used to illuminate the face, such that the direction of lighting is controlled, well solving the long-outstanding problem caused by illumination changes. Using active lights in the NIR spectrum, the method becomes non-intrusive to human eyes. A local feature representation of local binary pattern (LBP) [Ojala *et al.* (2002); Ahonen *et al.* (2004); Hadid *et al.* (2004)] extracted from the active NIR image is shown to be independent of illumination changes [Li *et al.* (2007)]. Thereby, a high performance face recognition solution is built. With years of development, NIR face recognition been in practical use with success and products have gone into the market.

However, there is still room to improve. One problem is caused by eyeglasses. One obvious issue is the occlusion by eyeglass frame. This is common in both NIR and conventional visual face recognition, when the target and probe face images differ by eyeglasses. Particular to the NIR face technology is strong specular reflections on eyeglasses caused by the active frontal NIR illumination while this is not so much a problem in the case of ambient lighting in conventional visual face recognition. Also a kind of occlusion, this may lead to inaccurate eye localization and noisy interference in extracted local features of the face, deteriorating the performance. NIR face recognition performance can be improved if this problem can be solved properly.

In this chapter, we introduce NIR face recognition as an emerging technology (Section 2). We then present a method [Yi and Li (2011)] for the eyeglass problem (Section 3), using a sparse representation learned in a local feature space of LBP. Experiments on a large NIR face database are reported, and advantages and the results are analyzed in comparison with other occlusion robust or synthesis based methods (Section 4).

2. NIR Face Recognition

Before discuss the eyeglass problem in NIR face recognition, we give some reviews about the NIR face recognition system proposed in [Li *et al.* (2007)]. The system consists of NIR imaging hardware [AuthenMetric Co. Ltd. (2003, 2004, 2005)] and NIR based face recognition algorithms [Li and and His Face Team (2005); Li *et al.* (2006a,b)]. A thorough overview is given in the following.

2.1. *NIR Face Imaging*

The goal of making NIR special-purpose hardware is to overcome the problem arising from uncontrolled environmental lights so as to produce face images of a good illumination condition for face recognition. By "a good illumination condition", we mean that the lighting is from the frontal direction and the image has suitable pixel intensities, *i.e.* having good contrast and not saturated.

We propose two strategies to control the light direction: (1) mount active lights on the camera to provide frontal lighting, (2) minimize environmental lighting. We set two requirements on the active lighting: (1) the lights should be strong enough to produce a clear frontal-lighted face image without causing disturbance to human eyes, and (2) the minimization of the environmental lighting should have minimum reduction of the intended active lighting.

Radiation spectrum ranges are shown in Figure 1. While far (thermal) infrared imaging reflects heat radiation, NIR imaging is more like normal visible light imaging though NIR is invisible by naked eyes. Ultraviolet radiation is harmful to the human body and cannot be used for face recognition applications.

Our solution for requirement (1) is to choose the active lights in the near infrared (NIR) spectrum between 780-1100nm, and mounted them on the camera. We use NIR light-emitting diodes (LEDs) as active radiation sources, which are strong enough for indoor use and are power-effective. A convenient wavelength is 850nm. Such NIR lights are almost invisible to human eyes, yet most CCD and CMOS sensors have sufficient response at this spectrum point.

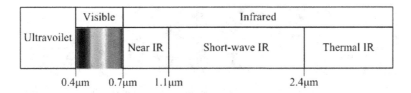

Fig. 1. Radiation spectrum ranges.

When mounted on the camera, the LEDs are approximately co-axial to the camera direction, and thus provide the best possible straight frontal lighting, better than mounted anywhere else; moreover, when the LEDs and camera are together, control of the lights can be more easily using a circuit in the box. The geometric layout of the LEDs on the camera panel may be carefully designed such that the illumination on the face is as homogeneous as possible.

Fig. 2. Active NIR imaging system and its geometric relationship with the face.

The strength of the total LED lighting should be such that it results in the NIR face images with good S/N ratio when the camera-face distance is between 50-100 cm, a convenient range for the user. A guideline is that it should be as strong as possible, at least stronger than expected environmental illumination, yet does not cause sensor saturation. A concern is the safety to human eyes. When the sensor working in the normal mode is not saturated, the safety is guaranteed.

Our solution for requirement (2) above is to use a long pass optical filter to cut off visible light while allowing NIR light to pass. We choose a filter such that ray passing rates are 0%, 50%, 88%, and 99% at the wavelength points of 720, 800, 850, and 880nm, respectively. The filter cuts off visible environmental lights (< 700nm) while allowing most of the 850nm NIR light to pass.

Figure 2 illustrates a design of the hardware device and its relationship with the face. The device consists of 18 NIR LEDs, an NIR camera, a color camera, and the box. The NIR LEDs and camera are for NIR face image acquisition. The color camera capture color face images could be used for fusion with the NIR images or other purposes. The hardware and the face are relatively positioned in a way that the lighting is frontal and NIR rays provide nearly homogenous illumination on face. The imaging hardware works at a rate of 30 frames per second with the USB 2.0 protocol for 640x480 images, and costs less than 20 US dollars.

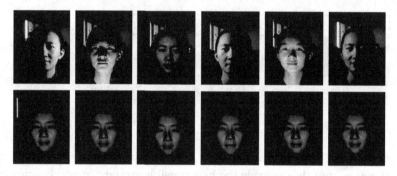

Fig. 3.   Color images taken by a color camera vs. NIR images taken by the present NIR imaging system. While unfavorable lightings are obvious in the color face images, they are almost unseen in the NIR face images.

Figure 3 shows example images of a face illuminated by NIR LED lights from the front, a lamp aside and environmental lights. We can see the following: (1) the lighting conditions are likely to cause problems for face recognition with the color images; (2) the NIR images, with the visible light composition cut off by the filter, are mostly frontal-lighted by the NIR lights, with minimum influence from the side lighting, and provide a good basis for face recognition.

2.2. *Illumination Invariant Face Features*

According to the Lambertian model, an image $I(x, y)$ under a point light source is formed according to the following

$$I(x, y) = \rho(x, y)\mathbf{n}(x, y)\mathbf{s} \tag{1}$$

where $\rho(x, y)$ is the albedo of the facial surface material at point (x, y), $\mathbf{n} = (n_x, n_y, n_z)$ is the surface normal, and $\mathbf{s} = (s_x, s_y, s_z)$ is the lighting direction (with magnitude). Here, albedo $\rho(x, y)$ reflects the photometric properties of facial skin and hairs, $\mathbf{n}(x, y)$ is the geometric shape of the face.

The top-most factor that affects the face recognition performance is the direction of the incidence lighting relative to the face surface normal. Assume $\mathbf{s} = \kappa\mathbf{s}^0$, where κ is the strength of the lighting and $\mathbf{s}^0 = (s_x^0, s_y^0, s_z^0)$ is a unit column vector of the lighting direction. A less restrictive modeling of constant κ would be a monotonic transform instead of a constant. When the active NIR lighting is from the (nearly) frontal direction (*cf.* Figure 2), *i.e.* $\mathbf{s}^0 = (0, 0, 1)$, the image can be approximated by

$$I(x, y) = \kappa\rho(x, y)n_z(x, y) \tag{2}$$

where κ is a multiplying constant due to possible changes in the strength of the lighting caused by changes in the distance between the face and the LED lights, and $n_z(x, y)$ is exactly the depth information that can be acquired by a range imaging system. An active NIR image $I(x, y)$ combines information about both depth map $n_z(x, y)$ and albedo map $\rho(x, y)$, and therefore provides the wanted intrinsic property about a face for face recognition.

LBP representation is usually used to compensate for the degree of freedom in κ or in a monotonic transform, to achieve an illumination invariant representation of faces for indoor face recognition applications. The basic form of the LBP operator is illustrated in Figure 4. The binary bits describing a local 3x3 subwindow are generated by thresholding the 8 pixels in the surrounding locations by the gray value of its center; the feature vector is formed by concatenating the thresholded binary bits anti-clockwise. There are a total of 256 possible values and hence 256 LBP patterns denoted by such an LBP code; each value represents a type of LBP local pattern. Such a basic form of LBP can be extended to multi-scale LBP $\mathrm{LBP}_{(P,R)}$ where R is the radius of the circle surrounding the center, and P is the number of pixels on the circle. An $\mathrm{LBP}(P, R)$ string is called uniform, denoted by $\mathrm{LBP}_{(P,R)}^{u2}$, if the neighboring bits (the circular sense) contains at most 2 bitwise transitions from 0 to 1 or vice versa (see [Ojala *et al.* (2002)] for details).

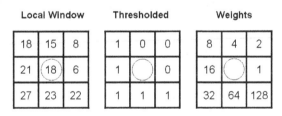

LBP String = (0001111)

LBP Code = 0+0+0+8+16+32+64+128=248

Fig. 4.   LBP code for 3x3 window.

As discussed in the above, the pixel intensities in an NIR image are subject to a multiplying constant κ due to changes in the distance between the face and the LED lights. This degree of freedom can be fixed by using an LBP based representation. To be less restrictive and more realistic, let us relax the effect of the multiplying constant to a *monotonic transform*, $\mathcal{T}$. Then, active NIR images can be modeled by

$$I(x, y) = \mathcal{T}(\rho(x, y)n_z(x, y)) \tag{3}$$

Let us be given an image $I'(x,y) = \rho(x,y)n_z(x,y)$ and a transformed image $I''(x,y) = \mathcal{T}(I'(x,y)) = \mathcal{T}(\rho(x,y)n_z(x,y))$. The ordering relationship between pixels in an image is not changed by any monotonic transform, namely, if $I'(x_1,y_1) > I'(x_2,y_2)$, then $I''(x_1,y_1) > I''(x_2,y_2)$. Therefore, the LBP codes generated from I'' are exactly the same as the ones generated from I'. From the analysis, we see that the NIR imaging and LBP features together lead to an illumination invariant representation of faces.

2.3. *NIR Face Classifiers*

While the initial LBP histogram feature set is of high cardinality, *e.g.* 748,592 for a 120×142 face image, the intrinsic dimension of the face pattern may not be so high. Some dimension reduction or feature selection methods are usually used to reduce the feature space, in an optimal sense, such as PCA, LDA, AdaBoost and etc. In this section, we present a AdaBoost+LDA method, for building face recognition classifiers. Given a training set of LBP features of faces subject to image noise, slight pose changes and alignment errors, such a learning method performs a transform to find most discriminative features and reduce their dimension further by LDA. We assume and ensure that a large set of training examples is available to sufficiently represent differences between individual faces, such that once trained, the classifier is able to recognize faces without need to be re-trained when a new individual client is added.

While AdaBoost [Freund and Schapire (1997)] procedure essentially learns a two-class classifier, we convert the multi-class problem into a two-class one using the idea of intra- and extra-class difference [Moghaddam *et al.* (1996)]. However, here the difference data are derived from the LBP Histogram features rather than from the images. A difference is taken between two LBP histogram feature sets, which is intra-class if the two face images are of the same person, or extra-class if not.

A training set of N labeled examples is given for two classes, $\mathbf{S} = (x_1,y_1),\ldots,(x_N,y_N)$, where x_i is a training example (which is the difference between two LBP histogram feature sets in this case) and $y_i \in \{+1,-1\}$ is the class label. The procedure learns a sequence of T weak classifiers $h_t(x) \in \{-1,+1\}$ and linearly combines it in an optimal way into a stronger classifier

$$H(x) = \mathrm{sign}\left(\sum_{t=1}^{T} \alpha_t h_t(x)\right) \qquad (4)$$

where $\alpha_t \in \mathbb{R}$ are the combining weights. We can consider the real-valued number $\sum_{t=1}^{T} \alpha_t h_t(x)$ as the score, and make a decision by comparing the score with a threshold. AdaBoost learning procedure is aimed to derive α_t and $h_t(x)$ so that an upper error bound is minimized [Freund and Schapire (1997)].

For LBP, a weak classifier is based on a single scalar, *i.e.* an LBP histogram bin value; a weak classification decision, represented by +1 or -1, is made comparing the scalar feature with an appropriate threshold, the threshold being computed to minimize the weighted error on the training set. In practice, we find about 3000 features are enough for face recognition task (in section 4 we use 3072 features). An AdaBoost strong classifier composed by 3072 weak classifiers can satisfy most of applications. But, the speed of AdaBoost classifier can not meet the requirements in large scale face recognition systems, *e.g.* 1,000,000 subjects in gallery.

To speed up the face matching, LDA are used on the selected LBP features to reduce the dimension further, *e.g.* from 3072 to 256. Finally, the similarity of lower dimensional features could be measure by normalized correlation or "cosine distance". By AdaBoost+LDA classifier, we can complete 1,000,000 matching per second on a P4 Duo Core 2.4GHz personal computer. Meanwhile, after processed by the AdaBoost and LDA procedures, the final features with 256 dimensions are very discriminative and robust to noise, nearly 99% recognition rate as reported in section 3.

3. Dealing with Eyeglasses

Eyeglass are very common in our daily life. In face enrollment and recognition, many people may wear, change or remove their eyeglass, so the status of eyeglass in gallery and probe face images are uncertain. Figure 5 shows a simple example to illustrate the influence of eyeglass problem, that the difference between intra (diagonal) and inter (off-diagonal) class similarity scores of boosted LBP [Li *et al.* (2007)] features are reduced by the eyeglass and specular highlight.

As stated in section 1, the eyeglass problem comes from two aspects: eyeglass frame and specular highlight. The first problem can be seen as a partial occlusion around the eyes, which usually reduce the recognition performance. The second problem of strong specular reflections may result in inaccurate eye localization, alignment and recognition. [Li *et al.* (2007)] has study the influence of specular highlight on eye localization, and proposes a simple-to-complex boosting detector to deal with this problem obtaining satisfactory localization precision. But the influence on recognition rate is less analyzed in that work. Compared with the first problem, specular highlight is usually of larger size and higher magnitude (*e.g.* 255), which can be seen as a more challenging case of occlusion.

3.1. *Related Work*

The eyeglass problem is closely related to occlusion, whether caused by eyeglass frame or specular highlight. As noticed in the early stage of face recognition [Turk and Pentland (1991)], occlusion could degrade recognition performance shapely, and this has not been well solved today. Much effort has been made to achieve occlusion robust face recognition.

In 2000, Leonardis *et al.* [Leonardis and Bischof (2000)] showed the impact of occlusion on object recognition and proposed a sub-sampling based PCA to

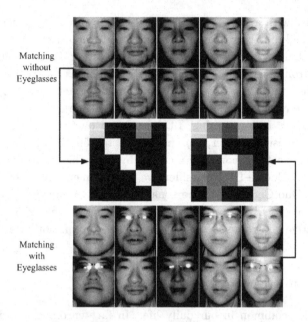

Matching
without
Eyeglasses

Matching
with
Eyeglasses

Fig. 5. Analysis the effects of eyeglass and specular highlight on similarity scores. Top: Face matching without eyeglass. Bottom: Face matching disturbed by eyeglass, images in each column belonging to a person. Middle: Matching score matrix.

deal with it. Martinez [Martinez (2002)] proposed a parted based method to solve occlusion problem in face recognition, in which each face are divided into many regions, analyzed in isolation and fused in a probabilistic way. In the last ten years, many works followed the direction of sub-sampling and part based method , such as [Fidler *et al.* (2006); Kim *et al.* (2005)]. These works are shown to improve recognition performance, but they are not quite practical due to their computational complexity or needing human intervention (*e.g.* choosing the layout of partition).

Sub-sampling and part based methods may be good to resist occlusions. We can divide each face images into several regions, *e.g.* four regions as shown in Figure 7. During recognition, the eye region can be discarded or fused with a weight according to the size of occlusion. Eye region is often one of the most important component in face recognition, *e.g.* nearly making 1/3 contribution in the example shown in Figure 7, so whether dropping it or choosing its weight is an obscure problem in real-world systems.

Image synthesis techniques have also been used to overcome the eyeglass occlusion problem in face recognition. In [Saito *et al.* (1999)] a group of face images without eyeglass was used to construct a PCA subspace. By projecting a new sample to the non-eyeglass face subspace, eyeglass in the face image could be removed with good quality. In 2005, Du *et al.* [Du and Su (2005)] improved the PCA method by using an iterated PCA, and achieve better results. Wu *et al.* [Wu *et al.* (2004)] proposed an automatic system based on Bayesian inference to remove eyeglass in face images consisting of three parts: eyeglass detection, eyeglass localization and eyeglass removal. In the FERET [Phillips *et al.* (1998)] and a private database, their method obtained nearly perfect results, but they didn't do any experiments about face recognition. In order to prevent artifacts in the synthesized face images, this kinds of methods need precise key points localization, and are easy to fail while existing specular highlight on eyeglass. In [Nayar *et al.* (1997)], the authors separate the diffuse and specular reflection components from a single image by using color and polarization information, but it need additional hardware and cannot apply to monochrome NIR image.

Synthesis based methods can obtain face image of good quality in the human visual sense, as shown in Figure 6, but don't take any additional information. As show in Figure 6, face images after eyeglass removal are usually lose some details in the eye region. Through our great efforts, we find synthesis based methods could not improve recognition performance.

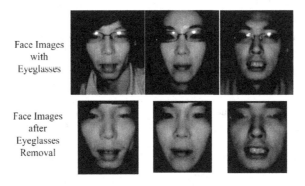

Fig. 6. eyeglass and highlight removal in face images. Top row shows face images with eyeglass and specular highlight of three different persons. Bottom row is their corresponding results after removing eyeglass and highlight. The eye region of the processed images become smoother than before.

Recently, sparse representation (SR) provides a solution for occlusion robust face recognition [Wright *et al.* (2009)]. A face image as probe could be sparsely represented by a linear combination of all face images in gallery plus a sparse error (*e.g.* partial occlusion). Through a L_1 minimization process [Chen *et al.* (1998)] , the sparse coefficient and occlusion error can be solved simultaneously. Owing to its sparse error model, SR can work well under large occlusion in theory and experiments. But as noted in [Wagner *et al.* (2009)], precise alignment is a key point to SR. Mis-alignment only in serval pixels may result in false non-sparse coefficient, and leading to false reject. Although the authors proposed several precise alignment algorithms in [Wagner *et al.* (2009)] and [Peng *et al.* (2010)], these algorithms don't suit for large scale face recognition, because of their high complexity. Therefore, we consider mis-alignment robustness is a problem to be solved in SR.

3.2. Problem Analysis

According to the distribution of eyeglass in face images, the difficulties of the eye-glass problem can be divided into the following situations from easy to hard:

(1) All images without eyeglass
(2) All images with eyeglass
(3) All images in gallery without eyeglass, and all images in probe with eyeglass, or vice versa
(4) Each person either "without eyeglass in gallery, with eyeglass in probe" or "with eyeglass in gallery, without eyeglass in probe"

The state-of-the-art method in [Li et al. (2007)] can achieve nearly perfect recognition rate in the first and second cases. In the third case, intra and inter similarity scores decrease consistently, thus the recognition rate just drop a little. The fourth case is the most challenging situation we meet in practical applications, in which similarity metric is disturbed by the nearly random eyeglass occlusion. Table 1 shows the influence of eyeglass under various situations on a large database including 292 persons (the database will be described in section 4). From the table we find the degradation rate of the fourth situation is the biggest one, especially in validation rate. To be more specific, the experiments in section 4 will focus on the hardest situation.

Table 1. The influence of eyeglass on recognition and validation rate at FAR=0.01 under various situations, which is evaluated using boosted LBP + LDA (cf. section 2).

Situations	Recognition Rate	Validation rate @ FAR=0.01
1	99.85%	99.61%
2	99.58%	99.28%
3	98.26%	87.43%
4	93.77%	70.21%

As shown in table 1 and reviewed in the prior subsection, we can see eyeglass is a very challenging problem must be conquered in NIR face recognition. Those noted methods all have their disadvantages in practicability, efficiency and robustness. In [Yi and Li (2011)], we propose an effective method for dealing with the eyeglass problem in NIR face recognition, inspired by sparse representation (SR). Unlike the existing sparse representation method that works on image pixels [Wright et al. (2009)], the sparse linear combination and sparse error model are learned from discriminative local feature space. Because of the locality, sparse error model can still hold when it moves from pixels to local features. Obviously, local features, as statistics in local regions, are more robust to small mis-alignment than pixels.

Figure 7 shows a eyeglass occluded NIR face image, a good NIR face image and its residual in pixel and LBP histogram bin feature [Ojala et al. (2002); Liao et al. (2007)]. From the figure, we can see the residual error are both sparse in pixel and feature domains, so local feature SR is a reasonable solution for eyeglass problem.

The principles and advantages of local feature SR will be discussed in the following sections.

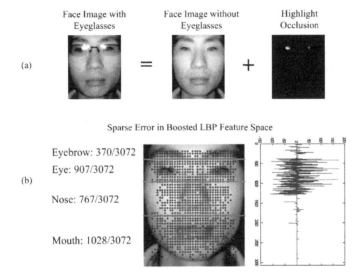

Fig. 7. The sparse error caused by eyeglass in pixel and LBP feature space. (a) A face image with eyeglass and specular highlight can be combined by its corresponding ideal (without eyeglass) face image plus sparse error; (b) The proportion of LBP features selected by boosting in each region (left), the position of boosted 3072 LBP features in the face image (middle), and the sparse error in LBP feature space (right).

In the following sections, we present two local feature SR methods, one based on multi-scale LBP histogram bins [Ojala *et al.* (2002); Liao *et al.* (2007)], one based on LBP and LDA. As discussed in section 2, LBP operator is invariant to monotonically transforming illumination, which achieves very high performance in NIR face recognition, so we use LBP as basic local feature.

LBP can be applied in various scales and sub-windows. For a face image, huge LBP histogram features are generated, with millions of dimensions. Such high dimensional features are infeasible for practical application and most of them are useless for face recognition actually. Therefore, a small but effective subset is selected from the over-complete feature set by using AdaBoost [Freund and Schapire (2001); Li *et al.* (2007)]. All the following steps will be applied on the boosted LBP feature set.

3.3. *Learning Sparse Representation of LBP*

Recently, SR (or compressive sensing) is popular in many fields [Candes and Tao (2005); Chen *et al.* (1998)]. One of the most interesting applications is occlusion robust face recognition [Wright *et al.* (2009)]. Assuming a occlusion in face image is not exceed a ratio (*e.g.* 60%), we can look the occlusion as a sparse error. A observed face image **y** can be represented by the following sparse linear combination and error model:

$$\mathbf{y} = A\mathbf{x} + \mathbf{e} \tag{5}$$

where **e** is sparse error (occlusion). A is a group of vectorized face images we want to recognize (*i.e.* gallery). **x** is sparse linear combination coefficients. In cooperative-user NIR face recognition, face images in probe and gallery are all frontal and illumination uniform, hence any face image **y** belonging to the ith person will be lie in the linear span of the gallery face images of the ith person. Ideally, **x**'s entries will be zeros except those associated with the ith person, and **e** will be the occlusion error.

Owing to the locality of LBP, sparse error model still can hold, while moving from pixels to local features. We denote the LBP feature of **y** as $\mathbf{y}_L$, the LBP feature of A as A_L, so the new model is

$$\mathbf{y}_L = A_L\mathbf{x} + \mathbf{e} \tag{6}$$

where **e** is the occlusion in LBP feature space, which is different from the pixel occlusion in Equ. (5). Merge the sparse coefficients and error, Equ. (6) can be written as

$$\mathbf{y}_L = \begin{bmatrix} A_L, I \end{bmatrix} \begin{bmatrix} \mathbf{x} \\ \mathbf{e} \end{bmatrix} = B\mathbf{z} \tag{7}$$

Given $\mathbf{y}_L$, A_L, solving **x** and **e** can be seen as the following L_1 optimization problem

$$\begin{aligned} \min \quad &\|\mathbf{z}\|_1 \\ \text{s.t.} \quad &\mathbf{y}_L = B\mathbf{z} \end{aligned} \tag{8}$$

The above constraint must be an under-determined system of linear equations. As concluded in [Yang *et al.* (2010)] , many existing algorithms can recover the sparsest solution of problem (8). Considering the overall performance of those algorithms, we use "Homotopy" [Osborne *et al.* (2000)] in this paper. When the solution is of high-sparsity (*e.g.* smaller than 40%), Homotopy is very fast and can achieve high performance.

As the entries of **x** encoding the identity information of the probe image **y**, we use it to do face recognition and validation. Reconstruction residual and sparsity concentration index (SCI) are defined in [Wright *et al.* (2009)] for face recognition and validation respectively. Without loss of generality, we use a modified SCI (mSCI) in face recognition and the original SCI for validation experiments. The definition of mSCI is slightly modified from SCI as follows

$$\mathrm{mSCI}_i(\mathbf{x}) = \frac{n\|\delta_i(\mathbf{x})\|_1 / \|\mathbf{x}\|_1 - 1}{n - 1} \in [0, 1] \tag{9}$$

where i is the index of each class; n is the number of samples in gallery. For each class i, $\delta_i(x)$ is the characteristic function that selects the coefficients associated with the ith class. The larger the corresponding $mSCI_i$ of a probe $\mathbf{y}$, that $\mathbf{y}$ is more similar with the ith class of the gallery. In experiments, (m)SCI is used to generate recognition rate (rank1) and ROC curves. Note that, reconstruction residual and mSCI have very similar recognition performance in face recognition. We use mSCI here because it's faster than reconstruction residual.

3.4. *LDA Extension*

LDA is widely used in face recognition for dimension reduction and discriminative feature extraction. Boosted local features are usually of relative higher dimensionality (*e.g.* 2000-10000), which would burden the next L_1 minimization step. To improve the performance in accuracy and computation efficiency further, we incorporate LDA into local feature SR. As we know, LDA is a holistic projection operation, it cannot reserve the sparsity of error, that a sparse error (occlusion) in LBP feature space will not remain sparse in the LDA subspace. Despite of this, we still can solve this problem by doing a small modification on Equ. (7).

Based on LBP features, we first train a LDA projection matrix P using intra-class and inter-class scatter matrix, then we project the probe and gallery features into the LDA subspace as $P\mathbf{y}_L$ and PA_L. Multiply matrix P on the two sides of Equ. (7), we get

$$P\mathbf{y}_L = \begin{bmatrix} PA_L, P \end{bmatrix} \begin{bmatrix} \mathbf{x} \\ \mathbf{e} \end{bmatrix} \tag{10}$$

Similar to Equ. (8), $\mathbf{x}$ and $\mathbf{e}$ can be solved by L_1 minimization. Note that $\mathbf{e}$ is still a sparse error in LBP feature space not in LDA subspace. (m)SCI of $\mathbf{x}$ is used for face recognition and validation.

4. Experiments

In this section, we aim to evaluate the performance of the described method in NIR face recognition with eyeglass problem, and compare with conventional local feature method [Li *et al.* (2007)] and SR [Wright *et al.* (2009)]. Five methods are evaluated as follows,

(1) Boosted LBP + LDA + nearest neighborhood (NN) (*cf.* section 2 and [Li *et al.* (2007)])
(2) Down-sampled image + SR [Wright *et al.* (2009)]
(3) LDA + SR [Wright *et al.* (2009)]
(4) Boosted LBP + SR (proposed)
(5) Boosted LBP + LDA + SR (proposed)

For equity, all methods are trained and tested on the same databases. To obtain more comprehensive results, all methods are tested in two scenarios: recognition and validation. For a given probe sample, we must first decide if it's a valid sample from one of the persons in gallery (validation), then judge which person the sample belongs to (recognition). These two properties are both crucial for real-world recognition systems.

4.1. Database Setup

A database containing 46351 NIR face images of 1925 persons are prepared for AdaBoost and LDA training. And we collect an independent database for testing including face images both with/without eyeglass. The testing database includes 9988 face images of 292 persons, 8-20 images with eyeglass and 8-20 images without eyeglass for each person. Some face images in training and testing set are shown in Figure 8. For the convenience of face recognition and validation, 4 subsets are constructed from the testing database as follows:

(1) Gallery for recognition: 292 persons, half of them with eyeglass, the other half without eyeglass;
(2) Probe for recognition: 292 persons, the other images not overlapped with subset 1);
(3) Gallery for validation: randomly selected 146 persons, half of them with eyeglass, the other half without eyeglass;
(4) Probe for validation: the other images of the same 146 persons in subset 3) as positive samples, all images of the other 146 persons not overlapped with subset 3) as negative samples.

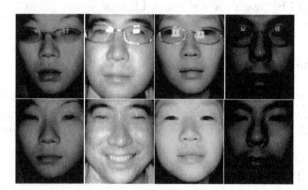

Fig. 8.   Some aligned face images according to eye coordinates in database.

The partition rules conforms to the most challenging situation described in section 3. Firstly, we detect the eye coordinates of face images by automatic face and eye detectors, and align them to 120×142 pixels. For every aligned face image, an over-complete LBP features of 1803072 dimension are extracted in multi-scale way. Then AdaBoost and LDA are both trained on the basis of the LBP features in training set. By using AdaBoost, the high dimensional LBP features can be reduced to 3072 dimensions. LDA further reduces the 3072 LBP features to 256 dimensions. The following experiments are conducted on 3072 or 256 dimensions.

4.2. *Face Recognition*

In this experiment, we test the rank1 recognition rate [Phillips *et al.* (1998)] of the five methods. Table 2 shows the rank1 recognition rates, where our proposed methods 4) and 5) achieve the best performance: 96.17% and 96.81%. The proposed methods are about 3% higher than conventional NIR face recognition method 1) and 16% higher than SR 3) and 4). Two conventional SR methods are poorer than the other methods, because of their sensitivity to mis-alignment and lack of discriminant. Although method 4) has nearly comparable recognition rate with 5), but is slower than 5).

The computational cost per query are also list in table 2. All methods are performed in MATLAB 7.8 on 64bits Windows 2003 with dual quad-core 2.66GHz Xeon processors and 16GB of memory. We can see that 1) is the fastest method among them, but it doesn't deal with occlusion explicitly, with lower recognition rate than the proposed methods. Considering the overall (recognition rate and speed) performance, 5) is the most appropriate choice for practical applications, 96.81% recognition rate and 300ms per query against a gallery with 4997 samples.

Table 2. Comparison of recognition rates and average run time per query for all methods.

No.	Method	Recognition Rate	Dimension	Run Time/Query
1	LBP+LDA+NN	93.77%	**256**	**0.5ms**
2	Down-sampled+SR	79.46%	24x28=620	460ms
3	LDA+SR	79.82%	620	460ms
4	LBP+SR	96.17%	3072	4100ms
5	LBP+LDA+SR	**96.81%**	**256**	300ms

4.3. *Face Validation*

Next, we demonstrate the performance of our methods for rejecting invalid probe images while accepting valid ones. As described above, the probe set for this experiment are composed by two parts, including 2528 positive samples and 4933 negative samples. By tuning the threshold τ in the range of SCI $[0, 1]$, we generate a series ROC curves for method 2)-5) in Figure 9. For comparison, the ROC curve of method 1) is generated according to the nearest neighborhood of "cosine" distance. From Figure 9 we can see the five methods have similar performance rank with recognition experiment, except for 2) and 3), that is about 5) $\simeq$ 4) > 1) > 2) > 3). The best methods 4) and 5) achieve 85.92% validation rate when FAR is 0.01, which has a big improvement (about 13%) compared with the current state-of-the-art method

1). The validation rates are list in table 3 at FAR=0.1, 0.01 and 0.001. Considering the overall performance in face face recognition, validation and computational complexity, method 5) LBP+LDA+SR is the best one.

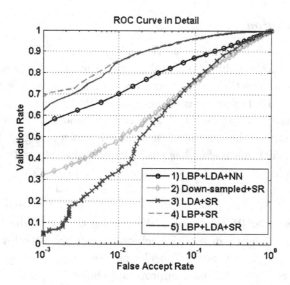

Fig. 9. ROC curves of comparison methods for face validation.

Table 3. Comparison of validation rates at FAR=0.1, 0.01 and 0.001 for all methods.

No.	Method	FAR=0.1	FAR=0.01	FAR=0.001
1	LBP+LDA+NN	87.18%	70.21%	54.31%
2	Down-sampled+SR	75.83%	49.17%	32.32%
3	LDA+SR	76.78%	33.98%	5.18%
4	LBP+SR	96.04%	**85.92%**	**68.83%**
5	LBP+LDA+SR	**96.52%**	**85.92%**	62.7%

5. Conclusion and Discussions

Although NIR face recognition already have many successful practical applications, the performance and user experience could be improved further by solving the eyeglass problem. Synthesis based methods could remove eyeglass in the face images with good visual quality, but they can not introduce any new information when compared with the original images. We have done many experiments based on those synthesized face images as like in figure 6, and get no improvement on recognition rate. SR provides a good theoretical and encouraging solution for occlusion problems, but it is prone to mis-alignment and usually of less discriminant. In this chapter, we introduce an effective method recently proposed in [Yi and Li (2011)], which reformulate SR in local feature (LBP) space and extend it by collaborating with LDA. The experiments show LBP reserves the advantages of SR in dealing with occlusion while becoming less sensitive to mis-alignment and more discriminant than SR derived from the image pixels. With respect to NIR face recognition with eyeglass and specular highlight, LBP+SR is obviously better than the conventional NIR face recognition method and original SR in both recognition and validation rate. Moreover, the LDA extension improves the recognition rate and computational speed further.

References

Ahonen, T., Hadid, A. and M.Pietikainen (2004). "Face recognition with local binary patterns", in *Proceedings of the European Conference on Computer Vision* (Prague, Czech), pp. 469–481.

AuthenMetric Co. Ltd. (2003). "A Method for Face Image Acquisition Using Active Lighting", Patent Application No.200310121340.1.

AuthenMetric Co. Ltd. (2004). "A Method for Face Image Acquisition and a Method and System for Face Recognition", Patent Application No.PCT/CN2004/000482.

AuthenMetric Co. Ltd. (2005). "An Image Acquisition Apparatus for Face Recognition", Patent Application No.200520022878.1.

Candes, E. and Tao, T. (2005). "Decoding by linear programming", *Information Theory, IEEE Transactions on* **51**, 12, pp. 4203–4215.

Chen, S. S., Donoho, D. L. and Saunders, M. A. (1998). "Atomic decomposition by basis pursuit", *SIAM Journal on Scientific Computing* **20**, 1, pp. 33–61.

Du, C. and Su, G. (2005). "Eyeglasses removal from facial images", *Pattern Recognition Letters* **26**, 14, pp. 2215–2220.

Fidler, S., Skocaj, D. and Leonardis, A. (2006). "Combining reconstructive and discriminative subspace methods for robust classification and regression by subsampling", *Pattern Analysis and Machine Intelligence, IEEE Transactions on* **28**, 3, pp. 337–350.

Freund, Y. and Schapire, R. (1997). "A decision-theoretic generalization of on-line learning and an application to boosting", *Journal of Computer and System Sciences* **55**, 1, pp. 119–139.

Freund, Y. and Schapire, R. (2001). "An adaptive version of the boost by majority algorithm", *Machine Learning* **43**, 3, pp. 293–318.

Hadid, A., Pietikainen, M. and Ahonen, T. (2004). "A discriminative feature space for detecting and recognizing faces", in *Proceedings of IEEE Computer Society Conference on Computer Vision and Pattern Recognition*, Vol. 2, pp. 797–804.

Kim, J., Choi, J., Yi, J. and Turk, M. (2005). "Effective representation using ica for face recognition robust to local distortion and partial occlusion", *Pattern Analysis and Machine Intelligence, IEEE Transactions on* **27**, 12, pp. 1977–1981.

Leonardis, A. and Bischof, H. (2000). "Robust recognition using eigenimages", *Comput. Vis. Image Underst.* **78**, 1, pp. 99–118.

Li, S. Z. and and His Face Team (2005). "AuthenMetric F1: A Highly Accurate and Fast Face Recognition System", ICCV2005 - Demos.

Li, S. Z., Chu, R., Liao, S. and Zhang., L. (2007). "Illumination invariant face recognition using near-infrared images", *IEEE Transactions on Pattern Analysis and Machine Intelligence* **26**, Special issue on Biometrics: Progress and Directions.

Li, S. Z., Chu, R. F., Ao, M., Zhang, L. and He, R. (2006a). "Highly accurate and fast face recognition using near infrared images", in *Proceedings of IAPR International Conference on Biometric (ICB-2006)* (Hong Kong), pp. 151–158.

Li, S. Z., Zhang, L., Liao, S. C., Zhu, X. X., Chu, R. F., Ao, M. and He, R. (2006b). "A near-infrared image based face recognition system", in *Proceedings of 7th IEEE International Conference Automatic Face and Gesture Recognition (FG-2006)* (Southampton, UK), pp. 455–460.

Liao, S., Zhu, X., Lei, Z., Zhang, L. and Li, S. Z. (2007). "Learning multi-scale block local binary patterns for face recognition", in *ICB*, pp. 828–837.

Martinez, A. (2002). "Recognizing imprecisely localized, partially occluded, and expression variant faces from a single sample per class", *Pattern Analysis and Machine Intelligence, IEEE Transactions on* **24**, 6, pp. 748–763.

Moghaddam, B., Nastar, C. and Pentland, A. (1996). "A Bayesian similarity measure for direct image matching", *Media Lab Tech Report* No.393, MIT.

Nayar, S. K., Fang, X.-S. and Boult, T. (1997). Separation of reflection components using color and polarization, *International Journal of Computer Vision* **21**, pp. 163–186.

Ojala, T., Pietikainen, M. and Maenpaa, T. (2002). "Multiresolution gray-scale and rotation invariant texture classification width local binary patterns", *IEEE Transactions on Pattern Analysis and Machine Intelligence* **24**, 7, pp. 971–987.

Osborne, M., Presnell, B. and Turlach, B. (2000). A new approach to variable selection in least squares problems, *IMA Journal of Numerical Analysis* **20**, 3, pp. 389–403.

Peng, Y., Ganesh, A., Wright, J., Xu, W. and Ma, Y. (2010). "RASL: Robust alignment by sparse and low-rank decomposition for linearly correlated images", pp. 763–770.

Phillips, P. J., Wechsler, H., Huang, J., and Rauss, P. (1998). "The feret database and evaluation procedure for face recognition algorithms", *Image and Vision Computing J* **16**, 5, pp. 295–306.

Saito, Y., Kenmochi, Y. and Kotani, K. (1999). "Estimation of eyeglassless facial images using principal component analysis", in *Proceedings of IEEE International Conference on Image Processing*, Vol. 4, pp. 197–201.

Turk, M. A. and Pentland, A. P. (1991). "Face recognition using eigenfaces", in *Proceedings of IEEE Computer Society Conference on Computer Vision and Pattern Recognition* (Hawaii), pp. 586–591.

Wagner, A., Wright, J., Ganesh, A., Zhou, Z. and Ma, Y. (2009). "Towards a practical face recognition system: Robust registration and illumination by sparse representation", pp. 597–604.

Wright, J., Yang, A. Y., Ganesh, A., Sastry, S. S. and Ma, Y. (2009). "Robust face recognition via sparse representation", *IEEE Transactions on Pattern Analysis and Machine Intelligence* **31**, pp. 210–227.

Wu, C., Liu, C., Shum, H.-Y., Xu, Y.-Q. and Zhang, Z. (2004). "Automatic eyeglasses removal from face images", *IEEE Trans. Pattern Anal. Mach. Intell.* **26**, 3, pp. 322–336.

Yang, A. Y., Sastry, S. S., Ganesh, A. and Ma, Y. (2010). "Fast l1-minimization algirhtms and an application in robust face recognition: A review", in *Proceedings of International Conference on Pattern Recognition*.

Yi, D. and Li, S. Z. (2011). "Learning sparse feature for eyeglasses problem in face recognition", in *Proceedings of IEEE International Conference on Automatic Face and Gesture Recognition*.

Zhao, W., Chellappa, R., Phillips, P. and Rosenfeld, A. (2003). "Face recognition: A literature survey", *ACM Computing Surveys* , pp. 399–458.

CHAPTER 2.3

THE PHASE ONLY TRANSFORM FOR UNSUPERVISED SURFACE DEFECT DETECTION

D. Aiger and H. Talbot

Université Paris-Est, Laboratoire d'Informatique Gaspard-Monge, Equipe A3SI, ESIEE Paris, 2 Boulevard Blaise-Pascal, 93162 Noisy-le-Grand Cedex France
{d.aiger,h.talbot}@esiee.fr

We present a simple, fast, and effective method for detecting defects on textured surfaces. Our method is unsupervised and contains no learning stage or information on the texture being inspected. The method is based on the Phase Only Transform (PHOT) which correspond to the Discrete Fourier Transform (DFT), normalized by the magnitude. The PHOT removes any regularities, at arbitrary scales, from the image while preserving only irregular patterns considered to represent defects. The localization is obtained by the inverse transform followed by adaptive thresholding using a simple standard statistical method. The main computational requirement is thus to apply the DFT on the input image. The method is also easy to implement in a few lines of code. Despite its simplicity, the methods is shown to be effective and generic as tested on various inputs, requiring only one parameter for sensitivity. We provide theoretical justification based on a simple model and show results on various kinds of patterns. We also discuss some limitations.

1. Introduction

Vision-based inspection of surfaces has many real-world applications, for instance industrial wood, steel, ceramic and silicon wafers, fruits, aircraft surfaces and many more. It is in high demand in industry in order to replace the subjective and repetitive process of manual inspection. A comprehensive survey on recent developments in vision based surface inspection using image processing techniques, particularly those that are based on texture analysis methods, was proposed by Xie.[1] According to this work, one can divide the methods for surface defect detection into four categories, namely: statistical approaches, structural approaches, filter-based methods, and model-based approaches. A significant differentiating factor in visual inspection approaches is that of *supervised* classification versus *novelty* detection. For applications where both normal and defective samples can be easily obtained, supervised classification based approaches are usually favored. However, when defects are unpredictable and defective samples are unavailable, novelty detection is more desirable.

Texture is one of the most important characteristics in identifying defects or flaws. Much effort was invested in extracting useful texture features.[2–4] Statistical texture analysis methods investigate the spatial distribution of pixel values. In structural approaches, texture is characterized by primitives or texture elements, and the spatial arrangement of these primitives.[5] The goals of structural approaches are to extract texture primitives, and to model the spatial arrangement. Filter based approaches share a common characteristic, which is to apply filter banks on the image and compute the energy of the filter responses. These methods can be divided into spatial domain, frequency domain, and joint spatial/spatial-frequency domain techniques. Model based methods include, among many others, fractal models,[6] autoregressive models,[7] random field models,[8] and the *texem* model.[9]

In a novelty detection task, the task is to identify whether an input pattern is an expected part of the data or unknown. As for defect detection, it involves assigning a normal or abnormal label to a pattern (e.g. a surface or a pixel). In contrast to supervised classification, novelty detection only needs the normal samples for training purposes and usually uses a distance measure and a threshold for decision making. Recently, Markou and Singh[10,11] gave a detailed review of novelty detection approaches, using statistical and neural network based approaches. Statistical parametric approaches are commonly used in visual inspection.[12–15] A fundamental assumption is that the data distribution is Gaussian in nature, thus, it can be easily statistically modeled by means and covariances.

Working in the frequency domain is closely related to our contribution. Many methods apply filtering in the frequency domain, particularly when no straightforward kernel can be found in the spatial domain. The image is transformed into the Fourier domain, multiplied with the filter function and then re-transformed into the spatial domain. Coggins and Jain[16] used ring filters and orientation filters for feature extraction. D'Astous and Jernigan[17] used peak features, such as strength and area, and power distribution features, such as power spectrum eigenvalues and circularity, to discriminate textures. Tsai and Heish[18] used the Fourier transform (FT) to reconstruct textile images for defect detection. Chan and Pang[19] extracted harmonic peaks from horizontal and vertical power spectrum slices. The phase of the DFT was used for matching images by the Phase Only Correlation (POC).[20] It was also used for image coding.[21]

The main focus of this chapter is to develop an unsupervised method for defect detection. We concentrated on a method that does not require any prior information or learning stage. This solves the practical problem of collecting usable examples of good patterns and defective ones. In addition, in many inspection applications, the pattern of the inspected surface is not known *a-priori*. As we work in the frequency domain, we have the advantages of a global view, solving the problems of selecting good kernel sizes. On the other hand, our method localizes defects without the need for any post processing. The idea is very simple: instead of trying to detect peaks in the Fourier transform of the input images, it simply removes all regularities in the image of various sizes and patterns at once by normalizing the FT of the input image by its magnitude. Through this operation, only the phase information remains while all regular patterns at all scales are removed. The localization of the defects is achieved by transforming back to the spatial domain. We show that since no analysis is being performed in the Fourier domain, removing regularities by normalizing the magnitude serves as a multiscale regularity removal, therefore, random textures are effectively removed as well.

2. The importance of the phase

In their important paper, Oppenheim and Lim investigated the importance of phase in signals.[22] We review here some of their insights. In the Fourier representation of signals, spectral magnitude and phase tend to play different roles and in some situations, many of the important features of a signal are preserved even if only the phase is retained. This is not true in general for the spectral magnitude. This observation about phase has been made in a number of different contexts and applications. In general, reconstructing an object from the magnitude only is not of much value in representing the original object, whereas reconstructions from the phase only have many important features in common with the original objects. A phase-only image has Fourier transform phase equal to that of the original image and a FT magnitude of unity. Figure 1 show reconstruction of Lenna with magnitude only and phase only. It shows that although the rebuilt image contains the same frequency terms as the original input, the magnitude-only transform bears no resemblance to the original image, it does not allow recognition due to the missing phase. Conversely, a phase-only reconstruction shows a picture similar to the shape of the original Lenna. It appears that the phase contains significant information, especially regarding the edge location. In addition, the phase only transform removes periodicity and regularity so it does more than just preserving the edges. In Section 3.2 we model the problem, give some theoretical justifications and explain why it works.

Fig. 1. The Phase only inverse Vs. Magnitude Only inverse - left: original, middle: magnitude only, right: phase only.

3. The new method

3.1. *Applying the Phase Only Transform*

In the context of this chapter, our goal is not to reconstruct a signal from its FT. In fact, we wish to do almost the opposite, namely, to filter out most patterns. We wish to eliminate some parts of the signal while preserving others. The above discussion provides a way to do it for our application of defect detection in images. Our purpose is to eliminate the so-called regular patterns while preserving correspondingly "rare" events in the image which can be considered to be defects. The fact that phase-only reconstruction preserves much of the correlation between signals would suggest that the location of events tends to be preserved. If we assume that in our application, a non defective region contains non-localized structures (e.g. regular patterns or homogeneous regions) and that a defect is well localized, it is reasonable to use the phase to filter all non localized patterns (see Section 3.2).

Our algorithm is very simple and can be implemented in a few lines of code. The first step is to apply the DFT on the input (real) image. The discrete two-dimensional Fourier transform of an image array $F(u, v)$ is defined in series form as:

$$\mathcal{F}(u,v) = \frac{1}{N} \sum_{j=0}^{N-1} \sum_{k=0}^{N-1} F(j,k) \exp\{\frac{-2\pi i}{N}(uj + vk)\}$$

where $i = \sqrt{-1}$. The indices (u, v) are called the spatial frequencies of the transformation. The result is a matrix of complex numbers in the frequency domain,

$$\mathcal{F}(u,v) = \mathcal{R}(u,v) + i\mathcal{I}(u,v)$$

or in magnitude and phase-angle form,

$$\mathcal{F}(u,v) = \mathcal{M}(u,v) \exp\{i\Phi(u,v)\}$$

where

$$\mathcal{M}(u,v) = \sqrt{\mathcal{R}^2(u,v) + \mathcal{I}^2(u,v)}$$

and

$$\Phi(u,v) = \arctan\{\frac{\mathcal{I}(u,v)}{\mathcal{R}(u,v)}\}$$

By normalizing every complex number by dividing both the real and imaginary parts by $\mathcal{M}(u, v)$ we essentially remove all regular patterns at every scales at once. Note that we don't have to analyze the Fourier image. This normalization works on all frequencies at once and eliminates the regularities. The algorithm of the first stage can now be written (algorithm 1):

Algorithm 1 The Phase Only Transform.

Require: input image $I(u, v)$
  compute $\mathcal{F}(u, v)\{I\}$
  for all (u,v) **do**
    $\mathcal{F}(u, v) = \frac{\mathcal{F}(u,v)}{\mathcal{M}(u,v)}$
  end for
  $O(u, v) = \mathcal{F}^{-1}(u, v)$

The resulted image is $O(u, v)$. The spectral magnitude of images tends to fall off at high frequencies, the phase-only signal will experience a high-frequency emphasis which will accentuate narrow events without modifying their position. It is reasonable to identify $\frac{1}{\mathcal{M}(u,v)}$ as generally emphasizing high frequencies over low frequencies.

3.2. *Theoretical justification*

We justify here why the Phase-Only Transform (PHOT) works. For this we express the fact that for texture, the range of value taken by the phase is small compared to that for a defect in the texture.

3.2.1. *Phase of regular texture*

We limit ourselves to 1-D, as the discussion extends readily to n-D due to the FT separability, and we carry out the discussion in the continuous domain for simplicity. We need some definitions to start:

Definition 1 (projection). *Let ϕ be a function defined over a 1-D domain $\mathcal{D}$. The projection p of ϕ is the indicator function of the domain of $\phi(y)$ for all y.*

$$p_{\mathcal{D}}[\phi(x)](y) = \begin{cases} 1 & \text{if there is an } x \text{ over } \mathcal{D} \text{ such that } \phi(x) - y = 0 \\ 0 & \text{otherwise.} \end{cases}$$

We note that if $\mathcal{D}$ is bounded and ϕ takes discrete values, i.e. if ϕ is zero almost everywhere, then $p_{\mathcal{D}}[\phi(x)]$ is also zero almost everywhere. Conversely, if ϕ is monotonic and non-constant, then it Lebesgue measure is non-zero.

Definition 2 (Phase integral excursion). *The integral excursion $\mathcal{E}$ of the phase of a real signal is a measure of the range of values that the phase of its FT actually takes. More precisely, let $S(x)$ be a real signal. Let*

$$\mathcal{F}[S](\omega) = \int_{-\infty}^{+\infty} S(x)e^{-i\omega x} dx = a_S(\omega)e^{i\phi_S(\omega)}$$

be its FT with ω real belonging to $[0, 2\pi]$. Assuming $\phi_S(x)$, the phase of S, to be either monotonic or discrete on that domain, we define

$$\mathcal{E}[S] = \int_0^{2\pi} p_\mathcal{D}[\phi(x)](y)dy. \tag{1}$$

Our phase integral excursion is a measure of the density of the values of $\phi_S(\omega)$ projected onto the y-axis. We limit ourselves to phases that are monotonic or discrete as this is sufficient for our discussion, and as in these cases, we can define $p_\mathcal{D}$ implicitly. To continue, we need a simple definition of texture.

Definition 3 (regular texture). *We defined a regular texture as a signal representable by a convergent Fourier sum:*

$$S(x) = \sum_{n=-N}^{+N} c_n \exp(inx), N \geq 0. \tag{2}$$

We know that Fourier sums can represent any square-integrable bounded periodic signal with arbitrary precision almost everywhere, which is what we require to represent regular textures. We now have then the following theorem:

Theorem 1 (Phase integral excursion of a regular texture). *The phase integral excursion of any regular texture over a finite domain is zero.*

Proof: The FT of S reduces to a superposition of a finite number of Dirac peaks, therefore $\phi(x \in \mathcal{D})$ is zero almost everywhere, and so is its projection. Finally, so is its phase integral excursion, as the integral of a function with only countably non-zero values is zero. $\square$

3.2.2. *Phase of a defect*

A defect in a texture can be defined as an abrupt change in its regularity. To study this we model it by a box function.

Definition 4 (Random box function). *A random box function $H(a, b)$ is a function which has the following form:*

$$H(a,b)[x] = \begin{cases} 0 \ if x < a \\ 1 \ if a \leq x \leq b \\ 0 \ if x > b \end{cases}$$

We use the following well-known properties of the Fourier transform to derive the expression of the FT of $H(a, b)$.

- Translation invariance : $\mathcal{F}[f(x - x_0)](\omega) = e^{-ix_0\omega}\mathcal{F}[f(x)](\omega)$.
- Scale invariance : $\mathcal{F}[f(ax)](\omega) = \frac{1}{|a|}\mathcal{F}[f(x)](\frac{\omega}{a})$.
- The expression of the centered box function : $H(-\frac{1}{2}, \frac{1}{2}) = \text{sinc}(\frac{\omega}{2})$

The expression of the FT of the random box function is therefore :

$$\mathcal{F}[H(a,b)](\omega) = e^{-i(\frac{a+b}{2})\omega} \left[\text{sinc}(\frac{\omega}{2(b-a)}) \right]. \tag{3}$$

The phase of this FT is simply

$$\phi_{H(a,b)}[\mathcal{F}[H(a,b)]](\omega) = -(\frac{a+b}{2})\omega. \tag{4}$$

We now have the following theorem :

Theorem 2 (Phase excursion of the random box function). *The phase excursion of the random box function is almost surely non-zero.*

Proof: Ignoring phase wraparound over 2π, If $a+b \neq 0$, then $\phi_{H(a,b)}$ is monotonic and non constant, and so, even including phase wraparound, $p_D[\phi(x)]$ is 1 on a measurable set. Its integral over the range of ϕ is therefore non-zero, and so is the integral phase excursion. We note that since a and b are random, the probability of $a+b=0$ is zero. $\qquad\square$

Let us assume a regular texture on the one hand, and a regular texture with a defect in the other. Theorem 1 tells us that the former has a FT with a phase composed of only a few different values. The latter might be viewed as a superposition of a regular texture and a random box function with random values for a and b. Theorem 2 tells us that its FT features a phase composed of uncountably infinite different values.

We now show that the phase-only transform can readily distinguish between these two cases even in the discrete setting.

3.3. 1D examples

In this section we show a few examples on $1D$ signals and give some insights about the behavior of the Phase Only Transform. We refer to the PHOT here, as the signal that is transformed back to the spatial domain, after being normalized by the magnitude. As already shown by the 2D example, most of the information on edges and sharp peaks is contained in the phase. If a signal contains a single peak or edge and a flat region, the phase part of the FFT must be significant, because the sum of many trigonometric functions is needed to construct the flat part. On the other hand, if a signal is constructed of a sum of pure sine or cosine functions of various frequencies with zero or little phase content, the PHOT will be almost zero. This is true not only for signals that are periodic within a finite support. Figure 2 shows such a signal. In Figure 3 we see a sharp peak that requires large phase content. We conclude that signals (not necessarily periodic) that have a small phase content would yield a smooth PHOT, while those with large phase content representing a peak or an edge yield a large peak in the PHOT which corresponds to the location of the peak or edge in the input signal. Assuming that a defected

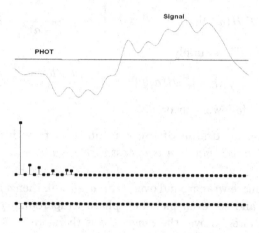

Fig. 2.   A signal with little phase content - the PHOT is almost flat. Top: signal and its PHOT. Middle: magnitude of frequencies. Bottom: phase of frequencies.

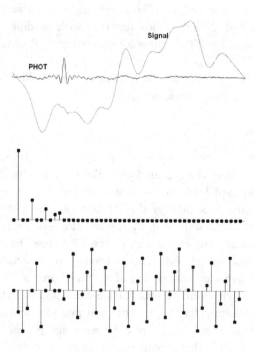

Fig. 3.   A small defect in a sum of sine curves. Top: signal and its PHOT. Middle: magnitude of frequencies. Bottom: phase of frequencies.

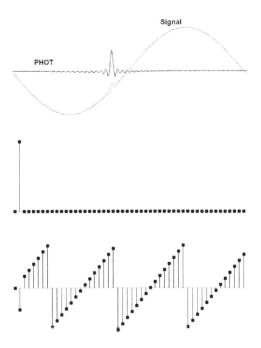

Fig. 4. A defect in a single sine curve. Top: signal and its PHOT. Middle: magnitude of frequencies. Bottom: phase of frequencies.

signal is composed of sum of sine function of various frequencies and a peak, the result of the PHOT is a collection of peaks in the spatial domain that are localized in the original defect location while the part that is corresponding to the first term is eliminated. Figure 4 shows a small defect (peak), composed with a sine (or cosine) wave. In Figure 5 we show another example on a signal that appears non-periodic due to the limited domain, yet, is composed from a sum of trigonometric functions which are all removed, while the defect remains.

Our model of an input signal is thus composed of two terms, a *non-defected* term, $A(x)$ which is a sum of sine or cosine functions with relatively small phase content, and a *defect* term, $B(x)$ which is assumed to be a peak or step edge, thus contains large phase content:

$$S(x) = A(x) + B(x)$$

Since the PHOT eliminates the sum of (low phase content) sines, we are left mainly with $B(x)$, as expected from Section 3.2. The inverse transform then yields the localization of the defect in the spatial domain.

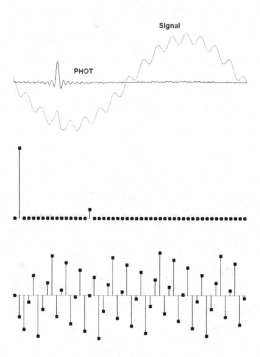

Fig. 5. Non periodic (but with little phase) signal and a defect (large phase content). Top: signal and its PHOT. Middle: magnitude of frequencies. Bottom: phase of frequencies.

3.4. *Thresholding using Mahalanobis distance*

In order to be able to use a totally unsupervised method with no learning component, we have to assume that for each input image the majority of the image pixels are intact (see Section 4). In this case, we can use simple statistics. We use the result of the PHOT as a probability map of a pixel being a defect. As commonly used, we assume a Gaussian distribution and use the Mahalanobis distance. We compute the mean and variance of the distribution from the image obtained by the PHOT. Since we normalize each of the FFT basis when we reconstruct the PHOT image, the global mean and standard deviation of the image are now both $1/N$ where N is the number of pixels. However, since the noise can be significant, we first smooth the PHOT image by a Gaussian filter and only then compute the mean and variance (we have used Gaussian of $\sigma = 3.0$). The user provides a value in sense of Mahalanobis distance. We threshold the PHOT result such that every pixel with a distance larger than this value is considered as a defected pixel. Figure 6 shows an input image, the PHOT result interpreted as Mahalanobis distance from the mean and the thresholding result using a Mahalanobis distance of 4.0. Of course more sophisticated statistical methods can be used instead.

4. Characteristics and limitations of the Phase Only Transform

The most appealing characteristic of the PHOT is that it removes any regularities from the image without the need to identify peaks in the Fourier domain. Only spikes that do not correspond to a sum of trigonometric functions inside the image domain are left. Note that the regularities should not be presented in the entire image. Every large enough regular patterns are removed by the transform by normalizing the resulted complex number by its magnitude. In this sense, our method is different from those that work only on periodic patterns. Figure 7 shows an example of image that has several subpatterns that are regular but the entire image is not. The only parameter in the threshold on the Mahalanobis distance and it is exactly the same in Figures 6 and 7. The result shows that the PHOT has no difficulty in detecting defects in this image. The results look very similar to the human perception of "novel pattern". The entire image is not regular but contains patterns that in some way similarly perceived. We should note here that this can be also considered as a limitation of the method, since large defects can be viewed as regular subpattern, thus might be removed by the PHOT.

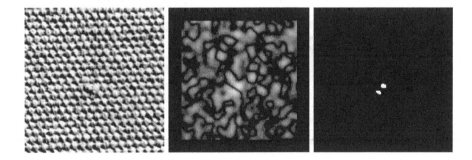

Fig. 6. Image in the process of defect detection: left - input image, middle - Mahalanobis distance from the mean (multiply by 30 for visualization), right - thresholding using distance 4.0.

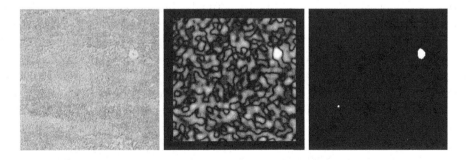

Fig. 7. Non regular patterns: left - input image, middle - Mahalanobis distance from the mean (multiply by 30 for visualization), right - thresholding using distance 4.0.

Fig. 8.　Limitation of 2D transform: Scratches could not be detected as they are 1D regular.

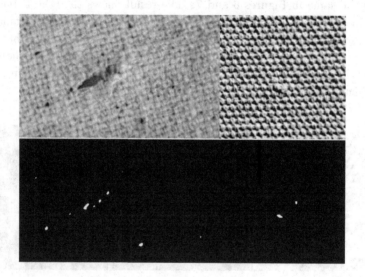

Fig. 9.　Multiple patterns: top - input image, bottom - thresholding using distance 4.0.

As can be expected, if we use 2D FFT on the image, every periodicity or regularity (or homogeneity) is removed by the PHOT. This contains also large defected patterns and 1D structures. For example, a defect structured as a line or scratch in the image, would not be well detected as can be seen in Figure 8. On the other hand, the same characteristic, can be used (to our advantage) to obtain defect detection on multiple patterns where nothing has to be known by the algorithm in advance ("blind" defect detection). In Figure 9 the results of our algorithm on a image that contains two totally different regularities are shown. It can be observed that the boundary regions between regularities were removed by the PHOT. This means that 1D long defected patterns may not be detected. A way to solve this problem is to apply the PHOT on lines instead on the entire 2D image. This would work however only in a highly regular patterns. We will investigate this direction in the future.

5. Complexity and real time performance

In many inspection system that apply defect detection algorithms for quality assurance, the time performance of the algorithm is critical as it might be used in a real manufacturing process. As can be easily concluded from our algorithm, the complexity is $O(n \log n)$ where n is the number of pixels in the input image. This, of course, comes from the DFT that we have to apply. The further processing and statistics is obviously linear with n. For very large or continuously inspected patterns, one can apply the algorithm on partial sub-windows without affecting the detection performance substantially . It is also very simple to implement the algorithm on parallel machines by decomposing the input. We successfully implemented the algorithm on a GPU (Graphics Processing Unit) using the Nvidia CUDA language. The FFT is also quite fast in practice and effective parallelization exists using Intel's SSE2 and SSE3 instructions, as well as on DSPs.

6. Results

We implemented the algorithm using C++ and Visual Studio. the results on a large set of images are shown in Figures 10. All the results were obtained using the same parameter for thresholding the Mahalanobis distance (4.0). No other parameter is needed for our algorithm. The sensitivity of the algorithm can be changed by the user by altering the Mahalanobis threshold.

6.1. *Multiple sub-patterns and arbitrary patterns*

As already mentioned in Section 4, our method does not require that the entire inspected pattern be regular. It can process many sub-patterns simultaneously. In fact, the PHOT is a detector for *novel* patterns. It emphasizes patterns that do not appear much in the image. It is worth noting that we do not assume anything about the size of the pattern, so it can vary. In Figure 11 an image containing many texture patches of different size and regularities is proceeded and the result (using Mahalanobis threshold 4.0) is shown on the right. The synthetic defect almost invisible by eye in the image is detected since it is novel. Another spike on top of the image is also detected. In Figure 12 a scene that contains a house with a textured roof is shown. The image contains textures as well as homogeneous and irregular regions. The synthetic defect as well as the *novel* pattern of the lamp on the right are well detected.

6.2. *Images with no defects*

We tested our simple adaptive threshold on input images which are texture patches without any defect. The purpose of this test is to verify that the method does not produce false positives. We used exactly the same parameter as in all other tests,

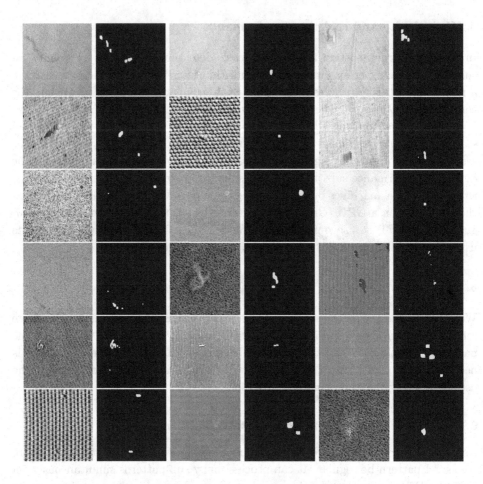

Fig. 10. Results on various patterns: in each of the three columns, left - input image, right - results by thresholding using distance 4.0.

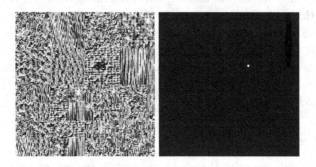

Fig. 11. Multiple textures of various size and regularities and a synthetic defect: top - input image, bottom - thresholding using distance 4.0.

Fig. 12.   Arbitrary scene with synthetic defect: top - input image, bottom - result.

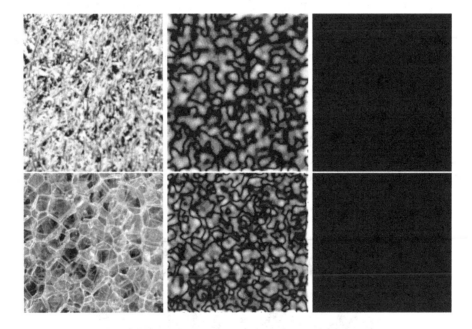

Fig. 13.   Images with no defects: left - input image, middle - PHOT result (multiplied by 30 for visualization), right - result using threshold of 4.0.

namely, a Mahalanobis distance $= 4.0$. In Figure 13 we show two texture patches which are not quite regular (to make the test more difficult), their PHOT results and the output using threshold equal to 4.0. It can be seen that no false positive defects were produced for either inputs. It can be observed in the PHOT result (middle), how the strength of the response is related to the perception of "novelty". Although no pixel exceeds distance 4.0, some regions have larger response correlated to the measure of their regularity.

7. Other potential applications

The main application of the Phase Only Transform presented in this chapter is defect detection, however, as the PHOT detects novel patterns in an image it can be also used for other applications. Salient regions are generally regarded as the candidates of attention focus in human eyes, which is the key stage in object detection. The phase spectrum plays a key role for saliency detection.[23] The saliency map can be calculated by the image's Phase spectrum of Fourier Transform alone. It was shown, similarly to the analysis in this chapter, that phase information specifies where each of the sinusoidal components resides within the image. The locations with less periodicity or less homogeneity in either the vertical or horizontal orientation show were the object candidates are located. In,[23] each pixel of the image is represented by a quaternion that consists of color, intensity and motion feature. The Phase spectrum is then used to obtain the spatio-temporal saliency map, which considers not only salient spatial features like color, orientation and etc. in a single frame but also temporal feature between frames like motion. Two examples from[23] is shown in Figure 14.

Another possible application is to measure the amount of "rectification" in images containing repeated patterns (like textures) that were taken in perspective. This can subsequently allow a rectification algorithm that maximizes this measure. For example, in Figure 15, the left image contains more homogeneity than the unrectified image to the right, thus a measure that is based on, say, the integration of the PHOT of the image would be much larger for the image to the right. Minimizing this measure (maximizing the homogeneity) would achieve rectification. The effect of repeated pattern on the PHOT is clearly observed.

Fig. 14. Results from[23] on saliency: bottom input image with main objects, top: saliency obtained by their method using the PHOT.

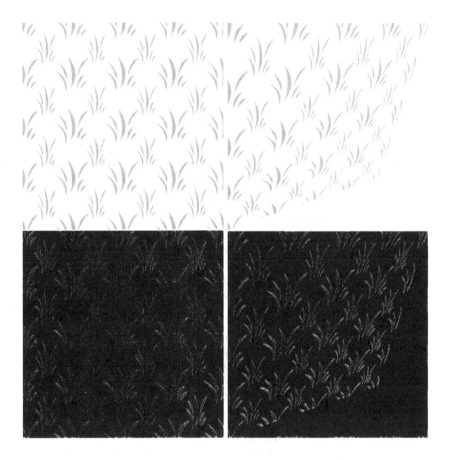

Fig. 15. Perspective and rectified textures. Top: rectified (left) and unretified (right) images with repeated patterns. Bottom: their corresponding PHOT.

8. Conclusions

A novel method for defect detection on surface patches was presented. The main advantage of the new algorithm is its extreme simplicity (it consists manly of a standard forward and inverse FFT), its generality to work for various pattern without prior knowledge and the fact that it is unsupervised. We gave theoretical justification for a reasonable model. We show results on a large set of inputs and the results are very similar to the perception of defects where no prior information is given. The new algorithm has only one parameter which is the sensitivity of the algorithm. It is an advantage in real inspection systems, where ease of use is important. The algorithm is also fast in practice and can be used in real time systems. Moreover, parallelization of the algorithm can be easily obtained by simply subdividing the input.

References

1. X. Xie, A review of recent advances in surface defect detection using texture analysis techniques, *Electronic Letters on Computer Vision and Image Analysis.* **7**, 1–22, (2008).
2. R. Haralick, Statistical and structural approaches to texture, *Proceedings of the IEEE.* **67**(5), 786–804, (1979).
3. T. Reed and J. Buf, A review of recent texture segmentation and feature extraction techniques, *Computer Vision, Image Processing and Graphics.* **57**(3), 359–372, (1993).
4. M. Tuceryan and A. Jain, Texture analysis, *In Handbook of Pattern Recognition and Computer Vision.* pp. 235–276, (1998). chapter 2, World Scientific, 1998.
5. F. V. R. Nevatia and K. Price, Structural analysis of natural textures, *IEEE Transactions on Pattern Analysis and Machine Intelligence.* **8**, 76–89, (1986).
6. W. Freeman, The fractal geometry of nature. (1983).
7. M. Comer and E. Delp, Segmentation of textured images using a multiresolution gaussian autoregressive model, *IEEE Transactions on Image Processing.* **8**(3), 408–420, (1999).
8. S. Li, Markov random filed modeling in image analysis. (2001). Springer.
9. X. Xie and M. Mirmehdi, Texem: Texture exemplars for defect detection on random textured surfaces, *IEEE Transactions on Pattern Analysis and Machine Intelligence.* **29**(8), 1454–1464, (2007).
10. M. Markou and S. Singh, Novelty detection: A review part 1: Statistical approaches, *Signal Processing.* **83**, 2481–2497, (2003).
11. M. Markou and S. Singh, Novelty detection: A review part 2: Neural network based approaches, *Signal Processing.* **83**, 2499–2521, (2003).
12. J. E. R. N. M. Millan and J. Pladellorens, Detection of local defects in textile webs using gabor filters, *Optical Engineering.* **37**(8), 2297–2307, (1998).
13. A. Kumar and G. Pang, Fabric defect segmentation using multichannel blob detectors, *Optical Engineering.* **39**(12), 3176–3190, (2000).
14. A. B. M. Bennamoun and S. Latham, Optimal gabor filters for textile flaw detection, *Pattern Recognition.* **35**, 2973–2991, (2002).
15. A. M. M. Mirmehdi and B. Thomas, Restructured eigenfilter matching for novelty detection in random textures, *British Machine Vision Conference.* pp. 637–646, (2004).
16. J. Coggins and A. Jain, A spatial filtering approach to texture analysis, *Pattern Recognition Letter.* **3**, 195–203, (1985).
17. F. D'Astous and M. Jernigan, Texture discrimination based on detailed measures of the power spectrum, *International Conference on Pattern Recognition.* **3**, 83–86, (1984).
18. D. Tsai and C. Heish, Automated surface inspection for directional textures, *Image and Vision Computing.* **18**(1), 49–62, (1999).
19. C. Chan and G. Pang, Fabric defect detection by fourier analysis, *IEEE Transactions on Industry Applications.* **36**(5), 1267–1276, (2000).
20. C. D. Kuglin and D. C. Hines, The phase correlation image alignment method, *IEEE Int. Conf. on Cybernetics and Society.* pp. 163–165, (1975).
21. L. M. G. C. Z. W. X. Luo and H. Lu, Phase-only transform based shaping for error concealment coding, *MobiMedia.* **11**, (2006).
22. A. V. Oppenheim and J. S. Lim, The importance of phase in signals, *IEEE, Proceedings.* **69**, 529–541, (1981).
23. C. Guo, Q. Ma, and L. Zhang. Spatio-temporal saliency detection using phase spectrum of quaternion fourier transform. In *CVPR*, (2008).

CHAPTER 2.4

CELLULAR AUTOMATA AS A TOOL FOR IMAGE PROCESSING

Paul L. Rosin and Xianfang Sun

School of Computer Science & Informatics, Cardiff University,
Cardiff, CF24 3AA, UK
{Paul.Rosin,Xianfang.Sun}@cs.cf.ac.uk

An overview is given on the use of cellular automata for image processing. We first consider the number of patterns that can exist in a neighbourhood, allowing for invariance to certain transformation. These patterns correspond to possible rules, and several schemes are described for automatically learning an appropriate rule set from training data. Two alternative schemes are given for coping with gray level (rather than binary) images without incurring a huge explosion in the number of possible rules. Finally, examples are provided of training various types of cellular automata with various rule identification schemes to perform several image processing tasks.

1. Introduction

Cellular automata (CA) consist of a regular grid of cells, each of which can be in only one of a finite number of possible states. The state of a cell is determined by the previous states of a surrounding neighbourhood of cells and is updated synchronously in discrete time steps. The identical rule contained in each cell is essentially a finite state machine, usually specified in the form of a rule table with an entry for every possible neighbourhood configuration of states.

Cellular automata are discrete dynamical systems, and they have been found useful for simulating and studying phenomena such as ordering, turbulence, chaos, symmetry-breaking, etc, and have had wide application in modelling systems in areas such as physics, biology, and sociology.

Over the last fifty years a variety of researchers (including well known names such as Stanislaw Ulam [Ulam (1962)] and John von Neumann [von Neumann (1966)], John Holland [Holland (1970)], Stephen Wolfram [Wolfram (1994)], and John Conway [Gardner (1970)]) have investigated the properties of cellular automata. Particularly in the 1960's and 1970's considerable effort was expended in developing special purpose hardware (e.g. CLIP) alongside developing rules for the application of the CAs to image analysis tasks [Preston and Duff (1984)]. More recently there has been a resurgence in interest in the properties of CAs without focusing on massively parallel hardware implementations, i.e. they are simulated on standard serial computers. By the 1990's CAs could be applied to perform a range of computer vision tasks, such as:

- calculating distances to features [Rosenfeld and Pfaltz (1968)],
- calculating properties of binary regions such as area, perimeter, convexity [Dyer and Rosenfeld (1981)],
- performing medium level processing such as gap filling and template matching [de Saint Pierre and Milgram (1992)],
- performing image enhancement operations such as noise filtering and sharpening [Hernandez and Herrmann (1996)],
- performing simple object recognition [Karafyllidis *et al.* (1997)].

Cellular automata have a number of advantages over traditional methods of computations:

- Although each cell generally only contains a few simple rules, the combination of a matrix of cells with their local interaction leads to more sophisticated emergent global behaviour. That is, although each cell has an extremely limited view of the system (just its immediate neighbours), localised information is propagated at each time step, enabling more global characteristics of the overall CA system.
- This simplicity of implementation and complexity of behaviour means that CA can be better suited for modelling complex systems than traditional approaches. For example, for modelling shell patterns, CA were found to avoid the considerable numerical problems inherent with partial differential equation based models, and were also substantially faster to compute [Kusch and Markus (1996)].
- CA are both inherently parallel and computationally simple. This means that they can implemented very efficiently in hardware using just AND/OR gates and are ideally suited to VLSI realisation [Chaudhuri *et al.* (1997)].
- CA are extensible; rules can easily be added, removed or modified.

2. Relating the Number of Cell States to the Number of Rules

For a 3×3 neighbourhood with cells taking 256 possible intensities there are 256^8 possible neighbourhood patterns (not considering the central cell's value). However, for many image processing tasks this number can be reduced by considering symmetries, e.g. the same rule should apply even if the pattern is rotated. To determine the number of distinct patterns after removing equivalent symmetric versions the Pólya-Burnside counting lemma [Roberts and Tesman (2005)] can be applied. If G

$$
\begin{array}{ccc}
\begin{array}{ccc} A & B & A \\ B & & B \\ A & B & A \end{array}
&
\begin{array}{ccc} A & B & C \\ D & & D \\ C & B & A \end{array}
&
\begin{array}{ccc} A & B & A \\ C & & C \\ D & E & D \end{array}
\end{array}
$$

Fig. 1. Patterns of a 3 × 3 neighbourhood that remain invariant under ±90° rotation, 180° rotation, and mirror symmetry through a vertical line of reflection respectively.

is a set of permutations of a set A, then the number of equivalence classes is

$$
N = \frac{1}{|G|} \sum_{g \in G} |\operatorname{Fix}(g)| \tag{1}
$$

where $\operatorname{Fix}(g)$ is the number of elements of A that are invariant under g. Figure 1 shows the patterns which are invariant under examples of the following transformations: ±90° rotation, 180° rotation, and mirror symmetry through a vertical line of reflection respectively. Thus, the number of distinct patterns N in terms of the number of possible intensities n is

$$
N = \frac{n^8 + 2n^2 + n^4 + 4n^5}{8} \tag{2}
$$

where the terms in the numerator correspond to: the identity transformation (i.e. 0° rotation), two rotations (±90°), a single rotation (180°), and four rotations corresponding to mirror symmetry through horizontal, vertical, and diagonal lines of reflection.

Given that even after eliminating symmetries the number of patterns still scales as $O(n^8)$, it can be seen that considering all $n = 256$ intensities leads to a prohibitive number of possible rules ($N > 2 \times 10^{18}$). Therefore, much of the previous application of cellular automata to image processing has been restricted to binary images. In this case, there are only $2^8 = 256$ possible patterns or rules, which reduces to $N = 51$ rules after eliminating symmetries – see figure 2.

3. Threshold Decomposition

In order to extend binary image CA methods to apply to gray level images without incurring the combinatorial explosion in the number of rules, Rosin [Rosin (2006)] proposed using threshold decomposition, a technique used extensively in image processing for rank order filtering [Fitch *et al.* (1984)]. This involves decomposing the gray level image into the set of binary images obtained by thresholding at all possible gray levels.[1] If a filter has the "stacking property" then it can be applied to

---

[1] Since threshold decomposition using all intensity levels incurs a substantial computation cost, a subset of intensity levels can be used instead to produce a faster approximation.

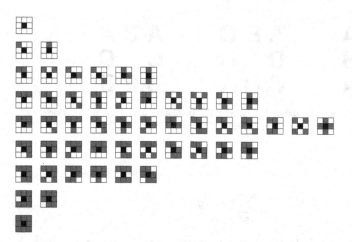

Fig. 2.   The complete rule set for a 3 × 3 neighbourhood of binary values with a central black pixel contains 51 patterns after symmetries and reflections are eliminated. The black central pixel is flipped to white after application of the rule. The neighbourhood pattern of eight white and/or black (displayed as gray) pixels which must be matched is shown.

each binary image, and when the set of processed binary images are summed then the result is identical to applying the filter to the original intensity image. A set of CA rules do not in general satisfy the stacking property[2], nevertheless applying this methodology is still useful since it allows intensity images to be processed. There is no equivalence of the results using the binary CA to those of a full intensity CA, but experimental results showed that results were still good.

In Rosin's initial work the CA were trained to perform denoising on a *single* binary image and subsequently the learnt rule set was reapplied to a gray level image using threshold decomposition. This idea was subsequently developed [Rosin (2010)] to take the more computationally expensive approach of training the CA on gray level images. That is, a search is made for a set of rules that when applied to the elements of the threshold decomposed input image and reconstructed produces a gray level image that provides a good match to the gray level target image. This has the advantage that it directly optimises the desired error function unlike the first approach which does not use threshold decomposition in the training phase but only during the subsequent application phase.

4. A 3-State Representation

An alternative approach to reducing the number of cell states was proposed by Rosin [Rosin (2010)] to enable more efficient training and application of CA to intensity images. It is based on the texture unit texture spectrum (TUTS) method

---

[2]Since CA can perform rank order filtering [Jagadish and Kailath (1989)] then it follows that at least some CA rules do possess the stacking property.

of texture analysis [Wang and He (1990)]. This involves using a pixel's value as a threshold for its eight neighbours. That is, for a central pixel value v_c its neighbours v_i are thresholded as

$$v_i' = \begin{cases} 0 \text{ if } v_i < v_c \\ 1 \text{ if } v_i = v_c \\ 2 \text{ if } v_i > v_c \end{cases} \tag{3}$$

Each neighbourhood can then be represented by a code formed from the eight ternary values: $\sum_{i=1}^{8} v_i' 3^i$, and histograms of the $3^8 = 6561$ different texture unit codes (the so called texture spectrum) make up the textural description of an image (or sub-image).

This local thresholding approach can be used in the context of our CA to reduce the large number of neighbourhood patterns. The basic idea is to maintain at each cell the image intensity as its primary state, but during the rule matching phase to consider the ternary pattern of the neighbourhood determined relative to the central cell's state. The CA rules are defined in terms of the 3 states rather than 256 states, and so from Eqn. (2) we find that 954 patterns need be considered.

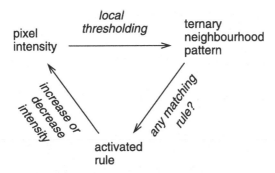

Fig. 3.   The CA is represented using both pixel intensities for cell states as well as ternary neighbourhood patterns. Transition rules involve changing a cell's intensity sufficiently such that the ternary neighbourhood pattern changes.

The scheme is as follows, and is illustrated in figure 3. The rules are applied to all the matching cells in parallel at each time step. This involves 1/ first generating at each cell its ternary neighbourhood pattern by thresholding the neighbourhood according to Eqn. (3), 2/ at each cell check for any rule that matches its ternary neighbourhood pattern, 3/ if a rule matches then apply it to update the central cell. In a standard binary CA, application of a rule inverts the cell's state. In the 3-state system despite using ternary rules it is the cell's intensity that needs to be updated. This is done by modifying its value such that its ternary neighbourhood changes. The minimal modification in either direction is to either increase its intensity to its closest neighbourhood intensity value, or alternatively to decrease it to its closest neighbourhood value. Thus for each pattern there are two possible rules (i.e. increasing or decreasing types). Note that for each type of rule there are certain neighbourhood patterns that preclude the use of that rule type. That is, there are 51 of the 954 basic patterns without any neighbourhood pixel intensity greater than the central pixel, and the same number without a smaller value. Therefore, the total number of rules in this CA system is $2 \times (954 - 51) = 1806$.

Two variations to the scheme are possible. The first allows all 1806 rules to be learnt independently. Thus, if the results of processing image I are $f(I)$, there is no constraint that $f(-I) = -f(I)$, which can be useful in some situations. The second variation enforces the constraint by treating the "increasing" and "decreasing" versions of a rule, $R_{TU_i}^{inc}$ and $R_{TU_i}^{dec}$, as equivalent when applied to inverted neighbourhood intensities. That is, for a neighbourhood N, $R_{TU_i}^{inc}(N) \equiv R_{TU_i}^{dec}(-N)$ and $R_{TU_i}^{inc}(-N) \equiv R_{TU_i}^{dec}(N)$, and so there are only 903 rules to be learnt.

One of the features of the TUTS (and LBP) schemes is that they provide a textural description that is invariant to a wide range of gray level transformations. Thus, for the 3-state CA this means that $f(\mathcal{T}(I)) = \mathcal{T}(f(I))$ where $\mathcal{T}$ is a monotonic, non-linear mapping that is information preserving in the sense that separate intensities are not collapsed into single values. In some situations this could be beneficial, but in others a lot of important information is lost by discarding quantitative information, and so the 3-state CA is not suited to solve certain quantitative tasks, e.g. computing edge magnitudes.

5. Training Strategies

Most of the literature on cellular automata studies the effect of applying manually specified transition rules. However, this is not a convenient way in which to build an image processing system, and a more automated approach is required. Ideally, it should be possible to automatically learn the rules given 1/ a set of training images, 2/ a set of corresponding target (i.e. ideal) output images, and 3/ an objective function for evaluating the quality of the actual images produced by the CA, i.e. the error between the target output and the CA output. However, the inverse problem of determining appropriate rules to produce a desired effect is hard [Ganguly *et al.* (2003)]. In general an optimal selection of rules cannot be guaranteed without an exhaustive enumeration of all combinations [Cover and Campenhout (1977)], and this is clearly generally not feasible. We shall describe three more practical approaches for learning appropriate rule sets from training data.

5.1. *Genetic Algorithms*

The most common approach in the literature is to use evolutionary algorithms, and in particular genetic algorithms (GAs). Most of this work is applied to a single, somewhat artificial, example which is a version of the density classification problem on a 1D grid. Given a binary input pattern, the task is to decide if there are a majority of 1s or not, i.e, a single binary outcome. For CAs with rules restricted to small neighbourhoods this is a non-trivial task since the 1s can be distributed through the grid, and so it requires global coordination of distant cells that cannot communicate directly. Early work by Mitchell *et al.* [Mitchell *et al.* (1994)] encountered several difficulties with the GA learning: 1/ breaking of symmetries in early generations for short-term gains, and 2/ the training data became too easy for the CAs in later generations of the GA. Juillé and Pollack [Juillé and Pollack (1998)] tackled the latter problem using GAs with co-evolution. To encourage better learning the training set was not fixed during evolution, but gradually increased in difficulty. Thus, once initial solutions for simple versions of the problem were learnt, they would be extended and improved by evolving the data to become more challenging. Instead of GAs Andre *et al.* [Andre *et al.* (1996)] used a standard genetic programming framework. Since this was computationally expensive it was run in parallel on 64 PCs. Extending the density classification task to 2D grids, Jiménez Morales *et al.* [Morales *et al.* (2001)] again applied standard GA to learn rules.

Applying GAs remains a dominant theme in research into learning CA rules [Bull and Adamatzky (2007); Terrazas *et al.* (2007)], although progress is limited. There are still papers attempting to solve the trivial task of binary image boundary detection using genetic algorithms and CA [Batouche *et al.* (2006); Slatnia *et al.* (2007)]. Another example at the same level attempts (with limited success) to generate simple shapes such as a square, circle, etc., again using genetic algorithms [Chavoya and Duthen (2006)]. Craiu and Lee [Craiu and Lee (2006)] use a minimum description length criterion to automatically learn both neighbourhood size and rules for stochastic CA to perform the task of synthesising binary patterns. However, the system exhaustively considers first all neighbourhood sizes and then all rule parameters, and so they could only demonstrate results on very small CA examples.

5.2. *Greedy Selection*

In comparison to such evolutionary methods, deterministic feature selection methods are extensively used for building classifier systems. In particular, we describe a popular approach called the sequential floating forward search (SFFS) [Pudil *et al.* (1994)] which was found to perform best compared to fourteen other feature selection algorithms (including a genetic algorithm) [Jain and Zongker (1997)]. SFFS was first used for training CA for image processing tasks by Rosin [Rosin (2006)]. The advantages of deterministic feature selection methods over evolutionary methods are that

- they tend to be extremely simple to implement
- their runtime is significantly less
- the quality of their results is as good or better

- being deterministic rather than randomised means that the results are repeatable (which is particularly helpful for other researchers attempting to duplicate published results)
- they do not require the many parameters necessary for genetic algorithms.

The SFFS algorithm can be described as follows. Let $\mathcal{R}_i$ denote the rule set at iteration i and its score be $J(\mathcal{R}_i)$. In our case, $J(\mathcal{R}_i)$ is computed by applying the CA with the rule set $\mathcal{R}_i$ to the input image, and returning the error computed by one of the objective functions described in the next section. The initial rule set $\mathcal{R}_0$ is empty. At each iteration i all rules are considered for addition to the rule set $\mathcal{R}_{i-1}$. Only the rule giving the best score is retained, to make $\mathcal{R}_i$. This process is repeated until no improvements in score are gained by adding rules (an alternative termination rule is when a known desired number of rules has been found). This describes the sequential forward search, which is extended to the sequential floating forward search by interleaving between each iteration the following test. One at a time, each rule in $\mathcal{R}_i$ is removed to find the rule whose removal provides the candidate rule set $\mathcal{R}'_{i-1}$ with the best score. If this score is better than $J(\mathcal{R}_{i-1})$ then $\mathcal{R}_i$ is discarded, $\mathcal{R}_{i-1}$ is replaced by $\mathcal{R}'_{i-1}$, and the process continues with the addition of the i'th rule. Otherwise, $\mathcal{R}'_{i-1}$ is discarded, and the process continues with the addition of the $(i+1)$'th rule to $\mathcal{R}_i$.

5.3. *Identification Algorithms*

Genetic algorithms or greedy selection methods can be used to learn CA rules from training data pairs effectively, but the learning procedure is very time-consuming. If only input/output data pairs (representing start/end CA states) are available, and multiple steps of CA evolution from the start to the end states are required, the above introduced CA rule learning methods are still the state-of-the-art ones. However, if the start and the end together with their intermediate CA states are available, or only one step is needed to evolve from the start to the end states, fast CA learning methods exist. Recent development in CA rule extraction for the cases with intermediate CA states involves parameter estimation methods from the field of system identification [Adamatzky (1994); Billings and Yang (2003a); Billings and Mei (2005); Zhao and Billings (2007); Sun *et al.* (2011); Billings and Yang (2003b)].

The basic idea underlying identification algorithms for CA rule learning is as follows. CA rules are modelled as a linear-in-parameter model, and the model parameters are estimated based on some error minimisation criterion using the start, the intermediate, and the end states. The model together with the estimated parameter values is then used to retrieve the CA rules. Not only are such approaches fast, but are also effective for larger neighbourhood sizes than are practical for the previous methods. The current fastest CA identification algorithm was developed by Sun *et al.* [Sun *et al.* (2011)], and the following is a brief description of the algorithm.

Let $x_i(t)$ be the state value of cell c_i at evolution step t, and $x_i^l(t)(l = 1, \ldots, m)$ be the state values of the cells in c_i's neighbourhood at step t. The state value of c_i at step $t + 1$ is then given by

$$x_i(t + 1) = \sum_{j=0}^{2^m - 1} \theta^j Q_i^j(t) + \epsilon_i(t), \tag{4}$$

where $Q_i^j(t)$ is the value of jth neighbourhood pattern defined by

$$Q_i^j(t) = \prod_{l=1}^{m} b_j^l(x_i^l(t)), \tag{5}$$

and b_j^l is defined as the coefficient of 2^{l-1} in j when j is written as a binary number. θ^j is either 0 or 1, and $\theta^j = 0$ represents the CA rule that when the neighbourhood state combination is pattern $Q_i^j(t)$, $x_i(t + 1)$ takes value 0, while $\theta^j = 1$ means that $x_i(t + 1)$ takes value 1. $\epsilon_i(t)$ is a noise term, which is to be minimised in the estimation of the parameters.

Note that here m is the neighbourhood size and the number of neighbourhood patterns is 2^m. Although in the context of image processing in this chapter, the neighbourhood size is 9 (including the central cell), and the number of patterns is reduced from $2^9 = 512$ to $51 \times 2 = 102$ when considering rotation and mirror symmetry, we used all the 512 original patterns in Eqn. (4) in the identification experiments in this chapter. However, the readers can easily generalise Eqn. (4) to the case of 102 patterns.

When noise variance is used as the minimisation criterion, the parameter estimation problem is solved by

$$\{\hat{\theta}^j\} = \arg\min \frac{1}{TC} \sum_{t=1}^{T} \sum_{i=1}^{C} \left(x_i(t + 1) - \sum_{j=0}^{2^m - 1} \theta^j Q_i^j(t) \right)^2, \tag{6}$$

where T is the number of evolution steps, and C the number of cells.

Considering that $x_i(t + 1)$, θ^j and $Q_i^j(t)$ always take values 0 or 1, the solution of Eqn. (6) is reduced to

$$\hat{\theta}^j = \begin{cases} 1, & \text{if } r^j > 0, \\ 0, & \text{if } r^j \leq 0, \end{cases} \tag{7}$$

where

$$r^j = \frac{1}{TC} \sum_{t=1}^{T} \sum_{i=1}^{C} (x_i(t + 1) - \bar{x}_i(t + 1)) Q_i^j(t), \tag{8}$$

and $\bar{x}_i(t + 1)$ denotes the logical NOT of $x_i(t + 1)$.

Sun *et al.* [Sun *et al.* (2011)] also discussed automatic selection of the CA neighbourhood size and gave an incremental neighbourhood algorithm that uses the Bayesian information criterion. The readers are referred to the original paper [Sun *et al.* (2011)] for details.

5.4. *Objective Functions*

Whichever optimisation method is used, an objective function is required, and its quality obviously has a crucial effect on the final results. For binary images the simplest objective function is the Hamming distance, or for images with more intensity values then the root mean square (RMS) error between the input and target image is a straightforward measure.

However, it is well known that RMS (and related) values have limitations. In particular, given that they do not involve inter-pixel relationships they often do not capture perceptual similarity. One possible improvement which incorporates spatial information into the comparison of binary images A and B is to use the Hausdorff distance

$$H(A, B) = \max_{a \in A} \min_{b \in B} ||a - b||.$$

For grey level images a well known image similarity measure is the Structural Similarity index (SSIM) [Wang *et al.* (2004)] which takes luminance, contrast and structure into account:

$$SSIM(A, B) = \frac{(2\mu_A\mu_B + C_1)(2\sigma_{AB} + C_2)}{(\mu_A^2 + \mu_B^2 + C_1)(\sigma_A^2 + \sigma_B^2 + C_2)}.$$

where the constants are set to $C_1 = (0.01 \times 255)^2$ and $C_2 = (0.03 \times 255)^2$. The SSIM index is then applied locally using an 11×11 circular-symmetric Gaussian weighting function, and the mean over the image is used as the final similarity measure.

6. Applications of Cellular Automata to Image Processing

6.1. *Binary Image Denoising*

The effectiveness of the CA is first demonstrated on binary image denoising. To demonstrate the differences between the three training strategies described in section 5, each was applied to learn the rule sets for the five input conditions (corresponding to three types/levels of noise).

The GA used in our study was a simple steady state GA [Syswerda (1989); Mumford-Valenzuela *et al.* (2003)]. During crossover, the first parent is selected deterministically in sequence, but the second parent is selected uniformly, at random. If new offspring duplicate existing values for the objective function, then they automatically "die". The GA was run for a fixed number of generations (100) and the population size was set to 100. For each image, the GA was run ten times with different initialisations of the random number generator.

For the identification algorithm, we considered two cases: one uses the 3×3 neighbourhood with size $m = 9$ which includes neighbours $\{c(i, j) : |i - i_0|, |j - j_0| \leq 1\}$ corresponding to the central cell $c(i_0, j_0)$, and the other uses the neighbourhood with size $m = 13$ which includes neighbours $\{c(i, j) : |i - i_0| + |j - j_0| \leq 2\}$.

Table 1. RMS errors of filtered versions of the binary test image corrupted by salt and pepper noise with probabilities $p = 0.01, 0.1, 0.3$; black squares containing a white central pixel; and black 'T's and white bars. For the GA the mean RMS over the ten runs is given with the standard deviation in brackets. For the median filter the RMS is given for the number of iterations (which is indicated in brackets) that gave the best results on the *test* image.

noise type	original image	CA				median filter
		GA	SFFS	Identification		
				$m = 9$	$m = 13$	
S & P, $p = 0.01$	17.9	14.6 (0.3)	14.1	14.1	12.7	42.5 (1)
S & P, $p = 0.1$	57.0	41.2 (0.4)	32.4	32.3	30.6	45.0 (1)
S & P, $p = 0.3$	99.0	49.5 (1.0)	47.6	49.2	49.0	53.3 (2)
B & W square	55.7	48.9 (5.6)	32.0	32.0	25.8	55.7 (0)
'T's and bars	64.1	65.5 (1.9)	48.2	54.3	44.1	64.1 (0)

Two large binary images (1536 × 1024 pixels) were constructed, one each for training and testing, and consisted of a composite of several 256 × 256 subimages obtained by thresholding standard images. Different types and levels of noise were added, and a set of rules learnt for each. Salt and pepper noise was added with probabilities $p = 0.01, 0.1, 0.3$. The next two types of noise are non-standard, and were included to demonstrate how the CA can learn good rules as long as it has good training data. Black squares (3 × 3) containing a single white central pixel were added; the final noise type consisted of added black 'T's and white bars. Examples of these noise types are shown in figure 4.

In all instances the rules were run for up to 100 iterations. For comparison, results of filtering are providing using a 3 × 3 median filter with the optimal number of iterations determined for the *test* image, giving the median a favourable bias.

As table 1 shows, for small amounts of noise the median filter degrades the image rather than improves it. Likewise, it is unable to cope with the unusual noise types (squares and 'T's/bars). In comparison, the CA (when trained using SFFS or the identification algorithm) consistently succeeds in denoising the images, and also improves on the median filter's results in all cases. The results are demonstrated visually on a small image window in figure 4.

For the lowest level of salt and pepper noise ($p = 0.01$) the CA learns to use a single rule to remove isolated pixels: ▦. As the noise level increases the number of rules required increases. For $p = 0.1$ the rules are: ▦ ▦ ▦ and the rules for $p = 0.3$ are: ▦ ▦ ▦ ▦ ▦ ▦. For noise consisting of the black square with white central pixel the following rules were learnt: ▦ ▦ ▦ ▦ ▦ ▦ ▦ ▦ ▦. Unlike the salt and pepper noise in which the rules also applied in their inverted versions (i.e. swapping the roles of black and white), the learnt rules for square were (automatically) determined only to operate when the central pixel was black. For the black 'T's and white bars noise the rules learnt for patterns with a black central pixel were: ▦ ▦ ▦ ▦ ▦ ▦ and one further rule to be applied for white central pixels: ▦.

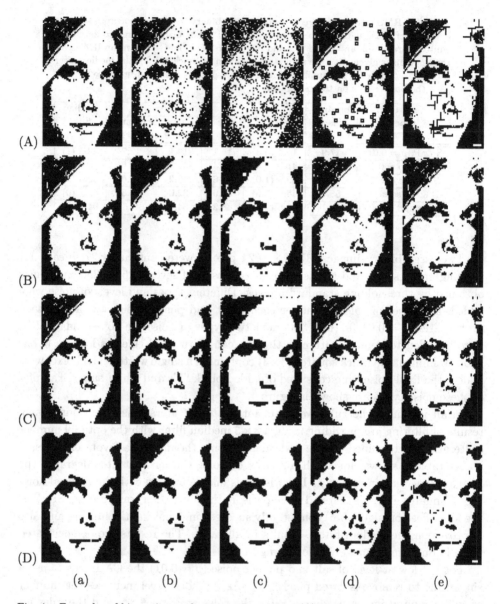

Fig. 4. Examples of binary image denoising. Rows show (A) the input images, and the results after denoising using (B) CA trained by SFFS , (C) CA generated by identification algorithm (D) median filtering. Columns (a)–(e) show different noise conditions: salt and pepper $p = 0.01$, salt and pepper $p = 0.1$, salt and pepper $p = 0.3$, added black square with white central pixel, added black 'T's and white bars. One iteration of the median filter was applied except for (c) for which two iterations were applied.

Table 1 shows that the GA was not competitive with the other training methods, generating much worse results. Furthermore, it can be seen from table 2 that its runtime is much greater. From these experiments it is evident that identification

Table 2. Timings (in seconds) for learning the rules for the binary image filtering tasks.

noise type	CA		
	GA	SFFS	Identification
S & P, $p = 0.01$	1427	46	8
S & P, $p = 0.1$	8717	290	8
S & P, $p = 0.3$	14230	843	8
B & W square	2776	2493	8
'T's and bars	25030	4483	8

algorithm generally performed the best out of the three training methods, both in terms of accuracy and computational efficiency.

6.2. *Gray Level Denoising*

Table 3. RMS errors of filtered versions of the gray level test image corrupted by Gaussian noise, single pixel salt and pepper noise, salt and pepper noise affecting 3×3 pixel blocks, and randomly recoloured stripes.

noise type	original	CA		shock filter	AM
		TD	3-state		
Gaussian, $\sigma = 25$	23.7	13.9	13.9	16.1	18.9
S & P, $p = 0.6$	113.5	25.0	20.9	58.5	18.8
3×3 S & P, $p = 0.01$	43.1	14.7	11.8	37.9	12.1
stripe, $p = 0.8$	83.3	54.0	35.4	62.7	65.1

Using either of the two techniques described in section 5 the CA for binary image denoising can be extended to perform gray level denoising. Results of experiments are shown here along with a comparison made with the complex shock filter [Gilboa *et al.* (2004)][3] and the adaptive median (AM) filter [Hwang and Haddad (1995)]. The CA was trained and tested on the grey level versions of the 1536×1024 image mosaics used for the binary image denoising experiments in section 6.1. The SFFS training strategy and the RMS objective function were used, and the number of rule iterations was set to 100.

Various types of noise were added: salt and pepper, and 3×3 blocks of salt and pepper. As with the binary image denoising a non-standard noise was included to show the capabilities of learning noise specific rules. This structured noise was created by randomly replacing (with probability p) each row or column in the image with a random intensity (constant along the stripe). Thus the probability of corrupting a pixel is $p - \left(\frac{p}{2}\right)^2$.

---

[3]Gilboa's code for his complex shock filter was used with its default parameters: number of iterations = 30, $|\lambda| = 0.1$, $\tilde{\lambda} = 0.2$ and $a = 2$.

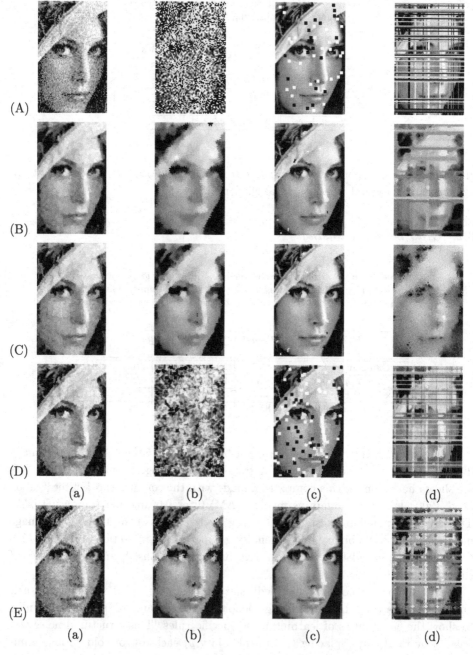

(A)

(B)

(C)

(D)

(a)   (b)   (c)   (d)

(E)

(a)   (b)   (c)   (d)

Fig. 5.   Examples of grey level image denoising. Rows show (A) the input images, and the results after denoising using (B) CA with threshold decomposition, (C) 3-state CA, (D) complex shock filter, and (E) the adaptive median filter. Columns (a)–(d) show different noise conditions: Gaussian $\sigma = 25$; salt and pepper $p = 0.6$, single pixel; salt and pepper, 3×3 block, $p = 0.01$; stripe $p = 0.8$.

The RMS errors resulting from denoising are listed in table 3, and examples of the results from applying the various methods are displayed in figure 5. The CA generally performs well, and the 3-state CA gives lower errors than the threshold decomposition CA. For the case of Gaussian noise it outperforms the shock filter and adaptive median, although there certainly exist other denoising methods that work better on this type of noise (although not so well on other noise types), e.g. the hidden Markov trees (HMT) applied to wavelet coefficients [Romberg *et al.* (2001)] that were previously compared against CA for denoising in [Rosin (2010)]. The adaptive median is well suited to salt and pepper noise (i.e. where the noise is at the extreme ends of the intensity range) but does not do so well for the stripe noise which consists of random intensities.

6.3. *Edge Detection*

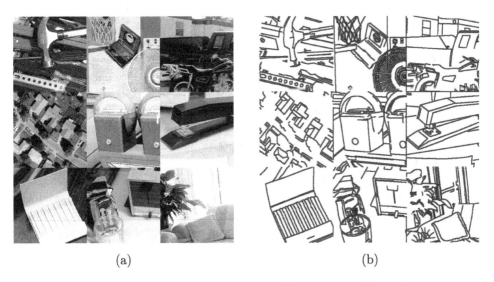

(a)   (b)

Fig. 6.   Training data for edge detection task. (a) input image, (b) target image.

We now demonstrate the training and application of CA to edge detection. To create a good set of training data a 750×750 image mosaic was created using sub-images from the University of South Florida data set which contains images along with manually generated ground truth edges – see figure 6. Since there is likely to be some positional error in the ground truth edges (which are one pixel wide) the target edge map was dilated twice, with the new edges set each time to an increasingly lower intensity. This process is similar to blurring the edge map whilst avoiding creating local intensity maxima at junctions.

Since the 3-state CA representation discards most of the intensity magnitude information there is insufficient information to compute edge magnitudes, and therefore this approach is not appropriate for performing edge detection. However, the

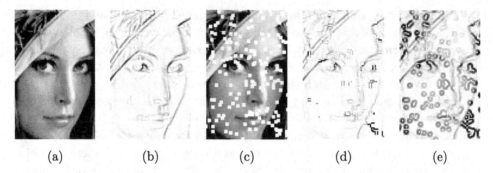

(a)   (b)   (c)   (d)   (e)

Fig. 7. Edge detection using the threshold decomposition CA. (a) input image, (b) results from CA, (c) noisy input image, (d) results of filtering figure 7c using a CA trained on corrupted version of figure 6a, (e) results of filtering figure 7c using the Sobel edge detector.

threshold decomposition method works well as demonstrated in figure 7b which used the SFFS training strategy and RMS objective function.

In fact, the rule set is exceedingly simple, consisting of a single rule, specifying that any white pixel in a 3×3 homogeneous (i.e. all white) neighbourhood is flipped. For each of the binary images that the input is decomposed into, this causes all white pixels to be replaced by black except for pixels adjacent to black pixels in the input image. Thus, a black image is formed containing a one pixel wide white strip along the original black/white transitions, which when summed at the reconstruction stage of the threshold decomposition produces the edge magnitudes.

An advantage of the CA methodology is that rules can be combined to perform multiple tasks. This is demonstrated by adding noise to the training image in figure 6a, specifically 3×3 blocks of white pixels. Retraining the CA produces rules that simultaneously perform denoising whilst detecting edges. The results of applying the new rules to the noisy image in figure 7c are shown in figure 7d. In comparison to the Sobel edge detector the CA is much more successful in being able to robustly detect the edges, the results being only minimally affected by the salt and pepper noise.

7. Conclusions and Discussion

As demonstrated in this chapter, it is possible to automatically learn rules for cellular automata that can effectively perform image processing tasks. Examples have been given here for image denoising and edge detection, and further examples (convex hull, connected set morphology, ridge detection) were shown in [Rosin (2006, 2010)]. A benefit of the CA approach is that it is flexible, and can easily be applied to a variety of tasks; the same architecture can be used, and the CA just needs retraining with new data.

It is crucial to have good methods for automatically learning the CA rules, and we have described those that are currently popular in the literature. Although genetic algorithms are the most commonly used they were shown to be the worst choice, being both slow and providing relatively poor solutions. In comparison, the greedy algorithms (such as sequential floating forward search) were faster and gave better results. Under certain circumstances, the system identification type algorithm was able to perform very well, with substantially reduced computation time. However, it assumes the availability of training data containing the desired outputs at intermediate iterations of the CA, which are unlikely to be available for the image processing tasks. Therefore, the system identification algorithm is most effective for image processing when the number of iterations required for the task is small (ideally one).

Another important factor is the objective function used for optimisation by the rule selection algorithms. Although several possibilities were considered, it has not been shown that the more sophisticated functions that incorporate spatial information provided significant benefits over the simpler functions such as root mean square error [Rosin (2006, 2010)].

There are several directions for future research:

- Although both the threshold decomposition and the 3-state approaches to extending the binary state cellular automata to operate on gray level images were effective, it would be worthwhile investigating alternatives that are better able to capture the gray level information while maintaining reasonable computational complexity.
- The system identification algorithm is very promising, and if it could be extended to cope with missing input/output data pairs this would greatly increase its usefulness.
- Most of the results described here and in the literature used small neighbourhoods (e.g. 3×3). It would be of interest to experiment with larger neighbourhoods which definitely provide the possibility of better results. However, this would also increase the cost in terms of storage requirements, computation time, and would need larger amounts of training data for effective rule identification.

References

Adamatzky, A. (1994). *Identification of Cellular Automata* (Taylor & Francis, London, UK), ISBN 0-7484-0172-5.

Andre, D., III, F. B. and Koza, J. (1996). Discovery by genetic programming of a cellular automata rule that is better than any known rule for the majority classification problem, in *Proc. 1st Conf. on Genetic Programming* (MIT Press), pp. 3–11.

Batouche, M., Meshoul, S. and Abbassene, A. (2006). On solving edge detection by emergence, in *Int. Conf. on Industrial, Engineering and Other Apps. of Applied Intelligent Systems*, Vol. LNAI 4031, pp. 800–808.

Billings, S. and Yang, Y. (2003a). Identification of probabilistic cellular automata, *IEEE Transactions on Systems Man and Cybernetics, Part B: Cybernetics* **33**, 2, pp. 225–236.

Billings, S. and Yang, Y. (2003b). Identification of the neighborhood and CA rules from spatio-temporal CA patterns, *IEEE Trans. on Systems, Man and Cybernetics, Part B* **33**, 2, pp. 332–339.

Billings, S. A. and Mei, S. S. (2005). A new fast cellular automata orthogonal least-squares identification method, *International Journal of Systems Science* **36**, 8, pp. 491–499.

Bull, L. and Adamatzky, A. (2007). A learning classifier system approach to the identification of cellular automata, *J. Cellular Automata* **2**, 1, pp. 21–38.

Chaudhuri, P., Chowdhury, D., Nandi, S. and Chattopadhyay, S. (1997). *Theory and Applications: Additive Cellular Automata* (IEEE Press).

Chavoya, A. and Duthen, Y. (2006). Using a genetic algorithm to evolve cellular automata for 2D/3D computational development, in *Genetic and Evolutionary Comp. Conf.*, pp. 231–232.

Cover, T. and Campenhout, J. V. (1977). On the possible orderings in the measurement selection problem, *IEEE Trans. on Systems, Man and Cybernetics* **7**, 9, pp. 657–661.

Craiu, R. and Lee, T. (2006). Pattern generation using likelihood inference for cellular automata, *IEEE Trans. on Image Processing* **15**, 7, pp. 1718–1727.

de Saint Pierre, T. and Milgram, M. (1992). New and efficient cellular algorithms for image processing, *CVGIP: Image Understanding* **55**, 3, pp. 261–274.

Dyer, C. and Rosenfeld, A. (1981). Parallel image processing by memory-augmented cellular automata, *IEEE Transactions on Pattern Analysis and Machine Intelligence* **3**, 1, pp. 29–41.

Fitch, J., Coyle, E. and Gallagher, N. (1984). Median filtering by threshold decomposition, *Acoustics, Speech and Signal Processing, IEEE Transactions on* **32**, 6, pp. 1183–1188.

Ganguly, N., Sikdar, B., Deutsch, A., Canright, G. and Chaudhuri, P. (2003). A survey on cellular automata, Tech. Rep. 9, Centre for High Performance Computing, Dresden University of Technology.

Gardner, M. (1970). The fantastic combinations of John Conway's new solitaire game "life", *Scientific American* , pp. 120–123.

Gilboa, G., Sochen, N. and Zeevi, Y. (2004). Image enhancement and denoising by complex diffusion processes, *IEEE Trans. on Pattern Analysis and Machine Intelligence* **26**, 8, pp. 1020–1036.

Hernandez, G. and Herrmann, H. (1996). Cellular automata for elementary image enhancement, *Graphical Models and Image Processing* **58**, 1, pp. 82–89.

Holland, J. (1970). Logical theory of adaptive systems, in A. Burks (ed.), *Essays in Cellular Automata* (University of Illinois Press).

Hwang, H. and Haddad, R. (1995). Adaptive median filters: new algorithms and results, *IEEE Trans. on Image Processing* **4**, 4, pp. 499–502.

Jagadish, H. and Kailath, T. (1989). Primitive cellular automata, threshold decomposition, and ranked order operations, *IEEE Trans. Comput.* **38**, pp. 148–149.

Jain, A. and Zongker, D. (1997). Feature-selection: Evaluation, application, and small sample performance, *IEEE Trans. on Pattern Analysis and Machine Intelligence* **19**, 2, pp. 153–158.

Juillé, H. and Pollack, J. (1998). Coevolving the ideal trainer: Application to the discovery of cellular automata rules, in *Proc. 3rd Conf. on Genetic Programming* (Morgan Kaufmann), pp. 519–527.

Karafyllidis, I., Ioannidis, A., Thanailakis, A. and Tsalides, P. (1997). Geometrical shape recognition using a cellular automaton architecture and its VLSI implementation, *Real-Time Imaging* **3**, pp. 243–254.

Kusch, I. and Markus, M. (1996). Mollusc shell pigmentation: cellular automaton simulations and evidence for undecidability, *J. Theor. Biol.* **178**, pp. 333–340.

Mitchell, M., Hraber, P. and Crutchfield, J. (1994). Evolving cellular automata to perform computation: Mechanisms and impedients, *Physica D* **75**, pp. 361–391.

Morales, F. J., Crutchfield, J. and Mitchell, M. (2001). Evolving two-dimensional cellular automata to perform density classification: a report on work in progress, *Parallel Computing* **27**, pp. 571–585.

Mumford-Valenzuela, C., Vick, J. and Pearl, Y. (2003). Heuristics for large strip packing problems with guillotine patterns: An empirical study, in D. Du and P. Pardalos (eds.), *Metaheuristics: Computer Decision-Making* (Kluwer Academic Press).

Preston, K. and Duff, M. (1984). *Modern Cellular Automata-Theory and Applications* (Plenum Press).

Pudil, P., Novovicova, J. and Kittler, J. (1994). Floating search methods in feature-selection, *Pattern Recognition Letters* **15**, 11, pp. 1119–1125.

Roberts, F. and Tesman, B. (2005). *Applied Combinatorics* (Pearson/Prentice-Hall).

Romberg, J., Choi, H. and Baraniuk, R. (2001). Bayesian tree-structured image modeling using wavelet domain hidden markov models, *IEEE Trans. on Image Processing* **10**, 7, pp. 1056–1068.

Rosenfeld, A. and Pfaltz, J. (1968). Digital distance functions on digital pictures, *Pattern Recognition* **1**, 1, pp. 33–61.

Rosin, P. (2006). Training cellular automata for image processing, *IEEE Trans. on Image Processing* **15**, 7, pp. 2076–2087.

Rosin, P. (2010). Image processing using 3-state cellular automata, *Computer Vision and Image Understanding* **114**, 7, pp. 790–802.

Slatnia, S., Batouche, M. and Melkemi, K. (2007). Evolutionary cellular automata based-approach for edge detection, in *Int. Workshop on Fuzzy Logic and Applications*, Vol. LNAI 4578, pp. 404–411.

Sun, X., Rosin, P. and Martin, R. (2011). Fast rule identification and neighborhood selection for cellular automata, *IEEE Transactions on Systems Man and Cybernetics, Part B: Cybernetics* **41**, 3, pp. 749–760.

Syswerda, G. (1989). Uniform crossover in genetic algorithms, in *Proc. Third Int. Conf. on Genetic Algorithms* (Lawrence Erlbaum Associates), pp. 2–9.

Terrazas, G., Siepmann, P., Kendall, G. and Krasnogor, N. (2007). An evolutionary methodology for the automated design of cellular automaton-based complex systems, *J. Cellular Automata* **2**, 1, pp. 77–102.

Ulam, S. M. (1962). On some mathematical problems connected with patterns of growth of figures, *Proc. Symp. Appl. Math.* **14**, pp. 215–224.

von Neumann, J. (1966). *Theory of Self-Reproducing Automata* (University of Illinois Press).

Wang, L. and He, D. (1990). Texture classification using texture spectrum, *Pattern Recognition* **23**, pp. 905–910.

Wang, Z., Bovik, A., Sheikh, H. and Simoncelli, E. (2004). Image quality assessment: from error visibility to structural similarity, *IEEE Trans. on Image Processing* **13**, 4, pp. 600–612.

Wolfram, S. (1994). *Cellular Automata and Complexity Collected Papers* (Addison-Wesley).

Zhao, Y. and Billings, S. (2007). The identification of cellular automata, *Journal of Cellular Automata* **2**, 1, pp. 47–65.

COMPUTER VISION IN MARINE FISHERY IDENTIFICATION

Bichuan Shen[1] and Chi Hau Chen[2]

University of Massachusetts Dartmouth, USA
[1]*bshen@umassd.edu*
[2]*cchen@umassd.edu*

This chapter focuses on marine fishery identification using computer vision techniques. Computer vision can help with the study of fish population and species by automatically counting and classifying live fish in situ, and is potentially useful for harvest and marketing. However, the tasks are quite challenging mostly due to large inter-class, and intra-class variations of fish species and accurate statistical models are usually not available. We propose a novel modeled approach to address these issues. Firstly we model the prior fish density as mixture of Gaussians (MOG) which is estimated by an expectation maximization (EM) based clustering algorithm. Secondly, we present a joint scheme to achieve the scale and deformation invariance by using a grid based shape descriptor to represent fish hypotheses and support vector learning. This is conducted by taking 32x32 cell grid of the hypothesized fish region and the shape pixels falling within each cell are counted as histogram values. Shape quantization in this 1024-dimension would have sufficient representation capability and enough allowance for non-rigid deformation and scalability. The modeled shape hypotheses are learned by support vector machine (SVM) classifiers and fish could thus be sorted out from non-fish objects and backgrounds. Two spatial metrics of the aspect ratio (length/width) and eccentricity are selected empirically to identify fish species. We also demonstrate that the spectral metrics can be employed to classify fish categories pretty well besides the spatial metrics. Experimental results show that our proposed methods are very promising.

1. Introduction

Marine fishing and the industry are closely related to our daily life and have big impacts on the economy and society. They provide us not only vital foods, but also many other important resources such as medicine and oil. Close study and deep understanding are thus highly desirable and necessary for many aspects including farming, monitoring, harvest, data collection, management, and stock assessment.

Up to present, however, manual grading, recording, measuring and sorting are still the major works performed onboard all commercial and research fishing vessels. These processes are quite tedious by involving intensive human efforts, it is therefore that the automation of these processes is highly desirable.

The aim of this chapter is to study automatic identification of fish and their species by jointly using computer vision approaches. Computer vision along with pattern classification techniques have been applied to the fishery study and in recent years. For example, [1] describes a computer vision system, CatchMeter, capable of identifying and measuring seven fish species which are placed on a conveyor belt in a well controlled illumination and projection, with a high sorting accuracy of 99.8% and a throughput of 30,000 fish/hour. [2] presents a fish sorting algorithm and has applied it to three fish species based on the moment-invariants (MI) of the shapes of the whole fish body, head and tail. Relevant works have also been reported for studying underwater fish stocks and habitats by taking advantage of widely available digital video cameras and analysis software [3].

Biologically, fish is a unique and distinctive species. In our study, however, we only focus on their visual appearance and photometric characteristics and study their dichotomy from computer vision perspectives. General inspection of fish species suggests the following observations,

- The class of fish is significantly distinctive from non-fish species and the backgrounds and fish species have different global and local characteristics that are different from each other.

- Shape and color are two prominent features which can be used for sorting fish, as most fish species have distinctive overall shape boundaries and contours, color distributions, and textures.

- The within-class intrinsic feature variations of the fish category and each fish species are quite large. Different fish species also have high between-class variations. And these variations are not always consistent.

- Fish objects on imagery show a lot of extrinsic variations in shape, pose and orientation, scale, color, illumination and contrast, occlusion and shadow.

We call the factors arising from the biological differences intrinsic, and extrinsic otherwise. Both factors affect the fish's luminescent and chromatic appearance. For example, catches in different temporal seasons and from spatial sea areas may show large variations in fish shapes, sizes, colors, and weights.

Indoor and outdoor lighting and illumination also play a significant role inducing variations. Fish variations of shape, size and color in a less-controlled setup tend to be much bigger and more complex than in a well-controlled environment. It is these large variations of fish species that cause a major challenge for the task of fish identification and classification. For example, fish can be placed at varying orientations on a moving conveyor belt, and the system is still required to identify and recognize their contours and shapes in different angles.

The objectives of our study include counting the number of fish and categorizing them into different fish species. This would potentially enhance the survey and population study of marine fishery. The fish imagery we collect includes fish species of redfish, haddock, pollock, codfish, yellowtail flounder, and winter flounder and fish piles. Shape and color are the two salient features and therefore can be used for identifying the fish from non-fish objects. They have been widely selected as the discriminative features in many object recognition and image retrieval systems. For example, the shapes of these fish are most likely to be smooth with a lot of curvatures on their belly and backs. Their heads and tails look like triangles.

To reach these goals, we have employed a bottom-up shape modeled framework for fish identification. The prior densities of fish species are modeled as Gaussian mixtures, and we can generate candidate regions through expectation maximization (EM) clustering. Over-segmentation and edge map may be combined to deal with neighboring and overlapping objects with occlusion which might exhibit similar colors, shapes, and spatial textures. Then the shape is extracted from the generated fish hypotheses and it is represented in a grid based model. Geometrically and empirically, fish shape is more robust and less sensitive to variations and noise as compared to other features. Additionally, fish scales can be normalized by a grid based description. This method relies on the external appearance of fish and not on its intrinsic biological characteristics which is beyond the scope of our study. And it might be sufficient to identify less complex fish species when they are well aligned and simply separable in feature space.

The chapter is organized as follows. Section 2 presents a mixture of Gaussian modeling of the fish prior distribution and the expectation maximization clustering. A grid modeled shape representation and description is given in Section 3. The support vector learning and case studies of fish hypotheses are provided in Section 4. Finally discussions and conclusions are presented in Section 5.

2. Fish identification and modeling

2.1. *Hypotheses generation by Gaussian mixtures*

We model fish's probability distributions as Gaussian mixtures and formulate the task of fish identification as finding a constrained maximum-likelihood solution. Fish hypotheses of region growing and grouping and its posteriori distribution can be jointly estimated by global context and local fish and non-fish object priors, a shape or color fish pattern model, and a measured fish pattern from a Bayesian perspective. The likelihood is then evaluated and its maximization is used to validate that a pattern belongs to a certain label.

According to the Bayesian theorem, the posteriori probability that a multi-dimensional feature $\mathbf{x}$ belongs to the fish pattern space Ω can be expressed as the product between the prior fish class probability $P(\Omega)$ and the class conditional probability $P(\mathbf{x}|\Omega)$. That is,

$$P(\Omega|\mathbf{x}) = P(\mathbf{x}|\Omega) P(\Omega) \tag{1}$$

The class conditional probability can be modeled as normal distribution $P(\mathbf{x}|\Omega) = N(\mathbf{x}|\mu, \Sigma)$, with the mean vector μ and covariance Σ. Let $\bar{\Omega}$ represent all the non-fish object patterns for a binary detection problem, then a simple decision for the fish hypotheses can be made

$$\text{H=fish: } P(\Omega|\mathbf{x}) > P(\bar{\Omega}|\mathbf{x}) \tag{2}$$

For fishery images taken in real environments, there may exist multiple fish objects with different colors and shapes and more than two fish species. Therefore, the probability distributions of these discrete patterns can be modeled as mixture of Gaussians (MoG) whose densities are a linear combination of each individual multivariate Gaussian cluster Ω_k, $k = 1, \cdots, K$, with each distribution described as $N(\mu_k, \Sigma_k)$. Therefore the likelihood can be written as

$$p(\mathbf{x}) = \sum_{k=1}^{K} p(\Omega_k) N(\mathbf{x}|\mu_k, \Sigma_k) \tag{3}$$

Here the ensemble is described by the parameter set $\theta = \{K, p(\Omega_k), \mu_k, \Sigma_k\}$ including the mixture number K, the class prior density $p(\Omega_k)$, the mean vector μ_k and the covariance matrix Σ_k of each class respectively.

By taking the logarithmic on both sides of equation (3), we can obtain the following ensemble log-likelihood of the Gaussian mixture model (GMM)

$$L(\mathbf{\theta}) = \sum_{k=1}^{K} \alpha_k \log\left(p_k\left(\mathbf{x}|\mathbf{\theta}_k\right)\right) \qquad (4)$$

Here α_k are the coefficients associated with the prior mixture densities.

Therefore, given the prior fish models and the estimated region grouping hypotheses, together denoted as $\mathbf{X}$, the objective of the fish identification and classification task is to find a maximum-likelihood solution

$$\text{H=fishclass: } \mathbf{\theta}^* = \arg\max_{\mathbf{\theta}} L\left(\mathbf{\theta}|\mathbf{X}\right) \qquad (5)$$

2.2. *Fish segmentation by expectation maximization clustering*

An analytical solution of the above maximum likelihood estimation tends to be hard and computationally prohibitive. In practice, these parametric estimation problems can be tackled by the expectation maximization algorithm, which includes the following two major steps: E-step and M-step [4].

Assume that there are a number of M mixing components and the data $\mathbf{z}_n$,

1. Initialization: Select randomly (or heuristically) $\Theta^{(0)}$. Set $i=0$. 2. Expectation step (E step): Using the observed data set $\mathbf{z}_n$ and the estimated parameters $\Theta^{(0)}$, calculate the expectations $\overline{\mathbf{x}}_n$ of the missing values $\mathbf{x}_n$. 3. Maximization step (M step): Optimize the model parameters $\Theta^{(i+1)} = \left(\hat{\alpha}_j, \hat{\mathbf{\mu}}_j, \hat{\mathbf{\Sigma}}_j\right)$ by maximum likelihood estimation using the expectation $\overline{\mathbf{x}}_n$ calculated in the previous step. 4. Stop if the likelihood did not change significantly in step 3. Else, increment i and go to step 2.

Theoretically, the expectation maximization algorithm is guaranteed to converge to the maximum of the objective log-likelihood function. After the optimal parameter set is converged, therefore, the identification of fish hypotheses would be that the feature vector $\mathbf{x}$ is assigned to class i if $p_i\left(\mathbf{x}|\mathbf{\theta}_i\right)/p_j\left(\mathbf{x}|\mathbf{\theta}_j\right) \geq 1$, for all $j \neq i$, it is assigned to class j otherwise.

In our fish study, segmentation has been employed as an important preprocessing approach to extract the prior fish knowledge and distribution. And it is implemented by the above expectation maximum clustering algorithm to estimate the mixture of Gaussian distributions. Segmentation has been widely used in many computer vision applications, medical image processing, and motion analysis. It essentially labels similar patterns in a feature space to the same group and dissimilar ones to different groups. The semantics of similarity among patterns could be defined in many metrics, such as the luminescent intensity, spectral color, spatial geometry, structural texture, and motion information. Two feature vectors in a multi-dimensional space are similar when their distance is less than a pre-defined threshold.

(a)

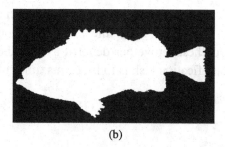

(b)

Fig. 1. Fish segmentation example. (a) Redfish. (b) EM segmented hypothesis.

We select an example of the original redfish and its segmented fish hypothesis in Figs. 1(a) and 1(b) respectively. In the right figure, the fish hypothesis region and non-fish background are labeled in different colors. As shown, the partition boundary is satisfactorily separated from the more homogeneous background area, as all the intrinsic redfish components and salient features, such as the jaw, gill, operculum, fins, scales, and tail have been successfully fused into one merged hypotheses, in consideration of their large spectral and spatial variations. This fish imagery is segmented in the RGB color space. Other spectral or spatial spaces can be used as well.

The advantage of using segmentation to acquire priori fish distribution is obvious. We can avoid dealing with the complicated issues such as identifying the complex spatial biological features around the fish eye, mouth, gill, fins, scales, textures and tail, and accommodating feature invariance to color variations, orientation and viewpoint changes, shape deformations, scale shrinkage and elongation at the early processing stages. This process also has the additional benefit to avoid local extrema and achieves better computational efficiency.

In the implementation, we have tested the clustering and compared several trained classifiers using the nearest neighborhood (knnc), linear (ldc), quadratic (qdc) and MoG (mogc) classifier models. Empirically, we have randomly selected 10% of the whole data set to train the clustering models, and used the rest for testing. The segmentation results in two-dimensional feature space are depicted in Figs. 2(a), 2(b), 2(c), and 2(d) for comparison.

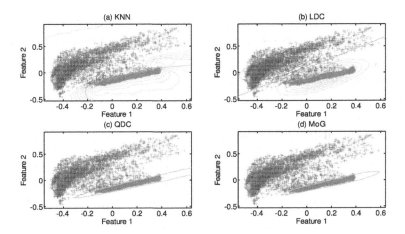

Fig. 2. Classifier comparison for fish segmentation in feature space.

Here the red markers * in feature space are more centralized with much smaller variance and correspond to the more uniform background area in Fig. 1(a). The blue markers + are more distributed and stand for the redfish region hypotheses. The equal density contours of each class are also plotted partially in the same color of red and blue as their feature samples respectively.

The green lines delineate the classification decision boundaries. As suggested from the plots, the decision boundaries of these different classifiers also show varying characteristics. The decision boundary of the nearest neighbor classifier seems to divide the feature space approximately at the middle. The partition boundary of the linear classifier is as straight as expected but shifts more to the side of blue markers. The last two between the quadratic classifier and the mixture of Gaussian classifier are very similar and both tightly bound the outer contours of the background distribution. Because the computational complexity of the quadratic classifier is much smaller than the mixture of Gaussian classifier, it is thus selected as the alternative model [4] for our fishery segmentation task.

The performance of fish detection and identification is closely related to the quality and accuracy of model based segmentation. And there exist several factors affecting the segmentation quality. For example, the initialization of cluster prototypes plays a big role in the convergence and performance of EM algorithm. Practically, the initial prototypes and the number of clusters have to be carefully chosen to avoid inconsistent segmentation and trapping in local extrema.

However, there are no general segmentation procedures for various problems and datasets and it may become unstable for complex cases. The fish objects with more uniform color and texture tend to be more easily segmented. For cases where inhomogeneous regions have large feature variations and adjacent fish may overlap and occlude each other, multiple salient features, over-segmentation and edge map can be employed jointly to insure the global optimality and consistency.

3. Fish shape model and descriptor

After the fish region hypotheses are generated from the segmentation preprocessing, the next problem is how to describe and represent the hypothesized fish and foreground. Conventional approaches and feature descriptors may be not sufficient to deal with large fish variations. We believe that shape is the most prominent and robust feature of fish generally speaking. Most shapes are highly curved and smooth around the fish body. In addition, the fish head is usually sharp and tail looks like triangle.

Shape is an important spatial feature and has played a vital role in many image retrieval and object recognition systems. Simple shape descriptors of using region metrics such as aspect ratio of parts [5], circularity, eccentricity, extent, solidity, Euler number are used in object [5][6]. They are easy to compute and intuitively meaningful. A lot of more complex descriptors are developed, such as Fourier descriptors, shape context [7], moment based shape descriptor, Zernike moment based shape descriptor. Shape descriptors in traditional low order moments are much easier to compute and have meaningful physical interpretation than higher order moments. Zernike moment based shape features are more difficult to compute by mapping shapes onto unit circle.

However, the deformation and complexity of fish appearance is quite large, and vary a lot according to different fish species, sizes, contours, orientations, poses and placements. Therefore, a rigid shape model such as simple ellipsoids cannot be well applied and a more flexible shape model is needed. Deformable template based approaches have been attempted by researchers to allow for inclusion of object-specific priors in the model. For example, [8] introduces a deformation term for a constrained parametric shape model to represent articulated object.

We have constructed a grid based shape descriptor to well represent the fish hypotheses. This is conducted by taking a 32x32 cell grid of the candidate fish hypotheses region. The continuous shape pixels delineated from EM segmentation are counted as histogram values falling within each cell. Then the grid based shapes of different fish sizes are converted to binary and normalized with respect to their minimum bounding regions to achieve scale invariance.

Shape quantization in this 1024-dimension would have sufficient representation capability and enough allowance for the non-rigid deformation and scalability. A fish region hypotheses is then identified by support vector classifiers and fish hypotheses could thus be sorted out from non-fish objects and backgrounds.

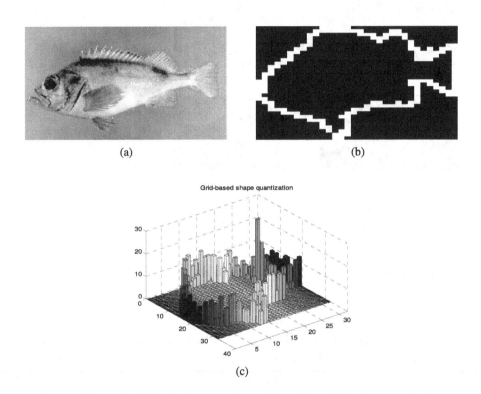

Fig. 3. (a) Recognized fish. (b) Shape model. (c) 3D grid-based fish shape quantization.

Figs. 3(a), 3(b), and 3(c) show the recognized fish, the normalized grid-based fish shape model, and the 3D fish shape quantization respectively. The redfish contour is plotted in red lines according to the segmentation result in Fig. 1(b). Each grid cell of the histogram bin in Fig. 3(c) represents the number of counted redfish shape pixels. In this representation, it is the spatial locations of these shape histogram bins really matter in fish identification, but not their absolute values. Fig. 3(b) is the 2D representation of Fig. 3(c).

Figs. 4(a)~(f) show more fish examples we have collected using the above approach to describe the shapes of fish species including codfish, haddock, pollock, redfish, winter flounder, and yellowtail flounder respectively. In the figures, they are generally similar looking, although not exactly the same. The representations of the tails and the wide bodies are the most similar with smaller variations, as compared to other fish features on parts such as the head and fins.

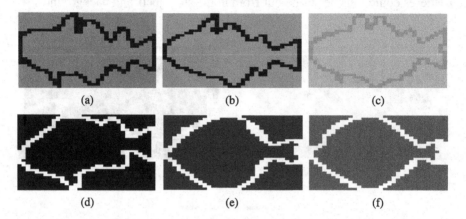

Fig. 4. Grid based shape model. (a) Codfish. (b) Haddock. (c) Pollock. (d) Redfish. (e) Winter flounder. (f) Yellowtail flounder.

4. Hypotheses learning of deformable fish shape

Using a Bayesian formulation, our approach is able to estimate the probabilistic hypotheses posterior likelihood given the fish priori distribution. In this section, we will now discuss learning the fish hypotheses based on the grid modeled shape, and identifying their categories. At first, we briefly present the support vector learning and our proposed approach for fish identification. A system block diagram is given. Then we present experimental results and case studies and demonstrate that our approach can integrate the hypotheses knowledge abstracted from the grid modeled shape and achieve good performance.

4.1. *Support vector learning*

Support vector machine has shown very good performance in classifying patterns for many applications such as object recognition, image retrieval, hyperspectral remote sensing classification, bioinformatics, and speech recognition. It is based on a supervised learning scheme which maximizes the distance between the closest points across the hyperplane separating two object classes. These points are called support vectors and the Euclidean distance is one of the most widely used metrics to measure similarity or dissimilarity.

The dimension of feature space is a key factor for general classification problems. That is, when clusters are difficult to separate in a low dimensional feature space, it is more easily separable in a high-dimensional feature space. However, the computational complexity increases very fast with much higher feature dimension. Support vector learning tends to avoid estimating probability densities directly by relying on margin maximization. This is the general idea of support vector machine learning. The advantages of support vector learning include efficiency in the learning process and its powerful capabilities in generalization which has been proved in literature.

We intend to use SVM to identify the shape modeled fish hypotheses. Let's assume a two-class problem $\{(\mathbf{x}_i, y_i)\}, i = 1, \cdots, l$, where each $\mathbf{x}_i \in R^n$ is the training data sample in the input space, n is the dimension of the input space, and $y_i \in \{-1, 1\}$ is the output label. A hyperplane separating y_i into positives and negatives will satisfy the following condition [9]

$$\mathbf{w} \cdot \mathbf{x} + b = 0 \tag{6}$$

Here $\mathbf{x}$ is the point lying on the hyperplane, and $\mathbf{w}$ is the weight normal to the hyperplane, with the perpendicular distance $|b|/\|\mathbf{w}\|$ from the hyperplane to the origin. Then these two classes could be discriminated as follows [9]

$$\mathbf{w} \cdot \mathbf{x}_i + b \geq +1 \quad \text{for } y_i = +1 \tag{7}$$

$$\mathbf{w} \cdot \mathbf{x}_i + b \leq -1 \quad \text{for } y_i = -1 \tag{8}$$

When the equalities hold, two hyperplanes $H_1 : \mathbf{w} \cdot \mathbf{x}_i + b = 1$ with the perpendicular distance $|1 - b|/\|\mathbf{w}\|$ to the origin and $H_0 : \mathbf{w} \cdot \mathbf{x}_i + b = -1$ with the perpendicular distance $|-1 - b|/\|\mathbf{w}\|$ to the origin are defined correspondingly. Thus the distance margin $2/\|\mathbf{w}\|$ between the two parallel hyperplanes could be maximized by minimizing $\|\mathbf{w}\|^2$.

After combining the above two equations together, we can get the following set of inequalities

$$y_i(\mathbf{w} \cdot \mathbf{x}_i + b) - 1 \geq 0 \quad \forall i \tag{9}$$

Fig. 5 depicts a binary support vector classification example using the Fisher Iris dataset. In the figure, the training and testing samples of both classes are denoted in different colors and a solid linear decision boundary equally divides the hyperplanes H_0 and H_1. Three support vectors marked in small circles are lying on these two hyperplanes respectively. It is therefore the distance margin between these two hyperplanes that SVM tends to maximize to discriminate the two classes.

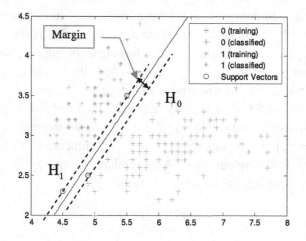

Fig. 5. Support vector classification example.

4.2. *Fishery dataset and hypotheses identification*

We have built a fishery database including multiple images of single fish and fish piles. All the fish on the imagery were caught in seven fishing trips made on commercial vessels. The fishery images were taken carefully under well-controlled illumination and angles in a well setup environment, and field fish geometries were also measured. The single fish images are all aligned in one direction and preprocessed to correct the color balance and contrast. Each individual fish image represents one species and totally there are six fish species. Note that these fish images and species are not intended to be typical or eigen-fish and we simply use some of them as training samples. The second fishery datasets include still images of fish piles in which some fish are manually placed and separated on a flat wood board, while others are mixed together with multiple species in random orientations and poses. There are totally over 30 fish inside each fish pile image.

We have conducted experiments and applied our proposed methods onto our field marine fishery image datasets. Fig. 6 shows the system block diagram of our proposed methods for fish identification. In our implementation, it might be necessary to correct contrast and brightness for better analysis and recognition purposes. So we have employed a preprocessing step by reducing cross-boundary overlaps and ambiguities and adjusting spectral contrast to prepare for the hypotheses segmentation. The segmentation labeled results are post-processed further to remove spurious regions which are below certain threshold according to the prior fish knowledge.

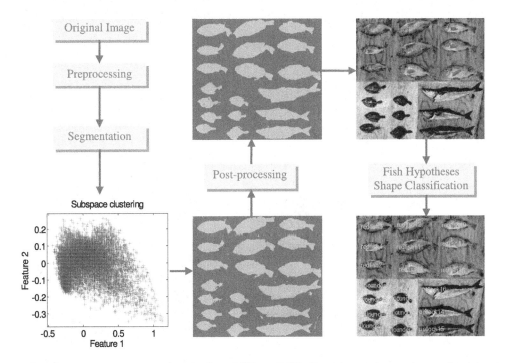

Fig. 6. Block diagram of the fish identification system.

4.3. *SVM training and classification*

Fig. 7 depicts the SVM learning block diagram for fish identification. To train the SVM, we have selected a positive training fish subset of 15 redfish, 27 flounder, and 3 pollock and there are totally 45 training positives. Other feature vectors such as the histogram of oriented gradients (HOG) can be also used to represent the region hypotheses for identification. At the same time, we have constructed two non-fish object datasets as the training negatives. The first is a corel dataset which includes 100 randomly selected images of 10 different object classes, and the second is an object 266 dataset. Both datasets are public available over the internet and they are selectively shown in Figs. 8(a) and 8(b) respectively. We have built all the training positives and negatives in the 32x32 grid based shape model respectively and formed an array of high-dimensional feature vectors. For example, the training vector dimension is 1025 which is equal to the dimension of the grid-based shape vector 1024 plus the label. There are totally 145 training samples with 100 negatives for the Corel100 dataset. And the number of training samples is 311 with 266 negatives for the Object266 dataset.

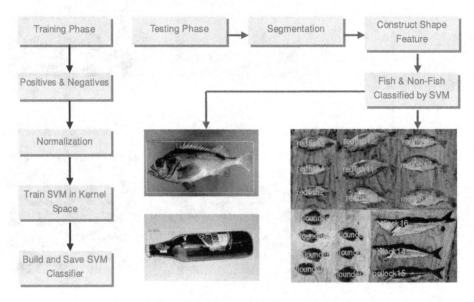

Fig. 7. SVM learning block diagram.

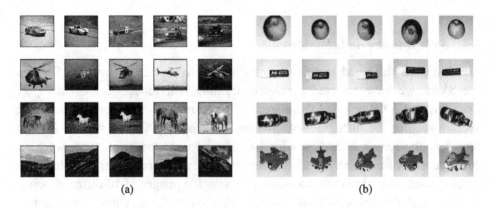

(a)   (b)

Fig. 8. (a) Corel100 dataset. (b) Object266 dataset.

4.4. *Fish species identification and case studies*

We have conducted multiple case studies of fish hypotheses identification and learning using more complex fishery imagery. Figs. 9(a), 9(b), 10(a) and 10(b) demonstrate fish identification for the well-separated fish case. The flounder hypotheses in Fig. 9(b) are extracted by an empirical area threshold of 5000, and the redfish in Fig. 10(b) by 1000. All of the hypotheses segmentation is performed by using the Bayesian quadratic classifier.

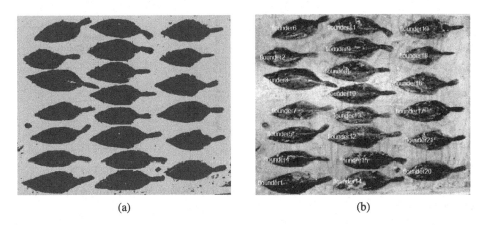

(a)   (b)

Fig. 9. (a) Filtered EM segmented flounders. (b) Recognized flounders.

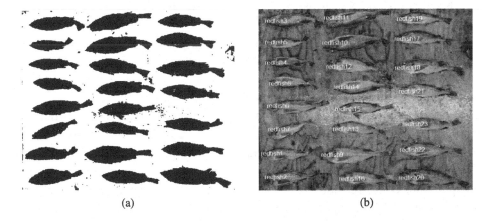

(a)   (b)

Fig. 10. (a) Filtered EM segmented redfish. (b) Recognized redfish.

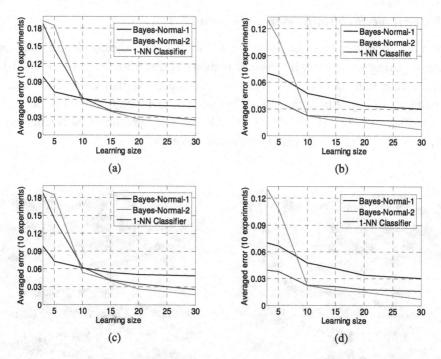

Fig. 11. Classifier learning curves for segmentation. (a) Codfish. (b) Redfish. (c) Fish pile 1. (d) Fish pile 2.

Figs. 11(a)~11(d) show the learning evaluation curves of three classifiers we select for hypotheses segmentation of codfish, redfish, fish pile 1 and 2 respectively. The horizontal axis is the learning size of the feature vector, and the vertical axis is the averaged error. The three classifiers of Bayes-Normal-1, Bayes-Normal-2, and 1-NN are the ldc, qdc, and knnc classifier we mention at the above respectively. It is clear that the averaged error of the quadratic Bayes-Normal-2 classifier decreases the fastest with respect to the learning size, as compared to the Bayes-Normal-1 and 1-NN classifiers.

Fish hypotheses detection and the effectiveness of modeled identification learning are tested and evaluated using SVM classifiers of various kernels. Throughout our experiments, SVMs with linear and quadratic kernels have been tested. After fish are classified from non-fish objects and backgrounds, we compute and compare multiple local spatial metrics of fish region hypotheses with respect to different fish species. For instance, the aspect ratio, eccentricity, circularity and extent are shown in Figs. 12(a)~12(d) respectively. In Figs. 12(a) and 12(c), we observe that the values of the aspect ratio (length/width) and eccentricity associated with the respective fish classes (flounder, redfish and pollock) are highly separable. Therefore we are able to classify these three fish species. However, these empirical results are fishery data dependent. The case is similar for the metric of circularity in Fig. 12(b) except that the circularity of redfish 15 is an outlier (in green circle) and not separable from those of pollock. This is due to the fact that one fin of redfish 15 is unfolded on the original fishery image but fins of the other redfish are not shown up in Fig. 10(b). In Fig. 12(d), the metric of extend is totally non-separable for all three species.

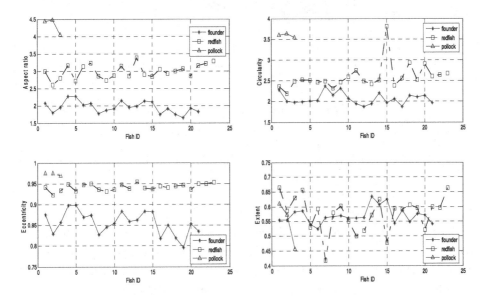

Fig. 12. Spatial metrics. (a) Aspect ratio. (b) Circularity. (c) Eccentricity. (d) Extent.

We have manually synthesized three fish images together by relatively keeping their aspect ratios constant, and applied our approaches for multi-class fish recognition. Fig. 13(a) show different fish species of seven flounder, three pollock and nine redfish instances respectively. As given on the above, it is preprocessed by the EM segmentation and further processed with local constraints. The SVM classifies all the 19 fish instances correctly. Simple rule based thresholds of aspect ratio are then employed to classify these three fish categories. For our cases, we set a threshold of 2.5 to differentiate the flounder from the redfish and 4.0 between the redfish and the pollock. All fish are truthfully classified with the recognition rate of 100%. Note that the number following the fish name in the figure is the index of fish number. Fig. 13(b) depicts the three fish categories in three colors respectively.

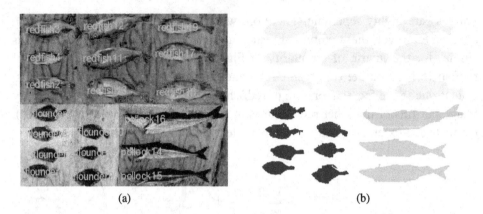

(a)   (b)

Fig. 13. Multi-class fish recognition. (a) Recognized fish. (b) Labeled fish categories.

Table 1. Spectral metrics of fish categories.

ClassID	Rmean	Gmean	Bmean	Rratio	Gratio	Bratio
1	96.87	72.35	52.76	0.44	0.33	0.24
2	98.67	88.94	79.63	0.37	0.33	0.30
3	162.06	93.50	72.11	0.49	0.29	0.22

Further more, we have calculated the spectral statistics of the three fish categories and the 19 fish instances. The results of fish categories are given in Table 1. The ClassID 1, 2, and 3 correspond to the flounder, pollock and redfish respectively. The color metrics of Rratio, Gratio, and Bratio denote the color ratios of the red, green and blue channels and are computed using $R/(R+G+B)$, $G/(R+G+B)$, and $B/(R+G+B)$ respectively. The results demonstrate that the spectral metrics including Rmeans and Rratio can be employed to classify these fish categories pretty well besides the spatial metrics.

We have used the unsupervised K-means algorithm to separate the three fish species based on both the spectral metrics of red channel ratio and the spatial metrics of aspect ratio and eccentricity. Figs. 14 (a) and (b) illustrate the classification results with different markers. Note that the numbers next to the class markers indicate the fish index. Clearly the 100% distinctiveness of these three fish species in these orthogonal spaces is demonstrated. These two figures also show that they are separable along the horizontal and vertical axes respectively.

To demonstrate the effectiveness of our proposed approach, we have also extended the scope of our testing to more complex cases. Two examples of identifying fish species on overlapping fish piles are depicted in Fig. 15.

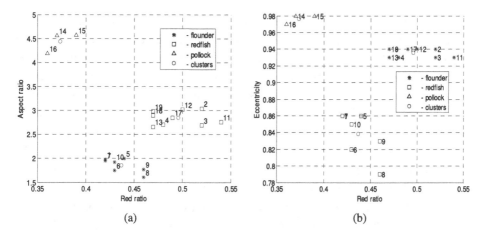

(a)   (b)

Fig. 14. Fish classification based on spectral and spatial metrics. (a) Aspect ratio with respect to red ratio. (b) Eccentricity with respect to red ratio.

(a)   (b)

Fig. 15. Overlapping fish pile. (a) Recognized redfish. (b) Recognized redfish.

5. Conclusion and discussions

In this chapter, we have presented a novel modeled approach to fish detection and species identification. In our study, the prior fish knowledge and distribution are modeled as mixture of Gaussian and estimated by using the distinctive fish color. We take an EM based clustering for the implementation. Fish is represented by grid based shape descriptors with robust invariance to scaling and deformation. We demonstrate 100% fish identification rate using spectral metrics of the red ratio and spatial metrics of aspect ratio and eccentricity. These numerical findings surpass the existing results reported by [1] and coincide well with the fish's biological characteristics. And the results show that the SVM based fish hypotheses learning is quite effective. We have constructed an extensive marine fishery dataset and conducted numerous experiments by comparison with both natural and man-made non-fishery objects. We demonstrate that various strategies can be employed for segmenting, generating, selecting, grouping, modeling, and classifying the fish hypotheses respectively.

The computational complexity of our proposed approach is another advantage. The description and learning based on the grid modeled shape are quite efficient as it usually takes several minutes to identify dozens of fish instances. This would especially speedup the identification processing when there are a lot of fish object regions and their scales are quite large.

The identification performance and accuracy of our proposed approach largely rely on the selection of segmentation parameters and thresholds. Fine tuning is usually necessary according to the test fishery images on hand. Hierarchical segmentation can be performed to separate overlapped and occluded fish regions with similar colors and textures. In addition, the edge map and local priors may be used to partition adjacent hypotheses and constrain the segmented hypotheses regions. The experimental results demonstrate our methods are quite promising and can be potentially applied onto other object recognition problems.

Acknowledgements

We thank Glenn Chamberlain of the School for Marine Science and Technology at the University of Massachusetts Dartmouth for providing the fishery images used in this chapter.

References

[1] D.J. White, C. Svellingen, and N.J.C. Strachan, "Automated measurement of species and length of fish by computer vision," *Fisheries Research,* 80, pp. 203–210, 2006.

[2] B. Zion, A. Shklyar, and I. Karplus, "Sorting fish by computer vision," *Computers and Electronics in Agriculture,* 23(3):17.

[3] D. Somerton and C. Glendhill (editors), "Report of the national marine fisheries service workshop on underwater video analysis," *U.S. Depart. of Commerce, NOAA Tech. Memo. NMFS-F/SPO-68,* 69 p. May 2005.

[4] Ferdinand van der Heijden, Robert Duin, Dick d. de Ridder, and David M. J. Tax, Classification, "Parameter estimation and state estimation: an engineering approach using MATLAB," *John Wiley & Sons Ltd,* 2004.

[5] L. Gorelick, M. Galun, E. Sharon, R. Basri, and A. Brandt, "Shape Representation and Classification Using the Poisson Equation," *IEEE Trans. on Pattern Analysis and Machine Intelligence,* vol. 28, no. 12, pp. 1991–2005, Dec. 2006.

[6] Dengsheng Zhang and Guojun Lu, "Review of shape representation and description techniques," *Pattern Recognition,* vol. 37, 1–19, 2004.

[7] S. Belongie, J. Malik, and J. Puzicha, "Shape matching and object recognition using shape contexts," *IEEE Trans. on Pattern Analysis and Machine Intelligence,* vol. 24, no. 4, pp. 509–522, April 2002.

[8] Stan Sclaroff and Lifeng Liu, "Deformable shape detection and description via model-based region grouping," *IEEE Trans. on Pattern Analysis and Machine Intelligence,* vol. 23, no. 5, pp. 475–489, May 2001.

[9] C. J. C. Burges, "A tutorial on Support Vector Machines for pattern recognition," *Data Mining and Knowledge Discovery,* 2, pp. 121–167, 1998.

CHAPTER 2.6

VEHICLE DETECTION IN REMOTE SENSING SATELLITE IMAGERY

Bichuan Shen[1,*], Chi Hau Chen[1,†] and Giovanni Marchisio[2,‡]

[1]*University of Massachusetts Dartmouth, USA*
[2]*DigitalGlobe, USA*
[*]*bshen@umassd.edu*
[†]*cchen@umassd.edu*
[‡]*gmarchis@digitalglobe.com*

Geophysical remote sensing satellite equipped with high-resolution hyperspectral sensors can provide us extremely large amount of data, which can be relayed to end nodes for processing and have many potential important applications in traffic and road network management, city planning and environmental monitoring. For instance, the commercial satellite WorldView-2 can provide 8-band multispectral imagery at high resolution of 0.5m, by scanning large area of 550,000 square km per day. How to extract useful information and knowledge from these large data repository is an active research topic. In this chapter, we aim to study the detection of vehicles from remote sensing satellite imagery. Although vehicles may be geometrically represented as rectangular shapes in limited resolution, the task is not trivial as most of vehicle details and characteristic features are lost during the image capturing process, and different vehicle types, colors, sizes, heights and orientations cause a lot of variations during projection. Additionally, wide area coverage may include very similar non-vehicle features due to building roofs and parts, trees, walls, shadows, and other objects. We present a Gaussian and maximum likelihood framework to model the local appearance densities and a machine learning approach using SVM and neural networks to classify vehicles from non-vehicle buildings, objects, and backgrounds. We show that linear edge features of vehicles can be well extracted by Prewitt and Canny edge detectors during the preprocessing stage. We also demonstrate that it is more advantageous to use median filtering to remove backgrounds and shadows. Global contextual information such as road networks and vehicle distributions can be used to enhance the recognition. Our method takes a spectral clustering perspective to estimate the parameters of centroid, orientation, and extents for local densities. We demonstrate a high detection rate 94.8%, with a missing rate 5.2% and a false alarm rate 5.3%. Experimental results show that our method is quite effective to model and detect vehicles.

1. Introduction

Remote sensing plays a significant role in applications such as urban planning, traffic management, military reconnaissance, geographical and environmental administration. Satellite equipped with high-resolution hyperspectral sensors can provide us extremely large amount of data, which can be relayed to end nodes for processing. The commercial satellite WorldView-2 can provide 8-band multispectral imagery at high resolution of 0.5m, by scanning large area of 550,000 square km per day. However a lot of human hours are needed to extract useful information from the huge amount of data, and the automatic target identification and recognition becomes attractive.

The focus of this chapter is on vehicle detection from satellite images. Object detection, representation, and recognition have inspired a lot of research interests in computer vision. However the general techniques are not adequate to address the vehicle detection in satellite imagery. One critical issue is that most vehicles appear as tiny blobs and most of their details and characteristic features are lost during acquisition by high resolution sensors mounted on remote sensing satellite, which is normally operating at hundreds of kilometers above ground. This resolution limit constrains the task of detecting and describing the vehicles in satellite images.

There have been many studies carried out on detecting vehicles from aerial imagery [1-8]. [1] describes an approach to vehicle detection and extraction from high resolution aerial imagery. The system uses mean-shift clustering based blob analysis to extract candidate vehicle hypotheses, by fusing the chromatic and geometric features in both the color LUV and Euclidean spaces. The density and symmetry of blob-like car appearance and shape are considered jointly. At the verification stage, a log-polar shape descriptor is then used for measuring similarity to remove the ambiguous outliers and reduce the detection cost. [2] illustrates an online boosting approach for car detection through exhaustive search and post-processed by mean shift clustering. Three car features including Haar-like feature, orientation histogram and local binary pattern are employed in the study. [6] introduces an appearance 3D model for vehicle detection based on line features. [8] presents a 3D object model and a Bayesian network learning framework for vehicle recognition, by integrating features of boundaries by shaped edge filter and shadows. [3][4][5] also present a wire-frame 3D model to describe the geometric and radiometric features of vehicle including their shadow region for aerial imagery. This work uses both local description and global knowledge to find and evaluate the vehicle support. Nonetheless, the resolution constraint of vehicles in satellite imagery makes the 3D modeling hardly applicable as prominent features are not available and their correspondences can not be easily identified.

Segmenting is attempted as a straight-forward approach or preprocessing step to separate the objects of interest from the background. In remote sensing imagery, however, vehicles are small objects and the sizes are usually tens of pixels. Different vehicle types, colors, sizes, heights and orientations cause a lot of variations during projection. Illumination and viewing angle also play a significant role leading to both cross-boundary ambiguity and variations. Additionally, wide area coverage may include very similar non-vehicle features due to building roofs and parts, trees, walls, shadows, and other objects. Thus segmentation is very difficult for satellite imagery with extensive detail variety.

The task of vehicle detection can be also complicated by similar non-vehicle segments, which are mostly found on building roofs, corners, walls and parking lots. Buildings and roads are common objects besides vehicles and usually have rectilinear edges. They are significantly different in shapes and sizes. Roads may have long and parallel lines and curves with almost equal distance. Additionally, buildings, trees and walls may cast shadows onto the ground in proportion to the projection angle and their heights. Shadows cause ambiguity in the cast regions and information might get lost. Shadows of cars and buildings contribute to edges and create a lot of false alarms. Much research has been carried out on shadow detection from the photogrammetric point of view and 3-D modeling. Street layers have been reported by researchers to increase the detection and decrease the false alarm.

As the orientations of vehicles can vary at any direction in the imagery with changing sizes, the property of rotational and scale invariance has been a core issue for recognition. Much work has been done by researchers using rotational features to achieve rotational invariance. For example, [2] employs rotational invariant features including Haar-like, orientation histogram and local binary pattern for describing cars.

We address vehicle detection using the model based approach. Vehicle hypotheses are fitted by elliptical Gaussians and the bottom-up features are grouped by Gabor orientation filtering based on multiscale analysis and distance transform. In consideration of the problem complexity due to dense and overlapping vehicle distributions, partial occlusion and clutters by building, shadows, and trees, we employ a spectral clustering strategy jointly combined with bootstrapped learning. The promising detection results and evaluations demonstrate the effectiveness of the vehicle detection algorithm.

This chapter is organized as follows. We first give an overview of our vehicle detection approach in Section 2. Section 3 presents vehicle modeling and detection. Results and evaluation are provided in Section 4. Finally section 5 presents the discussions and conclusions.

2. Overview

In our approach, we have taken a bottom-up hypothesize and verify scheme for vehicle detection. Considering the fact that the scales and shapes of vehicles are small and more uniform and those of the other object categories including buildings and roads are significantly large and varying, we first employ a multiscale wavelet based segmentation and analysis by taking the redundant wavelet decomposition. Our approach makes use of distance transform to provide important structural and geometric information. A distance transform is performed on the median filtered difference image, and candidate hypotheses are generated from the minima on the distance map. Prior distribution and knowledge of the roads and background are then aggregated with global and contextual information and local cues and evidences to improve the accuracy of the hypothesize processing step. Gabor filters are employed to identify the hypotheses orientation by convolving with the imagery at each pixel location. When the orientation of Gabor filter is well aligned with that of vehicle blobs, a local Gaussian distribution can be identified at the peaks and used to represent the hypothetical ellipses. We have used 12~18 orientations to sample a half of the unit circle, which correspond to 10 and 15 degree in angle. Fig. 1 shows a block diagram of the vehicle detection algorithm.

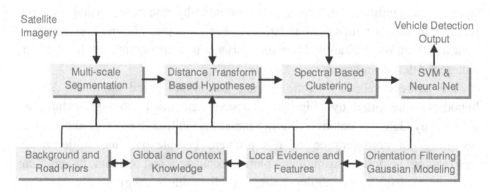

Fig. 1. Block diagram for vehicle detection.

Although our approach shares some similarities to the method proposed in [1] by following sequences of blob detection, feature description and vehicle detection, our system employs median filtering as one essential preprocessing step to efficiently detect extrema of candidate vehicle densities and effectively remove most of backgrounds and shadows. Background and shadow modeling used in many object detection and recognition systems therefore can be omitted, and this reduces the system complexity and decrease false alarms. Our method also differs in building feature description and representation by combining spectral clustering of local density with distance transform based orientation filtering and global contextual knowledge. The final difference is that the proposed approach uses machine learning to classify vehicles.

3. Vehicle modeling and detection

In this section, we intend to present model based vehicle description and representation. We start from brief descriptions of vehicle cues. This is followed by a study of orientation modeling and filtering. A Gaussian distribution model is presented to estimate the vehicle features in a probabilistic maximum likelihood framework. Then machine learning strategies using SVM and neural networks for vehicle recognition are given.

3.1. *Imagery dataset and vehicle cues*

The dataset we use is acquired by the commercial satellite WorldView-2, which provides 0.5 m resolution and 8-band multispectral imagery. It is operating at the altitude of 770 km and covers a wide field of views. It has a large-scale collection capacity nearly 1 million km^2 of high-resolution panchromatic imagery per day, and offers average revisit times of 1.1 days around the globe. However, only the grayscale intensity is considered in this study, without using the multispectral information.

The primary task of vehicle detection is to find strongly correlated evidence, either in local and global, to support the hypotheses. For the case of high resolution aerial image whose resolution may reach as low as 10 cm [3], vehicle scales are large and can have prominent cues and salient features. However, vehicles in the remote sensing imagery are usually much smaller, and look like blob structures with most of their details and characteristic features lost during the image capturing process. The typical vehicle appearance and distribution can be summarized as follows:

- Vehicles usually show up as a bright blob compared to the local neighborhood, and it is statistically symmetric with respect to its major axis, if not shadowed or cluttered;

- Vehicles are mostly found on roads and their orientations may align with the road direction. They can also appear on parking lots and open lands, but unlikely on the roofs of buildings;

- Vehicle characteristic features including windshields, license plates, tires, bodies are not visible in satellite imagery, and blobs appear as solid rectangles or ellipses;

- The scales of vehicle blobs are less variant than those of road segments, building lines, land patches, and other background objects. Small vehicles may contain about 2x4 pixels and long trucks can takes 3x6 pixels. The ranges can be slightly lower or higher, and they are affected significantly by illumination and reflectance variations;

- The blob sizes of the same type vehicles are almost constant. For example, passenger cars have almost the same dimensions. One thing we need mention is that orientation and spatial sampling do have significant impact on the blob sizes because of the limited resolution.

All of these factors have impacts on the task of vehicle detection, and drive us to look for salient vehicle evidences. For example, vehicle shadows are used as an important contextual feature [8]. Additionally, the prior knowledge of vehicle distribution on various land coverage will definitely help. Identification and classification of urban area and land usage have been studied by many researchers for many remote sensing applications. Although an exact knowledge is not always available, simple preprocessing with straightforward objectives can classify and segment the scene into foreground and background layers.

We have selected two views of neighboring city sectors from the imagery dataset, which are displayed at Fig. 2 (a) and (b). Because of its high spatial and spectral resolution, many fine objects above ground like roads, parking lots, buildings and roofs, walls, trees and plants, shadows, and other objects besides vehicles are highly visible in both scenes. Moreover, they may also contain very similar non-vehicle features for wide area coverage and make the vehicle detection task more complicated. Building sides near the projection angle appear as bright edges, and they look very much like vehicle evidences. Some roof structures and walls also resemble vehicle cues too. Fig. 3 shows a zoomed building area inside the left scene. Clearly, shadows are visible by the side of neighboring buildings and trees, and their dimensions are proportional to their heights.

(a)   (b)

Fig. 2. (a) City view 1. (b) City view 2.

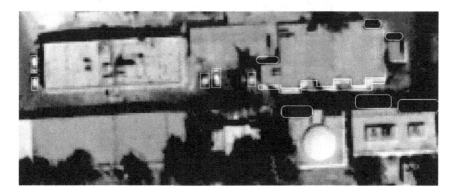

Fig. 3. Vehicles (white blobs), bright building edges (polygons) and shadows (round rectangles).

3.2. *Orientation filtering and modeling*

The property of rotational and scale invariance has been a central one in object recognition. The rational is the same for vehicle detection, as vehicle orientations of interests can be at any angle between 0 and 360 degree. Here we give an example of varying orientations in Fig. 4, which include 24 vehicles instances and 6 non-vehicle cues. The first and third rows show six different vehicles in both the horizontal and vertical directions respectively. The second row includes other vehicles with orientation close to 45 degree. The orientations of vehicles between 90 to 180 degree are illustrated at the fourth row.

To detect and classify these instances into binary object categories, we believe that orientation modeling and estimation is very critical for the task. Orientation estimation based on averaging squared gradient and principal component analysis (PCA) has been discussed in [9]. Both approaches have limitations to estimate vehicle orientation. For a local neighborhood containing background noise, the first method needs to take gradient and this would amplify high frequency noise and an estimate becomes less accurate. This is the case for vehicle signals with limited resolution in remote sensing imagery. The second method with PCA can use a subspace spanned by eigenvectors of Hessian matrix [5]. The major orientation axis corresponds to the long eigenvector whose eigenvalue and the variance are the largest. In the case of large scale variations and densely neighboring vehicles with variety of details, how to specify a search window is also not an easy task.

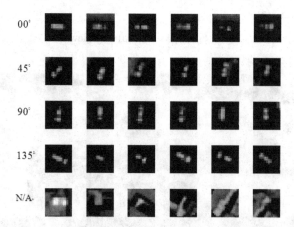

Fig. 4. Vehicle instances and non-vehicle cues (N/A means non-vehicle objects).

Fig. 5 (a) shows 15 vehicle objects in a local area. Vehicles in different orientations are manually labeled in different colors of rectangular boxes. In addition, we mark on top left region with two overlapping rectangles in different orientation. To estimate the probability distribution density accurately, a maximum likelihood estimation is employed jointly with an elliptical Gaussian model as shown in Fig. 5 (b).

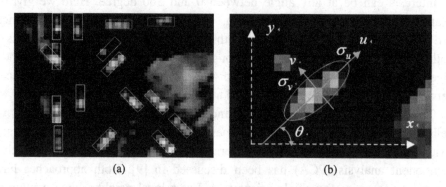

(a)   (b)

Fig. 5. (a) Vehicles in different orientations. (b) Gaussian model.

Assume that σ_x^2 and σ_y^2 are the variances along the horizontal and vertical coordinates. We can compute the directional variance based on this Gaussian model, where σ_u^2 and σ_v^2 are the variance along the major and minor axis. The Gaussian variances are given by equations [10]

$$\sigma_x^2 = \sigma_u^2 \cos^2 \theta + \sigma_v^2 \sin^2 \theta \tag{1}$$

$$\sigma_y^2 = \sigma_u^2 \sin^2 \theta + \sigma_v^2 \cos^2 \theta \tag{2}$$

From which the directional variances can be derived

$$\sigma_u^2 = |D|^{-1} \left(\sigma_x^2 \cos^2 \theta - \sigma_y^2 \sin^2 \theta \right) \tag{3}$$

$$\sigma_v^2 = -|D|^{-1} \left(\sigma_x^2 \sin^2 \theta - \sigma_y^2 \cos^2 \theta \right) \tag{4}$$

where D is the determinant of the system matrix. Thus a local variance can be estimated as $\sigma^2 = \sigma_u^2 + \sigma_v^2$.

Gabor wavelets and filter banks have also been widely applied to multi-resolution signal and image analysis. For example, they have been used to enhance ridges in fingerprint image which are well aligned with the orientation and spatial frequency [11]. Researchers have also attempted other applications including texture analysis. The adaptive frequency and resolution properties make it a good candidate for local orientation estimation and coding. Another attractive feature is that there is no need to compute gradient which is easily subject to both noise and outliers.

A complex Gabor filter consists of an even-symmetric cosine and an odd-symmetric part

$$h(x, y) = h_e(x, y) + j h_o(x, y) \tag{5}$$

$$h_e(x, y) = \exp\left(-\left(\frac{x^2}{\sigma_x^2} + \frac{y^2}{\sigma_y^2} \right) \right) \cos(\omega x) \tag{6}$$

$$h_o(x, y) = \exp\left(-\left(\frac{x^2}{\sigma_x^2} + \frac{y^2}{\sigma_y^2} \right) \right) \sin(\omega x) \tag{7}$$

where σ_x and σ_y are the variance and spread of the filter along the x and y coordinates. For vehicle detection in satellite imagery, the typical range is usually between 0.1~0.5. And the ω is the spatial frequency controlling the filter support, where a number around 4~10 can be set.

As most vehicle evidences are shown as edges, line segments, and blobs, orientation is another important parameter which must be considered. Vehicle signals and cues are essentially band-limited and have certain spatial frequency. Thus Gabor bandpass filter banks can be applied to each blob centroid. A multi-scale symmetry metric is identified using magnitude weighted difference between the Gabor even and odd filter response [1]. In contrary to their method, we have applied Gabor filtering at all the extrema found by median filtering at all the 12 orientations and 4 scales, which are shown the Fig. 6.

For each scale, we vary the orientation of the even Gabor filter and convolve with the distance thresholded image. The orientation can be estimated by identifying convolution peaks. Fig. 7 illustrates selected convolution results at 30 degrees of a local region containing multiple vehicles in varying orientation with Gabor filter banks in orientations from 0 to 180, with a step of 15 degrees.

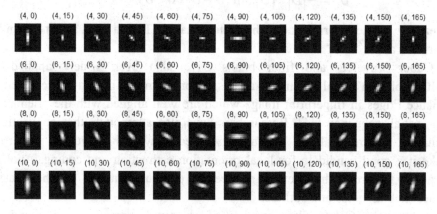

Fig. 6. Gabor orientation filter bank.

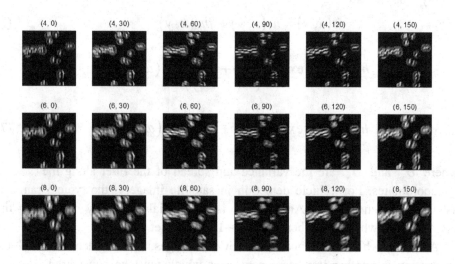

Fig. 7. Gabor filter vehicles.

3.3. *Gaussian modeling of vehicle densities*

The methods that hypotheses are verified by using local vehicle features and global constraints are essentially bottom-up. There has been much research using this paradigm. [1] uses a circular log-polar shape descriptor to represent and verify the candidate vehicle blob. [5] reconstructs the shape boundary by searching the maximum gradient along lines in a distance-angle space around the blob centroid. Most works involving blob detection group global evidences and constraints besides local cues for vehicle hypotheses to reduce the over segmentation and false alarm rate.

Assuming that Ω_m is the target pattern model in the pattern space $\Omega = \{\Omega_m\}$, $0 \le m \le M - 1$, with the pattern mean $\mathbf{u}_m$ and variance $\boldsymbol{\sigma}_m$, the probability of local neighborhood being the object pattern Ω_m at each pixel (i, j) can be formulated as Gaussian

$$P_{ij}\left(\mathbf{x}|\Omega_m : \mathbf{x} \in R^d\right) = \frac{1}{\sqrt{(2\pi)^d |\boldsymbol{\sigma}_m|}} \times \exp\left(-\frac{1}{2}(\mathbf{x}-\mathbf{u}_m)^T \boldsymbol{\sigma}_m^{-1}(\mathbf{x}-\mathbf{u}_m)\right) \quad (8)$$

Here $\mathbf{x}$ is a feature vector within the hyper-dimensional feature space $\mathbf{X} \in R^d$. It is generated with the dimension d inside a local neighborhood centered at the pixel.

Based on the estimated densities of hypothesis cues, we can formulate the problem of vehicle detection in a Bayesian reasoning framework

$$P_{ij}\left(\Omega_m|\mathbf{x}\right) = \frac{P_{ij}\left(\mathbf{x}|\Omega_m\right)P_{ij}\left(\Omega_m\right)}{\sum_{k=1}^{M} P_{ij}\left(\mathbf{x}|\Omega_k\right)P_{ij}\left(\Omega_k\right)} \quad (9)$$

Without losing generality, we can denote the object vehicle class as Ω_0, and all the other object classes $\overline{\Omega}_0 = \{\Omega_1, \cdots, \Omega_{M-1}\}$, for the case of our study. Then we can follow the Bayesian criterion to make decision for the vehicle hypotheses

$$H = \frac{P_{ij}\left(\Omega_0|\mathbf{x}\right)}{P_{ij}\left(\overline{\Omega}_0|\mathbf{x}\right)} > 1 \quad (10)$$

After inserting equations (8) and (9) into (10) and simplification, we can derive a general estimation problems based on the parametric model of mixture of Gaussians, which can be solved by optimization techniques, such as the expectation maximization algorithm using the maximum likelihood (ML) principle.

3.4. *Learning and recognition of vehicle descriptions*

As the object appearance, shape and other cues may be used to describe and represent the probabilistic density, likelihoods and distances of Gaussian mixtures can be employed to model both the object class and background. Besides these parametric modeling techniques, nonparametric methods also receive increasing research attention.

Much works on neural networks have been carried out during the past two decades, and we have witnessed a wide range of applications including function fitting, clustering, speech and pattern recognition, and texture classification [12], as well as car detection [13]. Neural networks refer to parallel and probabilistic connection models, which are designed to simulate mathematically the human brain perceptrons and functionalities. The building element is called neuron which consists of input, weights, bias, and transfer function. The typical sigmoid function is chosen with the characteristic that its derivative is Gaussian. The network architecture and training converged weights of neuron ensemble decides the overall function designed to perform. We have employed both SVM and neural networks for vehicle detection in this study.

4. Results and evaluation

Vehicle hypotheses are generated from features in a hierarchical manner, after the candidate extrema and blobs are detected. The feature hierarchy can include local geometric properties of symmetry and solidity, and radiometric statistics such as the mean and directional variance, which is closely related to the vehicle orientation as described in the previous section. Feature matching and selection is performed on the basis of grouping sufficient vehicle evidence and verifying the hypotheses. False hypotheses can thus be eliminated from queues without further processing.

This section shows first a sequence of preprocessing steps, including edge operators to detect linear features, median filtering to extract hypothesis peaks and remove backgrounds and shadows, and Gaussian blurring to improve the recognition accuracy and reduce the computational complexity. Then, we describe a SVM and neural networks based learning scheme for verifying the vehicle hypotheses. Finally, the detection results are presented and performance constraints are highlighted.

4.1. *Foreground filtering and analysis for vehicle detection*

We have conducted tests using various detection strategies. The first is based on foreground filtering and analysis. This approach is best used for the case of simple scene, static video analysis or when multiple scene views are available. The assumption is that background can be perfectly modeled and separated from the foreground. To perform this task, we have applied both the Canny and Prewitt edge filters to extract the lines and edge fragments (Fig. 8 (b) and (c)) and show processing results from (d) to (f). Because most vehicle and building can be represented by rectilinear contours and shapes, these edge maps can remove much of background and preserve the candidate hypotheses. This is followed by an intermediate Gaussian blurring (Fig. 8 (d)) for blob detection. Figs. 8 (e) and (f) demonstrate the final blobs based on simple thresholding and detected local extrema respectively.

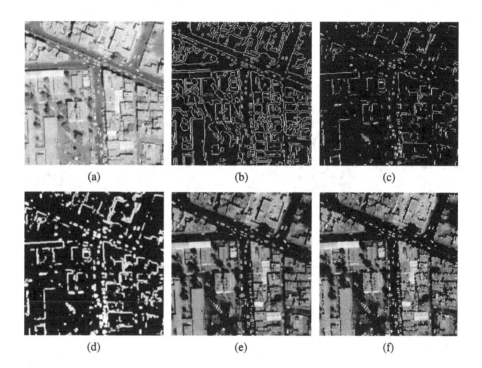

Fig. 8. (a) Original image. (b) Canny edge map. (c) Prewitt edge map. (d) Prewitt edge blur by Gaussian. (e) Blob detection. (f) Local extrema.

4.2. *Detection based on blob analysis*

Let us take the crude method of detection and learning by SVM and neural networks directly from the output of blob analysis to illustrate its effectiveness and the problem complexity. We have applied a two-phase recognition process. In the first phase recognition is learned by SVM with 8x8 feature vector arrays.

We have manually selected 42 (13.6 Percent) training positives, and 266 (86.4 Percent) negatives. Among a total of 28445 extrema, the SVM is able to classify 7723 of them as vehicles, and 20722 as non-vehicles. Fig. 9 (a) illustrates the SVM results of phase 1, where red color denotes the detected positives and green the negatives. Clearly there are too many false positives, caused mainly by building edges. In phase 2, a neural network is trained and asked to do further testing on the detected results from the first SVM detection phase. The feature vectors are formed by combining 8x8 local neighborhood with its mean and variance, and utilizing the knowledge from estimated background and canny edge. We have trained the network with 28 (63.6 Percent) positives and 16 (36.4 Percent) negatives, and detected 2820 vehicles, and 4903 non-vehicles. Fig. 9 (b) shows the results.

(a)   (b)

Fig. 9. (a) SVM results (phase 1). (b) Neural networks results (phase 2).

4.3. *Bootstrapped learning for vehicle detection*

To verify the vehicle hypotheses, we have employed a bootstrapped learning scheme to obtain negative vehicle hypotheses. We first perform a greedy vehicle detection by setting a high threshold to select positive training hypotheses. In this case, all of them can be trusted as truly positives, and we treat them as the pseudo-ground truth. Then we use the crude blob extraction method to randomly select non-vehicle hypotheses which resemble vehicle. This is based on the prior knowledge that if the locations of hypotheses are on the segmented layers other than roads, then they are unlikely to be vehicles. This learning strategy emphasizes on the dissimilarity between the vehicle and non-vehicle objects for training SVM and neural networks, and automates the training process without manual selection and subjective judgement.

Based on the pseudo-ground truth with high confidence, we have assembled a total of 366 (59.2%) training positives and randomly select 252 (40.8%) negative hypotheses. Fig. 10 (a) and (b) illustrate the idea. To obtain a good performance, we can repeat the process with another randomly selected ensemble of negatives and features sets to train both the SVM and the neural networks to improve the detection rate and reduce both the missed detection and false alarm rates.

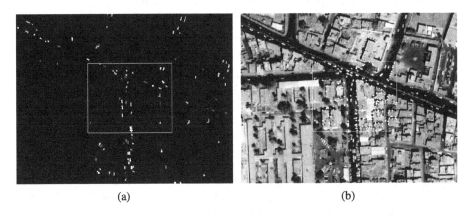

(a)   (b)

Fig. 10. (a) Greedy training positives. (b) Bootstrapped training negatives.

4.4. *Hypotheses generation based on multiscale wavelet decomposition*

The first component of our vehicle detection approach is nonlinear median filtering. Its design has dual objectives: extract the approximate locations of candidate vehicles and remove the background and shadows. As most vehicles in satellite imagery show up as tiny bright spots with a few tens of pixels and have generally less variant pixel intensity, the median filter is superb to isolate vehicle pixels from neighboring objects, buildings, and the background. It is also invariant to location, scale, and orientation.

The local neighborhood of median filtering has to be selected carefully, however, in order to match well the dimensions of vehicles, and sift out non-vehicle objects in larger scales. We have proposed a selection heuristics to estimate the filtering performance according to the number of found extrema for vehicle evidence and the overall distribution across the view. Practically we can segment the field of view into multiple layers, for example, a three-layer composition is made from roads, buildings and background, and shadows. This work can be performed by an expectation maximization based clustering approach. Because remote sensing satellite such as WorldView-2 can cover vast ground areas in one single image which provides a lot of spatial and structural information, this method is quite useful for scene segmentation that reduces the searching space to indexed layers and separates objects from the background due to urban scene complexity and variety.

The heuristic implementation results are illustrated in Fig. 11 (a)~(c). The middle figure (b) shows a processed clean background of the original scene (a). To keep the buildings, open lands, and vegetations well and remove almost all of the vehicles, we employ a multiresolution analysis by taking the redundant wavelet decomposition. The advantages of wavelet analysis include decreased computational complexity and very good noise performance. By selecting a lower resolution band which primarily corresponds to large scales of objects including buildings, roads, and lands, most vehicles in a smaller and constant scale are sifted. The right figure (c) demonstrate the three EM segments of roads, background, and shadows.

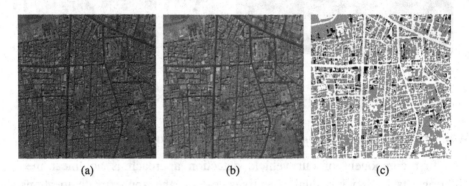

      (a)   (b)   (c)

Fig. 11. (a) Original image. (b) Background. (c) Segmented layers.

Fig. 12 (a) and (b) demonstrates the detected vehicle candidate hypotheses and distribution statistics with respect to the segment layers. The horizontal coordinate denotes the size of local neighborhood, for example, 4x4, in both (a) and (b). The vertical coordinate is the number of detected vehicle extrema. Clearly, median filter 2x2 generate too many false extrema and larger neighborhood sizes 3x3 and beyond have much better results between 2508 and 3570. Figure (b) plots the distribution percentage of the detected extrema for median filter ranging from 3x3 to 10x10. As can be seen from (b), vehicles have the largest distribution between 69.3% and 80.2% on roads overall. While for the background segment, the concentration of vehicles is much smaller between 19.3% and 28.4%. The smallest distribution of vehicles is on the shadow segment, between 0.4% and 2.3%. Overall, the 4x4 median filter produces totally 2670 extrema with the distribution of 77.8%, 21.6% and 0.6% on roads, background and shadows respectively and generates the best compromise between the detected extrema and the outliers.

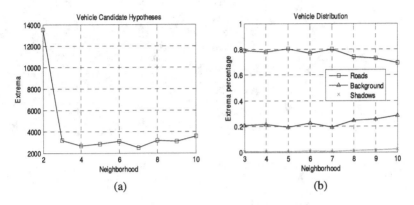

(a) (b)

Fig. 12. (a) Vehicle candidate hypotheses. (b) Vehicle distribution.

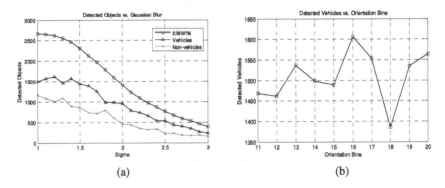

(a) (b)

Fig. 13. (a) The relationship between the detected hypotheses versus sigma. (b) The orientation distributions of detected vehicles.

Considering median filtering as a nonlinear operation, we have also employed Gaussian blurring on the difference image to minimize the candidate hypotheses. We have conducted multiple tests with respect to the parameter of blur spread. The relationship between the detected objects versus the sigma is shown in Fig. 13 (a). Thus we empirically set the sigma equal to 2.0.

The discrete Gabor orientation bins also affect the detection performance. We have performed tests by varying the orientation bins from 11 to 20, and the results are displayed in Fig. 13 (b). Clearly the case that the orientation bins is equal to 18 generates the best vehicle hypotheses.

4.5. *Distance transform based modeling and learning*

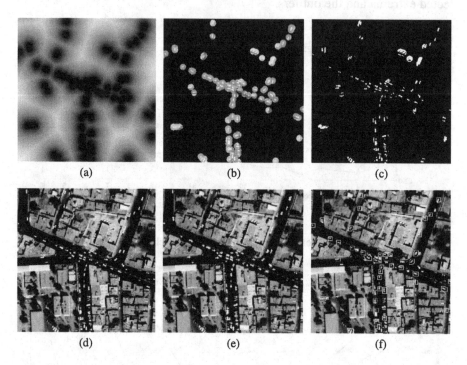

Fig. 14. (a) Distance transform. (b) Distance thresholded image. (c) Hard thresholded image. (d) Region perimeters. (e) Bounding ellipses. (f) Spectral clustered hypotheses.

As observed from the satellite imagery, most vehicle neighborhoods are near circular symmetric and have fixed distance between parallel linear segments. Road and building edges are also parallel with broader distances. These priors suggest that we can take a distance transform, to match between object modes. Fig. 14 (a) illustrate this idea and shows a distance map computed from the median filtered difference image. Local minimums which look like black holes are clearly visible. Further away from the centroids, the possibility of being vehicle pixels decreases with respect to the increasing orthogonal distance. Thus, the problem of vehicle detection can be approached from a spectral clustering perspective by fusing evidences from radiometric intensities, geometrical locations and regional features, and structural contexts. And the objectives of vehicle detection are to find the optimal solution based on cluster distances in a hyper-dimensional feature space.

Fig. 14 (b)~(f) illustrate the processing results based on distance transform. Figure (b) shows a circular neighborhood extracted around each vehicle candidate with an empirical distance threshold set to 7 pixels. Vehicles are well extracted and outliers are much less than the direct blob based approach. (c) displays a binary image that is simply thresholded at the value of 0.3. The bounding perimeters of the thresholded areas are drawn in (d) and a minimum bounding ellipse in (e). (f) shows the spectral clustered results.

4.6. *Neural network results*

To arbitrate the grouped vehicle hypotheses, we have constructed a two-layer feed-forward neural network that contains 25 hidden neurons, and multiple input nodes of features formed as hyper-dimensional vectors. The vehicle distribution is learned by presenting examples of aggregated vehicle cues to the network which is optimized by the error back-propagation algorithm. The number of training data and their representation capability in the object feature space affect the network generalization and performance. In practice, best tradeoff to avoid both overfitting and underfitting can be achieved by balancing the network complexity and the training goal. Numerous tests have been conducted using the constructed neural network. Fig. 15 (a) and (b) illustrate the relationship between the number of epochs and time for training the neural network versus the number of hidden neurons respectively. Totally we have detected 1355 vehicles by the selected neural network model shown in Fig. 16 (a). Vehicle orientations are marked in different color ellipses according to their binning in Fig. 16 (b).

The detection rate is 94.8% and the missing rate is 5.2% based on the approximate ground truth 1429 vehicles estimated in the 2000x2000 imagery. Then within the 1355 vehicles, we have verified that there are 1279 detected true positives, whose false alarm rate is 5.3%.

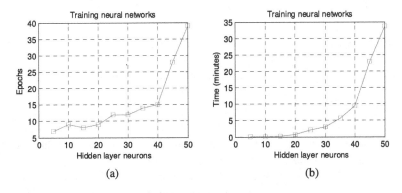

Fig. 15. (a) The neural network training epochs. (b) Neural networks training time.

(a)   (b)

Fig. 16. (a) Neural networks results. (b) Detected vehicles.

5. Discussion and conclusion

In this chapter, we have explored the challenging vehicle detection problem in satellite imagery from different perspectives. We have identified and presented a vehicle modeling and learning approach based on the multiscale analysis and the distance transform. The generation and representation of vehicle hypotheses in small scales and limited resolution are grouped from local evidences and global contextual priors. The hypothesis densities are fitted by elliptical Gaussians which have the advantage that various optimization techniques can be applied. Our method combines a bootstrapped learning mechanism with the adaptive orientation modeling by Gabor filters and hyperspectral feature clustering to handle dense and overlapping vehicle distributions. The high detection results and evaluations demonstrate the effectiveness of these models. And we believe the generalized models can be improved and applied to other computer vision tasks as well.

Acknowledgements

We thank DigitalGlobe for providing the satellite imagery used in this chapter.

References

[1] J.Y. Choi and Y.K. Yang, "Vehicle detection from aerial images using local shape information," *PSIVT 2009, LNCS 5414,* pp. 227-236, 2009.

[2] H. Grabner, T.T. Nguyen, B. Gruber, and H. Bischof, "On-line boosting-based car detection from aerial images," *ISPRS Journal of Photogrammetry & Remote Sensing,* 63(3):382-396, 2008.

[3] S. Hinz and A. Baumgartner, "Vehicle detection in aerial images using generic features, grouping, and context," *Pattern Recognition 2001 (DAGM Symposium 2001), Lecture Notes on Computer Science 2191,* Springer Verlag, 2001.

[4] S. Hinz, "Integrating local and global features for vehicle detection in high resolution aerial imagery," *International Archives of Photogrammetry, Remote Sensing and Spatial Information Sciences,* vol. 34(3W/8), pp. 119-124, 2003.

[5] S. Hinz and U. Stilla, "Car detection in aerial thermal images by local and global evidence accumulation," *Pattern Recognition Letters,* vol. 27, pp. 308-315, Mar. 2006.

[6] Z. Kim and R. Nevatia, "Expandable Bayesian networks for 3-D object descriptions from multiple views and multiple mode inputs," *IEEE Trans. Pattern Analysis and Machine Intelligence,* vol. 25, no. 6, pp. 769-774, 2003.

[7] T.T. Nguyen, H. Grabner, B. Gruber, and H. Bischof, "On-line boosting for car detection from aerial images," *Proc. of The 5th IEEE International Conference on Computer Sciences,* RIVF'06, 2006.

[8] T. Zhao and R. Nevatia, "Car detection in low resolution aerial image," *Proc. IEEE IVC(21),* no. 8, pp. 693-703, August 2003.

[9] A.M. Bazen and S.H. Gerez, "Systematic methods for the computation of the directional fields and singular points of fingerprints," *IEEE Trans. on Pattern Analysis and Machine Intelligence,* vol. 24, no. 7, pp. 905-919, July 2002.

[10] F. van der Heijden, R. Duin, D.d. de Ridder, D.M.J. Tax, "Classification, parameter estimation and state estimation: an engineering approach using MATLAB," *John Wiley & Sons Ltd,* 2004.

[11] L. Hong, Y. Wan and A. Jain, "Fingerprint image enhancement: algorithm and performance evaluation," *IEEE Trans on Pattern Anal Mach Intelligence,* vol. 20, no. 8, pp. 777-789, Aug. 1998.

[12] O. Garcia-Pineda, B. Zimmer, M. Howard, W. Pichel, X. Li, and I. R. MacDonald, "Using SAR images to delineate ocean oil slicks with a texture-classifying neural network algorithm (TCNNA)," *Canadian Journal of Remote Sensing,* vol. 35, no. 5, pp. 411-421, Oct. 2009.

[13] C. Goerick, D Noll and M Werner, "Artificial neural networks in real time car detection and tracking applications," *Pattern Recognition Letter,* vol. 17, pp. 335-343, 1996.

CHAPTER 2.7

APPLICATION OF POLAR HARMONIC TRANSFORMS TO FINGERPRINT CLASSIFICATION

Manhua Liu*, Xudong Jiang[†], Alex Chichung Kot[†] and Pew-Thian Yap[‡]

*Department of Instrument Science and Engineering,
Shanghai Jiao Tong University, Shanghai, China
[†]School of Electrical and Electronic Engineering,
Nanyang Technological University, Singapore
[‡]Department of Radiology and Biomedical Research Imaging Center (BRIC),
University of North Carolina at Chapel Hill, U.S.A.
Corresponding Author: Pew-Thian Yap

Accurate object classification is often confounded by the variation of the pose of the object. We show in this chapter how a set of recently introduced transforms, called the Polar Harmonic Transforms (PHTs), can be used to produce a set of features for rotation invariant fingerprint representation. For accurate classification, fingerprint images often need to be corrected for rotational differences. Determining an orientation reference for achieving this, however, often results in ambiguity and is hence not always reliable. PHTs allow rotation invariant representation of the fingerprint images and hence discard the need for reference detection. Experimental results indicate that the proposed PHT-based classification scheme yields results that are comparable with state-of-the-art methods.

1. Introduction

The ability to recognize an object irrespective of its orientation is often highly desirable for pattern classifiers. While this ability is born innate in human beings, it is non-trivial and not always easily achievable for machines. One method to achieve this is to train a number of classifiers that work in parallel to cater for a finite number of orientations. Another is to extract directly from the image a set of representation features that are invariant to object orientation. In this chapter, we will focus on the latter approach. We will demonstrate how a recently introduced set of image descriptors, which are obtained via the polar harmonic transforms (PHTs) [Yap et al. (2010)], can be employed for fingerprint classification.

Moments, as image descriptors often invariant to image transformations, are powerful tools that find vast applications in computer vision. Various types of moments have been extensively investigated for invariant image representation. Among them is the popular Zernike moments (ZMs) [Teague (1980)], and also the often mentioned-together pseudo Zernike moments (PZMs) [Teh and Chin (1988)]. These moments have been applied successfully in a variety of contexts including, but not limited to, modeling of corneal surface [Iskander *et al.* (2001)], watermarking [Kim and Lee (2003)], face recognition [Foon *et al.* (2004)], character recognition [Kan and Srinath (2002)], fingerprint recognition [Qader *et al.* (2007)], multi-spectral texture classification [Wang and Healey (1998)], and edge detection [Ghosal and Mehrotra (1994)]. Despite their wide usefulness, these two sets of moments are often faced with the problem of high computation cost and numerical instability, especially when high order moments are concerned. The computational problems associated with these moments are related to the fact that many factorial terms are involved in the computation of the respective moment basis functions.

Some other examples of moments invariant include rotational moments (RMs) [Teh and Chin (1988)] and complex moments (CMs) [Abu-Mostafa and Psaltis (1984, 1985)]. But these moments are not orthogonal, as are ZMs and PZMs. Non-orthogonality implies lower information compactness in each of the computed moments. On the other hand, orthogonality of the basis functions implies that an image is projected onto a set of pairwise orthogonal axes, and the difficulty of the task of analyzing the image can be substantially reduced. The basis functions of ZMs and PZMs can be shown to be the outcome of applying Gram Schmidt orthogonalization on the CM basis functions [Abu-Mostafa and Psaltis (1984)]. Refer to [Shu *et al.* (2007); Coatrieux (2008a,b); Mukundan and Ramakrishnan (1998)] for a comprehensive survey on moments.

In this chapter, we demonstrate the effectiveness of PHTs [Yap *et al.* (2010)] — a recently introduced set of transforms capable of generating rotation invariant features — in fingerprint classification. We will first give a brief introduction on PHTs and explain why they pose much less computation complication than ZMs and PZMs. We will then show how PHTs can be applied to orientation fields of fingerprints to extract compact and rotation invariant features for use in fingerprint classification. Preliminary experimental results are included as a proof of concept.

2. Polar Harmonic Transforms (PHTs)

For a square-integrable function $f(x, y)$, a transform of order $(p, q) \in \mathbb{Z}^2$ with respect to the basis functions $\psi_{p,q}(x, y)$, generating the transform coefficients $M_{p,q}$ can be defined as

$$M_{p,q} = \int_x \int_y \psi_{p,q}^*(x, y) f(x, y) dx dy, \qquad (1)$$

where the superscript * denotes the complex conjugate. The basis functions are typically designed to have some properties that are useful for the task at hand. Moments, for instance, can yield representations that are invariant to rotation, scaling and translation of $f(x, y)$. For digital images defined on a discrete domain, equation (1) can be discretized and written in the form

$$M_{p,q} = \sum_x \sum_y \psi_{p,q}^*(x,y) f(x,y) \Delta x \Delta y. \tag{2}$$

If the set of basis functions is complete, the image can be completely characterized by the corresponding set of transform coefficients $\{M_{p,q}\}$.

To appreciate the rotation invariance of a transform, it is helpful to express the transform in polar coordinates (r,φ), where $r = \sqrt{x^2 + y^2}$ is the radius of a circle with $(0,0)$ as its origin and (x,y) as one of its points, and $\varphi = \arctan(y/x)$ is the angle between the line joining the origin and the point (x,y) and the x-axis. The basis function $\psi_{p,q}(x,y)$ and the image function $f(x,y)$ can then be written as $\psi_{p,q}(x,y) = \psi_{p,q}(r\cos\varphi, r\sin\varphi) \equiv \psi'_{p,q}(r,\varphi)$ and $f(x,y) = f(r\cos\varphi, r\sin\varphi) \equiv f'(r,\varphi)$, respectively. A set of polar harmonic basis functions are introduced in [Yap *et al.* (2010)] for the purpose of generating rotation invariant features. These basis functions are mutually orthogonal and can be separated into the radial and angular components: $\psi'_{p,q}(r,\varphi) = R_p(r)\Phi_q(\varphi)$. Various types of basis functions are introduced in [Yap *et al.* (2010)] for achieving various polar harmonic transforms (PHTs). We will only use the simplest form where the radial and angular components are defined using complex exponentials as

$$R_p(r) = e^{i2\pi p r^2}, \tag{3}$$

$$\Phi_q(\varphi) = e^{iq\varphi}, \tag{4}$$

where $(p,q) \in \mathbb{Z}^2$. Compared with the well-known Zernike moments, the computation of these basis functions is extremely simple with no numerical stability issues. Given an image function $f(x,y)$, the PHT of order (p,q), expressed in polar coordinates, is defined as

$$M_{p,q} = \int_0^{2\pi} \int_0^1 R_p^*(r)\Phi_q^*(\varphi)f'(r,\theta)r\,dr\,d\theta \tag{5}$$

where $R_p(r)$ and $\Phi_q(\varphi)$ are defined according to equations (3) and (4). Visual illustrations of the basis functions are given in Fig. 1.

3. Fingerprint Classification Using Invariant Image Representation

In general, fingerprint recognition systems work in two modes: one-to-many identification and one-to-one verification. Identifying a fingerprint from a large data set is more difficult due to the complex database search involved. Verifying a fingerprint, on the other hand, needs to be performed based only on one-to-one matching. Fingerprint classification is important for indexing the fingerprints to speed up database search in a large scale fingerprint identification system. Fingerprint classification, widely studied for more than a century, aims to classify each fingerprint into one of the pre-specified classes in an accurate and consistent manner. Most classification

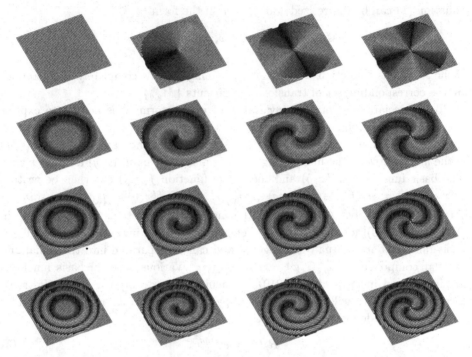

Fig. 1.   The PHT basis functions $\psi'_{p,q}(r,\varphi)$. The plots show the phase angle; the magnitude is unity for all cases. Top to bottom: $p = 0, 1, 2, 3$; left to right: $q = 0, 1, 2, 3$.

algorithms are based on the Galton-Henry classification scheme [Henry (1900)], classifying fingerprints into five common human-interpretable classes (see Fig. 2): arch, tented arch, left loop, right loop and whorl. Consistent and reliable fingerprint classification, however, remains a challenging problem due to the intrinsic complexities of the human fingerprint.

One important difficulty in fingerprint classification involves extracting representative features that are invariant to translation and rotation. The orientation field of a fingerprint is a form of coarse-level representation that is widely used for fingerprint classification and indexing [Cappelli *et al.* (1999a); Lumini *et al.* (1997); Cappelli *et al.* (1999b); Nagaty (2001); Jiang *et al.* (2006)]. Fig. 3 (a) shows the orientation field of a right loop fingerprint. For invariant representation of the orientation field, fingerprints are usually corrected for translation and rotation using a reference-based scheme. For this purpose, singular points, i.e., core and delta points shown in Fig. 3 (b), are important landmark points that can be used as reference points for alignment. Many methods have been proposed for the detection of these points [Karu and Jain (1996); Nilsson and Bigun (2003)]. For translational alignment, core points are usually used. Rotational alignment, however, is not as straightforward since it is generally difficult to detect a reliable reference orientation for alignment.

Fig. 2. The five commonly used fingerprint classes: arch, tented arch, left loop, right loop and whorl (from left to right and top to bottom).

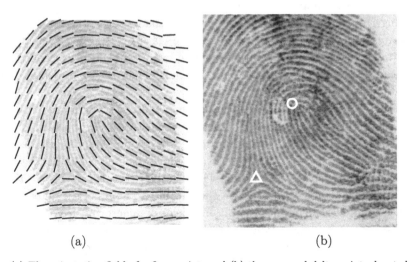

(a)   (b)

Fig. 3. (a) The orientation field of a fingerprint, and (b) the core and delta points denoted with ∘ and △, respectively.

In what follows, we will illustrate how PHTs can be applied for rotation invariant representation of the orientation field. We will show how the extracted PHT features can be employed for fingerprint classification, performed using support vector machines (SVMs) [Gunn (1998); Ma *et al.* (2002)], which are supervised classifiers widely used for data analysis and pattern classification. Experimental results performed on the NIST special fingerprint database 4 (NIST-DB4) [Watson and Wilson (1992)] are then presented to demonstrate the effectiveness of PHTs in fingerprint classification.

3.1. *Representing the Fingerprint Orientation Field Using PHTs*

The orientation field describes the global flow patterns of the ridges on a fingerprint. It provides coarse, but robust, features that are widely used for fingerprint classification. The orientation field is commonly obtained by dividing the fingerprint image into blocks of the same size and the orientation of each block is estimated using a gradient based averaging method [Hong *et al.* (1998); Bazen and Gerez (2002)]. In this work, a block size of 16×16 is used. The orientation is often represented by a gradient phase angle θ. This representation, however, often causes problems, owing to its periodicity and discontinuity. This is caused by the fact that local ridge-valley structures remain unchanged when rotated 180 degrees. Opposite gradient vectors will hence cancel each other, although they indicate the same ridge-valley orientation. To avoid this problem, one popular approach is to double the phase angle θ and construct an orientation field O with each field element represented using complex value $o = e^{i(2\theta)} = \cos(2\theta) + i\sin(2\theta)$. This angle-doubled vector field is used in this work to represent the orientation field.

A fingerprint, when captured under different conditions, varies in terms of translation, rotation and segmentation. A reliable feature extraction should hence be robust to these variations. For this purpose, we discussion in the following how PHTs can be employed to generate an invariant fingerprint representation of the orientation field.

Firstly, we note that image translation may shift the origin used in the computation of PHTs and results in undesirable changes to the computed coefficients $M_{p,q}$. The orientation fields have to be shifted to a common location to correct for translational differences. To achieve this, a unique reference point consistent to all fingerprint types needs to be located. The method described in [Liu *et al.* (2004)] is employed for this purpose. Figure 4 shows some examples of the detected unique reference point overlaid on the orientational fields of fingerprints from the five different classes. Given an orientation field $O(k', l')$, computed on a discrete domain with $(k', l') = [0, K] \times [0, L]$, we shift the orientation field to a padded discrete domain $(k, l) \in [-K, K] \times [-L, L]$ with the reference point placed at the origin $(0, 0)$.

Secondly, we note that the image background of a fingerprint usually contains uninformative noisy areas without apparent ridges and valleys. These areas should be excluded from the computation of PHTs for more reliable feature extraction.

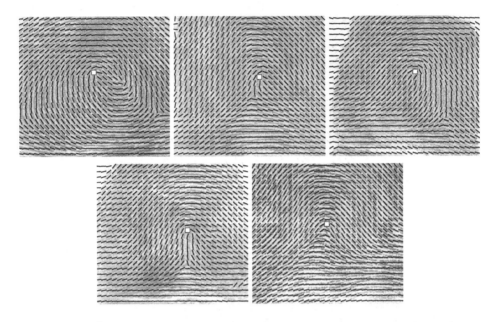

Fig. 4. The unique reference point, marked by the white squares, detected on fingerprints from five different classes. From left to right and top to bottom: whorl, right loop, left loop, tented arch and arch.

For this purpose, we use the fingerprint segmentation method described in [Bazen and Gerez (2001)] to separate the foreground (with clear ridges and valleys) from the background. The segmentation outcome is represented by a 2D binary matrix W. Matrix element $w(k,l)$ is set to one if the image block (k,l) is determined to be a part of the foreground; otherwise, it is set to zero, indicating a background block. Only the foreground image blocks are used in the computation of PHTs.

Thirdly, we extract rotation invariant features from the fingerprint orientation field using PHTs based on equation (2). The discrete domain (k,l) of the translation-corrected orientation field, $O(k,l)$, is first mapped to a new domain $(x_k, y_l) \in [-1,1] \times [-1,1]$ with $x_k = k/K$ and $y_l = l/L$. Noting that $\Delta x = 1/K$, $\Delta y = 1/L$, the PHT coefficient of order (p,q), $M_{p,q}$, can then be computed in the transformed domain as

$$M_{p,q} = \frac{1}{KL} \sum_{k=-K}^{K} \sum_{l=-L}^{L} \psi_{p,q}^*(x_k, y_l) O(k,l) w(k,l) \tag{6}$$

with the basis function $\psi_{p,q}(x_k, y_l) = \psi_{p,q}(r\cos\varphi, r\sin\varphi) = e^{i2\pi pr^2} e^{iq\varphi}$, where $r = \sqrt{x_k^2 + y_l^2}$ and $\varphi = \arctan(y_l/x_k)$. For better grasp of what is actually captured by the PHT coefficients, we inspect the orientation patterns that are represented by the basis functions when $p = 0$ and $q = \pm 1$. The equivalent local orientations $\theta_{k,l}$ represented by the basis functions $e^{\pm i\varphi}(q = \pm 1)$ are shown in Figs. 5 (a) and (b).

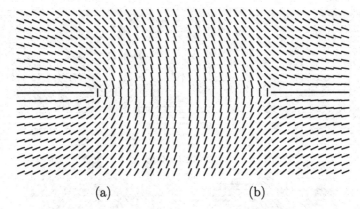

(a)   (b)

Fig. 5. The equivalent local orientation patterns represented by the basis functions (a) $e^{i\varphi}$ and (b) $e^{-i\varphi}$.

It can be observed that these orientation patterns are very similar to those typically given by the core and delta points of a fingerprint.

Rotation invariant features can be extracted easily from the PHT coefficients. If the orientation field $O(k, l)$ is rotated clockwise at an angle ϕ to become $O'(k', l') = O(k, l)e^{-i(2\phi)} = e^{i(2(\theta_{k,l}-\phi))} = \cos(2(\theta_{k,l} - \phi)) + i\sin(2(\theta_{k,l} - \phi))$, the new set of PHT coefficients, denoted by $\{M'_{p,q}\}$, is related to $\{M_{p,q}\}$ by

$$M'_{p,q} = M_{p,q}e^{-iq(2\phi)}, \tag{7}$$

noting here that $M'_{p,q}$ and $M_{p,q}$ are complex-valued. Many forms of rotation invariant representations can be derived from this equation based on phase cancellation [Yap et $al.$ (2010)]. The simplest way of canceling the effect of the rotation factor, i.e., $e^{-iq(2\phi)}$, is by taking the absolute values on both sides of equation (7), resulting in $|M'_{p,q}| = |M_{p,q}|$. This indicates that the magnitudes $\{|M_{p,q}|\}$ of the PHT coefficients $\{M_{p,q}\}$ are invariant to image rotation. We form a feature vector by concatenating the magnitudes of the PHT coefficients. Limiting $|p| \leq M, |q| \leq N$, the length of feature vector is fixed at $(2M + 1) \times (2N + 1)$. The merit of each PHT feature is application dependent and should be evaluated based on an application-specific objective function. In our case, we have simply set $M = N = 10$ to capture sufficient information from the orientation field, but at the same time truncating higher order PHT coefficients, which capture high-spatial-frequency information, but are more susceptible to noise.

Compared with the commonly used 1680-element representation of the orientation field [Lumini et $al.$ (1997)], the dimensionality is significantly reduced, by using PHT features, to: $21 \times 21 = 441$. For further reduction, the Karhunen-Loeve Transform (KLT) is employed to map the feature vectors into a lower dimensional space to obtain a more compact representation for fingerprint classification. KLT, which aims to minimize information loss, is one of the most widely used statistical framework for dimensionality reduction. It guarantees the best Euclidean distance preservation among the unitary transformations for dimensionality reduction [Lumini *et al.* (1997)]. Applying KLT, the length of the feature vector is eventually reduced to 80.

There are several advantages in utilizing the PHT coefficients instead of the raw orientation fields for fingerprint classification. Firstly, the PHT derived feature vector is intrinsically invariant to image rotation without requiring rotational alignment. Detection of a reference for orientation correction consistent to all types of fingerprints is not an easy task, especially for images with poor quality. A small deviation may result in large changes in the final computed features. Secondly, the required number of PHT coefficients is generally small, hence reducing the computation and memory costs. Thirdly, the raw orientation data are intrinsically periodic with its representation using sine and cosine functions. This may affect the performance of KLT since the transform is best suited for data with Gaussian-like distributions [Wang *et al.* (2007)]. The PHT coefficient magnitudes are real numbers that do not exhibit obvious periodic patterns.

3.2. *Fingerprint Classification*

The PHT feature vectors discussed in the previous section are fed as input to the support vector machines (SVMs) for fingerprint classification. SVMs [Gunn (1998); Ma *et al.* (2002)] are an effective machine learning technique that is widely used for classification and regression analysis. SVMs are found to be an effective classifier for fingerprint classification [Li *et al.* (2008); Hong *et al.* (2008)]. A SVM constructs a hyperplane or set of hyperplanes in a high or infinite dimensional space that has the largest distance to the nearest training data points of any class, since in general the larger the margin the lower the generalization error of the classifier. The points used to determine the hyperplane are called support vectors. If the data is not linearly separable in the input space, a non-linear transformation $\Phi(\cdot)$ can be used to map the data points $x \in \mathbb{R}$ to a Hilbert space $\mathbf{H}$. The mapping $\Phi(\cdot)$ is represented by a kernel function $K(\cdot, \cdot)$ that defines an inner product in $\mathbf{H}$, i.e., $K(x,t) = \Phi(x) \cdot \Phi(t)$. The decision function of the SVM has the form:

$$f(x) = \sum_{i=1}^{n} \alpha_i c_i K(x_i, x) + b \tag{8}$$

where n is the number of data points, and $c_i \in \{-1, 1\}$ is the class label of training point x_i. Coefficients α_i can be found by solving a quadratic programming problem with linear constraints. It is non-zero when x_i is a support vector that composes the hyperplane; otherwise it is zero. The kernel function $K(x, x_i)$ can be easily computed by an inner product of the non-linear mapping function. Table 1 shows some representative kernel functions of SVMs, including the linear, polynomial, Gaussian, and sigmoid functions. One of the main advantages of using SVMs for pattern classification is that they are capable of learning in sparse, high-dimensional spaces with very few training examples. In the current work, the publicly available

Table 1. Typical kernel functions for SVMs.

Linear	Polynomial	Gaussian	sigmoid
$(x \cdot x_i)$	$(x \cdot x_i + \gamma)^d$	$\exp(-\frac{\|x-x_i\|^2}{2\sigma^2})$	$\tanh(x \cdot x_i + \gamma)$

OSU SVM Classifier Matlab Toolbox (version 3.00) [Ma *et al.* (2002)] is used. The linear kernel function is employed. The penalty factor of the error term $C > 0$ is selected based on cross-validation using the training dataset, and is finally set to 20.

3.3. *Experimental Evaluation*

The NIST special fingerprint database 4 (NIST-DB4) is one of the most important benchmarks to test fingerprint classification algorithms. Similar to most published fingerprint classification algorithms, we evaluated our algorithm based on this database. The five common fingerprint classes (see Fig. 2) — plain arch, tented arch, right loop, left loop and whorl fingerprints — are evenly distributed in NIST-DB4. This database contains 4,000 fingerprint images of size 480×512 pixels, taken from 2,000 fingers with two instances per finger. The first fingerprint instances are numbered from f0001 to f2000 and the second fingerprint instances are numbered from s0001 to s2000. All fingerprints in this database were used in our experiment. This fingerprint database is divided into the training set and the testing set. The training set contains the 2,000 fingerprints from the first 1,000 fingers (f0001 to f1000 and s0001 to s1000). The remaining 2,000 fingerprints (f1001 to f2000 and s1001 to s2000) are included in the testing set.

The performance of a fingerprint classification algorithm is often measured in terms of the classification error rate. The error rate is computed as the ratio of the number of misclassified fingerprints to the total number of fingerprints in the testing set. Each fingerprint in the NIST-DB4 was assigned by human experts to one or two of the five common classes. There are about 17% ambiguous fingerprints that have two classes assigned to them. For simplicity, we made use of only the first class label to train the classifier. For testing, however, the classifier output was considered correct if it matches any of the human-specified fingerprint classes. This approach is commonly adopted by other researchers when evaluating the classification accuracy of their algorithms using NIST-DB4 [Candela *et al.* (1995); Karu and Jain (1996); Jain *et al.* (1999); Jain and Minut (2002); Yao *et al.* (2001); Zhang and Yan (2004); Hong *et al.* (2008); Li *et al.* (2008)].

The confusion matrix for the five fingerprint classes is shown in Table 2, where W, L, R, T and A represent the whorl, left loop, right loop, tented arch and arch fingerprint classes, respectively. It can be computed from the table that the error rate of our approach is 10.3%. We note here that the arch and tented arch fingerprints are very similar and it is sometimes difficult to distinguish between the two. Some researchers do not differentiate between these two classes and combine

Table 2. Confusion matrix for five fingerprint classes.

Actual	Predicted Class				
Class	W	L	R	T	A
W	360	7	26	0	3
L	7	358	1	14	6
R	2	0	339	40	9
T	0	8	8	338	40
A	0	3	2	31	398

them into a single class, which is simply called the arch class. For this four-class classification problem, the error rate of our approach is 6.6%.

We compared our approach with some representative approaches that have been published previously in the literature. The results reported by other researchers, along with our own, for both four- and five-class problems are shown in Table 3. All results are based on NIST-DB4. The features and classification techniques used in these approaches are also listed in the table. Singularities, i.e., the singular points, are the most widely used features in fingerprint classification [Karu and Jain (1996); Zhang and Yan (2004); Hong *et al.* (2008)]. The approach proposed by Hong et al. [Hong *et al.* (2008)] yields the best performance in both five- and four-class problems. In their approach, several popular fingerprint features such as singularities and pseudo ridges. They also utilize a combination of naïve-bayesian and SVM-based classifiers, which contributed to better classification accuracy. It should also be noted that in [Jain *et al.* (1999); Hong *et al.* (2008)] 1.8% of the fingerprints that are of poor quality were rejected in the testing set, resulting in higher classification accuracy. It can be observed that our fingerprint classification method performs better than those using only one type of feature, such as the orientation field, singularities or Gabor filters based features, indicating that PHT features are well-suited for fingerprint representation. Future work will be directed to incorporating other complementary information and more sophisticated classification approaches to further improve classification performance.

It is often difficult to classify fingerprint images that are of poor quality, even by human experts. We introduced a rejection scheme to exclude some fingerprints with poor quality from the testing set. Fingerprints were rejected for testing when their proportions of valid orientation elements were smaller than a threshold. For various levels of rejection, the threshold was set to different values. As can be seen from Figure 6, the classification error rate, as expected, decreases with the increase of the rejection rate. The results for the case of the raw orientation features, based on the same rejection scheme and SVM classifiers, are also included for comparison. In all cases, PHT features compare favorably against the raw orientation features. Future work entails designing classifiers that are more tailored to the properties of PHTs.

Table 3. Error rates (%) of various fingerprint classification approaches, evaluated based on NIST-DB4.

Algorithm	Feature Type(s)	Classifier	5-Class	4-Class	Testing Set
[Candela *et al.* (1995)]	Orientation field	NN	–	11.4	2nd-hv
[Karu and Jain (1996)]	Singularities	Rules	14.6	8.6	whole
[Jain *et al.* (1999)]	Gabor filters	k−NN & NN	10	5.2	2nd-hv 1.8% reject
[Jain and Minut (2002)]	Ridge kernel	Rules	–	8.8	whole
[Yao *et al.* (2001)]	Gabor filters	SVMs	10.7	6.9	2nd-hv
[Zhang and Yan (2004)]	Singularities & ridge	Rules	15.7	7.3	whole
[Park and Park (2005)]	Fourier transform	NDA	9.3	6.0	2nd-hv
[Hong *et al.* (2008)]	Singularities & Garbor filters	SVMs & NB	9.2	5.1	2nd-hv 1.8% rejection
[Li *et al.* (2008)]	Orientation field	SVMs	12.6	–	2nd-hv
Our approach	PHT features	SVMs	10.3	6.6	2nd-hv

where NN: Neural Network; k−NN: k−Nearest Neighbor; SVMs: Support vector machines; NDA: Nolinear Discriminant Analysis; NB: Naïve Bayes; 2nd-hv: the second half portion of the NIST-DB4 is used for testing.

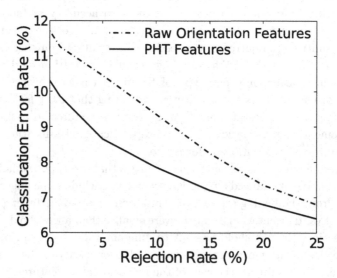

Fig. 6. Reduction in classification error rate with the increase in rejection rate for the five-class problem.

4. Conclusion

In this chapter, we demonstrate that polar harmonic transforms (PHTs) can serve as an effective representation for fingerprints. We show that PHT features are especially useful in the case where the reference for rotation correction cannot be determined reliably. The PHT features allow rotation invariant representation of the fingerprint orientation fields to generate a set of effective features for the purpose of classification. We have also shown that classification performance of the PHT-based classification scheme is comparable to state-of-the-art methods.

5. Acknowledgments and Appendices

This work was supported by the National Natural Science Foundation of China (No. 61005024), the Doctoral Fund of Ministry of Education of China (No. 20090073120019) and the Medical and Engineering Foundation of Shanghai Jiao Tong University (No. YG2010MS74).

References

Abu-Mostafa, Y. and Psaltis, D. (1984). Recognitive aspects of moment invariants, *IEEE Trans Pattern Analysis Machine Intelligence* **PAMI-6**, pp. 698–706.

Abu-Mostafa, Y. and Psaltis, D. (1985). Image normalization by complex moments, *IEEE Trans Pattern Analysis Machine Intelligence* **PAMI-7**, pp. 46–55.

Bazen, A. M. and Gerez, S. H. (2001). Segmentation of fingerprint images, in *Proceedings of ProRISC2001, 12th Annual Workshop Circuits, Systems and Signal Processing*.

Bazen, A. M. and Gerez, S. H. (2002). Systematic methods for the computation of the directional fields and singular points of fingerprints, *IEEE Transactions on Pattern Analysis and Machine Intelligence* **24**, 7, pp. 905–919.

Candela, G. T., Grother, P. J., Watson, C. I., Wilkinson, R. A. and Wilson, C. L. (1995). PCASYS - a pattern-level classification automation system for fingerprints, *Technique Report: NIST TR 5647*.

Cappelli, R., Lumini, A., Maio, D. and Maltoni, D. (1999a). Fingerprint classification by directional image partitioning, *IEEE Transactions on Pattern Analysis and Machine Intelligence* **21**, 5, pp. 402–421.

Cappelli, R., Maio, D. and Maltoni, D. (1999b). Fingerprint classification based on multi-space KL, in *Proceedings of Workshop on Automatic Identification Advanced Technologies*, pp. 117–120.

Coatrieux, J.-L. (2008a). Moment-based approaches in imaging. part 2: Invariance, *IEEE Engineering in Medicine and Biology Magazine* **27**, 1, pp. 81–83.

Coatrieux, J.-L. (2008b). Moment-based approaches in imaging. part 3: Computational considerations, *IEEE Engineering in Medicine and Biology Magazine* **27**, 3, pp. 89–91.

Foon, N. H., Pang, Y.-H., Jin, A. T. B. and Ling, D. N. C. (2004). An efficient method for human face recognition using wavelet transform and Zernike moments, *Proceedings of International Conference on Computer Graphics, Imaging and Visualization (CGIV)*, pp. 65–69.

Ghosal, S. and Mehrotra, R. (1994). Detection of composite edges, *IEEE Transactions on Image Processing* **3**, 1, pp. 14–25.

Gunn, S. (1998). Support vector machines for classification and regression, *University of Southampton.*

Henry, E. R. (1900). Classification and uses of fingerprints.

Hong, J.-H., Min, J.-K., Cho, U.-K. and Cho, S.-B. (2008). Fingerprint classification using one-vs-all support vector machines dynamically ordered with naïve bayes classifiers, *Pattern Recognition* **41**, 2, pp. 662–671.

Hong, L., Wan, Y. and Jain, A. K. (1998). Fingerprint image enhancement: Algorithm and performance evaluation, *IEEE Transactions on Pattern Analysis and Machine Intelligence* **20**, 8, pp. 777–789.

Iskander, D., Collins, M. and Davis, B. (2001). Optimal modeling of corneal surfaces with Zernike polynomials, *IEEE Transactions on Biomedical Engineering* **48**, 1, pp. 87–95.

Jain, A. K. and Minut, S. (2002). Hirerarchical kernel fitting for fingerprint classification and alignment, in *Proceedings of International Conference on Pattern Recognition (16th)*, Vol. 2, pp. 469–473.

Jain, A. K., Prabhakar, S. and Hong, L. (1999). A multichannel approach to fingerprint classification, *IEEE Transactions on Pattern Analysis and Machine Intelligence* **21**, 4, pp. 348–359.

Jiang, X. D., Liu, M. and Kot, A. C. (2006). Fingerprint retrieval for identification, *IEEE Trans. on information forensics and security* **1**, 4, pp. 532–542.

Kan, C. and Srinath, M. D. (2002). Invariant character recognition with zernike and orthogonal fourier-mellin moments, *Pattern Recognition* **35**, 1, pp. 143–154.

Karu, K. and Jain, A. K. (1996). Fingerprint classification, *Pattern Recognition* **29**, 3, pp. 389–404.

Kim, H. S. and Lee, H.-K. (2003). Invariant image watermark using Zernike moments, *IEEE Transactions on Circuits and Systems for Video Technology* **13**, 8, pp. 766–775.

Li, J., Yau, W.-Y. and Wang, H. (2008). Combining singular points and orientation image information for fingerprint classification, *Pattern Recognition* **41**, 1, pp. 353–366.

Liu, M., Jiang, X. D. and Kot, A. C. (2004). Fingerprint reference point detection, *International Conference on Biometric Authentication (ICBA)* **3072**, pp. 272–279.

Lumini, A., Maio, D. and Maltoni, D. (1997). Continuous versus exclusive classification for fingerprint retrieval, *Pattern Recognition Letter* **18**, 10, pp. 1027–1034.

Ma, J., Zhao, Y. and Ahalt, S. (2002). OSU SVM Classier Matlab Toolbox (Version 3.00), (http://eewww.eng.ohiostate.edu/ maj/osu svm/).

Mukundan, R. and Ramakrishnan, K. (1998). Moment functions in image analysis – theory and applications, *World Scientific Publishing* **Singapore**.

Nagaty, K. A. (2001). Fingerprints classification using artificial neural networks: A combined structural and statistical approach, *Neural Networks* **14**, pp. 1293–1305.

Nilsson, K. and Bigun, J. (2003). Localization of corresponding points in fingerprints by complex filtering, *Pattern Recognition Letters* **24**, pp. 2135–2144.

Park, C. H. and Park, H. (2005). Fingerprint classification using fast fourier transform and nolinear discriminant analysis, *Pattern Recognition* **38**, 4, pp. 495–503.

Qader, H. A., Ramli, A. R. and Al-haddad, S. (2007). Fingerprint recognition using zernike moments, *The International Arab journal of information technology* **4**, pp. 372–376.

Ren, H., Liu, A., Zou, J., Bai, D. and Ping, Z. (2007). Character reconstruction with radial-harmonic-Fourier moments, *Proceedings of the Fourth International Conference on Fuzzy Systems and Knowledge Discovery 2007 (FSKD07)* **3**, pp. 307–310.

Sheng, Y. and Duvernoy, J. (1986). Circular Fourier-Radial Mellin descriptors for pattern recognition, *J. Opt. Soc. Am. A* **3**, 6, pp. 885–888.

Sheng, Y. and Shen, L. (1994). Orthogonal Fourier-Mellin moments for invariant pattern recognition, *J. Opt. Soc. Am. A* **11**, 6, pp. 1748–1757.

Shu, H., Luo, L. and Coatrieux, J.-L. (2007). Moment-based approaches in imaging. part 1: Basic features, *IEEE Engineering in Medicine and Biology Magazine* **26**, 5, pp. 70–74.

Teague, M. (1980). Image analysis via the general theory of moments, *Optical Soc. Am.* **70**, 8, pp. 920–930.

Teh, C. and Chin, R. (1988). On image analysis by the method of moments, *IEEE Transactions on Pattern Analysis and Machine Intelligence* **10**, 4, pp. 496–513.

Wang, L. and Healey, G. (1998). Using Zernike moments for the illumination and geometry invariant classification of multispectral texture, *IEEE Transactions on Image Processing* **7**, 2, pp. 196–203.

Wang, Y., Hu, J. and Phillips, D. (2007). A fingerprint orientation model based on 2d fourier expansion (fomfe) and its application to singular-point detection and fingerprint indexing, *IEEE Transactions on Pattern Analysis and Machine Intelligence* **29**, 4, pp. 573–585.

Watson, C. I. and Wilson, C. L. (1992). NIST special database 4, fingerprint database, Tech. rep., National Institute of Standards and Technology.

Yao, Y., Frasconi, P. and Pontil, M. (2001). Fingerprint classification with combination of support vector machine, in *Proceedings of 3rd International Conference on Audio-and Video-based Biometric Person Authentication*, pp. 253–258.

Yap, P.-T., Jiang, X. and Kot, A. C. (2010). Two dimensional polar harmonic transforms for invariant image representation, *IEEE Transactions on Pattern Analysis and Machine Intelligence* **32**, 7, pp. 1259–1270.

Zhang, Q. and Yan, H. (2004). Fingerprint classification based on extraction and analysis of singularities and pseudo ridges, *Pattern Recognition* **37**, 11, pp. 2233–2243.

CHAPTER 2.8

OBJECT CORRESPONDENCE NETWORKS FOR UNSUPERVISED RECOGNITION OF IDENTICAL OBJECTS

Minsu Cho, Young Min Shin and Kyoung Mu Lee

Department of EECS, ASRI, Seoul National University, 151-742, Seoul, Korea
chominsu@gmail.com; shinyoungmin@gmail.com; kyoungmu@snu.ac.kr

Humans do not have a difficulty in discovering and matching objects from even a single image without any specific prior knowledge about the objects, whereas most of object recognition methods in computer vision require strong or weak supervision for object modeling. In this chapter, we introduce an unsupervised object detection and segmentation problem, which goes beyond the conventional limitations of one-to-one object correspondences or model-test settings between images, and propose an effective method to solve it based on an advanced match-growing framework. Unlike the previous approaches, the method can detect and segment identical objects directly from a single image or a handful of images without any supervision, and establishes 'object correspondence networks' that connect matching objects from the given images. To detect and segment object-level correspondences, a multi-layer match-growing framework is developed that starts from initial local feature matches and explores the images by intra-layer expansion and inter-layer merge. Experiments demonstrate a robust performance of our method on challenging datasets. The resultant object correspondence networks can provide useful information for further applications such as structural pattern analysis, object reconstruction, and scene understanding.

1. Introduction

Despite significant progress in object recognition researches, most approaches usually require strong or weak supervision for object modeling.[1-6] For example, uncluttered model images are required or bounding boxes are adopted for learning each target object. Although recent object categorization methods based on latent topic models[7-9] have presented unsupervised or weakly supervised approaches, they require a decent amount of training images for learning and do not deal with direct object-based image matching between severely cluttered images. In real-world images, however, objects of similar appearance often show up simultaneously in a single view. For example, an image can contain not only replicas, but also multiple shots of identical objects. Thus, for the general purpose, unsupervised object-level matching within *a single image* should be investigated. Interestingly, this problem is directly related to *many-to-many* object matching across images. Given two or more images, let us imagine a combined image consisting of the given images. Then,

Fig. 1.   A result of unsupervised object detection and segmentation. Without any specific object model, our method detects all identical object sets (middle) with dense correspondences (right). Each object set constitutes an 'object correspondence network' that connects the identical objects.

the problem of detecting many-to-many object correspondences across the original multiple images is actually solved by detecting multiple sets of identical objects within the combined one. Although several researches[10,11] have dealt with feature-level many-to-many matching, many-to-many object matching has not been focused on in literature. In this chapter *, we address unsupervised object detection and segmentation that goes beyond the conventional assumptions, that is, *one-to-one* object matching constraints or *model-test* settings between images. As shown in Fig. 1, our method can detect and segment identical objects directly from a single image or a handful of images. It estimates geometric relations between object entities and establishes *object correspondence networks* that connect matching objects. This work is the first attempt to direct object matching within a single image and many-to-many object matching across images in an unsupervised way.

To explore the given images for matching objects, we take a match-growing approach expanding initial seeds of local feature matches to neighboring regions.[4,13] The approach has been developed in several computer vision tasks to boost initial putative matches and eliminate the matching ambiguities.[4,5,14–16] Recently, Cho *et al.*[13] and Kannala *et al.*[17] extended it to demonstrate unsupervised object matching beyond the limitation of conventional model-test settings in image matching and registration. In particular, Cho *et al.* explicitly established the *co-recognition* problem,[13] which recognizes and segments common objects directly from cluttered image pairs, and extended the concept to action recognition in video pairs.[18] All these previous methods based on match-growing[4,5,13–15,17] as well as most of local feature-based methods, however, commonly rely on the restrictive assumptions of one-to-one constraints both in feature matching and in object matching to make their problems feasible. Our work generalizes the concept of $co - recognition$[13] beyond the one-to-one assumptions, and advances a novel match-growing approach to resolving the challenges of many-to-many feature matching due to the presence of multiple identical objects or similar patterns.[12] We cast the problem as a maximum a posteriori (MAP) task on a generative model in a Bayesian framework, and propose a multi-layer match-growing algorithm that detects and segments all

---

*This chapter is extended from our paper presented in CVPR 2010.[12]

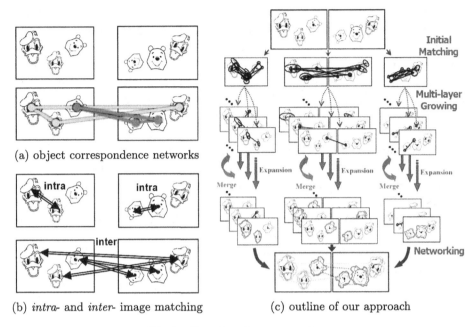

(a) object correspondence networks

(b) *intra-* and *inter-* image matching

(c) outline of our approach

Fig. 2.   Overview of our approach.

the object-level pairwise correspondences in the given images. Experiments demonstrate robust performance of the proposed method on challenging datasets of complex real-world images. The resultant object correspondence networks can provide useful information for further applications such as structural pattern analysis, object reconstruction, and scene understanding.

1.1. *Overview of Our Approach*

Supposing two images are given, relations of matching objects can be viewed as networks connecting the matching object regions as illustrated in Fig. 2 (a). Each network of connected objects represents an identical object set from the given images. This concept still holds when a single or multiple images are given; the more images we consider, the larger networks we are likely to obtain. These networks naturally reveal many-to-many object correspondences across images. As shown in Fig. 2 (b), a region pair of an object correspondence lies either within an image or across two images, thus object correspondences can be divided into two kinds, intra-image correspondences within each image and inter-image correspondences across images.

Our goal is to construct the *object correspondence networks* from the given images. Figure 2 (c) illustrates the outline of our approach. First of all, we divide the whole matching problem with N images into sub-problems consisting of N intra-matching problems and $\binom{N}{2}$ inter-matching problems. Although the problem can be considered as one intra-matching problem using a combined image, the decomposition increases efficiency by reducing complexity in the subsequent match-growing process. Second, based on local feature detectors, we obtain sufficient initial matches allowing multiple matching features for each feature. This one-to-many initial matching is required to consider the multiple identical patterns in images. Third, the initial matches are progressively enlarged by our multi-layer growing algorithm. At the start, each initial match forms a singleton *match cluster* and is endowed with its *expansion layer* which provides the match cluster with space for expansion. Those match clusters grow through *intra-layer expansion* and *inter-layer merge* until the posterior probability model of object correspondences is maximized. Finally, reliable object correspondences are chosen from the match clusters and connected into networks based on a region overlap criterion.

The remainder of this chapter is organized as follows. In Sec. 2, a set of object correspondences is described based on local region matches, and formulated in a Bayesian framework. In Sec. 3, our algorithm to detect object correspondences and connect them into networks is explained in detail. Then, experimental evaluations and comparisons are carried out in Sec. 4. Finally, in Sec. 5, conclusions are drawn with some discussions.

2. Bayesian Model for Object Correspondences

The object correspondence networks are composed of pairwise object correspondences, each of which relates two matching regions of an object. To model the object correspondences in an unsupervised way, we introduce a perceptually meaningful entity, termed the *Maximal Region Correspondence* (MRC) [†] which is defined as follows: (1) An MRC is a semi-global region pair, composed of local region matches (i.e. *match cluster*). (2) The region pair of an MRC should be mutually consistent in geometry and photometry. (3) An MRC should be maximal in its size.

Supposing the appearances of matching objects are not severely deformed in given images, an MRC from given images can be considered as an object correspondences under two conditions: (1) Two regions of each object correspondence lie on different backgrounds in photometry. (2) Object correspondences appear distinguishable from each other in geometry. With the first condition unsatisfied, an MRC includes a portion of a common background. With the second unsatisfied, multiple object correspondences can be recognized as one. For example, if two objects always appear aligned next to each other, we cannot discern them as separate objects without additional information. These are common intrinsic ambiguities of unsupervised object discovery.

Here, we formally describe MRCs which build the object correspondence networks. A set of MRCs Θ consists of MRCs Γ_i and is expressed by

$$\Theta = \{\Gamma_1, \Gamma_2, ..., \Gamma_K\}, \tag{1}$$

[†]This concept is an extension of the Maximal Common Saliency in *co-recognition*.[13]

| (a) a set of MRCs | (b) an MRC | (c) a local region match |

Fig. 3. A model of 'object correspondence networks'. (a) The object correspondences can be represented by a set of Maximal Region Correspondences (MRCs). (b) Each MRC consists of local region matches. (c) Each local region match is expressed by two affine regions and the homography between them. See the text.

where K denotes the number of MRCs in Θ. Each MRC Γ_i is represented by a cluster of local region matches (match cluster) as follows.

$$\Gamma_i = \{\mathcal{M}_{m|i} \mid m = 1, ..., N_i\}, \tag{2}$$

$$\mathcal{M}_{m|i} = \{(\mathcal{R}_{m|i}, \mathcal{R}'_{m|i}), \mathbf{H}_{m|i}\}, \tag{3}$$

where $\mathcal{M}_{m|i}$ denotes a member match that consists of a small local region pair $\mathcal{R}_{m|i}$, $\mathcal{R}'_{m|i}$, and affine homography $\mathbf{H}_{m|i}$ between them. N_i denotes the number of member matches in Γ_i. As shown in Fig. 3 (b)-(c), each MRC has a pair of corresponding object regions, and we refer to the larger region as the *head* and the smaller one as the *tail*. $\mathcal{R}_{m|i}$ and $\mathcal{R}'_{m|i}$ represent the local regions belonging to the head and the tail, respectively. In our model, a local region $\mathcal{R}$ is represented by an affine region centered on point $\mathbf{c}$, and is formulated using an ellipse equation

$$\mathcal{R} = \{\mathbf{x} \mid (\mathbf{x} - \mathbf{c})^{\mathrm{T}}\mathbf{A}(\mathbf{x} - \mathbf{c}) \leq 1\}, \tag{4}$$

where the 2×2 shape matrix $\mathbf{A}$ corresponds to the parameters of the ellipse equation. Since the transformation that projects the elliptical region to a unit circle with equal eigenvalues is given by the square root of the shape matrix $\mathbf{A}^{\frac{1}{2}}$, the points in $\mathcal{R}$ can be projected into a normalized unit circle by $\mathbf{A}^{\frac{1}{2}}(\mathbf{x} - \mathbf{c})$.[19] This transforms the intensity pattern of $\mathcal{R}$ to a normalized pattern. This normalized pattern has its characteristic orientation, which can be estimated by the gradient histogram.[1] Denoting the rotation matrix to the orientation by $\mathbf{R}$, the neighborhood points $\mathbf{x}$ of region $\mathcal{R}$ is further normalized in orientation by

$$T(\mathbf{x}) = \mathbf{R}^{-1}\mathbf{A}^{\frac{1}{2}}(\mathbf{x} - \mathbf{c}). \tag{5}$$

As illustrated in Fig. 3 (c), the intensity pattern in the normalized frame captures the photometric information of region $\mathcal{R}$ invariant to affine deformation. Applying Eq. (5) to the region pair of match $\mathcal{M}_{m|i}$, that is $\mathcal{R}_{m|i}$ and $\mathcal{R}'_{m|i}$, it holds that $T_{m|i}(\mathbf{x}) = T'_{m|i}(\mathbf{x}')$ for two corresponding points $\mathbf{x} \in \mathcal{R}_{m|i}$ and $\mathbf{x}' \in \mathcal{R}'_{m|i}$. Thus, we can easily show that $\mathcal{R}_{m|i}$ and $\mathcal{R}'_{m|i}$ are related by an affine homography $\mathbf{H}_{m|i}$ as

$$\mathbf{H}_{m|i} = \mathbf{A}'^{-\frac{1}{2}}_{m|i}\mathbf{R}'_{m|i}\mathbf{R}^{-1}_{m|i}\mathbf{A}^{\frac{1}{2}}_{m|i}, \tag{6}$$

(a) an MRC   (b) DT neighborhoods   (c) deform measure of a match

Fig. 4.   A neighborhood system in an MRC, and a deformation measure for a match. (a) Member matches in an MRC are shown for illustration. In two corresponding regions of the MRC, the blue region (right) and the red region (left) denote the head and the tail, respectively. (b) A neighborhood system of the MRC is constructed by Delaunay triangulation on the head region. (c) Based on this neighborhood system, geometric deformation of a member match is defined as the average transfer error of neighbor points on the normalized domain. For details, see the text.

$$\mathbf{x}' = \mathbf{H}_{m|i}(\mathbf{x} - \mathbf{c}_{m|i}) + \mathbf{c}'_{m|i}. \tag{7}$$

Based on the concepts and notations described so far, we design a Bayesian model of MRCs. The posterior probability $p(\Theta|\mathcal{I})$ is the probability of Θ being a set of MRCs given image set $\mathcal{I}$. In the Bayesian framework, it is decomposed as

$$p(\Theta|\mathcal{I}) \propto p(\mathcal{I}|\Theta)p(\Theta). \tag{8}$$

In our model, the prior probability $p(\Theta)$ evaluates the geometric coherency and maximality of MRCs, and the likelihood probability $p(\Theta|\mathcal{I})$ represents the photometric coherency of MRCs. All these terms are defined in the following subsections.

2.1. *Dynamic Neighborhood System*

In order to design a robust prior model, we construct a neighborhood system of each MRC Γ_i. It defines the neighborhood relation of each match $\mathcal{M}_{m|i}$ in Γ_i, and dynamically varies along evolution of Γ_i. Suppose that Γ_i is endowed with an undirected graph structure $G_i = (\mathcal{V}_i, \mathcal{E}_i)$ where the member matches of Γ_i consists in node set $\mathcal{V}_i = \{\mathcal{M}_{m|i}\}$. Edge set $\mathcal{E}_i$ is constructed by the Delaunay triangulation (DT)[20] on the center points $\mathbf{c}_{m|i}$ of head regions $\mathcal{R}_{m|i}$ as shown in Fig. 4 (b). Then, the neighbor set of $\mathcal{M}_{m|i}$ is defined as $\mathcal{N}_{\mathcal{M}_{m|i}} = \{\mathcal{M}_{n|i} \mid (\mathbf{c}_{m|i}, \mathbf{c}_{n|i}) \in \mathcal{E}_i\}$. Since DT has the low computational complexity of $\mathcal{O}(n \log n)$ and can be updated incrementally,[21] this DT-based dynamic neighborhood system provides an efficient and adaptive neighborhood system along expansion/merge moves.

2.2. *Geometric Prior*

To evaluate the geometric consistency of an MRC Γ_i, we utilize the DT-based neighborhood system described above. Consider a local region match $\mathcal{M}_{m|i}$ in Γ_i and its neighbor nodes $\mathcal{N}_{\mathcal{M}_{m|i}}$. As shown in Fig. 4 (c), we project the centers of $\mathcal{N}_{\mathcal{M}_{m|i}}$ into the normalized domain from both head $\mathcal{R}_{m|i}$ and tail $\mathcal{R}'_{m|i}$, and measure the distances between the corresponding points. Using Eq. (5), let $T_{m|i}$ and $T'_{m|i}$ be the transformations that map $\mathcal{R}_{m|i}$ and $\mathcal{R}'_{m|i}$ onto the unit domain, respectively. Then, the local deformation measure of $\mathcal{M}_{m|i}$ is defined as

$$
\begin{aligned}
d_{\mathrm{G}}(\mathcal{M}_{m|i}) &= \frac{1}{|\mathcal{N}_{\mathcal{M}_{m|i}}|} \sum_{\mathcal{M}_{n|i} \in \mathcal{N}_{\mathcal{M}_{m|i}}} |T_{m|i}(\mathbf{c}_{n|i}) - T'_{m|i}(\mathbf{c}'_{n|i})| \\
&= \frac{1}{|\mathcal{N}_{\mathcal{M}_{m|i}}|} \sum_{\mathcal{M}_{n|i} \in \mathcal{N}_{\mathcal{M}_{m|i}}} |\mathbf{R}_{m|i}^{-1} \mathbf{A}_{m|i}^{\frac{1}{2}}(\mathbf{c}_{n|i} - \mathbf{c}_{m|i}) - \mathbf{R'}_{m|i}^{-1} \mathbf{A'}_{m|i}^{\frac{1}{2}}(\mathbf{c}'_{n|i} - \mathbf{c}'_{m|i})|,
\end{aligned}
$$
(9)

where $\mathbf{c}_{n|i}$ and $\mathbf{c}'_{n|i}$ denote the center points of $\mathcal{R}_{n|i}$ and $\mathcal{R}'_{n|i}$, respectively. This measure represents the average transfer error of neighbors on the normalized domain. Summing up the all local deformation measures, the geometric energy of the MRC Γ_i is defined as

$$
E_{\mathrm{G}}(\Gamma_i) = \sum_{m=1}^{N_i} d_{\mathrm{G}}(\mathcal{M}_{m|i}).
$$
(10)

This provides robust geometric prior not only for all co-planar regions but also for smoothly deformed surfaces. Due to the neighborhood system, it dynamically adapts to the changes in the MRC configuration so that it generates more flexible deformation measures with much lower computational cost than geometric constraints used in the previous methods.[4,13]

2.3. *Maximality Prior*

To grow MRCs to larger ones satisfying the maximality in their sizes, rewards for both expansion and merge are required in the prior model. Thus, we design the maximality energy as

$$
E_{\mathrm{M}}(\Theta) = \sum_{i=1}^{K} \left(-N_i - |\Delta_i| \right),
$$
(11)

where N_i and $|\Delta_i|$ denote the number of member matches and the number of Delaunay triangles of MRC Γ_i, respectively. Since expansion of any MRC increases the total number of local matches, the first term encourages MRCs to expand. On the other hand, while merge of two MRCs does not increase the total number of local matches, it makes additional triangles of DT in a merged MRC. Thus, the second term provides a reward for merge of MRCs.

2.4. *Photometric Likelihood*

The likelihood probability $p(\mathcal{I}|\Theta)$ evaluates the photometric coherency based on intensity patterns of images $\mathcal{I}$. We define the dissimilarity[13] of two local regions in a member match $\mathcal{M}_{m|i}$ by

$$d_{\mathrm{P}}(\mathcal{M}_{m|i}) = 1 - \mathrm{NCC}(\mathcal{R}_{m|i}, \mathcal{R}'_{m|i}) + \frac{\mathrm{dRGB}(\mathcal{R}_{m|i}, \mathcal{R}'_{m|i})}{100}, \qquad (12)$$

where $\mathrm{NCC}(\cdot, \cdot)$ is the normalized cross-correlation between the gray patterns of two regions, while $\mathrm{dRGB}(\cdot, \cdot)$ is the average pixel-wise Euclidean distance in a normalized RGB color-space. Assuming that local matches are mutually independent in photometric coherency, the photometric energy of MRC $\varGamma_i$ is measured by

$$E_{\mathrm{P}}(\varGamma_i) = \sum_{m=1}^{N_i} d_{\mathrm{P}}(\mathcal{M}_{m|i}). \qquad (13)$$

Finally, using Eqs. (10)-(13), we formulate the overall prior and the likelihood in the posterior model as follows:

$$p(\varTheta) \propto \exp\left(-\sum_{i=1}^{K} E_{\mathrm{G}}(\varGamma_i) - E_{\mathrm{M}}(\varTheta)\right), \qquad (14)$$

$$p(\mathcal{I} \mid \varTheta) \propto \exp\left(-\lambda_{\mathrm{P}} \sum_{i=1}^{K} E_{\mathrm{P}}(\varGamma_i)\right), \qquad (15)$$

where λ_{P} controls the balance between the likelihood and the prior in the model.

Hence, given images $\mathcal{I}$, an optimal set of MRCs $\varTheta^*$ is obtained by maximizing the posterior probability $p(\varTheta|\mathcal{I})$ as

$$\varTheta^* = \arg\max_{\varTheta} p(\mathcal{I}|\varTheta)p(\varTheta). \qquad (16)$$

3. Construction of Object Correspondence Networks

In this section, we explain our algorithm for constructing the object correspondence networks. As briefly illustrated in Fig. 2, the proposed algorithm consists of three main phases: initial matching, multi-layer growing, and object correspondence networking. We will explain them one by one in the following subsections.

3.1. *Initial Matching*

To make seeds for MRCs, initial matches are established using local affine invariant feature detectors[22] and feature descriptors.[23] Affine feature pairs having similar descriptors consists in the initial matches. In our experiments, we adopted Harris-affine[24] and MSER[3] detectors, and the SIFT descriptor[1] for the purpose. Under the one-to-one matching assumption, various nearest neighbor (NN) matching schemes are widely used in the literature.[1,4,5,13,17] The NN methods, however, severely degrade inlier detection in the presence of multiple identical objects or similar patterns since the features have multiple matching features beyond the NN features. To the contrary, unlike usual initial matching schemes, we allow each feature to pair with

| (a) an image of multiple identical objects | (b) Lowe's NN[1] (True: 6 / Total: 50) | (c) loose thresholding (True: 18 / Total: 317) |

Fig. 5. An example of initial matching given an image of multiple identical objects. (b) While providing the higher true match ratio, the NN method misses a large amount of inliers in the presence of multiple identical patterns. (c) Loose thresholding preserves even more true matches at the cost of allowing high outlier ratio.

multiple features for the general cases. In our loose thresholding method, all the matching feature pairs exceeding a loose similarity threshold are simply collected among the possible feature pairs. Matching a feature with itself is avoided in the case of intra-matching within one image. As shown in the example of Fig. 5, although the initial matching by loose thresholding usually provides a low true match ratio, it prevents a large amount of true matches from being eliminated by conventional NN methods in the presence of multiple identical objects or repetitive patterns. In other words, it increases the absolute number of inlier seeds at the cost of the inlier ratio. Considering that inliers of an object correspondence become outliers for another object correspondence, inlier ratio for each object correspondence can be significantly lower than the conventional matching problem with one-to-one assumptions. [‡] The challenging problem of such a low inlier ratio is solved by boosting inliers in the following multi-layer growing.

3.2. Multi-Layer Growing

For growing initial seed matches to object correspondences, we propose a multi-layer growing algorithm driven by expansion/merge moves. The growing procedure is summarized in Fig. 6. First, multi-layering is performed using the initial seed matches obtained from initial matching. Each initial seed match forms an initial singleton MRC having its own expansion layer that provides space for expansion. As shown in Fig. 7 (a), each expansion layer consists of an overlapping circular grid of regular local regions that covers the entire image domain.[4,13] Then, the iterative growing loop starts. At each iteration, expansion is proposed with probability Q_E or merge with $1 - Q_E$. In an expansion proposal illustrated in Fig. 7 (b), a match in an MRC is selected as a *supporter* match (shown as the black solid line), and a local region around the selected supporter match is chosen as a *target* region on

---

[‡]Of course, a more advanced initial matching step can improve the initial matching result; however, this is not considered in this work.

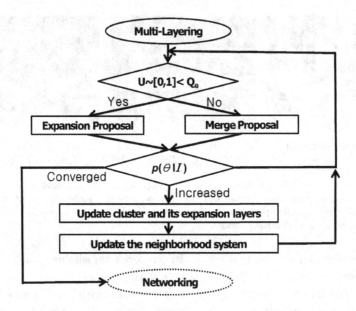

Fig. 6.   Multi-layer iterative growing.

the expansion layer of the MRC (shown as the blue circle). Then, the target region establishes a new match (the blue dotted arrow) propagated by the support match. If the expansion proposal is accepted, the propagated match is included in the MRC and the target region is eliminated from the expansion layer. On the other hand, in a merge move proposal illustrated in Fig. 7 (c), two geometrically similar MRCs are selected and proposed to be merged. If the merge move proposal is accepted, the expansion layers are also combined into their intersection. Our algorithm accepts the expansion/merge proposal moves when it increases the posterior probability $p(\Theta|\mathcal{I})$. The growing is iterated until $p(\Theta|\mathcal{I})$ no longer increases.

Note that although the multi-layer approach multiplies expansion layers at the start, merge moves gradually reduce the number of layers and guide expansion moves to concentrate on potential region. Likewise, expansion moves also guide merge moves to find compatible clusters by gradually growing them. Through these cooperative moves, our algorithm efficiently finds object correspondences in spite of significant outliers in initial matches. The details for expansion/merge proposals will be described in the following.

3.2.1. *Multi-Layer Expansion Move*

The expansion move is proposed by the following stochastic procedure. First, an MRC Γ_i is chosen from the current K clusters with probability $p(\Gamma_i) \propto N_i$, which reflects preference for larger MRCs. Second, among the members in the MRC a *supporter* match $\mathcal{M}_{m|i}$ is selected with probability $p(\mathcal{M}_{m|i}|\Gamma_i) \propto$

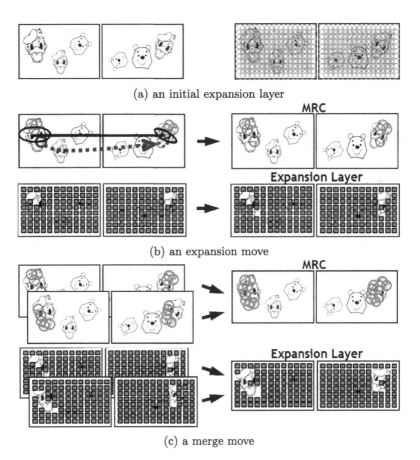

(a) an initial expansion layer

(b) an expansion move

(c) a merge move

Fig. 7. Expansion and merge moves in multi-layer growing. (a) Each expansion layer consists of overlapping circular regions covering the image domain. (b) In an expansion proposal, a cluster attempts to expand a region (blue circle in the bottom) using a support match (solid black line). When the proposal is accepted, the MRC and its expansion layer is updated. (c) In a merge proposal, two clusters attempt to merge into one. When the proposal is accepted, the MRC and their expansion layers are combined into one. Note that in (b) and (c) local regions in expansion layers are represented by red squares for better visualization.

$\sum_{\mathcal{R} \in \Omega_i} \exp\left(-|\mathbf{c} - \mathbf{c}_{m|i}|^2 / 2\sigma_{\mathrm{E}}^2\right)$ where Ω_i represents the expansion layer of Γ_i and $\mathbf{c}$ denotes the center points of $\mathcal{R}$. This means that the supporter match $\mathcal{M}_{m|i}$ is likely to have more unoccupied regions $\mathcal{R} \in \Omega_i$ in close vicinity. Finally, a target region $\mathcal{R}_{\mathrm{T}} \in \Omega_i$ is chosen with probability $p(\mathcal{R}_{\mathrm{T}}|\mathcal{M}_{m|i}, \Gamma_i) \propto \exp\left(-|\mathbf{c}_{\mathrm{T}} - \mathbf{c}_{m|i}|^2 / 2\sigma_{\mathrm{E}}^2\right)$ which favors regions near the supporter $\mathcal{M}_{m|i}$. Using the homography relation of the supporter match $\mathcal{M}_{m|i}$ formulated by Eq. (6) and (7), the target region $\mathcal{R}_{\mathrm{T}}$ is paired with the propagated region $\mathcal{R}'_{\mathrm{T}}$ as represented by the blue dotted arrow in Fig. 7 (b). Then, this propagated match $\mathcal{M}_{\mathrm{T}} = \{(\mathcal{R}_{\mathrm{T}}, \mathcal{R}'_{\mathrm{T}}), \mathbf{H}_{m|i}\}$ is further refined to adapt to propagation errors and object deformation. The refinement is done by locally varying the ellipse parameter of $\mathbf{H}_{m|i}$ and the center position of $\mathcal{R}'_{\mathrm{T}}$

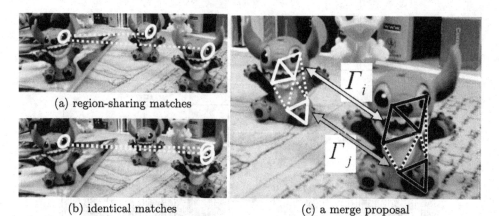

(a) region-sharing matches

(b) identical matches

(c) a merge proposal

Fig. 8. Region-sharing matches, identical matches, and a merge proposal. (a) Region-sharing matches in different MRCs are allowed to detect multiple object correspondences. (b) Identical matches are not allowed to avoid inconsistent or redundant region correspondences. (c) Merge proposals reflect the compatibility of newly emerging triangles (dotted lines) with the current triangles (solid lines). For details, see the text.

to minimize the dissimilarity $d_P(\mathcal{M}_T)$ of Eq. (12).[4] Finally, the expansion proposal gives $\Gamma_i \leftarrow \Gamma_i \cup \{\mathcal{M}_T\}$.

Since each MRC has its own expansion layer, multiple local region matches occupying the same region, called *region-sharing matches*, can be established in our multi-layer framework if they do not belong to the same MRC. As shown in Fig. 8 (a), this enables different pairwise object correspondences to share the same image region. However, if the matches are almost identical in their region pairs as depicted in Fig. 8 (b), such *identical matches* are not allowed since two different object correspondences do not share identical matches in images. Thus, when an expansion proposal of MRC Γ_i produces a match identical to an existing match in another cluster Γ_j, the expansion proposal is combined with contraction of its identical match in Γ_j. Let us denote the identical match in Γ_j by $\mathcal{M}_I$. Then, in this case, the expansion proposal gives $\Gamma_i \leftarrow \Gamma_i \cup \{\mathcal{M}_T\}$ (expansion) and $\Gamma_j \leftarrow \Gamma_j - \{\mathcal{M}_I\}$ (contraction). This implies that MRCs Γ_i and Γ_j compete to have the identical match. In this work, we define the identical matches as the matches overlapping over 50% both in their heads and in their tails.

3.2.2. *Multi-Layer Merge Move*

To propose a merge move of geometrically similar cluster pairs, we exploit the area ratio of the head to the tail in each cluster. For a candidate cluster pair Γ_i and Γ_j, we perform the Delaunay triangulation on the head regions of the combined cluster $\Gamma_i \cup \Gamma_j$ just as we construct the neighborhood system. As illustrated in Fig. 8 (c), the obtained triangles are divided into two sets: the triangles $\Delta_{i\cup j}$ (solid lines) that are already present in Delaunay triangles of Γ_i or Γ_j, and the newly added triangles $\Delta_{i\cup j}^{\text{new}}$ that are not included in $\Delta_{i\cup j}$ (dotted lines). If two cluster Γ_i and Γ_j belong to the same object correspondence, the area ratio of the head (black) to the tail (white) in $\Delta_{i\cup j}$ and that in $\Delta_{i\cup j}^{\text{new}}$ is likely to be similar because the new triangles $\Delta_{i\cup j}^{\text{new}}$ lie between Γ_i and Γ_j. Thus, denoting the head-tail area ratio of $\Delta_{i\cup j}$ by $\rho_{i\cup j}$, and that of $\Delta_{i\cup j}^{\text{new}}$ by $\rho_{i\cup j}^{\text{new}}$, the candidate cluster pair is proposed with probability $p(\Gamma_i, \Gamma_j) \propto \exp\left(-|\log(\rho_{i\cup j}/\rho_{i\cup j}^{\text{new}})|^2/2\sigma_{\text{M}}^2\right)$. This proposal reflects smoothness on the corresponding regions of the merged MRC. Finally, the merge proposal gives $\Gamma_i \leftarrow \Gamma_i \cup \Gamma_j$.

When Γ_i and Γ_j have region-sharing matches that have the same regions in common, the merge proposal is combined with a split of the region-sharing matches as another MRC. Let us denote the set of region-sharing matches in Γ_j by Γ_{S}. Then, the merge proposal gives $\Gamma_i \leftarrow \Gamma_i \cup \Gamma_j - \Gamma_{\text{S}}$ and $\Gamma_j \leftarrow \Gamma_{\text{S}}$.

3.3. *Algorithmic Aspects of Multi-Layer Growing*

While the multi-layer growing algorithm maximizes the posterior probability of MRCs in a greedy manner, its expansion/merge proposals efficiently drives the solution to a good local optimum. The algorithm is extended and modified from the single-layer growing algorithm of co-recognition[13] which is based on the data-driven Markov chain Monte Carlo (DDMCMC). In the MCMC theory, each move should be reversible for backtracking and satisfy the detailed balance condition.[25] It requires that in each move proposal the transition probability between the current state and the proposed state is computed considering all possible transitive paths. Since it is often computationally expensive, various approximation shortcuts are adopted in recent works.[13,26,27] In the sense of optimization in practice, however, reverse moves are required only because forward moves are likely to be trapped at bad local minima. Therefore, instead of constructing computationally expensive MCMC algorithm with reversible moves, we designed a stochastic algorithm with strong forward moves (expansion/merge) partly embedding their reverse moves (contraction/split) in themselves. Utilizing these strong expansion/merge moves, our greedy algorithm explores the solution space efficiently without explicit reverse moves.

The computational complexity mainly depends on two factors: the number of initial matches N_I forming multi-layers and the number of regions N_E in the expansion layer. The worst case complexity is $\mathcal{O}(N_I N_E^2 \log N_E)$. However, the practical complexity is even lower than it because the proposed algorithm progressively reduces the layers by merge moves and also rejects all expansion moves beyond matching regions. Concerning the complexity of updating an object correspondence with N_i member matches, the proposed algorithm takes significantly less cost of $\mathcal{O}(N_i \log N_i)$ than $\mathcal{O}(N_i^3)$ of the previous growing methods.[4,13]

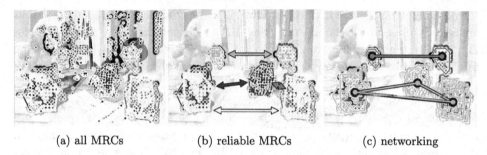

(a) all MRCs   (b) reliable MRCs   (c) networking

Fig. 9.   Networking step. (a) Each MRC obtained from the growing step is represented by convex hull region pairs of the same color. Many small and spurious MRCs are observed arising from outliers of initial matches. (b) Based on the reliability criteria, MRCs are selected as object correspondences. (c) Reliable MRCs are connected by SL-HAC based on region overlaps, and constitutes the object correspondence networks.

3.4. *Networking of MRCs*

In this final step, the object correspondence networks are assembled from the set of MRCs Θ^* obtained by the growing algorithm. Since the set of MRCs usually includes trivial MRCs arising from outliers of initial matches, we first eliminate such unreliable MRCs from Θ^*. Typically, object region correspondences are likely to grow larger and have distinctive textures in their regions. Thus, we evaluate the reliability of an MRC to be an object correspondence by its expanded area at the tail region and its mean variance of the intensity patterns, and discard MRCs not satisfying the threshold values. In our experiments, we set the thresholds for reliable MRCs as the expanded area over 3% of the given image and the mean intensity variance over 0.005.

Then, the reliable MRCs are considered as object correspondences and connected to construct the object correspondence networks according to their overlapping regions. In this work, we use a simple and popular algorithm of the Single-Link Hierarchical Agglomerative Clustering (SL-HAC)[28] to group the MRCs. Similarity between two MRCs is defined as the ratio of overlapping area to the area of the smaller MRC region, and SL-HAC sequentially connects the most similar MRC pairs until the similarity becomes less than 0.8. As shown in Fig. 9, the procedure assembles MRCs into the object correspondence networks, each representing connected object correspondences, and the detected object regions are all classified into sets of identical objects. Notably, this networking scheme does not require the pre-determined number of identical object sets, and is robust to missing object correspondences since others can provide indirect pathways. Of course, any other clustering algorithm could be adopted for this purpose as well. In particular, the partial linkage HAC[29] or the noise-robust spectral clustering[30] could make the networks robust to falsely detected MRCs.

4. Experiments

In this section we present three experiments that demonstrate the proposed method on unsupervised object detection and segmentation tasks. The experimental results show that the method can solve general matching problems of many-to-many object matching, and also outperforms the previous methods[13,17] on the restricted problem under one-to-one constraints. All experiments were performed under the following settings. In the initial matching step, affine region detectors of MSER[3] and Harris-affine[24] are used for extracting features, and the SIFT descriptor[1] to measure the similarity between the features. The threshold value for loose matching was set to 0.4 of distance in the SIFT descriptor space. The multi-layer growing was performed with $Q_E = 0.9, \lambda_P = 2.0, \sigma_E = 0.01l, \sigma_M = 0.2l$, where l denotes the diagonal length of the image.

4.1. *Common Object Detection and Segmentation*

In the first experiment, we compared the proposed method to the previous common object (one-to-one) matching and segmentation methods. First, Fig. 10 shows a comparative example on segmentation of deformed object with the co-recognition algorithm[13] and the quasi-dense matching method.[17] Their results are borrowed from their papers. As can be observed in the detailed segmentation boundaries, the proposed method shows superior performance in accuracy compared to other methods. While both co-recognition and quasi-dense matching fail to segment severely deformed regions, e.g. the corner part of the deformed magazine, the proposed method (OCN) discovers the deformed common parts with better accuracy. It

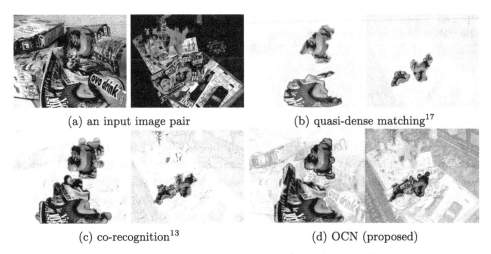

(a) an input image pair   (b) quasi-dense matching[17]

(c) co-recognition[13]   (d) OCN (proposed)

Fig. 10. Common object detection and segmentation. Compared to results of co-recognition[13] and quasi-dense matching,[17] the proposed method (OCN) shows better robustness to non-rigid deformation so that it covers the whole deformed part more accurately.

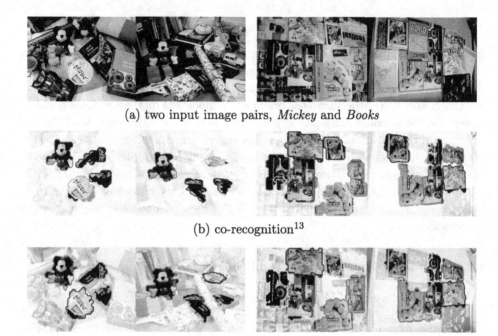

(a) two input image pairs, *Mickey* and *Books*

(b) co-recognition[13]

(c) OCN (proposed)

Fig. 11. Two example results on the image pairs from the SNU dataset. The proposed method shows better or comparable accuracy than the co-recognition method[13] in segmentation.

Table 1. Segmentation performance on the SNU dataset.[13]

Data	Mickey's	Minnie's	Jigsaws	Toys	Books	Bulletins	**Avg.**	Avg. co-recognition[13]
Hit (%)	84.4	87.8	81.6	85.2	97.2	93.8	**88.3**	85.5
Bk (%)	19.4	31.1	19.4	20.8	11.3	15.2	**19.5**	22.4

shows that our geometric prior based on the dynamic neighborhood system effectively adapts to deformed surfaces of the objects. For quantitative evaluation and comparison, we applied the proposed method on the SNU dataset § organized for common object matching and segmentation, and compared the results to those of co-recognition.[13] In detection accuracy, both methods perfectly found all the common objects. The results of segmentation accuracy are reported in Table 1, where the higher hit ratios (Hit) and the lower background ratios (Bk) represents the better results. The hit ratio is computed as the intersection area of ground truth and detected region divided by the total area of ground truth. The background ratio is measured as the background area of detected region divided by the total area of detected region. The average performance of co-recognition[13] is included at the last column for comparison. Figure 11 shows two examples among the results where detected common objects are delineated by the same color and the background is

§http://cv.snu.ac.kr/~corecognition

faded. While the proposed method can deal with more general matching problems and has lower computational complexity, it outperforms the previous method[13] in segmentation accuracy as well. For more detailed segmentation, object boundaries could be further refined by seeded segmentation algorithms[31-33] using the MRC regions as seeds.

4.2. *Object Matching and Segmentation on Real-World Images*

In the second experiment, we tested the proposed method on a more general problem beyond the one-to-one matching assumption, which the previous method cannot deal with. For evaluation, we constructed a dataset each image of which has several identical objects with occlusion and clutter. This aimed at demonstrating unsupervised identical object matching and segmentation on various real-world images. The results on single images and image pairs are shown in Fig. 12 and Fig. 13, respectively. The object boundaries are determined based on the local regions of MRCs, and delineated by lines of different colors according to which network they belong to. The object correspondence networks are represented by thick lines connecting detected object regions. In the third column of each result, object regions of each identical object are displayed in groups to better understand the object matching results. Our method provides impressive detection and segmentation results in spite of deformation, occlusion, and clutters. Although some false pairwise object correspondences were detected and some were missing, most of the identical objects were correctly localized and classified into identical object groups by the networking step. For evaluating the detection accuracy of the pairwise object correspondences, we computed recall and precision rates for our dataset considering all the possible pairwise correspondences. The average precision and recall rates are reported in Table 2. For the detection criterion, we used the standard detection criterion of 50% area overlap with the ground truth. The recall rate is particularly low for image pairs due to the challenging cases of densely stacked objects as shown in Fig. 13 (e.g. the sixth and eighth rows). In those results, several identical objects often merged into one, thus object correspondences were not distinguishable within the merged MRCs. The last row in Fig. 12 shows additional results on images with repetitive patterns. In these cases, a large number of MRCs gradually grew inside the repetitive pattern region at the growing step, and all the MRCs connected into a single network at the networking step. Thus, the network finally covered all repetitive patterns in the image as shown in the results, where the second column represents local matches of the network and the third shows its segmentation. The

Table 2.   Recall and precision of the pairwise object matching on our dataset. Some of the results are visualized in Fig. 12 and Fig. 13.

	Single Image	Image Pair	Total Average
Recall	0.89	0.61	0.75
Precision	0.94	0.89	0.92

Fig. 12. Unsupervised object matching and segmentation on single images. For each result, the input images, segmentation with object correspondence networks, and their classified regions are shown. Each color illustrates the identity of an object. In the last row, additional results are shown on images with repetitive patterns. For explanation, see the text.

network information could be further used to analyze the properties of the patterns such as the type and period of repetitions.

4.3. *Object Correspondence Networks from a Multi-Shot Image*

In the third experiment, we selected five complex test images from the ETHZ dataset¶, and generated a penta-tiled image as shown in Fig. 14. We then applied our method to detect and segment multiple sets of identical objects from the image. We did not use any information regarding boundaries of the original images and considered it as one image. The previous methods[13,17] cannot be applied to this problem as in the previous experiment. The penta-tiled image includes 8 sets of recurrent objects with clutter and occlusion under severe changes of viewpoint, scale, and illumination. As shown in Fig. 14, the proposed method successfully detected and segmented seven sets of objects missing one set (leopard puppet). Although one false set (dice hole) is included, the detected regions appear to be an

¶http://www.vision.ee.ethz.ch/~calvin/datasets.html

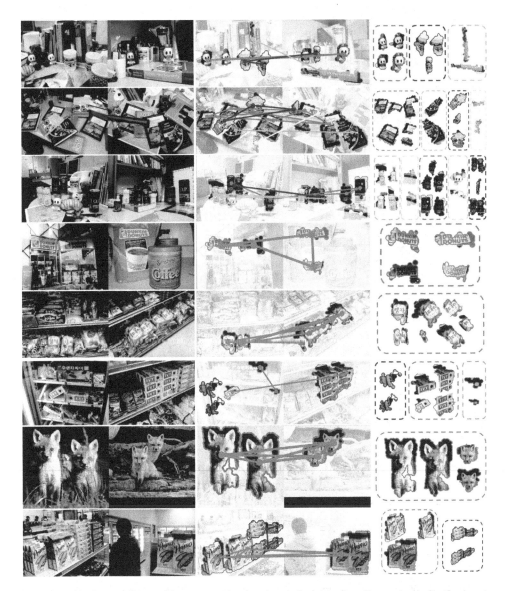

Fig. 13.   Object matching and segmentation results on image pairs. For each result, the input images, segmentation with object correspondence networks, and their classified regions are shown.

object as shown at the bottom-right of Fig. 14 (d). For more information, refer to our project site: http://cv.snu.ac.kr/~objectnetworks/

5. Discussion and Conclusion

To process arbitrary real-world images taken from a variety of environments, fully unsupervised analysis of images is required and has remained as one of the most

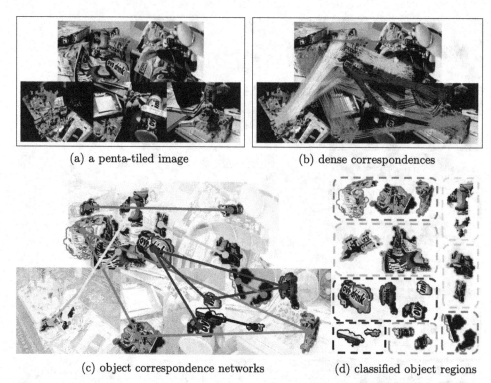

(a) a penta-tiled image   (b) dense correspondences

(c) object correspondence networks   (d) classified object regions

Fig. 14.   Object correspondence networks from a penta-tiled image. Note that we used it as one large image without any information about boundaries of original images. Eight sets of identical objects were recognized with one false set (dice hole) detected, one true set (leopard puppet) missing.

challenging issues in computer vision. In this chapter, a novel approach to unsupervised object detection and segmentation, based on the concept of the object correspondence networks, was introduced. The proposed method can detect and segment multiple sets of identical objects even from a single image or multiple images in an unsupervised way. The basic idea is to grow initial matches for object correspondences in a multi-layer match growing framework, and to analyze their relations by connecting them. Experiments show a robust performance of our method on challenging complex scenes which none of the previous methods has dealt with. The proposed match-growing framework also can be directly extended for a variety of unsupervised vision applications, such as symmetry analysis,[34] action detection,[18] and multiple object reconstruction.

While the proposed approach provides impressive results, the current method still has some limitations. As we noted in Sec. 2, an MRC can detect an object correspondence under the geometric and photometric distinctiveness of the object. The conditions, however, are not strictly satisfied in usual images. As the consequences, in the presence of photometrically similar backgrounds, the detected object regions are over-expanded or non-object regions are falsely detected. To suppress the background regions or non-objects, the *foregroundness*[35,36] or *objectness*[37] could be additionally incorporated for better results. In the case of geometrically similar arrangements of objects such as stacked objects, the image has intrinsic ambiguity between repetitive patterns of one object and dense arrangements of several objects. This problem requires a statistical inference on MRCs using various scenes of the objects. Thus, further analysis on large object correspondence networks need to be investigated to infer the objects and their whole-part relations. Pursuing the direction, the object correspondence networks could be extended for unsupervised scene understanding based on mutual relations of various objects.

References

1. D. G. Lowe, Object recognition from local scale-invariant features, *IEEE International Conference on Computer Vision*. (1999).
2. S. Obdržálek and J. Matas, Object recognition using local affine frames on distinguished regions, *British Machine Vision Conference*. (2002).
3. J. Matas, O. Chum, M. Urban, and T. Pajdla, Robust wide baseline stereo from maximally stable extremal regions, *British Machine Vision Conference*. (2002).
4. V. Ferrari, T. Tuytelaars, and L. Gool, Simultaneous object recognition and segmentation from single or multiple model views, *International Journal of Computer Vision*. **67**(2), 159–188, (2006).
5. I. Simon and S. M. Seitz, A probabilistic model for object recognition, segmentation, and non-rigid correspondence, *IEEE Conference on Computer Vision and Pattern Recognition*. (2007).
6. M. Cho and K. M. Lee, Partially occluded object-specific segmentation in view-based recognition, *IEEE Conference on Computer Vision and Pattern Recognition*. (2007).
7. J. Sivic, B. C. Russell, A. A. Efros, A. Zisserman, and W. T. Freeman, Discovering object categories in image collections, *IEEE International Conference on Computer Vision*. (2005).
8. S. Todorovic and N. Ahuja, Unsupervised category modeling, recognition, and segmentation in images, *IEEE Transactions on Pattern Analysis and Machine Intelligence*. **30**(12), 2158–2174, (2007).
9. L. Karlinsky, M. Dinerstein, D. Levi, and S. Ullman, Unsupervised classification and part localization by consistency amplification, *European Conference on Computer Vision*. (2008).
10. M. F. Demirci, A. Shokoufandeh, Y. Keselman, L. Bretzner, and S. Dickinson, Object recognition as many-to-many feature matching, *International Journal of Computer Vision*. **69**(2), 203–222, (2006).
11. M. Carcassoni and E. R. Hancock, Correspondence matching with modal clusters, *IEEE Transactions on Pattern Analysis and Machine Intelligence*. **25**(12), 1609–1615, (2003).
12. M. Cho, Y. M. Shin, and K. M. Lee, Unsupervised detection and segmentation of identical objects, *IEEE Conference on Computer Vision and Pattern Recognition*. (2010).
13. M. Cho, Y. M. Shin, and K. M. Lee, Co-recognition of image pairs by data-driven monte carlo image exploration, *European Conference on Computer Vision*. (2008).
14. M. Lhuillier and L. Quan, Match propagation for image-based modeling and rendering, *IEEE Transactions on Pattern Analysis and Machine Intelligence*. **24**(8), 1140–1146, (2002).

15. K. L. Steele and P. K. Egbert, Correspondence expansion for wide baseline stereo, *IEEE Conference on Computer Vision and Pattern Recognition.* (2005).

16. A. Vedaldi and S. Soatto, Local features, all grown up, *IEEE Conference on Computer Vision and Pattern Recognition.* (2006).

17. J. Kannala, E. Rahtu, S. Brandt, and J. Heikkila, Object recognition and segmentation by non-rigid quasi-dense matching, *IEEE Conference on Computer Vision and Pattern Recognition.* (2008).

18. Y. M. Shin, M. Cho, and K. M. Lee, Co-recognition of actions in video pairs, *International Conference on Pattern Recognition.* (2010).

19. K. Mikolajczyk and C. Schmid, Scale and affine invariant interest point detectors, *International Journal of Computer Vision.* **60**(1), 63–86 (Oct., 2004).

20. F. Preparata and M. Shamos, *Computational Geometry.* (Springer-Verlag, 1985).

21. O. Faugeras, *Three-Dimensional Computer Vision: a Geometric Viewpoint.* (MIT Press, 1993).

22. T. Tuytelaars and K. Mikolajczyk, Local invariant feature detectors: a survey, *Foundations and Trends® in Computer Graphics and Vision.* **3**(3), 177–280, (2008).

23. K. Mikolajczyk and C. Schmid, A performance evaluation of local descriptors, *IEEE Transactions on Pattern Analysis and Machine Intelligence.* **27**(10), 1615–1630, (2005).

24. K. Mikolajczyk and C. Schmid, An affine invariant interest point detector, *European Conference on Computer Vision.* (2002).

25. P. Green, Reversible jump markov chain monte carlo computation and bayesian model determination, *Biometrica.* **82**(4), 711–732, (1995).

26. F. Han and S. C. Zhu, Bottom-up/top-down image parsing by attribute graph grammar, *IEEE International Conference on Computer Vision.* (2005).

27. H. Y. Jung, K. M. Lee, and S. U. Lee, Window annealing over square lattice markov random field, *European Conference on Computer Vision.* (2008).

28. A.K.Jain and R.C.Dubes, *Algorithms for Clustering Data.* (Prentice Hall, 1998).

29. M. Cho, J. Lee, and K. M. Lee, Feature correspondence and deformable object matching via agglomerative correspondence clustering, *IEEE International Conference on Computer Vision.* (2009).

30. Z. Li, J. Liu, S. Chen, and X. Tang, Noise robust spectral clustering, *IEEE International Conference on Computer Vision.* (2007).

31. C. Rother, V. Kolmogorov, and A. Blake, Grabcut -interactive foreground extraction using iterated graph cuts, *ACM SIGGRAPH.* (2004).

32. V. Lempitsky, P. Kohli, C. Rother, and T. Sharp, Image segmentation with a bounding box prior, *IEEE International Conference on Computer Vision.* (2009).

33. T. H. Kim, K. M. Lee, and S. U. Lee, Nonparametric higher-order learning for interactive segmentation, *IEEE Conference on Computer Vision and Pattern Recognition.* (2010).

34. M. Cho and K. M. Lee, Bilateral symmetry detection and segmentation via symmetry-growing, *British Machine Vision Conference.* (2009).

35. L. Itti, C. Koch, and E. Niebur, A model of saliency-based visual attention for rapid scene analysis, *IEEE Transactions on Pattern Analysis and Machine Intelligence.* **20**, 1254–1259, (1998).

36. X. Hou and L. Zhang, Saliency detection: A spectral residual approach, *IEEE Conference on Computer Vision and Pattern Recognition.* (2007).

37. B. Alexe, T. Deselaers, and V. Ferrari, What is an object?, *IEEE Conference on Computer Vision and Pattern Recognition.* (2010).

Part 3

Modern Systems and Technology in Computer Vision

CHAPTER 3.1

THE USE OF CONSTRAINTS FOR CALIBRATION-FREE 3D METRIC RECONSTRUCTION: FROM THEORY TO APPLICATIONS

B. Boufama and S. Bouakaz*

School of Computer Science, University of Windsor, Windsor, ON, Canada N9B 3P4
boufama@uwindsor.ca; bouakaz@liris.cnrs.fr

The recovery of three-dimensional metric geometry of a scene from its two-dimensional images is a central problem in computer vision. Despite the many advances achieved so far, most, if not all of the proposed solutions are undermined by the requirement for an explicit or self calibration of the cameras. This chapter investigates the use of geometric and other constraints for the calculation of three-dimensional metric reconstruction of a scene, without resorting to any camera calibration. This is a different alternative solution for the 3D metric reconstruction, not widely investigated, with the potential to overcome the drawbacks of both pattern-based and self calibration methods. Three different kinds of constraints, used in three different methods, are presented here. In particular, it was shown that using only pixel correspondence, the 3D metric geometry of the scene can be retrieved either by using a few geometric constraints or, by using approximate values of the intrinsic parameters or, by combining a single geometric constraint with a search for the best solution. The results of our experiments, carried out on both synthetic and real data, have validated the proposed methods and have demonstrated their potential to be a substitute for calibration-based methods.

1. Introduction

For over three decades, researchers in the computer vision community have been working to enable machines to perceive the three-dimensional structure of our surrounding environment, a naturally three-dimensional entity. Although 3D cameras, based on different technologies, have been appearing recently,[1,2] their success is still very limited as they cannot replace the high resolution, color-rich and, very cheap digital cameras. In particular, the vast majority of the cameras being used nowaday are two-dimensional (2D) sensors delivering 2D arrays of color pixels as images. However, it is well known that the 3D structure of a scene cannot be retrieved using a single 2D image of that scene, with no a priori information

---

*Laboratoire d'Informatique en Images et Systèmes d'information, Université Lyon1, Lyon, France.

available. At least two images of a rigid scene are required in order to recover the 3D reconstruction (geometry) of that scene. Such 3D reconstruction can be either metric (Euclidean up to a scale factor), affine or projective, depending on the method used and on whether or not the cameras have been calibrated.[3–5] In particular, many methods exist for the projective 3D reconstruction using only pixel correspondences in two images.[6,7] However, such 3D structure does not carry any metric information and is therefore, of limited use for most applications. On the other hand, the 3D metric structure remains the most useful information in computer vision with applications in navigation, recognition, medical imaging, etc. The problem of 3D metric reconstruction is intimately related to the problem of camera calibration. In particular, when the cameras' intrinsic parameters are known, solving this problem becomes straightforward. The classical solution to solve the 3D reconstruction problem consists of two steps. First, the cameras are calibrated using a known 3D pattern, then the 3D coordinates are computed. However, the requirement of such an explicit calibration step makes the classical solution less attractive and even impossible in many situations. Hence, for over a decade, researchers have been proposing new methods, relying solely on point correspondences across a sequence of images, to calculate the camera's intrinsic parameters.[8,9] These solutions, known as camera self-calibration methods, have the advantage of avoiding the use of a known pattern and only require the 2D images as inputs. Most of the self-calibration methods rely on the calculation of virtual shapes in which the information on the camera parameters is embedded. For instance, early self-calibration methods are based on the calculation of the Dual of the Image of the Absolute Conic (DIAC) using the so-called Kruppa's equations.[10,11] Other methods rely on the calculation of the Absolute Quadric[12] that has the advantage of constraining both the DIAC and the plane at infinity on which the Absolute Conic lies. More recently, the so-called Absolute Line-Quadric was used for the case of a camera with varying parameters.[13]

Unfortunately, the accuracy of the estimated intrinsic parameters when using the above mentioned methods, and others, is undermined by the pixel correspondence problem and by the numerous degenerate motion configurations.[14] The problem of camera self-calibration, as any nonlinear problem, suffers from the difficulty of choosing the initial values for its parameters. Furthermore, the convergence to the desired solution of the numerical optimizations of the existing self-calibration techniques cannot be guaranteed, even if the parameters have been initialized close to the ground truth.

It is clear from all the above papers that the self-calibration problem has proven to be a very difficult one to be used in practice. As a consequence, it is very rare to find any commercial vision system that includes a self-calibration module. The difficulty to solve this problem mainly comes from its nonlinear nature because only pixel correspondences across the images are used as constraints. However, very often, known information about the scene or about the camera or about both is completely ignored. For example, knowing the approximate values of the camera's intrinsic parameters or knowing that some segments in the scene are perpendicular, could be used to get the metric reconstruction. In particular, the use of such constraints has the potential to overcome the drawbacks of both the explicit and self-calibration. In particular, these constraints typically consists of distances, angles and parallelism, if coming from the scene. When the constraints are from the cameras, they usually consist of approximate or partial knowledge of the intrinsic parameters. Although a few published works have been appearing on the use of constraints for metric reconstruction,[15–17] this promising research direction is still not widely investigated. Furthermore, these works[16,17] are mostly concerned with the theoretical part making their use in real-life situation difficult.

This chapter investigates the use of all kinds of constraints that can help retrieve the 3D metric structure of a scene from two images. First, we show that when approximate values of the intrinsic parameters are available, the 3D metric reconstruction is possible. By using normalized image coordinates instead of pixel coordinates, we show that even with large errors on the intrinsic parameters the reconstruction process yields a good quality 3D metric reconstruction. Then, we present a method that uses geometric constraints to map a 3D projective reconstruction, obtained from pixel correspondences only, into a metric space. Finally, a method that combines the use of approximate values of the intrinsic parameters with the perpendicularity constraint is presented.

This chapter is organized as follows. Section 2 presents the notations and background needed for the remainder of the chapter. Section 3 presents the method for obtaining the 3D metric reconstruction when only approximate values of the intrinsic parameters are available. Then, in Section 4 we show how geometric constraints can be used to map a projective reconstruction into a metric one and how one geometric constraint, perpendicularity, can be used to search for the best values of the intrinsic parameters leading to the best metric reconstruction. Finally, a conclusion is provided in Section 5.

2. Notations and Background

2.1. *Geometric transformations in projective space*

Geometric transformations in projective space, usually called projective transformations, play a central role in computer vision. In particular, the scene-image perspective projection is a projective transformation.

We present here a few definitions and a quick introduction to these transformations, necessary to understand the remainder of this paper. The interested reader is referred to the book by Hartley[3] for more information and for advanced use of geometry in computer vision.

Projective space   When considering the $(n + 1)$-dimensional space $I\!R^{n+1} - \{(0, \ldots, 0)\}$ together with the equivalence relationship given by

$$(x_1, \ldots, x_{n+1}) \sim (x'_1, \ldots, x'_{n+1}) \Longleftrightarrow$$
$$\exists \lambda \neq 0 \ \ such \ that \ \ (x_1, \ldots, x_{n+1}) = \lambda(x'_1, \ldots, x'_{n+1})$$

The projective space $I\!P^n$ is defined as the quotient space that is obtained from the above equivalence relation. In particular, $(x_1, \ldots, x_{n+1})$ and $(x'_1, \ldots, x'_{n+1})$, known as homogeneous coordinates, represent the same point in $I\!P^n$. When $x_{n+1} = 0$ for a point in $I\!P^n$, the point is said to be at infinity.

Hyperplane   A set of points in $I\!P^n$ whose homogeneous coordinates $X = (x_1, \ldots, x_{n+1})$ satisfy

$$\sum_1^{n+1} c_i x_i = 0$$

define a hyperplane, described by the vector $C = (c_1, \ldots, c_{n+1})$. In particular, the set of points with $x_{n+1} = 0$ define the hyperplane at infinity.

projective basis   Any set of $n + 2$ points of $I\!P^n$ such that no $(n + 1)$ of which define a hyperplane (linearly dependent), is a projective basis.

A widely used projective basis of $I\!P^n$, known as the canonical projective basis, is given by the following points:

$$e_1 = (1, 0, \ldots, 0), \ e_2 = (0, 1, 0, \ldots, 0), \ldots, e_{n+1} = (0, \ldots, 0, 1) \ and \ e_{n+2} = (1, \ldots, 1)$$

When considering the case of the 3D projective space $I\!P^3$, a projective basis can be any 5 points such that no 4 of which are linearly dependent. Therefore, when choosing 5 space points to form a basis in $I\!P^3$, one must make sure that no 4 of them are co-planar.

Projective transformations   Also called collineation, it is a linear mapping from $I\!P^n$ to $I\!P^m$ given

$$\begin{cases} I\!P^n \longrightarrow I\!P^m \\ P \ \ \longrightarrow Q \sim WP \end{cases}$$

where W is a nonsingular $(m + 1) \times (n + 1)$ matrix representing the projective transformation.

When $n = m$, these projective transformations are also called homographies and the matrix W becomes an $n \times n$ square matrix. The latter is completely determined by $(n + 2)$ point correspondences between the two projective spaces. In particular, a change of projective basis in $I\!P^n$ is a homography described by an $n \times n$ invertible matrix.

Affine transformations   It is a projective transformation that keeps the hyper-plane at infinity invariant.

For the case of the 3D space, an affine transformation is defined by the 4×4 matrix

$$\begin{pmatrix} M & T^T \\ 0\ 0\ 0 & 1 \end{pmatrix}$$

where M is a 3×3 linear matrix and T is a 3D translation vector.

2.2. *Camera model and image formation*

Image formation is the process by which a three-dimensional representation of a scene is reduced to a two-dimensional representation of that same scene, an *image*.

Using a pinhole model, a camera is attached to a reference frame with origin O, called *centre of projection*, and three axes, Ox, Oy, Oz (see Figure 1).

An image point (pixel), given by its image coordinates, is the result of a three-step transformation of a space point defined in a scene reference frame. These three steps are applied in a sequential order as follows:

(1) A 3D rigid displacement: the scene points, initially defined in a scene reference frame, are transformed so that they would be defined in the camera reference frame. This transformation is an Euclidean transformation consisting of a rotation and a translation given by the 3×4 matrix $D = (R|T)$, where R is the 3×3 rotation matrix and T is the 3D translation vector.

(2) A 3D-2D projection: 3D points defined in the camera reference frame are projected onto the image plane. This transformation is a projective transformation from $I\!P^3$ to $I\!P^2$ representing the perspective projection, described by the matrix:

$$I = \begin{pmatrix} 1\ 0\ 0\ 0 \\ 0\ 1\ 0\ 0 \\ 0\ 0\ 1\ 0 \end{pmatrix} \tag{1}$$

(3) A 2D-2D transformation: The obtained coordinates from the previous step, expressed in the scene's units, undergo a 2D affine transformation to become defined in pixels in the image plane. This affine transformation can be described by the matrix below:

$$A = \begin{pmatrix} \alpha_u & 0 & u_0 \\ 0 & \alpha_v & v_0 \\ 0 & 0 & 1 \end{pmatrix} \tag{2}$$

where:

- α_u *and* α_v are the 2 scale factors along the image Ox and Oy directions respectively.

- u_0 and v_0 are the coordinates, in pixels, of the intersection between the *optical axis* and the image plane. This point is also called the *centre of the image*.

The above three-step geometric transformation is usually described by a single matrix, called M, given by:

$$M = AID \qquad (3)$$

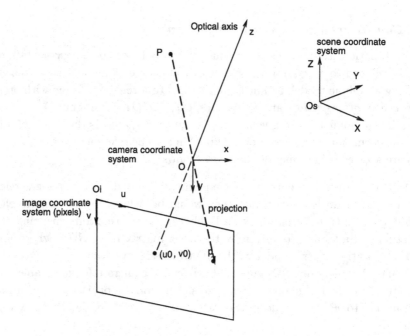

Fig. 1.   Image formation with a pinhole model for the camera.

3. Metric Reconstruction Using Approximate Values for the Intrinsic Parameters

Although accurate values for the intrinsic parameters of a camera are usually unknown without a calibration procedure, some kind of estimates of these parameters, not necessary accurate, are often available. Such estimates can be obtained either from the cameras' manufacturer's data or from some previous experiments with the same camera.

In this section, we assume that the intrinsic parameters are approximately known and we show that it is possible to recover a good quality 3D metric structure of a scene from two images. Simulations and real experiments have demonstrated this claim as the quality of the obtained 3D geometry was still good even with a 25%

errors on the intrinsic parameters. In addition, all the steps of the 3D reconstruction process are linear and easy to implement for practical applications.

3.1. *The eight-point algorithm*

Consider a space point P and its projections, p and p', on the first and second image respectively, it is clear from Figure 2 that the three vectors $\vec{Op}$, $\vec{Op'}$ and $\vec{O'O}$ are coplanar. Therefore, we have the relation:

$$\vec{O'p'} \cdot (\vec{O'O} \wedge \vec{Op}) = 0 \tag{4}$$

where $\cdot$ denotes the scalar product and $\wedge$ denotes the cross product.

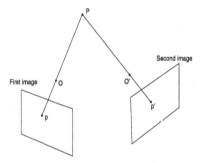

Fig. 2. $\vec{Op}$, $\vec{O'p'}$ and $\vec{O'O}$ are coplanar as they all belong to the plane defined by O, O' and P.

However, the above relation is only true when all vectors are defined in the same reference frame. Without loss of generality, let the coordinate system of the second camera be our reference frame. This will transform the above relation into:

$$\vec{O'p'} \cdot (\vec{O'O} \wedge R\vec{Op}) = 0 \tag{5}$$

where R is the 3×3 rotation matrix from the reference frame of the first camera to the one of the second camera.

Let's denote $\vec{O'p'}$, $\vec{O'O}$ and $\vec{Op}$ by p'^T, t and p respectively, equation (5) becomes:

$$p'^T \cdot ([t]_\times Rp) = p'^T T R p = 0 \tag{6}$$

where $t = (t_x, t_y, t_z)$, the translation vector, representing the coordinates of O in our reference frame and T is the 3×3 antisymmetric matrix given by

$$T = \begin{pmatrix} 0 & -t_z & t_y \\ t_z & 0 & -t_x \\ -t_y & t_x & 0 \end{pmatrix}$$

the above relation is usually written as

$$p'^T E p = 0 \qquad (7)$$

where E is a 3×3 matrix, called the *essential matrix*.

E describes the Euclidean motion, rotation and translation, between two cameras. Note that the *fundamental matrix* is a generalization of the *essential matrix* to the uncalibrated case (pixel coordinates).

The above relation is known as Longuet-Higgins' constraint.[18] E can be computed by solving a set of linear equations given eight or more point correspondences. The calculation of E using the linear equations from 7, known as the Eight-Point Algorithm, is excessively sensitive to pixel noise and it usually yields inaccurate motion and structure. However, it was shown in[19] that the high sensitivity to noise of this simple algorithm was mainly a numerical issue. Thanks to a simple data normalization on the pixel coordinates, the Eight-Point Algorithm performs just as good as the best nonlinear algorithm.

3.2. *Eight-point-based 3D reconstruction*

Calculating E is not enough for the 3D reconstruction process. First, R and T must be extracted from the calculated E. This is done by factoring E into the product TR of a skew-symmetric matrix and a rotation matrix.[20] Then, given R and T, the two projection matrices, M_1 and M_2, associated with the first and the second camera, respectively, can to be computed. Without loss of generality, the scene coordinate system can be assumed to be the coordinate system attached to the first camera. Therefore, we have

$$M_1 = A_1 I \qquad \text{and} \qquad M_2 = A_2 I D$$

where A_1, A_2 and I are the approximate intrinsic parameters and the perspective projection matrices respectively, D is a 4×4 displacement matrix made of R and T, extracted from E (see Eq. 3).

Once M_1 and M_2 are calculated, the 3D reconstruction can is done by triangulation.

3.3. *Experimental results*

We have carried out experiments on both simulated and real data to investigate how errors on the intrinsic parameters affect the quality of the 3D reconstruction. Our main goal here is to show that (1) a simple linear method can be used to calculate the metric geometry of a scene and (2) the calibration is not required as we are using coarse estimates of the intrinsic parameters.

Tests on simulated data   A 60 points points covering a $30cm \times 30cm \times 20cm$ volume was simulated. The virtual camera was located at about one meter off the scene with its intrinsic parameters set to $\alpha_u = \alpha_v = -1000$ and $u_0 = v_0 = 256$. The image resolution was 512×512. Two stereo images of this scene were generated with an inter-image displacement consisting of a rotation and a translation given by $(5, 20, 0)$ and $(-25, 12, 12)$, respectively. The units for the angles and for the translation were degrees and centimeters, respectively.

In these experiments, the errors added to the values of the intrinsic parameters ranged from 2.5% to 25%. In addition, we have considered 2 cases, a noise-free and a one-pixel uniform noise for the image coordinates. The relative errors are obtained by first applying a similarity transform (translation, rotation and a uniform scaling) to the reconstructed coordinates so that they get as close as possible to the ground-truth. Then, we calculate the simple difference between the ground-truth values and the computed ones. On the other hand, the absolute errors were obtained by first scaling the reconstructed points so that they are defined in the same units (cm), then by taking the simple difference between the ground-truth values and the computed ones. Table 1 summarizes these results.

Table 1. Results of the 3D reconstruction with errors, ranging from 2.5% to 25%, added to the intrinsic parameters for 2 cases: noise-free and 1-pixel noise in the image coordinates.

Noise on intrinsic parameters (%)	pixel noise 0 or 1	*Relative errors*				*Absolute errors*			
		ΔX (cm)	ΔY (cm)	ΔZ (cm)	average distance (cm)	ΔX (cm)	ΔY (cm)	ΔZ (cm)	average distance (cm)
2.5	0	0.011	0.020	0.012	0.026	0.081	0.319	0.229	0.426
	1	0.0841	0.097	0.207	0.244	0.133	0.298	1.163	1.251
5	0	0.016	0.030	0.017	0.039	0.432	0.787	0.372	1.057
	1	0.075	0.084	0.207	0.236	0.392	0.872	1.048	1.559
10	0	0.021	0.037	0.025	0.050	0.711	1.741	0.418	2.008
	1	0.094	0.112	0.209	0.256	0.624	1.734	1.420	2.503
15	0	0.057	0.077	0.056	0.111	4.337	3.066	1.104	5.528
	1	0.051	0.072	0.211	0.229	4.450	2.971	0.888	5.524
20	0	0.044	0.073	0.049	0.098	2.642	0.924	0.893	3.011
	1	0.086	0.115	0.209	0.254	2.725	0.983	1.248	3.271
25	0	0.218	0.387	0.266	0.518	4.797	1.553	4.519	7.374
	1	0.227	0.381	0.295	0.533	4.692	1.482	4.744	7.380

3.4. *Tests on real data*

Two scenes were used in these experiments with real images (see Figure 3).

(1) Pattern scene: a known calibration pattern consisting of three parallel planes covering a volume of $18cm \times 12cm \times 6cm$ was placed at about one meter off a stereo-head. A total of 23 interest points were used here. Note that as we know the 3D coordinates of these points, the accuracy of the reconstruction can be reliably estimated and compared.

(2) The house scene: this is a $40cm \times 40cm \times 25cm$ scene that was located at about one meter off the camera. A total of 38 points matched in two images were used. Unlike the pattern scene, the 3D coordinates are not perfectly known as they have been measured with a regular ruler.

Table 2 summarizes the results of the 3D metric reconstruction process for both scenes. Each row in this table provides the reconstruction errors for a given set of values assigned to the intrinsic parameters. In the first row, the intrinsic parameters were assigned the values from a previous explicit calibration procedure.

The pattern image pair   The house image pair

Fig. 3.   The two pairs of real images used in the experiments.

Table 2.   The results of the 3D metric reconstruction for real images where the intrinsic parameters were assigned different sets of values within an interval.

Intrinsic parameters used	Scene	Relative errors				Absolute errors			
		ΔX (cm)	ΔY (cm)	ΔZ (cm)	average distance	ΔX (cm)	ΔY (cm)	ΔZ (cm)	average distance
-1483 -1007	Pattern	0.010	0.002	0.016	0.019	0.430	0.216	0.115	0.498
242 239	House	0.051	0.054	0.035	0.082	2.906	4.678	0.651	5.589
-1500 -1000	Pattern	0.011	0.007	0.016	0.021	0.188	1.447	0.144	1.471
256 256	House	0.048	0.050	0.034	0.078	2.640	4.248	0.604	5.085
-1600 -1050	Pattern	0.008	0.006	0.017	0.020	0.852	0.827	0.155	1.221
276 246	House	0.055	0.056	0.039	0.088	2.269	4.484	0.644	5.128
-1600 -900	Pattern	0.029	0.027	0.025	0.047	1.564	0.680	0.772	1.967
236 240	House	0.041	0.045	0.032	0.069	2.941	4.722	0.618	5.648
-1400 -900	Pattern	0.022	0.019	0.022	0.037	0.791	1.405	0.618	1.834
236 260	House	0.036	0.040	0.028	0.061	3.067	4.160	0.561	5.233
-1700 -1100	Pattern	0.014	0.004	0.012	0.019	0.566	2.657	0.309	2.776
250 270	House	0.057	0.057	0.040	0.090	2.616	3.895	0.624	4.811

Discussion It was widely accepted that the simple Eight-Point Algorithm was too sensitive to noise and is numerically unstable, especially when the intrinsic parameters were not accurately known. However, the simple affine transformation (translation and scaling) on the image coordinates before the calculation of the essential matrix has overcome most of these problems. Furthermore, the relative reconstruction quality has become less sensitive to errors on the intrinsic parameters. On the other hand, the depth, being an absolute measurement, is poorly estimated when noise level gets high. This is in agreement with what is commonly known about the difficulty to accurately recover the depth information. The major conclusion of these experiments is that errors on the intrinsic parameters generate mostly a uniform shift that is easily canceled by a similarity transform. Both real and simulated tests have confirmed this conclusion.

We can also note from Table 2 that because the interest points in the pattern's images were extracted with a 0.1-pixel accuracy, the obtained 3D reconstruction was of very good quality even with large errors on the intrinsic parameters. On the other hand, the interest points extracted from the house's images had a 1-pixel (or more) accuracy, yielding a less accurate 3D metric geometry. This difference in pixel accuracy between the two scenes explains the differences in the obtained 3D errors. Furthermore, by looking closely at Table 2, one can see that the aspect ratio has a greater effect on the accuracy of the reconstruction. For example, by comparing rows 2,3 and 6 to rows 4 and 5, we clearly see that when the value of the aspect ratio is accurate (close to the value from the calibration process), we obtain better results than those with a less accurate aspect ratio. This is a good news since from our experiments with many digital cameras we found that the value of the aspect ratio is very stable and does not change significantly with camera settings. These experiments clearly demonstrate that a good 3D metric reconstruction can be easily calculated using two images and coarse estimates of the intrinsic parameters.

4. Metric Reconstruction Using Geometric Constraints

It has been known for over a decade that the projective three-dimensional structure of a scene can be easily computed using only a pair of images as inputs.[6,21] Several of these methods rely on computing the projection matrices then use them to get the projective 3D coordinates of the points. Without loss of generality, one can assume that the reference frame is attached to the first camera. Hence, the projection matrices for two images can be defined as follows:

$$M_1 = [I_D|O] \qquad \text{and} \qquad M_2 = [H|e]$$

where I_D is the 3×3 identity matrix, O is the null 3-vector, H is a 3×3 matrix representing an inter-image plane homography and e is the epipole in the second image.

The method in[22] can be used to get H and e using pixel matches only.

Once M_1 and M_2 are calculated, the projective 3D reconstruction can be easily obtained by triangulation. However, such a reconstruction is defined up to 3D projective transformation, described by a 4×4 nonsingular matrix, call it W. In fact, W is simply a basis-change in the 3D projective space, stating that any 3D projective reconstruction represents a set of 3D reconstructions that are related by a 3D collineation.

Unfortunately, this easy-to-get 3D projective reconstruction lacks any metric information, making its use very limited in practice. The next paragraphs present methods on how to use geometric constraints from the scene for obtaining a 3D metric reconstruction.

4.1. *The use of five known metric points*

Consider the case when we know the relative metric geometry of 5 scene points, $P_0, \ldots, P_4$, such that no 4 of which are co-planar. We can easily define a metric coordinate system by assigning these 5 points metric coordinates as follows. One point can be used as the origin, a second one as being on the X-axis, the third one as being on the X-Y plane and, by calculating the remaining coordinates with respect to this local coordinate system. Given that a 3D projective reconstruction is defined up to a 3D collineation, then a 3D metric reconstruction is related to the former by

$$P_i^m \sim W P_i^p$$

where P_i^p and P_i^m are the 3D coordinates in the original projective coordinate system and in the newly constructed metric coordinate system, respectively.

As every pair of such points provides three linear equations, W can be calculated with 5 pairs. Therefore, knowing the metric geometry of 5 points in space, allows to map a projective reconstruction to a metric one. This is the easiest way to get a metric reconstruction from a projective one as the problem turns into a simple change of basis in $I\!P^3$.

4.2. *The use of the plane at infinity*

It is not always possible to know the metric geometry of 5 points in space and in some cases, only the affine structure is sought. In this case, knowing three points (a plane) at infinity is enough to estimate a collineation W that maps a projective reconstruction into an affine one.

Note that the set of affine transformations is the subset of projective transformations that leaves the plane at infinity Π_∞ invariant. The affine space is defined from the projective space by choosing Π_∞. This plane has to be defined in the projective space by at least three points which are not collinear. The most common way to get such points is to consider lines that are known to be parallel and therefore intersect at infinity. There are two practical ways of finding such three points: (1) by extracting three noncollinear vanishing points from the images or, (2) by assuming the knowledge of three sets of parallel lines from the scenes (from walls, roads, corridors, etc). Parallel planes can also be used to obtain the plane at infinity. In particular, when we know that many parallel planes are present in the scene, then their automatic identification could be done using the method in.[23]

Given three points at infinity, call them P_1, P_2 and P_3, we can create a new reference frame with known affine geometry. As a consequence, we can calculate a collineation W, the same way as in the previous paragraph, that transforms a 3D projective reconstruction into an affine one. First, take any point from the reconstructed scene, call it P_0, and make it the origin of an affine coordinate system with $(0, 0, 0, 1)$ as its affine coordinates. Then, P_1, P_2 and P_3 can be used to define the three axes of this affine coordinate system by assigning them the coordinates $(1,0,0,0)$, $(0,1,0,0)$ and $(0,0,1,0)$, respectively. Finally, we can use any other point of the scene as a fifth point and assign it the affine coordinates $(1,1,1,1)$. This is a similar problem as the one in the previous paragraph, so we can easily calculate a 4×4 matrix W to transform a 3D projective reconstruction into an affine one.

4.3. *The use of geometric constraints*

The idea of using geometrical constraints[24] was first proposed over a decade ago. These constraints arise from geometric information about the observed scene such as points belonging to the ground, distances, parallelism, perpendicularity, etc. The main advantage in this case is that it is much easier to get this kind of information from an uncalibrated image than to get the 3D geometry of a set of scene's points. This is particularly true when the images represent scenes with man-made structures, such as walls, windows and roads. Furthermore, some of these geometric constraints could be automatically extracted in the near future.[23] As in the previous paragraphs, this problem consists of finding the appropriate collineation W, that maps points from the projective space into a metric one, where these geometric constraints are satisfied. Although W has 15 independent parameters, seven of them can be fixed as they correspond to a similarity transform. This can be done as follows. Any point from the scene can be set to be the origin with coordinates $(0,0,0,1)$. Then, another scene point can be used to define the X-axis and the metric unit with coordinates $(1,0,0,1)$. Finally, a third point can be used to fix one of the three planes of the coordinate system, for instance the X-Y plane, by setting its Z-coordinate to 0. By doing so, we get a total of seven linear constraints on W, and therefore, only 8 constraints, corresponding to 8 independent parameters, remain to be found. These 8 parameters together define the plane at infinity (3 parameters) and the absolute conic (5 parameters). This is in total agreement with the fact that the affine space is defined from the projective one by identifying the plane at infinity and that the Euclidean space is defined from the affine one by identifying the absolute conic.

In practice, each geometric knowledge about the scene is translated into a constraint on W. Here are a few examples of equations resulting from some constraints:

- A point A is known to belong to the plane $X = 0$ yields the equation

$$w_{11}x_A + w_{12}y_A + w_{13}z_A + w_{14}t_A = 0 \tag{8}$$

  where (x_A, y_A, z_A, t_A) are the known projective coordinates of the point A.
- Two segments S_1 and S_2 are known to be parallel yields the equation

$$w_{41}x_I + w_{42}y_I + w_{13}z_I + w_{14}t_I = 0 \tag{9}$$

  where (x_I, y_I, z_I, t_I) are the known projective coordinates of the intersection between the two lines defined by S_1 and S_2.

- The distance between A and B is known to be equal to d yields the equation

$$distance(WA, WB) = d \tag{10}$$

where A and B are known in the projective space, WA and WB are their corresponding unknown metric coordinates.

Note that the use of geometric constraints has the advantage of not assuming anything about the cameras. In particular, the intrinsic parameters can freely change from one image to the other.

4.4. *The use of the perpendicularity constraint*

Another possibility would be to combine some geometric constraints from the scene with constraints on the intrinsic parameters. The perpendicularity constraint is the easiest and most interesting one to be used in this case. This paragraph proposes a practical method to use this constraint in order to search for the best values of the intrinsic parameters. Assuming that we have a few segment pairs in the scene, known to be perpendicular, we can search for the best values of the intrinsic parameters yielding the best metric reconstruction. The required number of such pairs depends on the number of intrinsic parameters to be found. For instance, when using two cameras with very similar intrinsic parameters, the number of perpendicularity to be used must be greater or equal than 4 and not lying in the same plane. Note that this is not a self-calibration method as it is a simple linear search for the best values.

Consider the case of two images taken by the same camera whose unknown intrinsic parameters are given by the matrix A (see Paragraph 2.2). Assume that A' is a coarse estimate of A, available from some good guess. For each A', the 3D metric coordinates of the scene points are calculated using the simple Eight-Point Algorithm described in Section 3. Then, the dot product is calculated for each pairs of perpendicular segments. Since this product is supposed to be very close to zero for a correctly reconstructed metric scene, it is used to measure the quality of the obtained 3D metric reconstruction. Basically, the algorithm chooses the best A' that minimizes the sum of all available dot-products.

The method can be summarized by the pseudo-code below:

(1) Identify a few perpendicular segments in the scene
(2) for $(A = A' - \Delta_A;\ A = A' + \Delta_A;\ A = A + d_A)$

    (a) Get the 3D metric geometry of the scene using the Eight-Point Algorithm
    (b) Get the error = Sum of dot product of all perpendicular segments
    (c) If $(error < error_{min})$ then
        $error_{min} = error$
        $A_{min} = A$

(3) solution is A_{min} and its corresponding 3D metric geometry.

where Δ_A represents the search range, depending on our confidence about A', and d_A is the search resolution.

4.5. *Experimental results*

We have carried out experiments on real images only for this part as the goal here was to show that these ideas work in practice. Furthermore, we have also used some real scenes with known 3D metric coordinates as ground truth that has allowed us to compare the accuracy of the results.

3D reconstruction with no constraints   Two images of an outdoor scene were taken with an unknown camera and no geometric information about the scene was available. This is a worst-case scenario where in theory only the 3D projective geometry can be retrieved. A set of about a hundred feature points were extracted and matched between the images then, their 3D projective reconstruction was calculated. Five points of the scene were selected and were assigned approximate metric coordinates values, based only on visual clues. Although no constraints were used in this case, the results were surprisingly good as it can be seen from Figure 4. This

Fig. 4.   Two views of an outdoor scene and a top view of the 3D reconstruction of a set of points.

particular case shows that it is possible to retrieve a scene's 3D projective structure that is very close to the its metric counter-part. The accuracy of the latter depends only on the accuracy of the geometry of the five points used.

Metric reconstruction using geometric constraints   We have used an indoor scene here as it is easy to obtain geometric constraints and to compare the results with the measurements (see Figure 5).

After obtaining the 3D projective reconstruction of the feature points, we have used a variety of geometric constraints to estimate the collineation W, that maps the projective reconstruction into a metric one. Given that it was possible to measure the scene's coordinates using an ordinary ruler, we were able to estimate the errors on the reconstructed coordinates. The average distance between the measured and

Fig. 5. Two images of an indoor scene and a general view of the 3D metric geometry of a set of feature points.

the calculated coordinates was around $1.5mm$. This is an excellent accuracy as the feature points we have used were extracted with a 1-pixel accuracy.

The use of the perpendicularity constraint Two scenes were used for this case, an indoor scene with known ground truth and outdoor scene. For both cases, six perpendicularity constraints were used in the search for the best intrinsic parameters.

- **Indoor scene** The scene used here is the pattern scene given on Figure 3. Table 3 summarizes the obtained results for this scene where, the first row shows the reconstruction errors we have obtained when using the values of intrinsic parameters from an explicit calibration. The next 4 rows, case 1 through case 4, show the results for different cases depending on which intrinsic parameters were being searched (underlined ones). As it can be seen, all errors are very small compared to the size of the scene.

Table 3. 3D reconstruction errors for the pattern scene using the perpendicularity constraint.

case	best intrinsic parameters				3D errors		
	α_u	α_v	u_0	v_0	ΔX	ΔY	ΔZ
Calibration	-1483	-1007	242	239	0.0095	0.0087	0.0161
case 1	-1560	-1043	256	256	0.0105	0.0076	0.0154
case 2	-1528	-1044	242	239	0.0091	0.0034	0.01 39
case 3	-1483	-1007	182	222	0.0109	0.0079	0.0 133
case 4	-1583	-1075	282	239	0.0096	0.0042	0.0147

- **Outdoor scene (Figure 6)** A total of 40 interest points have been extracted and matched with about 1-pixel accuracy. Although we have a set of known values for the intrinsic parameters, coming from previously published works on self-calibration, we did not have any 3D metric measurement to compare with.

Therefore, we have used the perpendicularity errors, obtained by summing up the six dot products of the normalized perpendicular vectors, to compare our results with the ones obtained using the known intrinsic parameters.

Fig. 6.   The outdoor scene.

Table 3 summarizes the obtained results for this scene where, the first row shows the perpendicularity errors obtained when using the self-calibration values of the intrinsic parameters. The next 4 rows, case 1 through case 4, show the results for different cases depending on which intrinsic parameters were being searched, in which case they are underlined. As it can be seen, the error in the first row, obtained via self-calibration, is worse than all the other 4 errors.

Table 4.   Perpendicularity Errors for real images when some or all of the intrinsic parameters are searched.

case	calculated intrinsic parameters				Perpendicularity error
	α_u	α_v	u_0	v_0	(sum of dot products)
Published Intrinsic parameters	-682	-682	256	384	1.692
case 1	-781	-781	256	384	1.401
case 2	-782	-792	256	384	1.387
case 3	-682	-682	244	492	1.278
case 4	-887	-892	271	467	0.916

4.6. *Discussion*

Three objectives have been achieved in these experiments that have aimed at a calibration-free 3D metric reconstruction. First, using five scene points, with only their approximate metric geometry, it was possible to obtain an approximate 3D metric reconstruction of the scene. This 3D reconstruction is a true projective reconstruction with its metric geometry as good as the one of the five points used.

This is a very important result as one can always recover approximate 3D metric geometry from two images without any knowledge about the cameras. Second, geometric information about the scene can be translated into constraints. The latter, are used to calculate the collineation that maps a projective reconstruction into a metric one. The obvious advantages here are twofolds: (1) such geometric information is easily available for man-made environments and (2) information about the cameras is not needed at all. Third, it is possible to combine geometric constraints with information about the intrinsic parameters of the camera. The perpendicularity constraint was used as a quality measurement to search for the best values of the intrinsic parameters that yield the best metric 3D reconstruction of the scene. All that is needed is a number of perpendicular segments, very easy to pick from images, and a coarse estimate of the value ranges for the intrinsic parameters.

5. Conclusion

The recovery of three dimensional geometry of a scene from its two-dimensional images has attracted a lot of interests from numerous researchers for over three decades. Even if the theory for solving this problem has been thoroughly investigated, this problem is still considered unresolved in practice. This is mostly due to the camera calibration problem that is needed in order to obtain a metric reconstruction. In early solutions, the 3D metric reconstruction was obtained by performing a pattern-based camera calibration procedure first then the actual 3D reconstruction is calculated. However, the pattern-based camera calibration is neither realistic in practice nor always possible. This has led many researchers to investigate other possibilities to recover the 3D geometry of a scene without resorting to the explicit calibration. In particular, researchers have proposed new methods, relying solely on point correspondences across images, to retrieve the camera intrinsic parameters. This new solution, known as self-calibration, seemed to be attractive and promising as it has been seen as an alternative solution for the pattern-based calibration. Unfortunately, these self-calibration algorithms are very sensitive to pixel noise as they are nonlinear by nature, with no guaranteed convergence. Hence, self-calibration is rarely used in practice so far and still remains an active research topic. This chapter has proposed a different alternative to overcome the drawbacks of both pattern-based and self calibration methods. Instead of requiring the cameras to be calibrated, we have investigated other possibilities to obtain a calibration-free metric reconstruction. Three methods for such possibilities were proposed: (1) the use of approximate intrinsic parameters, (2) the use of geometric constraints and, (3) the use of both. In the first method, we have shown that when applying a simple affine transformation to the pixel coordinates, the 3D reconstruction process has become less sensitive to both pixel noise and errors on the intrinsic parameters. As a consequence, it was possible to retrieve a good quality metric 3D reconstruction even when using approximate values for the intrinsic parameters. This method could be used in many real situations, for instance, for robot navigation, where high 3D accuracy is not needed. In the second method, we have proposed the use of geometric constraints to calculate a collineation that maps a 3D projective reconstruction into a metric one. Such geometric constraints are easily obtained from images of man-made environments and are therefore, a good alternative that yield accurate 3D metric reconstruction. We have also shown that even in the absence of accurate geometric information about the scene, it was possible to calculate an approximate metric geometry using 5 space points, whose metric geometry was visually approximately from the images. When only the 3D affine geometry is sought, it can be calculated using the parallel geometric information. In the third proposed method, we have combined the use of one geometric constraint, the perpendicularity, with constraints on the ranges of the intrinsic parameters, to obtain the 3D metric geometry of a scene. The perpendicularity constraint was used as an accuracy measurement for the 3D metric reconstruction in order to search for the best intrinsic parameters.

Overall, we have proposed in this chapter three different alternative solutions for calibration-free 3D metric reconstruction. These solutions aim at bridging the gap between theory and application as they can be easily applied in real-life situations. In particular, should the extraction of geometric information, such as perpendicularity and parallelism, become fully automatic in the near future, the use of these geometric constraints may become the method of choice for 3D metric reconstruction. Therefore, future work in this direction will focus on the automatic extraction/identification of parallelism and perpendicularity from images. Some new methods are already being investigated for identifying parallel and perpendicular planes from images. Their early results are promising even though many challenges still remain to be overcome.

References

1. T. Oggier, F. Lustenberger, and N. Blanc. Miniature 3d tof camera for real-time imaging. In *Lecture Notes in Computer Science*, vol. 4021, pp. 212–216, (2006).
2. S. Soutschek, J. Penne, J. Hornegger, and J. Kornhuber. 3-d gesture-based scene navigation in medical imaging applications using time-of-flight cameras. In *Proc. IEEE Computer Vision and Pattern Recognition Worshops (CVPRW'08)*, pp. 1–6, (2008).
3. R. I. Hartley and A. Zisserman, *Multiple View Geometry in Computer Vision*. (Cambridge University Press, ISBN: 0521540518, 2004), second edition.
4. A. Heyden and F. Kahl. Direct affine reconstruction. In *Proc. Int. Conf. on Pattern Recognition (ICPR'00)*, vol. 1, pp. 885 – 888, (2000).
5. Y. Wang, B. Zhang, and T. Yao. A linear and direct method for projective reconstruction. In *Proc. IEEE Int. Conf. on Intelligent Computing and Intelligent Systems (ICIS'2009)*, vol. 4, (2009).
6. C. Rothwell, O. Faugeras, and G. Csurka, A comparison of projective reconstruction methods for pairs of views, *Computer Vision and Image Understanding*. **68**(1), 37–58, (1997).

7. A. Habed and B. Boufama. Three-dimensional projective reconstruction from three views. In *Proc. Int. Conf. on Pattern Recognition (ICPR'02)*, pp. 415–418, (2000).

8. M. Pollefeys and L. Van Gool, Stratified self-calibration with the modulus constraint, *IEEE Transactions on Pattern Analysis and Machine Intelligence.* **21**(8), 707–724, (1999).

9. A. Habed and B. Boufama, Camera self-calibration from bivariate polynomial equations and the coplanarity constraint, *Pattern Recognition.* **41**, 2484–2492, (2008).

10. O. D. Faugeras, Q. T. Luong, and S. J. Maybank. Camera self-calibration: Theory and experiments. In *Proc. of the European Conf. on Computer Vision (ECCV'92)*, pp. 321–334, (1992).

11. C. Lei, F. Wu, Z. Hu, and H. Tsui. A new approach to solving kruppa equations for camera self-calibration. In *Proc. Int. Conf. on Pattern Recognition (ICPR'02)*, vol. 2, pp. 308–311, (2002).

12. A. Fusiello, Uncalibrated euclidean reconstruction: a review, *Image Vision Computing.* **18**(2), 555–563, (2000).

13. A. Valdès, J. I. Ronda, and G. Gallego, The absolute line quadric and camera auto-calibration, *Int. J. of computer vision (IJCV).* **22**(10), 1199–1204, (2006).

14. P. Sturm, A case against kruppa's equations for camera self-calibration, *IEEE Transactions on Pattern Analysis and Machine Intelligence.* **22**(10), 1199–1204, (2000).

15. R. Mohr, B. Boufama, and P. Brand, Understanding positioning from multiple images, *Artificial Intelligence.* (78), 213–238, (1995).

16. D. Bondyfalat and S. Bougnoux. Imposing euclidean constraints during self-calibration processes. In *Lecture Notes in Computer Science*, vol. 1506, pp. 224–235, (1998).

17. J. I. Ronda and A. Valdés, Euclidean upgrading from segment lengths, *Int. J. Comput. Vision.* **90**(3), 350–368, (2010).

18. H. C. Longuet-Higgins, A computer program for reconstructing a scene from two projections, *Nature.* **293**, 133–135, (1981).

19. R. Hartley, In defence of the eight-point algorithm, *IEEE Transactions on Pattern Analysis and Machine Intelligence.* **19**(6), 580–593, (1997).

20. R. Hartley. Estimation of relative camera positions for uncalibrated cameras. In *Proc. of the European Conf. on Computer Vision (ECCV'92)*, pp. 579–587, (1992).

21. O. Faugeras. What can be seen in three dimensions with an uncalibrated stereo rig? In *Proc. of the European Conf. on Computer Vision (ECCV'92)*, pp. 563–578, (1992).

22. B. Boufama and R. R. Mohr, A stable and accurate algorithm for computing epipolar geometry, *Int. J. of Pattern Recognition and Artificial Intelligence.* **12**(6), 817–840, (1998).

23. A. Habed, A. Amintabar, and B. Boufama. Reconstruction-free parallel planes identification from uncalibrated images. In *Proc. Int. Conf. Pattern Recognition (ICPR'10)*, pp. 1828–1831, (2010).

24. B. Boufama, R. Mohr, and F. Veillon. Euclidean constraints for uncalibrated reconstruction. In *Int. Conf. on Computer Vision (ICCV'93)*, pp. 466–470, (1993).

AUCTION-BASED DYNAMIC CAMERA GROUPING WITH ACTIVE CONTROL

Bir Bhanu and Yiming Li

Center for Research in Intelligent Systems, University of California, Riverside
Engineering Unit II, Room 216, 900 University Ave., Riverside, CA 92521
{bhanu, yimli}@ee.ucr.edu

Video surveillance in a camera network requires the collaboration and competition among cameras. In this chapter, we present a novel auction-based approach to dynamically form coalitions of cameras to follow objects in a camera network. Active control (pan/tilt/zoom) is considered both theoretically and experimentally. One of the contributions of this chapter is that we formulate the bid price as a vector representation, such that when a camera is available to follow multiple objects, we consider the "willingness" of this camera to track a particular object. Most of the computation is decentralized by computing the bid price locally while the final assignment is made by a virtual auctioneer based on all the available bids, which is analogous to a real auction in economics. Thus, we can take advantage of distributed/centralized computation and avoid their pitfalls. The experimental results show that the proposed approach is effective and efficient for dynamically grouping cameras based on user defined performance metrics.

1. Introduction

The problem of efficient cooperation among multiple cameras has risen to the forefront of the video sensor networks. For example, in the video surveillance scenario, when there are multiple persons in the network, it is intuitive to follow these persons using all the available[a] cameras in the system instead of just using one. However, the problem is how these cameras cooperate with one another such that all the existing persons can be followed with an optimal group of cameras. There are several related questions. When active control (pan/tilt/zoom) of cameras is also available, how can we effectively control a group of cameras

---

[a] By available cameras, we mean those cameras that can currently "see" an object, i.e. have the object in its FOV for the current setting.

to have a better monitoring and understanding of the persons in the system? Also, how can we know in advance whether it is better to pan or tilt a camera to follow a person, who is originally not in its field-of-view (FOV), or how to use a currently available camera to follow that person? When a camera can "see" more than one person in its monitored range (all the areas that can be possibly covered by a camera by panning or tilting, even if it may not be covered for the current setting) how does a camera decide which one to follow? How can we form the groups of cameras based on the online information as the persons are walking in the camera network. All these questions are what we aim to address in this chapter.

There is a large amount of work done in the field of multi-camera multi-person tracking [1, 2, 3, 4, 5]. Only a few of the proposed approaches [4] focus on forming a group of active cameras dynamically for an object to track it collaboratively. Most of the work does not consider potentially available cameras to be involved in a group. Some existing approaches [5] use greedy algorithms, which prevents them from achieving the global optimum.

Auction-based approaches are used in multi-agent systems (multi-robot systems, manufacturing systems) for resource/task allocation problems. In an auction-based scenario, there is an *auctioneer* auctioning a good and all the potential buyers calculate their *bids* for the good locally. Finally, the auctioneer decides whom to sell the good based on the buyers' bids. This process, to a large extent, distributes the heaviest load of computation, the computation for bids, to each buyer, while the final decision is still optimal as long as there is a reasonable mechanism to make all the buyers *rational*. There is only very little work done to apply auction theory in camera dynamic control. [6] uses the ContractNet protocol for controlling active cameras. Our proposed work is different from it in (1) we formulate the bid as a vector to imply the willingness of a camera to be panned or tilted; (2) We test our approach using real data instead of simulated environments.

In this chapter, we model the process of grouping cameras to follow multiple objects in a camera network as the process of an economic auction. There is a virtual auctioneer holding an auction for each object to be followed and all the potential cameras bidding for it. By doing so, we benefit from the auction mechanism for distributed computation and consider the "willingness" of buyers (cameras). We choose from the top N bids to form a group and, thus, make the cameras with higher potentials to work collaboratively.

The rest of this chapter is organized as follows. Section 2 details some related work in camera networks and the existing applications of the auction-based techniques. The contributions of this chapter are also listed. Section 3 presents the mechanism how auction and bidding work for the camera grouping problem, the metrics design and the calculation of bidding price. Section 4 illustrates the effectiveness of the proposed approach by demonstrating various experimental. Section 5 concludes the chapter.

2. Related Work and Contributions

2.1. *Related Work*

In the field of tracking multiple persons in a camera network, there have been various approaches. There are many criteria to classify these approaches into different categories, such as whether an approach is distributed or centralized, whether it is under a calibrated environment or an uncalibrated one, whether active cameras are used or only static cameras are deployed. In Table 1, we compare the pros and cons in these different aspects while Table 2 compares our approach with some of the most related work.

Auction-based technique shows its effectiveness in solving many problems in multi-agent systems. For example, auction-based mechanism is established in [7] by He and Ioerger for computational grids. Gerkey and Mataric [8] use the auction method for dynamic task allocation for groups of failure-prone autonomous robots. Dias and Stentz [9] propose an opportunistic optimization approach for auction-based multirobot control. Leaders are used to do optimization within subgroups. Chen *et al.* [10] achieve single target tracking in wireless networks by deploying auction-based coalition. [6] applies ContractNet protocol for camera control, but the optimization part is done by formulating the problem as a constraint satisfaction problem. Our work differs from all of the above work. We formulated the whole problem as an auction process. The best camera is selected based on the bidding vector. This can provide the advantage of considering the potential availability of a camera, which outperforms the traditional bidding systems which usually treat the bids as scalar values.

2.2. *Contributions of this Chapter*

The proposed approach in this chapter aims at combining tracking in a camera network and the auction-based techniques together so as to solve the camera grouping problem with a new perspective.

The contributions of this chapter are:

1. We introduce the economic auction model for modeling the camera active control problem in a video network.
2. We make the bid a vector to consider the willingness of both currently available and potentially available cameras for camera selection.
3. We design the metrics and price function to calculate the bid locally.

Table 1. Comparisons of the pros and cons of various characteristics encountered in a camera network.

Properties	Advantages	Disadvantages
Fully distributed	Low bandwidth requirement; hard to die fully.	Lack of global cooperation; suffer from local optima.
Fully centralized	Easy for cooperation among cameras; easy to get the global optimum.	Bandwidth and computational time consuming; severe problem happens once the central server dies.
Calibrated	Help to know the topology of the camera network.	Pre-processing is required.
Uncalibrated	No computing for calibration is required.	Difficult to deduce the topology.
Active cameras	Provide a better view of objects; fewer cameras are required to cover a monitoring range.	Calibration may be required, Complex algorithms to account camera motions.
Static cameras	Easy to determine the topology; simpler algorithms than those for active cameras.	More (statitc) cameras are needed to have a full coverage.

Table 2. Comparisons of this paper with closely related approaches.

Authors	Approach	Comments
Tessens et al. [1]	Messages related to a distributed process are sent to a base station to determine the principal camera based on some score which is decided by experiments offline.	Use both distributed and centralized control. The criteria are determined offline so it is not suitable for active control. The principal view is complemented by helper views.
Li and Bhanu [2]	Models the camera assignment problem as a potential game. Applies a bargaining mechanism to negotiate among available cameras.	No active control is considered. The conflicts that arise in the camera assignment are not stated clearly. The theoretical model falls into a narrow category of a game.
Chen et al. [3]	Learn relationships between adjacent disjoint cameras by learning the spatio-temporal relationships and the brightness transfer function.	Handoff only takes places in the entry/exit zone; no active control; learning is needed.
Monari and Kroschel [4]	Determine the minimum number of sensors needed to relocate an object (even if the object is temporarily out of sight) based on computational geometry algorithms.	No active cameras are considered. This work aims to determine the minimum number of sensors instead of the best available ones.
Shen et al. [5]	Use greedy algorithm to compute the configuration of the cameras to meet the given quality of view requirement, while attempting to minimize the number of cameras.	It suffers from the drawback of greedy algorithm by losing the global optimal solutions.
This paper	Group cameras automatically and dynamically based on auction mechanisms. All the objects are followed by the top N available cameras.	Most of the computation is decentralized while the global optimal can still be achieved. Active control is considered as well by representing the bid as a vector.

4. We perform experiments with real data to corroborate the proposed approach.

3. Technical Approach

An illustration (Figure 1-a) as well as a bock diagram (Figure 1-b) in Figure 1 provide an overview of the proposed approach, which is detailed in this section. There are some cameras in the corridor and court yard, with some of the FOVs overlapped while some are non-overlapped. The virtual auctioneer announces the locations of the persons (L_1 and L_2) in the system and the cameras send out their bids (Figure 1-a). Most of the computation is distributed by calculating the bid prices locally and the group is automatically formed by choosing the top N cameras (Figure 1-b).

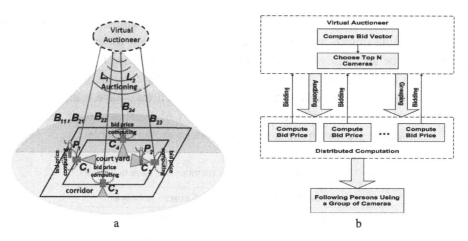

a   b

Figure 1. Overview of the auction-based approach. P_i: Person i. C_j: Camera j. B_{ij}: bid from Camera C_j for Person P_i. L_i: the location information of Person P_i. P_1 stands in the corridor where the person can only be covered by C_1, while P_2 stands on the court yard, where can be covered by C_3, C_4 and C_2 (by panning).

3.1. *Problem Formulation and Notations*

3.1.1. *Background*

An auction is the process of an auctioneer selling an item (goods or services) to many potential buyers, i.e., bidders. Typically, in the auction, the potential buyers first offer their prices (the price offer is also called a bid) [11, 12]. If the potential buyers bid for profitable trades only, we say that they are *rational*. Then, the auctioneer collects the bid prices (or, bids) information, and decides who wins the item and how much the winner has to pay. In the real world, there are many kinds of auction, which specify different bidding rules and different final payments of the winner. For example, in the first-price sealed-bid (FPSB) auction, all bidders simultaneously submit sealed bids so that no bidder knows the bid of any other participants. The bidder with the highest bid pays the submitted price. This is similar to the case in the proposed approach.

If any agent in a system cannot increase its well-being without damaging others' well-beings, we say that it is Pareto optimum [13]. The advantage of selling an item through auction method lies in the fact that in spite of asymmetric buying and selling information among bidders and auctioneer, the *Pareto optimum* can be achieved through auction under *Revelation Principle,* which rules out the possible inefficiency caused by asymmetric information [14].

3.1.2. *Problem Formulation*

The goal of the proposed approach is to form groups of cameras dynamically to follow multiple objects in the camera network. We want to select the cameras which have better *quality of views (QOV)* for an object, based on our pre-defined metrics, to form a group. This group may include the cameras which currently can "see" the object as well as those cameras which may have a high QOV by panning or tilting to somewhere else. The analogy of a real auction in economics and the grouping process in the camera network is shown in Table 3.

Table 3. The analogy of auction in economics and camera network.

Economics	Camera network
Auctioneer	Central Program
Goods	Objects
Bidder	Camera
Bid price	Camera's QOV of the object
Sale of the good	Group formation of cameras

A virtual auctioneer (a component that is not a real device like a camera, but something that is manipulated by the program) holds an auction for each of the objects in the system, i.e. objects are goods for sale. All the potentially available cameras are modeled as potential buyers for the object. There is a set of metrics according to which the cameras will evaluate their willingness to buy the good or not and if they decide to buy how much bid price they will provide. The auctioneer collects all this information and finally makes a decision on who should sell goods, i.e., to which camera(s) to use to follow objects.

3.2. *Auction Mechanism for Camera Network*

3.2.1. *System Assumptions*

Before describing the detailed approach, we first clarify some assumptions made in our system:

1. Homographies are calculated and the cameras' heights are known, so that we know the coordinate conversion between different camera images.
2. The camera's focal length is set to a fixed number such that the angle of view (the largest angle that a camera can cover without any active control) is $51.2°$. Each camera has 8 overlapping pre-defined pan settings to seamlessly cover 360 degrees. Also, there are three tilt settings, up $5°$, down $5°$ (or $-5°$) and no tilt ($0°$). So, there are 24 settings for each cameras. We will call these 24 settings for Camera C_j as $l = \{l_j^1, l_j^2, ..., l_j^{24}\}$ where l_j^1 is the current location of Camera C_j.
3. The cameras are rational and honest, i.e. they calculate their bid price solely based on the pre-defined metrics and they will only do the profitable trades.
4. There is no communication error.
5. There is no communication congestion.

    Based on the above assumptions, we propose an auction protocol to form groups of cameras automatically and dynamically to follow the objects in the network. For the convenience of the readers, some notations that are used in the following description are summarized in Table 4.

3.2.2. *Auction Protocol*

The auction protocol is described as follows:

1. ***Task announcement.*** A virtual agent (program running on a central server) holds an auction for each object to be tracked. An auction message is broadcast to the whole network. The message includes information such as the location of an object and camera IDs of those cameras which are in the same group to follow it. As illustrated in Figure 1-a, the virtual auctioneer broadcasts the locations of P_1 and P_2, L_1 and L_2, to the whole system. Note that we will initialize the location of the object by a motion detection module. The camera that first "sees" the object will be initialized as the *leader camera* in the group to follow this object. The object's location is initialized as the centroid location in the leader camera's image. After that, the leader camera is decided as the one with the highest bid price and the object's centroid in this leader camera will be broadcast.

Table 4. Symbols and Notations.

Symbols	Notations
P_i	Person i
C_j	Camera j
n_c	The number of cameras that can "see" P_i
l	Camera setting vector
l_j^k	The kth setting of C_j
B_{ij}	Bid price sent from camera C_j for person P_i
L_i	Location of P_i in the leader camera
$\boldsymbol{b}_{ij}$	Bid vector from C_j for P_i
b_{ij}^k	Intermediate bid from C_j for P_i at the setting l_j^k
B_i	$B_i = B_{i1} + B_{i2} + \cdots + B_{in_c}$
ρ	A percentage number decided by the user
N_i	The number of cameras that in the group to follow P_i
M_{ijm}	The mth metric score for P_i in C_j
w_m	Weights for different metrics
γ	Threshold for the size metric
(x, y)	Current location of the person in the camera image
(x_c, y_c)	Center of the camera image
α_k	Weight on k^{th} dimension in bid price function
λ	Elasticity of substitution between different dimensions in bid price function

2. **Bid price calculation.** The overall bid price B_{ij}, which is from camera C_j for person P_i, is decided by a 24-dimensional bid vector, $\boldsymbol{b}_{ij} = \{b_{ij}^1, b_{ij}^2, \ldots, b_{ij}^k, \ldots, b_{ij}^{24}\}, k \in [1,24]$. b_{ij}^k stands for the intermediate bid that the camera can get by panning or tilting to the setting l_j^k. If it cannot "see" an object at l_j^k, then b_{ij}^k is 0. Otherwise, b_{ij}^k is decided by the pre-defined metrics, such as the view, size and position of the object, which will be discussed in the next subsection. By modeling the bid as a vector, we are able to informing the auctioneer the willingness of a camera to follow an object, which is implied by the order of the elements in the bid vector $\boldsymbol{b}_{ij}$. We prefer to use a camera without any panning or tilting, since panning and tilting make some frames blurred and it takes time to have a sharp image. If an object is moving at a high speed, when the camera can have a sharp image after panning or tilting a large degree of angle, the object may already be out of the FOV again. However, the necessity of having this vector representation instead of by considering the current

location l_j^1 only lies in the fact that in some cases, all the cameras that can currently "see" the object have a back or side view of the object while if we pan or tilt some camera, which is currently unavailable for this object, it will have the object's frontal view, which can provide us more information of interest. Or, there might be the case when a camera pans or tilts to another setting, it will gain more welfare by following another object instead of continuously following the object currently assigned to it. For example, in Figure 1-a, although Camera C_2 cannot "see" the person P_2 at the moment, it still sends its bid for P_2, which is B_{22}. This vector representation helps to take into account the inclination of a camera, which, therefore, avoids the drawbacks of greedy algorithms. Finally, the overall bid price B is calculated as a function of all the intermediate bids in $\boldsymbol{b}_{ij}$, i.e.

$$B_{ij} = f(b_{ij}^1, b_{ij}^2, \ldots, b_{ij}^k, \ldots, b_{ij}^{24}).$$

This function is designed in the next subsection.

3. ***Bid submission.*** After evaluating the price for each object, all the related cameras send their bid prices for the object(s). As mentioned in the assumptions, the prices must be honest and can truly imply their willingness to follow an object.

4. ***Close of auction.*** Unlike in the traditional auction, where the auctioneer will sell the good to the buyer who provides the highest bid price, the virtual auctioneer in our system choose the top N_i cameras (whose bid prices are the top N_i ones) to form a group to follow an object. All the prices are sorted from high to low and then are summed up. Let $B_i = B_{i1} + B_{i2} + \cdots + B_{in_C}$. N_i is the minimum number such that

$$B_{i1} + B_{i2} + \cdots + B_{iN_i} \geq \rho\% \times B_i,$$

where ρ is a parameter decided by the user.

The whole auction process is described in Figure 1-b as a block diagram. Note that the highest computational load, the calculation of bid prices, is distributed to each camera node and, thus, done locally. Our works is different the work done in [6], because we do not solve the camera grouping problem using any other optimization mechanism, such as the constraint satisfaction. All the camera selection and grouping are done based on the bid vector and such, are very easy to calculate. No backtracks are needed.

3.2.3. *Optimality Discussion*

Intuitively, under the assumption that the cameras are rational and honest, all the cameras report their true evaluations of the object to be tracked to the virtual auctioneer. The virtual auctioneer can, thus, obtain the maximal benefit by "selling" the item (the object to be tracked) to those cameras that have the top N_i evaluations on the object. From the cameras' viewpoint, this transaction is optimal, since the camera which has the highest evaluation wins the right to track the object. Also, from the virtual auctioneer's standpoint, it can obtain the highest "payment" from the winner. The fact that the cameras always reveal their true evaluation of the object to be tracked validates that the Pareto optimality [13] of the camera grouping system is always achievable. Also, the optimal camera group is dynamically formed by this auction-based camera grouping process.

3.3. *Metrics and Price Function Design*

For the metrics used for evaluating the bids, we mainly consider the size of the person and the position of the person in the camera image, which are described as follows:

1. *The size of the tracked person*, measured by the ratio of the number of pixels inside the bounding box of the person to that of the size of the image. Assume that γ is the threshold for the best observation, i.e. when $r = \gamma$ this criterion reaches its peak value, where $r = \dfrac{\text{\# of pixels inside the bounding box}}{\text{\# of pixels in the image plane}}$.

$$M_{ij1} = \begin{cases} \dfrac{1}{\gamma} r, & \text{when } r < \gamma \\ \dfrac{1-r}{1-\lambda}, & \text{when } r \geq \gamma \end{cases} \tag{1}$$

2. *The position of the person in the FOV of a camera.* It is measured by the Euclidean distance that a person is away from the center of the image

$$M_{ij2} = \frac{\sqrt{(x-x_c)^2+(y-y_c)^2}}{\frac{1}{2}\sqrt{x_c^2+y_c^2}} \tag{2}$$

where (x, y) is the current position of the person and (x_c, y_c) is the center of the camera image plane.

Each intermediate bid b_{ij}^k is decided by the above metrics and is calculated

$$b_{ij}^k = \sum_{m=1}^{2} w_m M_{ijm} \tag{3}$$

where w_m is the weight for different metrics. The calculation of these M_{ijm} is described in the experimental part.

The final bid price B_{ij} is computed as

$$B_{ij} = \left(\alpha_1(b_{ij}^1)^\lambda + \alpha_2(b_{ij}^2)^\lambda + \cdots + \alpha_{24}(b_{ij}^{24})^\lambda\right)^{\frac{1}{\lambda}} \tag{4}$$

where $\alpha_1 + \alpha_2 + \cdots + \alpha_{24} = 1, \lambda \in (-\infty, +\infty)$.

The bid price function B_{ij} implies the utility that Camera C_j would obtain if it is assigned to follow Person P_j.

The parameter λ in equation (4) measures the degree of easiness in substitution among different dimensions in the intermediate bid vector $\boldsymbol{b}_{ij}$, i.e., when multiple setting of a camera can cover the object to be followed, to what extent we can use one of these available settings to substitute among one another in terms of the cost and benefit the camera can get. Figure 2 depicts the contour curves of the bidding function given different λ. For the purpose of illustration, the dimension of the intermediate bid is reduced to two (i.e. each camera has only two settings), which reduces the bid price function to

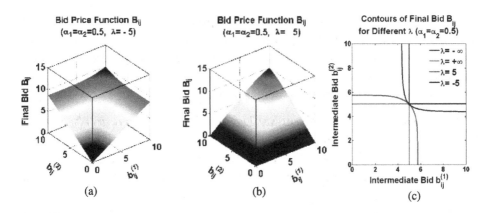

Figure 2. Effects of different λ on the bid price B_{ij}, Bid Price increases as the color changes from blue to red.

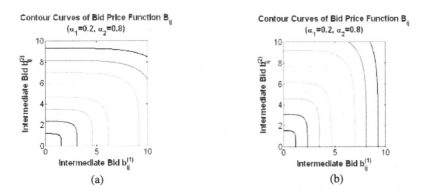

Figure 3. Contour Curves of B_{ij} (the effect of different α_k, Bid Price associated with contour curve increases as the color changes from blue to red.)

$$B_{ij} = \left(\alpha_1 (b_{ij}^1)^\lambda + \alpha_2 (b_{ij}^2)^\lambda\right)^{\frac{1}{\lambda}} \tag{5}$$

As λ approaches to negative infinity we have

$$\lim_{\lambda \to -\infty} B_{ij} = \min\{\alpha_1 b_{ij}^1, \alpha_2 b_{ij}^2\}$$

which means that B_{ij} is determined by the $\alpha_k b_{ij}^k$ with the *lowest* value, and the change of other b_{ij}^k cannot change the final bid B_{ij}, i.e. the camera's bid price solely depends on the setting that will give the worst result. Therefore, each dimension b_{ij}^k in the intermediate bid vector $\boldsymbol{b}_{ij}$ cannot be substituted by any other dimension, as shown by the green curve in Figure 2-c. On the other hand, if λ equals 1, the bid price function degenerates to a simple linear function

$$B_{ij} = \alpha_1 b_{ij}^1 + \alpha_2 b_{ij}^2 \tag{6}$$

which means that each dimension b_{ij}^k is a perfect substitution for any other dimension in $\boldsymbol{b}_{ij}$, i.e. each setting of the camera will give exactly the same result. Finally, as λ goes to positive infinity, the bid price function converges to the *max* function

$$\lim_{\lambda \to \infty} B_{ij} = \max\{\alpha_1 b_{ij}^1, \alpha_2 b_{ij}^2\}$$

which means that the bidder's utility level is determined by $\alpha_k b_{ij}^k$ with the *highest* value, and the change of other elements in $\boldsymbol{b}_{ij}$ cannot change the overall bid B_{ij}, i.e., B_{ij} solely depends on the setting that can provide the best result, as shown by the orange curve in Figure 2-c. The magenta and red curves in Figure 2-c show the contours of B_{ij} for two example cases that when $\lambda \in (-\infty, +\infty)$ and their actual functions are shown in Figure 2-a and Figure 2-b respectively. Intuitively, in the camera network scenario, it makes no sense to make $\lambda < 0$, since we will never use a camera with the worst result to follow an object. On the other hand, when a camera has more than one setting that can "see" an object, these settings will not be perfect substitution for one another, since panning or tilting the camera with different angles may cause different time delay and blur the image. In the parameterization of λ, we prefer the range $(1, +\infty)$, which means the camera's bid price B_{ij} depends largely on higher intermediate bids other than those lower ones.

In addition, α_k in the bid price function measures the camera's relative preference on b_{ij}^k to other b_{ij}^n ($n \neq k$). The larger the α_k is, the larger weight is put on b_{ij}^m in the bid price function B_{ij}. One extreme case is $\alpha_k = 1$, then the bid price function degenerates to $B_{ij} = b_{ij}^k$, which means that only b_{ij}^k contributes to the utility of camera C_j in following person P_i. Figure 3 describes the contour curves of B_{ij} bidding function under different parameterizations on α_i (the dimension of the intermediate bid is reduced to two for the convenience of illustration). In our experiments, we put the highest weight on α_1, which means that we prefer to use a camera to follow a person without any active control to avoid blurred images.

Note that if the dimensions of the camera setting vector l are non-overlapped with each other, then there is only one non-zero dimension in the bid vector b_{ij}. Thus, the bid vector simplifies to a scalar.

Although the bid price function B_{ij} can be picked up arbitrarily, we choose the form of equation (4) mainly because it has simple explanations of parameters, and also it provides considerable flexibilities in parameterization. This function is also known as the Constant Elasticity of Substitution (CES) in economics [12].

The zoom control is done when a person's frontal view is detected around the centroid of an assigned camera. We zoom in that camera (if more than one are available for the frontal view, then we zoom in the one that provides a higher bid) for 2 frames and then zoom out (in case that some other person will be lost when zooming in the camera).

The overall algorithm of the proposed approach is given in Algorithm 1.

Algorithm 1. Auction-based Approach for Camera Grouping

Input: Locations of the detected objects
Output: Group formation of cameras for each object
Initialization:
1.   Reset all the cameras to their default settings.
2.   Broadcast the location of the object in the camera that first detects it. Make this camera the leader camera for this object.
Do:
1. Hold an auction for each detected object.
2. Broadcast the location of this object in the leader camera.
3.   All the currently available cameras and potentially available cameras calculate their bid prices for this object based on the given metrics and submit their bids
4.   The virtual auctioneer selects the top N cameras to form a group to follow this object and make the camera with the highest bid the leader camera of the group.
5.   Whenever a person's face is detected in an assigned camera, we zoom that camera to have a close-up view and then zoom out that camera to the previous setting.
6.   Go to 1.

---

4. Experiments

4.1. *Data and Parameters*

We perform the experiments in our department building, where we have 37 outdoor cameras in a network. All the cameras are commercially available Axis

215 PTZ cameras. The map of the camera network is given in Figure 4. We choose 6 of the 37 cameras to perform our experiments. The selected FOVs are marked in red in Figure 4. We select 2 cameras in the corridor and 4 on the court yard, with 2 on each side. Some of the cameras' FOVs are overlapped (e.g., Camera 4 and Camera 5) while some are non-overlapped (e.g., Camera 2 and Camera 6).

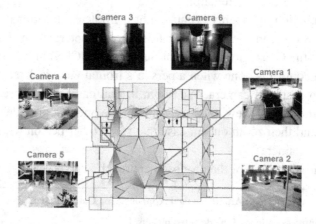

Figure 4. Map of the camera network.

We calculate the homographies for different settings of cameras such that we know the correspondence between each pair of cameras for any setting. The homographies are computed by picking up corresponding point pairs on the same ground plane. They are computed off-line and no detail camera calibration is needed. When a person's location in the image of the lead camera is known, we can use the homographies to predict the person's location in all possibly available settings for all the *potentially available* cameras. Then the height of this person is estimated by using homography and the height of camera placement, which is measured beforehand. Using the camera height, we are able to estimate the person's actual height in the world coordinates from his height in pixel in the image of the lead camera. Similarly, we can estimate the person's height in pixel in all those potentially available camera settings (from those cameras that cannot "see" the person currently but it is possible to "see" this person by panning or tilting) from the previously estimated actual height in world coordinates. M_{ij1}, the size metric of the person, is estimated by making a bounding box using the same ratio of height to width of a person as it is in the lead camera and calculate the area of the bounding box. M_{ij2}, the position metric of the person, is estimated by picking the center of the top and bottom (deduced from the height) as the centroid of the person. Error occurs when the person is not fully visible. There is also inaccuracy caused by the measurement in the world coordinates and the assumption for the flat ground.

We apply the particle filter tracker and use color as the feature. The face detection is done by applying the face detector in OpenCV around the top half of the bounding box. We choose a particle filter tracker because it is relatively robust to occlusions. It is to be noted that the focus of this chapter is not to design a robust tracker and face detector, but lies in how to form groups of cameras dynamically and integrate camera active control into this process.

The parameters in the experiments are set empirically. The threshold for the size of the person is $\gamma = \frac{1}{15}$. The weights for different metrics are selected as $w_1 = 0.6$ and $w_2 = 0.4$. The weights in the bid price function B_{ij} are given in Table 5.

From Values of α_k, we can note that using the camera with active control as little as possible is preferred, since it may cause blurring of images and the time delay may cause missing more objects. The elasticity of substitution parameter in the bid price function B_{ij}, $\lambda = 8$. The percentage based on which we decide the number of cameras to form a group, $\rho - 50$.

Table 5. Values of α_k.

α_k	Pan	Tilt	Value	α_k	Pan	Tilt	Value
α_1	$0°$	$0°$	0.150	α_{13}	$-90°$	$5°$	0.020
α_2	$45°$	$0°$	0.080	α_{14}	$135°$	$5°$	0.010
α_3	$-45°$	$0°$	0.080	α_{15}	$-135°$	$5°$	0.010
α_4	$90°$	$0°$	0.030	α_{16}	$180°$	$5°$	0.006
α_5	$-90°$	$0°$	0.030	α_{17}	$0°$	$-5°$	0.100
α_6	$135°$	$0°$	0.015	α_{18}	$45°$	$-5°$	0.060
α_7	$-135°$	$0°$	0.015	α_{19}	$-45°$	$-5°$	0.060
α_8	$180°$	$0°$	0.010	α_{20}	$90°$	$-5°$	0.020
α_9	$0°$	$5°$	0.100	α_{21}	$-90°$	$-5°$	0.020
α_{10}	$45°$	$5°$	0.060	α_{22}	$135°$	$-5°$	0.010
α_{11}	$-45°$	$5°$	0.060	α_{23}	$-135°$	$-5°$	0.010
α_{12}	$90°$	$5°$	0.020	α_{24}	$180°$	$-5°$	0.006

4.2. *Experimental Results*

In Figure 5, we show some typical frames in a simple case where we deploy 3 cameras (Camera 1, Camera 4 and Camera 5) and let 2 persons walk in the camera network. In frame a, although all the three cameras that can "see" the person in grey, camera 2's bid price takes up to 67% of the summation of all the bids. Therefore, there is only Camera 2 in the group that is assigned to follow the person in grey. The person in purple is in a similar situation. In frame b, although Camera 2 is potentially available for both of the two persons, it can "see" the person in purple pretty well without panning or tilting, thus, its bid for this person is higher and it forms a group with Camera 3 together for the person in purple while the person in grey is monitored by the group formed by Camera 1 and Camera 3. In frame c, the frontal view of the person in red is detected in camera 3. So, we zoom (1.5 times) in camera 3 to have a close-up view. In frame d, the person in grey can only be covered by camera 2. The process of bidding for the person in grey (Person 1) is shown in Figure 6. Note that some of the bids are zero value because Person 1 is not seen by that camera in those frames.

Figure 5. Some typical frames for the 3 cameras 2 persons case. The camera images from the cameras in the same group for a person are boxed in the same color as the person they are assigned to follow.

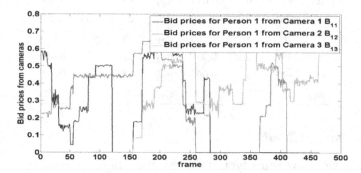

Figure 6. Bid prices for the person in grey (Person 1). See Figure 5 for the grouping results for the person.

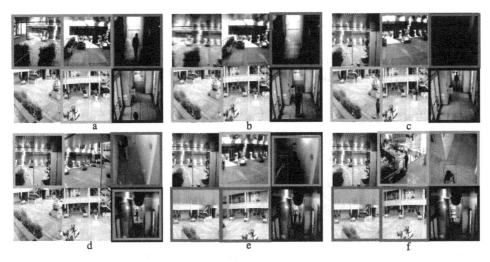

Figure 7. Experimental results in the 6 cameras 4 persons case . The camera images from the cameras in the same group for a person are boxed in the same color as the person they are assigned to follow.

In Figure 7, we show some typical frames in a more complicated case with 6 cameras and 4 persons. As stated previously, QOVs in the cameras influence the proposed camera grouping results. We can observe that, when there are more than one camera available for a person and none of them can dominate any other in terms of the tracking quality, then these cameras will form a group to follow this person, e.g. in frame a for the person in red and frame e and f for the person in brown. Otherwise, when there is a camera that has a much higher score for a person, then it's bid price will be greater than 50% of the overall summation of all the bids. In this case, we say that this camera dominate all the other cameras for that person. For example, for the person in green and blue, in all the presented frames in Figure 7, there is always one camera that can dominate all the other cameras, such that there is only one camera in the groups for tracking these two persons. This also includes the case when there is only one camera that is available for a person, e.g., in frame c.

The overall performance of the proposed approach is shown in Figure 8. We define the correct following rate of a person as the ratio of the number of frames that a person is successfully followed by at least one of the cameras when the person is visible in the network (although, sometimes, in other cameras, the person may be occluded) to the number of frames of the video sequence. Figure 8-a and Figure 8-c show the success following rate for all the persons in the 3 cameras/2 persons case and the 6 cameras/4 persons case respectively, where we ignore all the errors caused by various factors, not the proposed camera grouping algorithm itself, such as those come from the tracker, the face detector, the blurring image caused by panning and zooming, and the inaccuracy in measurements when calculating the homographies. If we count the errors caused by all these factors, then the correct following rates are shown in Figure 8-b and Figure 8-d.

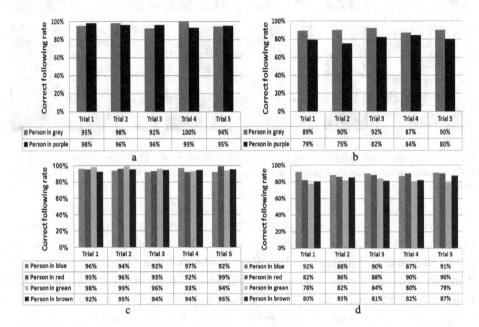

Figure 8. Correct following rate for the two cases in 5 trials.

5.   Conclusions and Future Works

We proposed a novel auction-based mechanism to form groups of cameras to follow objects in a camera network. This chapter introduced the auction concept into the camera network area and achieved promising results. We made the bid as a vector to take into account the cameras' willingness to follow an object or not. We show results for following various number of persons, active control of cameras and dynamic group formation. Experiments in real-time (15-19 fps) with real data are performed, which show the effectiveness of the proposed approach.

In the future, we will combine this auction-based system with more accurate tracking systems and scene analysis. For example, we can enhance the tracking performance by fusing multiple trackers to corroborate tracking in different scenarios. Scene analysis can be done more accurately by considering homography and epipolar geometry together.

References

1. L. Tessens *et al*. Principal view determination for camera selection in distributed smart camera networks. *ICDSC* 2008.
2. Y. Li and B. Bhanu. Utility-based dynamic camera assignment and hand-off in a camera network. *ICDSC* 2008.
3. K. Chen *et al*. An adaptive learning method for target tracking across multiple cameras. *CVPR* 2008.
4. E. Monari and K. Kroschel. A knowledge-based camera selection approach for object tracking in large sensor networks. *ICDSC* 2009.
5. Shen et al. A Multi-Camera Surveillance System that Estimates Quality-of-View Measurement. *ICIP* 2007.
6. F.Z. Qureshi, D. Terzopoulos, "Smart Camera Networks in Virtual Reality," *ICDSC 2007*.
7. L. He and T.R. Ioerger. Task-oriented computational economic-based distributed resource allocation mechanisms for computational grids. *ICAI*, vol. 1, 2004, pp. 462–468.
8. B.P. Gerkey and M.J. Mataric. Sold!: Auction methods for multirobot coordination. *IEEE Transaction on robotics and automation*, vol. 18, no. 5, October, 2002.
9. M.B. Dias and A. Stentz. Opportunistic optimization for market-based multirobot control. *IEEE/RSJ international conference on intelligent robots and systems*, 2002.
10. Chen et al. Auction-based dynamic coalition for single target tracking in wireless sensor networks. *The sixth world congress on intelligent control and automation, 2006.*
11. E. Wolfstetter. Auctions: an Introduction. *Journal of economic surveys*, vol. 10(4), pages 367–420, December, 1996.
12. R. P. McAfee and J. McMillan. Auctions and bidding. *Journal of economic literature*, Vol. 25, No. 2 pp. 699–738, 1987.
13. H. R. Varian. Microeconomic analysis (3rd edition). *W.W.Norton & Company*, New York,1992.
14. J. G. Riley and W. F. Samuelson. Optimal auctions. *The American economic review*, Vol. 71, No. 3 pp. 381–392, 1981.

CHAPTER 3.3

NOVEL PHOTOGEOMETRIC METHODS FOR CAPTURING STATIC AND DYNAMIC OBJECTS

Yi Xu

Computer Vision Lab, GE Global Research,
1 Research Circle, Niskayuna, NY, USA
xuyi@ge.com

Daniel G. Aliaga

Department of Computer Science, Purdue University,
305 N University Street, West Lafayette, IN, USA
aliaga@cs.purdue.edu

Obtaining high-resolution 3D models of real world static and dynamic objects is a very important and challenging task in computer vision. Using off-the-shelf hardware for active capture often requires careful calibration of the devices (e.g., projector-camera systems). In the case of moving objects, achieving high-resolution acquisition is difficult because information must be captured during a very short period of time. In this chapter, we discuss a family of *photogeometric* methods using only off-the-shelf hardware enabling self-calibration and quickly and accurately obtaining point-sampled colored models of real-world static and dynamic objects. Our system simultaneously obtains mutually registered surface position and surface normal information and produces a single high-quality model. Acquisition processing freely alternates between using a geometric setup and using a photometric setup with the same hardware configuration. Moreover, our approach generates reconstructions at the resolution of the camera and is not limited to the resolution of the projector.

1. Introduction

Robustly capturing 3D high-resolution models of real-world static and dynamic objects is critical to a wide range of applications, including virtual reality, telepresence, and entertainment. Acquisition methods that rely on off-the-shelf hardware (e.g., projector-camera systems) are appealing because building such a system, including both hardware and software, is relatively easy. However, there are several challenges associated with this type of system. First, accurate calibration of the projectors and cameras is required – obtaining such can be a

Moving and Deforming Object   **Highly Detailed Model**   **Multi-view Modeling**

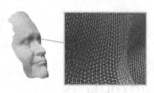   

- Three fixed Canon Realis SX6 projectors
- One fixed PTGrey Dragonfly Express camera
- Model captured at 640x480 pixels from single viewpoint
- Six images captured for one frame of motion

- Three fixed Optoma EP910 projectors
- One fixed Canon Digital Rebel XTi camera
- Model captured at 3888x2592 pixels from single viewpoint
- 120 images captured to obtain optimized model

- One Mitsubishi Handheld PK10 projector moved to 7 locations
- Three fixed PTGrey Flea cameras
- Model captured at 1024x768 pixels from multiple views for better coverage
- 756 images captured for multi-view modeling

Increasing hardware complexity and acquisition time.

Fig. 1. Examples using our Photogeometric Framework. The framework provides flexible designs using off-the-shelf hardware and enables trade-off between acquisition time and model accuracy.

tedious task. Second, the resolution of the capture is often limited by the device with the lowest resolution (e.g., the projectors). Third, it is also difficult to use off-the-shelf hardware to capture moving and deforming objects, especially given the goals of high-resolution and robustness. Acquisition must occur within only one or a few frame times of a typical camera. There are numerous methods that address some of the aforementioned problems, but none that address them all.

In this chapter, we present a novel photogeometric framework consisting of a family of methods using off-the-shelf projectors and cameras. Our method concurrently captures photometric and geometric information in order to produce a photogeometric modeling system. This combination enables self-calibration, which makes our system easy to build, deploy and use. Further, our framework exploits the high-visual details of photometric methods and supports capturing 3D models at the resolution of the camera. Our same fundamental framework also offers design flexibility and supports capture configurations ranging from coarse and fast acquisition engines to highly-detailed and optimized 3D capture systems. For example, the simplest configuration of our framework consists of one fixed camera and one projector moved to three different locations. It captures 3D models of static objects from a single viewpoint. Another configuration uses three fixed projectors and one camera to perform self-calibration method and to capture 3D models for moving and deforming objects. A multi-camera configuration acquires highly-detailed models consisting of several million point samples of real-world objects with a sampling resolution as fine as 0.1mm – still using only consumer hardware. Fig. 1 shows a few example designs that can be achieved using our framework.

The key observation behind our framework is that a digital projector can be simultaneously used as either an active light source or as a virtual camera. This permits acquiring per-pixel photometric and geometric observations of objects and enables self-calibration of the projectors acting as virtual cameras. Our photogeometric method differs from previous approaches in several ways. While Helmholtz reciprocity[27] and dual photography[18] methods have identified a projector as the dual to a camera, they do not explicitly obtain 3D models. Although previous efforts have captured photometric and geometric observations to build 3D models, they use separate *a priori* calibrated systems.[15, 16] Our photometric and geometric systems are one and the same; and thus acquisition processing is free to change from one setup to the other.

A photometric solution (e.g., shape-from-shading and photometric stereo) requires a small number of images and has the advantage of fast processing and high resolution but suffers from bad global shape recovery (e.g., General Bas Relief (GBR) transform[2]). A geometric solution (e.g., structured light) provides metric accuracy but often needs more compute time, requires more images and is at lower-resolution than a photometric solution. By combining the two sources of information, our photogeometric framework enables changing from a photometric solution, to a photogeometric solution, and then to a geometric solution. It is able to trade nonlinear geometric computations for faster linear computations and is able to incorporate the additional resolution and details of a photometric method. The combination also supports performing self-calibration of the intrinsic and extrinsic parameters of the used hardware. Moreover, in the case when only a small number of images are needed for a sparse geometric solution, by combining with photometric stereo, our framework enables capturing 3D models for moving and deforming objects.

Our framework uses one of several configurations of off-the-shelf digital projectors and digital cameras to obtain a large collection of single- or multi-view samples of the object each with geometric and photometric observations. In this chapter, we demonstrate 1) a self-calibration algorithm that uses the projectors as light sources and then as virtual cameras; 2) a photogeometric reconstruction algorithm that uses a linear optimization and produces very high-quality multi-viewpoint models; and 3) a special configuration that is capable of capturing moving and deforming objects. The main contributions of our framework are:

- a novel photogeometric framework that is self-calibrating and supports reconstructions at the resolution of the cameras,
- a photogeometric reconstruction algorithm that produces multi-view models best complying with both photometric and geometric measurements, without having to align various reconstructions, and
- a configuration of our framework that reconstructs 3D models for moving and deforming objects using a linear spatial-temporal photogeometric optimization of sparse geometric and dense photometric data.

2. Related Work

In this section, related work on active methods using photogeometric, geometric, and photometric based reconstruction are introduced. Approaches that exploit camera-light duality are also briefly discussed.

Photogeometric Modeling: Combining geometric and photometric modeling has helped obtain high quality models for static scenes. Rushmeier and Bernardini use two separate and pre-calibrated acquisition devices to obtain surface normals that are consistent with an underlying mesh.[16] Nehab et al. use the positional data obtained by dense structured-light acquisition and the normals measured by photometric stereo to perform a hybrid reconstruction of improved quality.[15] Compared to these methods, our method is both self-calibrating and multi-viewpoint. This makes the setup significantly more flexible, practical, and able to obtain more complete 3D models. We use the exact same equipment for both geometric and photometric acquisition. Our single capture system also has the benefit of removing the rotational alignment[15] or positional-and-rotational alignment[16] needed between photometric and geometric samples. Moreover, the two aforementioned methods did not explore the use of sparse geometric information, which in turns enables us to process moving and deforming objects.

Another group of approaches use specialized camera and lighting hardware to capture dynamic shapes using photogeometric methods. The University of Southern California (USC) Institute for Creative Technologies (ICT) Light Stage 5 operates at 1500Hz and projects 24 binary structured-light patterns and 29 basis lighting directions at 24Hz.[10] Vlasic et al. use Light Stage 6 with 1200 controllable light sources and 8 cameras to capture human performance.[21] By combining multi-view photometric stereo and silhouette-based visual hull reconstruction, this method obtains impressive results. However, these two methods do not explore sparse geometric sampling which in turn enables the use of much simpler hardware.

Geometric Modeling: Active methods using geometric-based acquisition produce detailed models but the process is often time consuming and does not necessarily produce smooth and accurate normals. Although some self-calibrating structured-light systems exist[5], typical structured-light approaches use *a priori* calibration and reconstruct mostly-Lambertian objects.[19] Camera self-calibration methods often rely on features and on either scene or geometry constraints to estimate parameters. In general, it is difficult to achieve convergence using such camera self-calibration.[20]

The reconstruction resolution of structured light usually depends on the number of patterns being projected onto the objects. Methods that only use a single pattern are suitable for acquiring moving scenes. However, they often obtain low-resolution geometric models.[12] State-of-the-art methods can achieve very good results, but the resolution is still limited by the projectors.[17] Our photogeometric framework augments geometric data using the normal information computed from photometric stereo and is able to exploit the higher resolution of the cameras. Further, some one-shot methods depend on recognizing intricate color patterns, which is difficult for an arbitrary object. In contrast, in our method for capturing moving and deforming objects (Sec. 4), we use a simpler geometric pattern (white dots) and white light photometric stereo; thus, our approach is robust, efficient, and able to handle full color objects.

In addition to the spatial coding of one-shot methods, temporal coding can also be used to enhance reconstruction resolution. Space-time stereo enhance traditional stereo by projecting rapidly changing stripe patterns and using oriented space-time windows for correspondence.[3, 26] Weise et al. present a fast 3D scanning system using phase-shifting patterns, a projector with the color wheel removed, and three cameras.[22] Both methods rely on stereo matching to obtain correspondence. Compared to stereo, the sparse geometric pattern used in our method for moving and deforming objects is more robust and can be computed more efficiently.

Photometric Modeling: As opposed to photometric stereo, our framework actively generates robust features and is able to overcome typical low-frequency deformations. In particular, we surmount the ambiguity of the GBR transform[2] and obtain 3D models up to a global scale factor. To improve upon ambiguity, some previous methods rely on feature tracking and/or on structure-from-motion refinement.[11, 13, 25] While this may reduce global deformations, it relies on passive feature correspondence. Another self-calibrating photometric option builds closed models of an object placed on a turntable and sequentially illuminated by three distant lights.[7] Many images of the object at constant rotational steps are captured. Silhouettes are extracted and used to form a 3D visual hull, which is perturbed to find matching photometric normals and mesh normals. This approach depends on silhouette detection, initial estimates for object orientation, and distant light sources (i.e., lights 3 to 4 meters away for a 15 cm object). For deforming and uniformly colored objects, Hernandez et al. simultaneously capture the appearance under three different lighting directions and each direction using a different color channel.[8] The algorithm requires a calibration object with the same material as the target object. Further, since the surface is integrated from normal maps, no globally-accurate geometry is acquired.

Camera-Projector Duality: Duality between cameras and light sources has been exploited for novel image generation and for surface reconstruction. For instance, a stereo method can use a projector as a virtual camera in the matching error metric formulation.[3] Dual photography demonstrates how the view from a light source (e.g., projector) can be obtained by transposing the light transport matrix from projector to camera; however, no object geometry is obtained.[18] Helmholtz stereopsis physically swaps a light source and camera so as to enable their co-location and to reconstruct an object without having to make assumptions about its bidirectional reflectance distribution function (BRDF).[27] While the initial Helmholtz stereopsis method requires calibration, it has been extended to an un-calibrated approach requiring known epipolar geometry[28] or using reciprocal image features[29]. Both still require co-location of the camera and light source and the latter method also depends on the presence of either texture features or specularity features.

3. Photogeometric Acquisition for Static Objects

In this section, we discuss our self-calibration algorithm and photogeometric reconstruction method for static objects. We use one of three fundamental configurations with $C \geq 1$ cameras and $R \geq 1$ projectors to obtain object point samples $S = \{S_i\}$ where $i \in [1, N]$ and N is desired to be large. An object point sample is defined as $S_i = (p_i, n_i)$, where p_i is the position and n_i is the normal of the sample. Each sample is computed from a set $G_i = \{q_{ij} \mid i \in [1, N], j \in [1, R]\}$ of at least two geometric observations needed for geometric/classical stereo and a set $P_i = \{c_{ij} \mid i \in [1, N], j \in [1, R]\}$ of at least three photometric observations needed for photometric stereo, where $q_{ij} = (u_{ij}, v_{ij})$ is the projection of sample i onto the image plane of projector j, and $c_{ij} = (c_{ij}^r, c_{ij}^g, c_{ij}^b)$ is the RGB color of point i under the illumination of projector j. Our configurations hinge on the notion that a digital projector can be either a virtual camera or a digitally controlled light. This leads to the following three possible configurations.

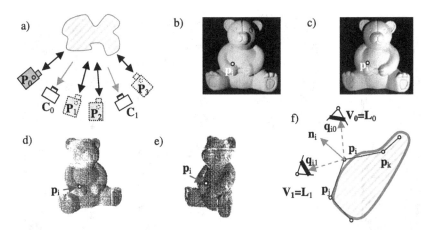

Fig. 2. Example Configuration. a) A multi-projector and multiple-camera configuration for acquisition. b) A photograph of an object to be captured. c) Photometric reconstruction using three images illuminated from three directions. d-e) Two corresponded virtual views between two projectors and the camera picture in b. f) Photogeometric setup surrounding point p_i.

Single-camera configuration (SC) has one static camera and one projector moved to several locations. A static camera-projector configuration provides the two viewpoints needed for a geometric reconstruction, but does not provide the multiple light sources for a photometric reconstruction. Moving the same single projector to several distinct locations increases both the number of viewpoints and the number of light sources. This configuration is the simplest for obtaining the minimum needed geometric and photometric observations per S_i. One disadvantage is that as the number of projector locations increases, the number of object points visible from the camera and all projectors decreases.

Multi-camera configuration (MC) has several static cameras and one projector moved to several locations. In this configuration, object points have correspondences in multiple cameras, and collectively sample a larger portion of the object as compared to single camera configuration.

Multi-projector Configuration (MP) has several identical projectors and one or more static cameras. This approach assumes the same focal length for all projectors and avoids moving the projector. Fig. 2a shows a MP/MC configuration. For static objects, MP can also be realized using a single projector.

3.1. *Image Capture*

For each projector, a sequence of $A + B$ patterns are projected where A patterns are used for photometric observations P_i and B patterns are used for geometric

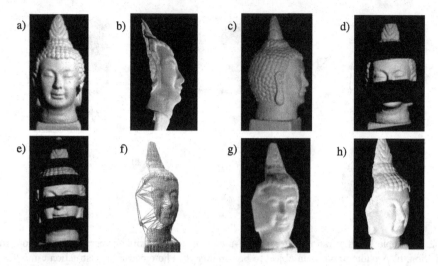

Fig. 3. Processing. a) Illuminated by a projector. b) Photometric reconstruction. c) Side-view photograph (compare to b). d-e) Structured light input images. f-g) Wireframe and filled view of coarse self-calibrated model. h) Final model after up-sampling and photogeometric optimization (rendered using synthetic lighting).

observations G_i. Changing the number of patterns enables different time-quality trade-offs. In the limit, using $B = 0$ yields a purely photometric capture while using $A = 0$ produces a purely geometric capture. The image data for the photometric observations P_i consists of the color intensity of the samples S_i as lit by the projectors (Fig. 3a). We assume a Lambertian object illuminated by each projector acting as a diffuse light source.

The image data for geometric observations G_i is captured using Gray code patterns[9], which are a set of coarse-to-fine multi-level binary patterns (Fig. 3d-e). Our method projects Q pairs of (horizontal and vertical) stripe patterns from each of the projectors and captures images for all cameras. To prevent determining surface albedos, we project the patterns and their inverses; resulting in $4Q$ patterns per projector. These patterns permit robustly corresponding about $(2^Q - 1)^2$ surface points between a camera and a projector. For a projector with 1400x1050 pixels, the maximum Q is 10; resulting in ~10^6 samples. To yield fewer points and faster processing, we can use smaller Q (e.g., 5). Other structured light patterns can also be used for point correspondence (e.g., grid pattern, color coded pattern, etc.). We choose to use temporally-coded Gray code patterns due to its robustness against different surface albedos and its ability to achieve high resolution. In addition, changing the levels of Gray code pattern can be used to easily control the density of geometric measurements.

3.2. *Self-Calibration*

After capturing all the images, self-calibration is performed using the sampled data. We first establish point correspondence between projectors and cameras and perform an un-calibrated photometric stereo reconstruction. The estimated light directions are used as initial values for the viewing directions of the projectors (i.e., virtual cameras) and are fed into the calibration optimization.

3.2.1. *Object Point Sampling*

To convert projectors to virtual cameras, we need to correspond pixels amongst as many projectors (and cameras) as possible. Gray code sequence provides camera-to-projector correspondence. We then re-sample the camera-to-projector correspondences to produce dense projector-to-projector correspondences.

Each camera has observed all Gray codes and has a set of point correspondences with the projectors. A camera and a projector each creates a 2D triangulation of the points they have in common. The camera then computes the barycentric coordinates for a regular grid of 2D points on the camera's image plane and uses the barycentric coordinates to compute corresponding 2D points on the projector's image plane. A newly created point in the projector's view is then corresponded with points on all the other cameras' regular grid provided the projector triangle is visible in the other camera. The result is a large set of points which at most are visible in all projectors and cameras and at the least equal to the correspondence between one camera and one projector. For example, point p_i in Fig. 2b-c is corresponded with the projector views in Fig. 2d-e. Points too close to others in all images are eliminated and the final outcome is a near-regular distribution of points corresponded between projectors. The projector-projector correspondences become geometric observations G_i and the projector-camera correspondence are used to extract photometric observations P_i.

3.2.2. *Un-calibrated Photometric Stereo*

Assuming three distant point sources of illumination, we use a Lambertian un-calibrated photometric stereo approach to compute the magnitude of and the angles between light directions, which is up to an unknown global rotation R, and a surface normal field N.[23] A surface height field can then be calculated from N up to rotation R and a GBR transform $G = [1\ 0\ 0; 0\ 1\ 0; \mu\ \nu\ \lambda]$.[2] Since typically the same projector is used at multiple locations, we assume equal light source intensity and simplify the GBR transform to $(\lambda, 0, 0)$. Using an arbitrary object, we manually estimate λ, which transforms the photometric surface to a shape similar to the real object. Registration does not need to be accurate and $\lambda = 0.3$ works well with all our objects. We determine a 3D rotation to bring the lighting setup into registration with the xy image plane using a simple user interface.

We then integrate the photometrically-estimated normals to construct a surface height field $z(x, y)$ using the method of Ref. 4 (Fig. 3b-c). Although depth discontinuities cause problems when integrating normals, our method does not assume the integrability of the surface. Our geometric self-calibration algorithm is able to handle a certain amount of self-occlusion by using a rough photometric estimate as shown later.

3.2.3. *Geometric Modeling*

Our self-calibration algorithm uses a photometrically-estimated surface $z(x, y)$, approximate lighting directions, and tailored reprojection equations to obtain a reconstruction of a subset S' of the object point samples $\{S_i\}$. The size of S' affects the time and quality of the resulting reconstruction. Our approach estimates both the focal length and pose of the projectors, acting as virtual cameras. We do not correct for radial distortion and thus assume long focal lengths and/or high-quality lenses. For all configurations, the projectors will be necessarily fully calibrated but the physical cameras are only optionally calibrated. We seek to minimize re-projection error expressed by the well-known nonlinear system of equations

$$\Sigma_j \Sigma_i \left(\frac{1}{h_{ij_z}} \begin{bmatrix} h_{ij_x} \\ h_{ij_y} \end{bmatrix} - \begin{bmatrix} u_{ij} \\ v_{ij} \end{bmatrix} \right)^2 \text{ where } h_{ij} = F_j(R_j p_i + T_j) \tag{1}$$

and R_j, T_j, and F_j are the unknown 3x3 rotation matrix, 3D translation vector and 3x3 perspective projection matrix, respectively.

Initialization: During initialization, a sparse and uniformly-distributed subset of object point samples of $S'$are used to estimate the distance from each projector to the object's center as well as the global projector focal length f. Initial values for p_i come from the photometrically-estimated surface $z(x, y)$. The calculated lighting (or virtual viewing) directions l_j and an assumed up-vector of $w = [0 \ 1 \ 0]^T$ define an initial oriented orthogonal coordinate system for each projector: $M_j = [l_j \times w \quad (l_j \times w) \times l_j \quad -l_j]$. The free parameters are the focal length f and distances z_j along l_j from the origin to each projector j. To bring the re-projection of the object points into rough alignment with the observed projections (u_{ij}, v_{ij}), we optimize the following simplified nonlinear system of equations of only $R + 1$ unknowns (f and z_j for $j \in [1, R]$ and where $\hat{p}_{ij} = M_j p_i$:

$$\Sigma_j \Sigma_i \left(\frac{\hat{p}_{ij_x} f}{\hat{p}_{ij_z} + z_j} - u_{ij} \right)^2 + \left(\frac{\hat{p}_{ij_y} f}{\hat{p}_{ij_z} + z_j} - v_{ij} \right)^2 . \tag{2}$$

Optimization: Next, our method optimizes for a linear correction to each projector location and performs a global bundle adjustment. In particular, in Eq. 1, F_j is replaced by a perspective projection matrix parameterized by f, T_j is replaced by $\begin{bmatrix} 0 & 0 & z_j \end{bmatrix}^T$ and R_j is replaced by $Q_j M_j$. Each matrix Q_j is computed using the following linear system of equations in the 8 unknowns of the matrix (i.e., $q_{33} = 1$)

$$\Sigma_i \begin{bmatrix} q_{ij} - u_{ij}(q_{ij} + z_j)/f \\ q_{ij} - v_{ij}(q_{ij} + z_j)/f \end{bmatrix} \text{ where } \hat{q}_{ij} = Q_j M_j p_i . \tag{3}$$

Using an iterative process, we then include all the remaining object point samples of S' and optimize the projector poses, object points, and remove outliers. First, we fix projector pose parameters and use a sparse bundle-adjustment optimization of all object points in Eq. 2. Second, we use sparse nonlinear bundle adjustment to refine both projector pose and all object points. Image-space and world-space point culling criteria are also applied. The culling criteria are based on the inter sample distance in the image plane and in world space. Thus, the same criterion is applied to similar situations (e.g., similar object size, camera/projector to object distances, and resolutions.). The optimization and culling repeats until convergence; M_j is updated to the final pose matrix.

3.3. *Photogeometric Reconstruction*

After self-calibration, our system combines photometric and multi-view geometric data in a single linear optimization. This approach enables a time-quality tradeoff whereby a variable amount of geometric modeling is performed and an approximation of the missing details is obtained from the faster photometric processing. Effectively, the traditional nonlinear modeling of multi-million point samples is converted into a fast and specialized nonlinear optimization of a small set of points followed by a linear up-sampling and linear multi-view optimization of all points.

3.3.1. *Up-sampling*

First, we increase the sampling density of the point samples to that of the camera. While the relative pose of the photometric reconstruction with respect to the geometric reconstruction is unknown, the known image-space correspondence defines a piecewise linear mapping because both observation types are from the

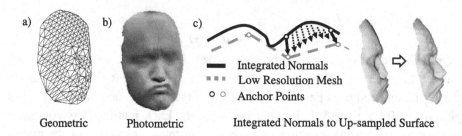

Fig. 4. Up-sampling. a) A low resolution triangulated mesh. b) A high-resolution normal map using photometric stereo. c) A 2D illustration and rendering that show the up-sampling process. The integrated normals are warped to a low-resolution geometrically-reconstructed surface.

same viewpoint (i.e., a projector). Thus points from the photometric surface can be warped to the geometric surface.

To perform the up-sampling, our method computes a 2D triangulation of all geometric surface points and then the barycentric coordinates $(\alpha_i, \beta_i, \gamma_i)$ of all photometric surface points within this triangulation. To warp a photometric point p_{P_i} to the geometric surface, $(\alpha_i, \beta_i, \gamma_i)$ and the corresponding vectors formed by pairing the vertices of the photometric surface triangle with the corresponding vertices of the geometric surface triangle are used to compute a displacement. The paired points are called anchor points. A new geometric surface point p_{G_i} corresponding to p_{P_i}, is computed by

$$p_{G_i} = p_{P_i} + \alpha_i\big(a_{G_i} - a_{P_i}\big) + \beta_i\big(b_{G_i} - b_{P_i}\big) + \gamma_i\big(c_{G_i} - c_{P_i}\big) \qquad (4)$$

where $(a_{P_i}, b_{P_i}, c_{P_i})$ is the photometric surface triangle surrounding p_{P_i} and $(a_{G_i}, b_{G_i}, c_{G_i})$ are the corresponding geometric surface points. The pixel observation of the up-sampled point is determined by interpolating the triangle's pixel observations. Fig. 4 demonstrates the process.

3.3.2. *Multi-view Optimization*

Our optimization alters object points so as to best match both photometric and geometric measurements. We search for a solution that follows the properties of

(i)   minimizing re-projection error onto the projectors,

(ii)   keeping a similar relative position of points, and

(iii)   reducing the difference between photometrically- and geometrically-computed normals.

An important aspect of our multi-view optimization method is to prevent undesired "flipping" and "self-intersecting" of the mesh of object points. One option is to keep each object point restricted to lie on the corresponding projector ray emanating from the center of projection and passing through its observation on the image plane.[15] However, such an approach is not suitable for multi-view processing. Only letting the point move along one or more projector rays does not support freely compensating for inaccuracies in the estimated pose (and focal length). In our method, each object point moves freely in 3D space and is able to accommodate larger corrections to the surface. The equations that attempt to ensure property (i) keep points near the correct geometric locations. The equations for property (ii) assist in yielding a distribution of points over the object's surface that is similar to the original one. Property (iii) guides the up-sampled points to an arrangement resembling the additionally captured photometric detail and provides a smoothly reconstructed surface.

The surface normals computed via the initial photometric processing are now updated to better represent the configuration. We upgrade to point lights because it more accurately imitates the true setup: projectors are kept relatively close to the objects. Surface normal n_i is now computed as

$$n_i = \frac{L_i^{-1}\left[c_{ij_1} \cdots c_{ij_R}\right]^T}{\left\|L_i^{-1}\left[c_{ij_1} \cdots c_{ij_R}\right]^T\right\|} \text{ where } L_i = \begin{bmatrix} [M_{j_1}^{-1}[0\ 0\ 0\ ]^T - p_i] \\ \cdots \\ [M_{j_R}^{-1}[0\ 0\ 0\ ]^T - p_i] \end{bmatrix}. \tag{5}$$

L_i is a per-point light direction matrix computed by linking the centers of projections of the projectors to the current estimate of 3D location of p_i. Other non-Lambertian photometric methods can also be used with our approach to produce a better photometric solution for non-Lambertian surfaces.[6, 14]

Formulation: The linear equation that satisfies the aforementioned triple of properties and that we wish to minimize is

$$e_t = (1 - \alpha)(1 - \beta)\kappa_g e_g + \beta\kappa_r e_r + \alpha\kappa_p e_p \rightarrow 0 \tag{6}$$

$$e_g = \Sigma_j \Sigma_i \begin{bmatrix} \hat{p}_{ij_x} - \left(\frac{u_{ij}\hat{p}_{ij_z}}{f}\right) \\ \hat{p}_{ij_y} - \left(\frac{v_{ij}\hat{p}_{ij_z}}{f}\right) \end{bmatrix} \tag{7}$$

$$e_r = \Sigma_i \delta_{ik}((p_i - p_k) - d_{ik}) \tag{8}$$

$$e_p = \Sigma_i \delta_{ik}(n_i \cdot (p_i - p_k)) \tag{9}$$

where d_{ik} is the initial distance between p_i and p_k, δ_{ik} is 1 when p_k is considered a neighbor of p_i and 0 otherwise. The unknowns are the 3D coordinates of each p_i. Eq. 7-9 correspond to the properties (i-iii), respectively. To determine the neighbors of a point p_i and to define δ_{ik}, we create a local Delaunay triangulation using a set of neighboring points projected onto the tangent plane. Eq. 6 can be written as $Ax = b$ and solved using over-constrained sparse linear least squares. Given K equal to the average number of neighbors per object point, $N(2R + K) \geq 3N$ ensures the number of equations is larger than or equal to the number of unknowns.

To control the tradeoff between geometric error, photometric error, and relative distance error, we scale the individual error terms to the range [0 1] (using $\kappa_g, \kappa_p, \kappa_r$) and optimize a weighted linear combination of them. The error terms are combined using α and β; e.g., a low value for α implies low photometric importance and high geometric importance; a low value for β implies lack of importance of keeping the same relative distance between points.

4. Photogeometric Acquisition for Moving and Deforming Objects

Our photogeometric reconstruction algorithm enables a time-quality tradeoff that supports capturing moving and deforming objects. When a certain amount of geometric modeling is performed, the missing details are obtained from photometric processing. When the geometric information is very sparse, it can be acquired in one or a small number of frames. Meanwhile, dense photometric information can be acquired in three frames using photometric stereo. By efficiently combining sparse geometric and dense photometric information using a nonlinear weighting scheme based on expected accuracies, we can build dense and precise models of moving and deforming objects at the resolution of the camera. Furthermore, unlike typical structured-light patterns, the diffuse light sources (e.g., projectors) used for photometric processing illuminate the scene uniformly and under the same conditions. This enables using optical flow based motion compensation amongst photometric images.

4.1. *Image Capture*

We use a single-camera and multi-projector configuration. A calibration step is performed beforehand either using our self-calibration method or using a traditional calibration package. For each frame, a pattern is rendered to one of the projectors. The camera, which is synchronized to the projectors, captures video frames of the object under the illumination of temporally-coded patterns.

The number of geometric patterns needed to obtain a desired quality is object-dependent. We found using three geometric patterns to yield a good balance of motion compensation and final quality. In the results section we explore the reconstruction quality when varying the number of geometric points. Furthermore, the inter-frame distance needed for the optical flow algorithm used in motion compensation equals the number of unique patterns. Hence, for our 60Hz camera, six unique patterns implies being able to detect optical flow for

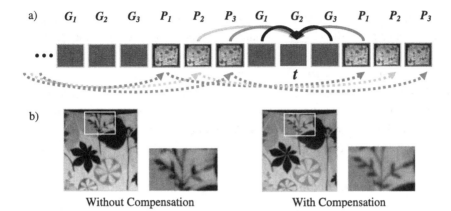

Fig. 5. Motion Compensation. a) The bottom arrows show the optical flows between photometric frames using the same light source. For the current frame *t*, a set of six frames (including itself) are warped to *t* (top arrows). b) For visualization, we store the three closest photometric frames into RGB channels. Without motion compensation, the channels are not aligned (double impression on the left). With compensation, the three frames observe a virtually static object and lead to a clean composite.

motions sampled at 10Hz, a frame rate that we do not want to go below. Since the majority of point samples are initially reconstructed using photometric stereo, we seek to minimize the amount of compensation for the photometric frames by placing the three photometric patterns temporally together; thus, resulting in a repeating six frame pattern sequence of three geometric patterns followed by three photometric patterns.

4.2. *Motion Compensation*

The reconstruction starts by warping the surrounding three geometric frames and three photometric frames to the current frame using motion compensation. We use an optical-flow based motion compensation method to bring all the desired frames into alignment with any frame *t*. Motion compensation is necessary because the object motion leads to misalignment between the frames used for reconstruction (Fig. 5b). Fast alternating patterns violate the illumination constancy assumption of traditional optical flow algorithms. However, the photometric frames of the same projector are captured under constant illumination conditions every six frames. Moreover, the constant illumination is white light which does not significantly interfere with scene colors. Hence, these frames are suitable for optical flow calculations.

Since the motion between a pair of adjacent photometric frames of the same projector can be large, we rely on the robustness of the optical flow calculation.

We compute point features and track them using OpenCV's pyramidal implementation of the Lucas-Kanade optical flow method. Per-pixel dense optical flow is interpolated using barycentric coordinates of the three surrounding features. Then, photometric frames are directly warped to frame t using their own flow fields. Geometric frames are warped to frame t by using an average of the three flows that pass through them (Fig. 5a). In this way, we compute a set of six frames that captures a virtually static scene and use them to model the non-rigid moving and deforming object.

4.3. *Geometric and Photometric Processing*

Geometric processing robustly obtains a sparse set of 3D positional measures. We project a 2D array of white dots for each geometric frame. The three dot patterns are projected by the same projector using shifted versions of the same dot array. Although the patterns could come from any of the projectors, using only one enables to control the sampling of the dots and intentionally produce a nearly uniform point sampling on the object's surface.

The dot array is constructed so that it yields disjoint and well-separated epipolar line segments on the camera's image plane for a chosen scene depth range (Fig. 6a). This property avoids ambiguity and enables very robust camera-to-projector ray correspondence. The resolution of the dot array is limited by the depth range and the camera resolution. The dot array consists of $d_X \times d_Y$ dots and is rotated around the image center by d_R degrees. We optimize for a set of d_X, d_Y, and d_R that maximizes the number of dots and meets a minimum inter-segment distance requirement. Typically, the resolution of the dot array is relatively low (e.g., 35x25); thus, using simple intensity thresholding is very robust as compared to other patterns using complicated geometric shapes and colors. The small number of dots is ameliorated by the use of multiple geometric frames and by the fact that missing details will be filled in using photometric information.

Since the camera and projectors are pre-calibrated, the corresponded camera and projector rays are triangulated to obtain a sparse 3D point sampling G of the moving and deforming object. If multiple dots are mapped to the same epipolar line segment, all of these dots are ignored to avoid outliers and depth discontinuities. The remaining points in G are meshed using 2D Delaunay triangulation from the camera's view (Fig. 6a right).

Unlike the un-calibrated photometric stereo used to initialize self-calibration, we use calibrated photometric stereo and use a point light source model similar to that used in Eq. 5. For each camera pixel, we find the 3D intersection between

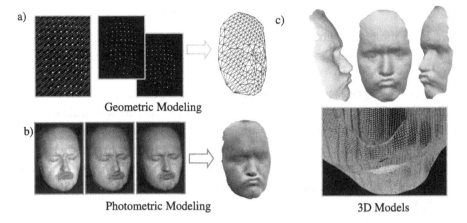

Fig. 6. Capturing Moving and Deforming Face. a) (left) A thresholded geometric pattern frame with epipolar line segments superimposed and two other frames. (right) A low resolution triangulated mesh. b) (left) Three photometric frames. (right) A high-resolution normal map using photometric stereo. c) Final optimized model rendered from three viewpoints using synthetic shading (top) and wireframe close-up (bottom).

the ray and the polygonal model and then re-project the 3D point to each of the three projectors. This operation gives us for each camera pixel i an initial estimate of the incident light directions. Light intensities from the three projectors are photometrically calibrated and equalized and Lambertian photometric stereo is used to compute a dense normal map (Fig. 6b right).

4.4. *Spatial-Temporal Photogeometric Processing*

The final reconstructed object is obtained by using an iterative algorithm and a system of linear equations. The objective is to find the surface that best satisfies a weighted combination of a sparse geometrically-computed surface, a dense photometrically-computed surface, and temporal smoothness constraints (Fig. 6c). We first compute the initial photogeometric surface using the up-sampling procedure discussed in Sec. 3.3.1 (Fig. 4c). The initial photogeometric surface provides a better approximation to the final solution than the low resolution polygonal mesh generated using geometric points. Thus, we update the per-pixel incident light directions using the new 3D position of each point; and re-compute per-pixel normals.

We extend the photogeometric optimization described in Sec. 3.3.2 by enforcing temporal coherence. Since we only use one camera, we restrict the 3D point of each pixel p_{ij} (j-th pixel in frame i) to lie along its camera ray and parameterize the pixel using only its depth value z_{ij}. In this formulation, we ignore for simplicity the error term e_r. Our new objective function is:

$$e = \alpha \cdot e_p + (1 - \alpha) \cdot e_g + \lambda \cdot e_s \qquad (10)$$

where e_p is the photometric error term, e_g is the geometric error term, and e_s is an additional temporal smoothness constraint. To optimize frame f, Eq. 10 is written as a linear least squares problem. The error terms are defined over a window of (e.g. 3) consecutive frames centered on frame f.

The use of very sparse geometric data necessitates a per-pixel weighting scheme to balance between photometric and geometric measurements. Only the anchor pixels, which are in both photometric and geometric surfaces, have accurate geometric measurements. The majority of pixels have approximations computed by the up-sampling process. Thus, we assign each pixel a weight that is defined as

$$w_{ij} = \begin{cases} 1 & p_{ij} \text{ is an anchor} \\ 1/\sqrt{s_{ij}} & \text{otherwise} \end{cases} \qquad (11)$$

where s_{ij} is the image-space distance between the pixel and the closest anchor in frame i. Hence, geometric measurements for pixels closer to anchor points are given higher weights since they are more accurate. The resulting error term that captures closeness to the geometric observations can now be written as

$$e_g = \Sigma_{i \in W(f)} \Sigma_j w_{ij} (z_{ij} - \hat{z}_{ij})^2 \qquad (12)$$

where $W(f)$ is a temporal window around frame f, and $\hat{z}_{ij}$ is the original depth value for the pixel.

To obtain best agreement between photometrically- and geometrically-computed normals and thus achieve spatial smoothness, we use a photometric error term e_p that minimizes the dot product between the surface tangents and surface normals, similar to Ref. 15. For locally smooth surfaces, tangents are approximated by vectors from point p_{ij} to each of its neighboring points. These vectors are represented as a linear combination of the depth values of p_{ij} and its neighbors. The resulting error term is:

$$e_p = \Sigma_{i \in W(f)} \Sigma_j \Sigma_{k \in N(j)} [(z_{ij} r_{ij} - z_{ik} r_{ik}) \cdot n_{ij}]^2 \qquad (13)$$

where $N(j)$ is the set of neighbors of pixel p_{ij}, and r_{ij} and r_{ik} are the ray directions of pixels p_{ij} and p_{ik}, respectively.

To ensure temporal smoothness, we assume locally linear motion and minimize the second-order derivatives of object points. The second-order difference is used to approximate the derivative. The smoothness term is

$$e_s = \Sigma_{\delta=1}^n \Sigma_j w_{fj} (z_{(f+\delta)j} r_{(f+\delta)j} + z_{(f-\delta)j} r_{(f-\delta)j} - 2 z_{fj} r_{fj})^2 \qquad (14)$$

where n is the temporal window half size, and $z_{(f-\delta)j}$, z_{fj}, and $z_{(f+\delta)j}$ are the same object point in three frames $f - \delta$, f, and $f + \delta$. The correspondence of points over time is established using the same dense optical flow employed for motion compensation. Since we only track sparse features and interpolate flow in between, we assign per-point weights w_{fj} that favor tracked features. The weights are computed by finding the closest tracked feature for each point.

Our new spatial-temporal photogeometric optimization is still a linear optimization and is fast to compute. Only the optimized results for the center frame f are stored. Since the photometric and geometric error terms are of different units, a weight α is used to control the balance. We have dense photometric samples and sparse geometric ones, thus α is usually small to favor geometric samples (e.g., $\alpha = 0.005$).

After photogeometric optimization, we re-compute point light sources and per-pixel normals and perform the optimization again until the change for one iteration is too small. The dense point clouds are meshed using 2D Delaunay triangulation from the camera's view. To enforce triangle consistency, we triangulate in the first frame and displace triangles to the next frame using optical flow. The edges of the displaced triangles are used to perform a constrained Delaunay triangulation in the next frame. Intersecting edges are ignored in order not to introduce new points. Thus, the same triangulation is used for as many frames as possible.

5. Results

We have implemented several systems. The first one uses a 800x600 Mitsubishi Handheld PK10 projector and a 1024x768 Point Grey Flea camera. The second uses three Optoma EP910 projectors and a 3888x2592 Canon XTi camera to capture very high-resolution models. The third one uses a Point Grey Dragonfly Express 640x480 camera and three Canon Realis SX6 projectors driven by a single PC through a Matrox® TripleHead2Go unit to capture moving and deforming objects. Table 1 summarizes the example data. For more results, the readers are referred to our previous publications on the topic.[1, 24]

The first four objects are static and we acquire them using our self-calibrating photogeometric approach. Total processing time for any model is less than 15 minutes: self-calibration takes a few minutes, up-sampling needs a few seconds, and photogeometric optimization requires from 1 to 10 minutes. All final calibration errors are less than one pixel. The last three objects are moving and deforming objects. We capture them using a pre-calibrated system. Modeling takes 12-37 seconds per frame, with 75% of the time usually used for 2-3 iterations of the spatial-temporal photogeometric reconstruction algorithm.

Table 1 Datasets. We show configuration, numbers of projectors, cameras, and points.

Data	Configuration	# Projectors	# Cameras	# Points
Bear	MC (PK10+FLEA)	4	2	215k
Beethoven	MP (PK10+FLEA)	7	3	311k
Buddha	SC (PK10+FLEA)	3	1	204k
Textured Pot	SC (EP910+REBEL)	3	1	3037k
Moving Face	SC (SX6+ Dragonfly)	3	1	34k
Moving Hand	SC (SX6+ Dragonfly)	3	1	34k
Moving Cloth	SC (SX6+Dragonfly)	3	1	63k

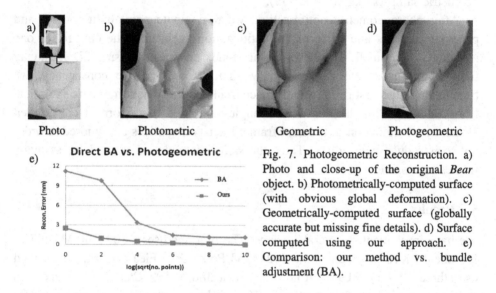

a)   b)   c)   d)

Photo   Photometric   Geometric   Photogeometric

e) Direct BA vs. Photogeometric

Fig. 7. Photogeometric Reconstruction. a) Photo and close-up of the original *Bear* object. b) Photometrically-computed surface (with obvious global deformation). c) Geometrically-computed surface (globally accurate but missing fine details). d) Surface computed using our approach. e) Comparison: our method vs. bundle adjustment (BA).

Fig. 7 demonstrates how our method captures the *Bear* object by combining a photometric surface (Fig. 7b) with that of a geometric surface (Fig. 7c) to yield a single higher quality surface (Fig. 7d). Fig. 7e shows the effectiveness of our method compared to a direct bundle adjustment (BA) solution. BA uses the photometric surface as initial guess and performs nonlinear optimization using Eq. 1 directly. We vary the number of geometrically-calibrated points by changing the level of Gray code patterns. The total number of reconstructed points remains the same (i.e., equal to the number of camera pixels). With the initial photometric surface having a reconstruction error of 15%, our approach reconstructs a surface with almost up to an order of magnitude less error as compared to BA. Error is shown as the difference, in percentile of the model diagonal, between the BA method and our best solution (i.e., all pixels used in self-calibration/photogeometric optimization).

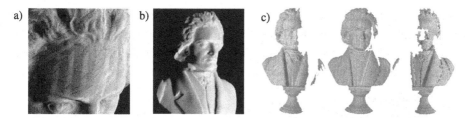

Fig. 8. Beethoven. a) A synthetically illuminated wireframe close-up view. b) A model rendered with light direction different from the actual one. c) Visible points for each of three projectors.

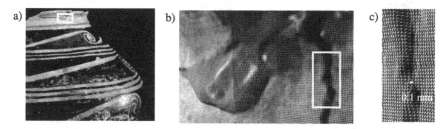

Fig. 9. Textured Pot. a) A textured model produced by using our method. b) A close-up view. c) An even closer view showing details beyond those possible using only geometric observations.

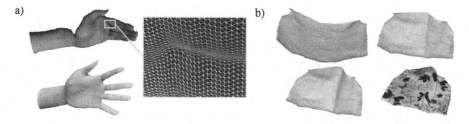

Fig. 10. Moving and Deforming Objects. a) Hand rendered with texture mapping and synthetic illumination. b) Novel views of a cloth for a static observer seeing the motion over time.

Fig. 8 shows the reconstructions of the *Beethoven* object. Fig. 8c visualizes the points visible from each of the three virtual cameras (i.e., projectors). As seen, the multi-view configuration provides better coverage of the object as compared to single-camera system. The processing pipeline and reconstruction results for the *Buddha* object were shown in Fig. 3. Fig. 9 demonstrates the case where a high-resolution digital SLR camera is used to capture a *Textured Pot*. Our approach uses the same hardware as standard structured-light but is fully self-calibrating and able to capture 3D models at the resolution of the camera (in this example camera resolution is 10x greater than projector resolution). The average triangle edge length is 0.09035 or about 0.1 mm.

Fig. 1-left and Fig. 10a show results of creating high quality models of moving and deforming facial expression and hand gestures. Our method uses only white dots and white photometric patterns; therefore, it is robust against colored and textured objects. Fig. 10b shows modeling results for a piece of deforming cloth. To provide color texture, we warp the closest photometric pattern frame from a chosen projector to the current frame.

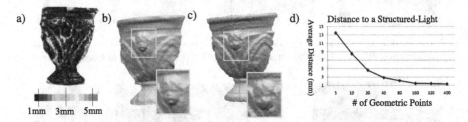

Fig. 11. Comparison with Structured Light (SL). a) Distance map between the photogeometric and SL reconstructions. b) Model using our method and c) using SL. d) Horizontal axis is the number of geometric points used. Vertical axis is the average distance to a SL reconstruction.

To show the effectiveness of combining sparse geometric data with dense photometric data, we implement a standard 16-frame Gray code dense structured-light method without sub-pixel optimization using the camera and one of the projectors. Then, we compare it with our six frame photogeometric method. For this setup and object distance, one camera pixel corresponds roughly to 0.8mm. The reconstructions using our method and the structured-light method are corresponded since they reside in the same camera. We visualize the per-pixel difference between points common to both reconstructions using a Jet color map (Fig. 11a). Most points reconstructed using our method are within 1mm of the positional measurements (near the limiting accuracy of the structured-light system). Moreover, details on the vase are better reconstructed using our method (Fig. 11b-c). To show the influence of the number of sparse geometric points, we compute photogeometric reconstructions of the vase using 5 to 400 geometric points and compare them to a structured light reconstruction. We plot the average errors distances in Fig. 11d. When the number of geometric points is small, the photogeometric reconstruction (close to a photometric-only reconstruction) has a big distortion as compared to structured light. The distortion is reduced when using more geometric points. Once 160 points (in this case) is surpassed, there is little benefit in using more. Hence, a small number of geometric frames might often be enough for high-quality modeling using our method.

6. Conclusions

We presented our self-calibrating and multi-view 3D photogeometric acquisition framework. Our technique uses the same hardware setup to obtain both photometric and geometric observations and to perform a unified acquisition and modeling effort. Our approach successfully extracts from the photometric data an increased amount of surface detail and extracts the metric accuracy from the geometric data in order to compensate for the global deformations typical of pure photometric methods. Our work can be applied to one of several configurations, including standard structured light but providing additional quality and avoiding the need for *a priori* calibration. Our work can also be used to capture moving and deforming objects by enforcing temporal smoothness in the photogeometric optimization. Since we use a robust method for both geometric and photometric processing, our system is fully automatic.

There are several current limitations to our method. First, our formulation for photometric reconstruction is restricted to capturing at most a hemisphere of the object. Second, the number of points reconstructed using our self-calibration is less than those of a stereo system because the points must be visible from the viewpoint of the camera and at least three projector viewpoints. Third, we use Lambertian photometric stereo with three lights. This leads to artifacts due to imperfect Lambertian reflectance and/or complex geometry (e.g., self-occlusion). Fourth, our method relies on features to compensate for motion when capturing moving and deforming objects. For future work, we would like to incorporate more sophisticated photometric methods (e.g., non-Lambertian photometric stereo[6, 14]), unobtrusive capturing (e.g., infrared), and real-time computing.

References

1. D. Aliaga and Y. Xu. A Self-Calibrating Method for Photogeometric Acquisition of 3D Objects, *IEEE Trans. on Pattern Analysis and Machine Intelligence*, 32(4):747-754, 2010.
2. P. Belhumeur, D. Kriegman, and A. Yuille. The Bas-Relief Ambiguity, *Intl. Journal of Comp. Vision*, vol. 35, no. 1, pp. 33-44, 1999.
3. J. Davis, D. Nehab, R. Ramamoorthi, and S. Rusinkiewicz. Spacetime Stereo: A Unifying Frame-work for Depth from Triangulation, *IEEE Trans. on Pattern Analysis and Machine Intelligence*, vol. 27, no. 2, pp. 296-302, 2005.
4. R. Frankot and R. Chellappa. A Method for Enforcing Integrability in Shape from Shading Algorithms, *IEEE Trans. on Pattern Analysis and Machine Intelligence*, 10(4):439-451, 1988.
5. R. Furukawa R and H. Kawasaki. Uncalibrated Multiple Image Stereo System with Arbitrarily Movable Camera and Projector for Wide Range Scanning, *In Proc. of Intl. Conf. on 3-D Imaging and Modeling*, pp. 302-309, 2005.
6. D. Goldman, B. Curless, A. Hertzmann, and S. Seitz S. Shape and Spatially-Varying BRDFs from Photometric Stereo, *In Proc. of IEEE Intl. Conf. on Comp. Vision*, pp. 341-348, 2005.
7. C. Hernandez, G. Vogiatzis, and R. Cipolla. Multi-view Photometric Stereo. *IEEE Trans. on Pattern Analysis and Machine Intelligence*, 30(3), 2008.

8. C. Hernandez, G. Vogiatzis, G. Brostow, B. Stenger, and R. Cipolla. Non-rigid Photometric Stereo with Colored Lights. *In Proc. of Intl. Conf. on Comp. Vision,* pp. 1-8, 2007.

9. S. Inokuchi, K. Sato, and F. Matsuda. Range Imaging System for 3-D Object Recognition, *In Proc. of Intl. Conf. on Pattern Recognition,* pp. 806-808, 1984.

10. A. Jones, A. Gardner, M. Bolas, I. McDowall, and P. Debevec. Simulating Spatially Varying Lighting on a Live Performance. *In Proc. of 3rd European Conf. on Visual Media Production,* pp. 127-133, 2006.

11. N. Joshi and D. Kriegman. Shape from Varying Illumination and Viewpoint. *In Proc. of Intl. Conf. on Comp. Vision,* pp. 1-7, 2007.

12. T. Koninckx and L. van Gool. Real-Time Range Acquisition by Adaptive Structured Light, *IEEE Trans. on Pattern Analysis and Machine Intelligence,* 28(3): 432-445, 2006.

13. J. Lim, J. Ho, M. Yang, and D. Kriegman. Passive Photometric Stereo from Motion, *In Proc. of Intl. Conf. on Comp. Vision,* pp. 1635-1642, 2005.

14. S. Mallick, T. Zickler, D. Kriegman, and P. Belhumeur. Beyond Lambert: Reconstructing Specular Surfaces Using Color, *In Proc. of IEEE Conf. on Comp. Vision and Pattern Recognition,* pp. 619-626, 2005.

15. D. Nehab, S. Rusinkiewicz, J. Davis, and R. Ramamoorthi. Efficiently Combining Positions and Normals for Precise 3D Geometry, *ACM Trans. on Graphics,* 24(3), 536-543, 2005.

16. H. Rushmeier and F. Bernardini. Computing Consistent Normals & Colors from Photometric Data, *In Proc. of Intl. Conf. on 3-D Imaging and Modeling,* pp. 99-108, 1999.

17. R. Sagawa, Y. Ota, Y. Yagi, R. Furukawa, N. Asada, and H. Kawasaki. Dense 3D Reconstruction Method using a Single Pattern for Fast Moving Object. *In Proc. of Intl. Conf. on Comp. Vision,* 2009.

18. P. Sen, B. Chen, G. Garg, S. Marschner, M. Horowitz, M. Levoy, and H. Lensch. Dual Photography, *ACM Trans. on Graphics,* 24(3):745-755, 2005.

19. D. Scharstein and R. Szeliski. High-Accuracy Stereo Depth Maps Using Structured Light, *In Proc. of IEEE Conf. on Comp. Vision and Pattern Recognition,* pp. 195-202, 2003.

20. P. Sturm. Critical Motion Sequences for the Self-calibration of Cameras and Stereo Systems with Variable Focal Length, *Image and Vision Computing,* 20(5-6):415-426, 2002.

21. D. Vlasic, P. Peers, I. Baran, P. Debevec, J. Popović, S. Rusinkiewicz, and W. Matusik. Dynamic Shape Capture using Multi-view Photometric Stereo, *ACM Trans. on Graphics,* 28(5), article 174, 2009.

22. T. Weise, B. Leibe, L. van Gool, Fast 3D Scanning with Automatic Motion Compensation, *In Proc. of IEEE Conf. on Comp. Vision and Pattern Recognition,* 2007.

23. R. Woodham, Y. Iwahori, and R. Barman. Photometric Stereo: Lambertian Reflectance and Light Sources with Unknown Direction and Strength, *UBC TR-91-18,* 1991.

24. Y. Xu and D. Aliaga. High-Resolution Modeling of Moving and Deforming Objects using Sparse Geometric and Dense Photometric Measurements. *In Proc. IEEE Conf. on Comp. Vision and Pattern Recognition,* 2010.

25. L. Zhang, B. Curless, A. Hertzmann, and S. Seitz. Shape and Motion under Varying Illumination: Unifying Structure from Motion, Photometric Stereo, and Multi-view Stereo, *In Proc. of Intl. Conf. on Comp. Vision,* pp. 618-625, 2003.

26. L. Zhang, B. Curless, and S. Seitz . Spacetime Stereo: Shape Recovery for Dynamic Scenes, *In Proc. of IEEE Conf. on Comp. Vision and Pattern Recognition,* pp. 367-374, 2003.

27. T. Zickler, P. Belhumeur, and D. Kriegman. Helmholtz Stereopsis: Exploiting Reciprocity for Surface Reconstruction, *In Proc. of European Conf. on Comp. Vision*, pp. 869-884, 2002.

28. T. Zickler, P. Belhumeur, and D. Kriegman. Toward a Stratification of Helmholtz Stereopsis, *In Proc. of IEEE Conf. on Comp. Vision and Pattern Recognition*, pp. 548-555, 2003.

29. T. Zickler. Reciprocal Image Features for Uncalibrated Helmholtz Stereopsis, *In Proc. of IEEE Conf. on Comp. Vision and Pattern Recognition*, pp. 1801-1808, 2006.

CHAPTER 3.4

MULTI-VIEW SURFACE RECONSTRUCTION BY QUASI-DENSE WIDE BASELINE MATCHING

Juho Kannala, Markus Ylimäki, Pekka Koskenkorva and Sami S. Brandt

Machine Vision Group, Computer Science and Engineering Laboratory,
University of Oulu, Finland
{jkannala,ylimakma,pkoskenk,sbrandt}@ee.oulu.fi

This chapter provides a review on correspondence growing techniques which have been used in multi-view stereo reconstruction problems. Typically these methods approach the problem of multi-view image matching by first determining a sparse set of feature correspondences between pairs of views and then iteratively expanding the matching regions. Sometimes such techniques are also referred by terms like match propagation, quasi-dense matching, or surface growing. Besides providing an overview of the research area, this chapter introduces a particular method, called quasi-dense wide baseline matching, which employs the best-first correspondence growing principle for matching pixels in views with substantially different viewpoints. In addition, the properties and performance of different methods are illustrated by examples and experiments with real images.

1. Introduction

Automatically acquiring a three-dimensional model of a scene from multiple photographic images is an important research area in computer vision. The basic geometric and computational principles for automatic multi-view reconstruction systems have been known for some time [20], and several such systems have already been built. The first systems of this kind were designed to use continuous video sequences as their input whereas some of the more recent approaches are able to acquire scene reconstructions from wide baseline image sets, where the views are captured at sparsely located viewpoints [36], or even unorganized image sets downloaded from Internet photo collections [48; 17].

However, despite the large number of previous research efforts and existing reconstruction systems, there are still many challenges and open problems in the field. In fact, image-based modeling continues to be an active research area. For example, one current focus area is the construction of large-scale reconstruction systems, which are able to reconstruct city-scale scenes from thousands or hundreds of thousands of images [1; 13; 12]. In addition, there have been efforts to improve the efficiency of video-based reconstruction pipelines in order to achieve real-time performance [42; 40; 39]. Also, one important research topic is to develop methods for representing three-dimensional models in a compact form that facilitates storage and transmission [32; 6].

The great interest towards image-based modeling is partly motivated by the recent trends in information and communication technology industry. For instance, in many applications and Internet-based services, there is an increased need for three-dimensional photorealistic visualization of scenes and objects, possibly with additional spatially localized data. Examples of such applications include navigation and driving assistance, entertainment and virtual tourism, architectural and environmental planning, and various forms of personal communication. In fact, image-based models are already utilized in many recent services by major software companies, such as Photosynth image stitching tool by Microsoft, and geographical mapping and visualization tools like Google Earth and Bing Maps.

In this article, we concentrate on image matching which is an essential part of any generic multi-view reconstruction system. Specifically, we focus on a particular correspondence growing method called *quasi-dense wide baseline matching* that was originally proposed in the articles [24] and [29] on which this chapter is partly based. This method tries to establish a large number of point correspondences, i.e. a quasi-dense set of matching points, between two or three views of a scene. As its input the algorithm takes a sparse set of corresponding regions between pairs of views and then iteratively expands these regions pixel-wise by using a best-first match propagation strategy similar to [33].

The quasi-dense approach can be used to match pairs of views of arbitrary possibly deforming scenes if the surfaces of the scene are sufficiently textured [25]. However, if the fundamental matrix is known for a pair of perspective views of a rigid scene, the method [24] can also utilize that to improve the reliability of matching. Further, in the three-view case, the method of [29] can be used if the trifocal tensor is known. In this chapter we mainly concentrate on cases where the cameras are perspective and their projection matrices are known. In fact, this is the usual problem setting in multi-view stereo [45], where the main task is to acquire a dense or semi-dense reconstruction by matching pixels between multiple calibrated views. Multi-view stereo is a key stage in a typical 3-D reconstruction pipeline after the camera motion and sparse scene structure have been estimated by structure from motion techniques [20; 16].

The structure of this chapter is as follows. First, in Section 2 we review related work on correspondence growing methods and multi-view reconstruction. Then, in Section 3, we describe the quasi-dense matching algorithms for image pairs and triplets and also discuss possible generalizations in order to match multi-view image sets with more than three views. Section 4 presents illustrative examples and experimental results with real images. The results are discussed in Section 5 and Section 6 concludes the chapter.

2. Related work

Matching multiple views of a scene in order to obtain a reconstruction of it is an old [20] but still timely problem in computer vision. Perhaps the most well studied topic in this area is two-view stereo matching. Much of the early research on two-view stereo was concentrated on the narrow baseline case [43] but some more recent works [24; 52] have focused on matching widely separated views by building upon the recent techniques in sparse wide baseline matching [38]. However, although some image-based modeling approaches use two-view matching as a basic building block and then combine point clouds from pairwise stereo depth maps into complete object models as a post process, there are also many multi-view stereo methods which are inherently designed to process more than two views [45; 14]. In the following, we aim to give an overview of two-view and multi-view stereo reconstruction methods with a particular focus on correspondence growing methods that are related to our approach.

2.1. *General overview*

Multi-view stereo methods can be characterized according to the scene representation that they use. Most approaches to represent three-dimensional scenes are based on voxels [30; 44], level-sets [11], polygonal surface meshes [10; 21], depth maps [49; 16; 15] or point clouds [14; 34]. Typically level set methods and voxel-based approaches represent scene geometry as a function on a regularly sampled 3-D grid. That is, in the case of level sets, the function encodes distance to the closest surface and, in voxel representation, it is a simple discrete occupancy function. The main problem with regular 3-D grids is their high memory usage which makes them inefficient for high-resolution representation of large scenes.

In contrast, polygonal meshes are efficient to store and render since they represent surfaces as a set of connected planar facets that may have unconstrained position in space. For example, triangular meshes have been used to model large scenes in some recent works [21]. Nevertheless, sometimes a depth map for each input view is the most convenient and natural representation for 3-D information. For example, in a real-time reconstruction system that operates from a moving vehicle, an efficient plane-sweeping method can be used to produce depth maps sequentially from successive views [15; 42]. However, multi-view depth maps are not necessarily mutually consistent and, in the end, some kind of a fusion process is often used to combine multiple depth maps into a single point cloud or surface mesh [37; 31]. In fact, besides meshes, point clouds (or clouds of surface patches) are representations that are also commonly used [14; 27]. Still, in many cases point clouds are eventually transformed into meshes in order to facilitate efficient storage and rendering [32; 6]. Finally, it is also common that systems use different 3-D representations in different stages of the reconstruction pipeline.

Besides scene representation, another essential characteristic of multi-view stereo methods is the type of reconstruction algorithm. In fact, reconstruction algorithms can be categorized into global and local methods. Global methods typically define a global cost function for shapes and then use some global optimization algorithm to recover a shape that minimizes the cost function. Examples of global methods include approaches based on volumetric Markov Random Field (MRF) models [3], which utilize graph cut optimization techniques [2], and variational approaches, which use convex optimization [28]. A common property of global methods is a relatively large memory usage and time complexity due to the use of a dense volumetric grid. Hence, such methods are not particularly suitable for large-scale scenes. However, there are also works which improve efficiency of graph cut based approaches by using adaptive grids [47; 31].

On the other hand, local matching methods often allow faster reconstruction with smaller memory requirements by, for example, dividing large image sets into subsets which may be processed separately and partly in parallel [42]. However, this may increase reconstruction errors in scene regions that are not particularly distinctive. That is, the global fitness of reconstruction may be compromised. Thus, the erroneously reconstructed parts need to be improved by some post-processing approach, using e.g. depth map fusion [42] or other ways of enforcing visibility consistency [14].

There are various local matching methods, including approaches based on the plane-sweep algorithm [8; 15] or correspondence growing [33; 14], for instance. Often these local approaches use scene representation based on surface patches, point clouds, or depth maps, which are flexible, as they can easily model various topologically complex structures and do not need a bounding box or a visual hull of the scene for initialization.

In addition, besides approaches that can be clearly categorized as local or global, there are semi-global methods [22], which do not directly aim in finding a global optimum of a global cost function but still consider certain long-range interactions between matched scene points (instead of just using their immediate neighborhood as in purely local methods). For example, the commonly used dynamic programming approach for narrow-baseline two-view stereo matching [43] can be seen as a semi-global method.

Overall, it should be noted that a complete multi-view reconstruction system may use different kinds of algorithms in different stages. For example, the recent approach [21] uses local matching of input views to extract a dense but possibly redundant point cloud, then applies a global approach for meshing the point cloud, and finally refines the resulting mesh model by iterative local optimization of a global cost function.

2.2. *Correspondence growing methods*

In the following sections we concentrate on a local image matching approach [24; 29], which we call quasi-dense wide baseline matching and which iteratively expands corresponding image regions by the match propagation algorithm [33]. However, somewhat similar ideas of correspondence growing have been used in other works as well, as briefly reviewed below.

One of the first correspondence growing methods is [41], which gradually expands matching image patches in a pair of views by using iterative alignment of patches [18] during each expansion step. Like [41], our approach uses the best-match-first growing stategy but we do not perform iterative refinement of the matched patches during growing. Also, [41] applies only for pairs of views whereas [29] extends the best-first matching approach to triplets of views if the trifocal tensor is known.

Our growing algorithm is similar to the match propagation algorithm of [33], which also avoids iterative patch refinements during growing and imposes a uniqueness constraint (i.e. one-to-one matching) and a disparity gradient limit simultaneously. However, unlike [33], our implementation [24] can be directly used for wide baseline image pairs. This is achieved by using an affine transformation model for the local patches instead of a translational model.

Another relatively recent correspondence growing method is [5], which is also inspired by [33]. However, unlike [33] and [24], [5] solves a global optimization task, which allows to reduce matching errors and prevents matching ambiguous structures, e.g. due to repetitive texture patterns. Nevertheless, the implementation in [5] requires a rectified stereo image pair as input and, hence, it can not be used for images of nonrigid scenes. Also, in practice, the ambiguities caused by repetitive patterns can be reduced by using more than two views, and even greedy local matching [24] may perform well in such cases [29].

Recently, correspondence growing algorithms have also been used for true multi-image matching with datasets that contain more than two or three views [19; 14]. These methods expand surface patches in 3-D space so that the patches are directly matched between multiple views. The advantage of such patch-based approaches is their flexibility in modeling both small, compact objects and large, complex scenes, and even crowded scenes where moving obstacles appear in multiple images of a static structure of interest [14].

Perhaps one of the most widely used patch-based multi-view stereo methods is [14], which is publicly available in source form and uses repeated match expansion and filtering stages for reconstruction. The method has produced good results with benchmark datasets [45] and it has also been used for large datasets [13]. However, [14] does not use a best-first growing strategy as we do. In fact, our results indicate that the best-first strategy allows to acquire good matches with a single growth stage and without repeated expansion and filtering steps. Hence, one key aspect in this chapter is to discuss the possibilities to further advance the use of correspondence growing methods for direct multi-view matching that is both accurate and efficient.

3. Quasi-dense wide baseline matching

In this section, we describe the two-view and three-view quasi-dense matching algorithms which were originally proposed in [24] and [29], respectively. The algorithms are based on the match propagation algorithm [33] which is extended to be applicable for wide baseline images whose viewpoints differ substantially.

3.1. *Two-view matching*

Given two views, $\mathcal{I}_1$ and $\mathcal{I}_2$, and optionally the associated fundamental matrix $\mathcal{F}_{21}$, our matching approach produces a quasi-dense set of point correspondences between the two views by growing a sparse set of seed matches, which are determined by matching affine covariant regions [24; 38].

Hence, the method contains two stages: the initial matching stage and the growth stage. The output of the initial matching stage is a set of seed matches $\{\mathbf{s}^i\}_i$, where each seed contains image coordinates $\mathbf{x}_a$ and $\mathbf{x}_b$, which denote the centroids of the matched regions [38], and an affine transformation matrix $\mathbf{A}_{ab}$, which approximates the local geometric transformation between the views. For each seed, the index $a \in \{1, 2\}$ indicates the reference view, and b is the other view. The reference view is determined so that the affine transformation $\mathbf{A}_{ab}$ from a to b is magnifying, i.e. $|\det \mathbf{A}_{ab}| \geq 1$ [24]. Further, each seed is associated with a texture similarity score s_{ab} and intensity variance score v. The zero-mean normalized cross-correlation (ZNCC) of geometrically normalized image patches is used as the similarity measure, and the score v is set equal to the minimum intensity variance of the two patches. The data structure for the two-view seeds is summarized in Definition 1. The same structure is used also for the grown matches.

In the growth stage, the seeds are sorted into a priority queue $\mathcal{Q}$ according to their similarity scores and then propagated by iterating the following steps:

(i) The seed $\mathbf{s}$ with the best score is removed from $\mathcal{Q}$.
(ii) New candidate matches are searched nearby $\mathbf{s}$ by using $\mathbf{s}.\mathbf{A}_{ab}$ for the geometric normalization of local image patches.
(iii) The candidates, which have a sufficiently high similarity and which satisfy the disparity gradient limit and the epipolar constraint (optional), are added to $\mathcal{Q}$ and to the list of matches after updating their affine transformation estimates. The corresponding pixels in the matching tables are marked as reserved.

The geometric normalization of patches and the update of affine transformations are detailed in [24]. Further, it should be noted that, for each grown match, the indices a and b are determined from the updated transformation matrix, i.e., the role of views 1 and 2 may be swapped during propagation. The outline of the growth stage is described in pseudo-code in Algorithm 1.

Updating affine transformation parameters for new matches allows the propagation to adapt to variations in the orientation and pose of surfaces as the matching

---

Definition 1: Data structure for two-view seed matches

```
struct twoviewseed{ int a, b;
                    double x_a, x_b, A_ab, s_ab, v; };
```

struct twoviewseed$\{$ int a, b;
double $\mathbf{x}_a$, $\mathbf{x}_b$, $\mathbf{A}_{ab}$, s_{ab}, v; $\}$;

Algorithm 1: Two-view match propagation

Input: images $\mathcal{I}_1$, $\mathcal{I}_2$, two-view seed matches $\mathcal{S}_{12}$, thresholds ϵ_d, ϵ_e, t, t_u, z, z_u,
   and, optionally, fundamental matrix $\mathcal{F}_{21}$

Output: list of matches $\mathcal{M}$, matching tables $\mathcal{J}_1$, $\mathcal{J}_2$

1 Initialize $n=0$, $\mathcal{M}=\emptyset$, $\mathcal{J}_k(\mathbf{p})=0$ for all k, $\mathbf{p}$

2 Compute pairwise similarity scores $\mathbf{s}.s_{ab}$ and uniformity
   scores $\mathbf{s}.v$ for all seeds $\mathbf{s}$, $[\mathbf{s}.s_{ab}, \mathbf{s}.v] = \text{sim}(\mathbf{s}, \mathcal{I}_{\mathbf{s}.a}, \mathcal{I}_{\mathbf{s}.b})$

3 Sort the seeds according to the scores $\mathbf{s}.s_{ab}$

4 Initialize priority queue $\mathcal{Q}$ with sorted seeds

5 while $\mathcal{Q}$ not empty

6   Draw the seed $\hat{\mathbf{q}} \in \mathcal{Q}$ with the best score $\hat{\mathbf{q}}.s_{ab}$

7   Set $a=\hat{\mathbf{q}}.a$ and $b=\hat{\mathbf{q}}.b$
     (In the following, $\mathcal{I}_a$ and $\mathcal{I}_b$ define new seeds)

8   for each new match $\mathbf{q}^i$ nearby $\hat{\mathbf{q}}$ which satisfies the disparity gradient limit ϵ_d
     and, optionally, the epipolar constraint ϵ_e

9   Set $\mathbf{q}^i.s_{ab} = -\infty$

10   if $\mathcal{J}_a(\text{round}(\mathbf{q}^i.\mathbf{x}_a))=0$ & $\mathcal{J}_b(\text{round}(\mathbf{q}^i.\mathbf{x}_b))=0$

11         $[\mathbf{q}^i.s_{ab}, \mathbf{q}^i.v] = \text{sim}(\mathbf{q}^i, \mathcal{I}_a, \mathcal{I}_b)$

12   end for

13   Sort matches $\mathbf{q}^i$ according to the scores $\mathbf{q}^i.s_{ab}$

14   for each $\mathbf{q}^i$ satisfying $\mathbf{q}^i.s_{ab} \geq z$ and $\mathbf{q}^i.v \geq t$

15   Set $n=n+1$

16   if $\mathbf{q}^i.s_{ab} \geq z_u$ and $\mathbf{q}^i.v \geq t_u$

17   Update $\mathbf{q}^i.\mathbf{A}_{ab}$, and thereafter $\mathbf{q}^i.a$ and $\mathbf{q}^i.b$

18   Set $\mathcal{Q}=\mathcal{Q}\cup\{\mathbf{q}^i\}$ and $\mathcal{M}=\mathcal{M}\cup\{\mathbf{q}^i\}$

19   Set $\mathcal{J}_a(\text{round}(\mathbf{q}^i.\mathbf{x}_a))=n$, $\mathcal{J}_b(\text{round}(\mathbf{q}^i.\mathbf{x}_b))=n$

20   end for

21 end while

---

expands further from seed regions. The update of the affine transformation matrix is implemented via a simple non-iterative update rule which is based on local second-order intensity moment matrices of the images [24]. In addition to intensity moments, the update rule requires a pair of corresponding directions in the images. Such directions can be obtained from the fundamental matrix or, if it is not available, they can be estimated from local image gradients [25]. In the latter case, the match propagation may even be used to track non-rigid deformations of surfaces [25]. The frequency of affine adaptation is controlled by parameters z_u and t_u in Algorithm 1 which restrict the update to occur only for textured image patches whose similarity scores exceed a threshold z_u. If $z_u > 1$, the update is omitted always and the algorithm operates in a non-adaptive mode.

In summary, as shown in Algorithm 1, the result of two-view match propagation is a list of grown matches $\mathcal{M}$ and two matching tables $\mathcal{J}_1$ and $\mathcal{J}_2$, which have the same size as images $\mathcal{I}_1$ and $\mathcal{I}_2$, respectively. The nonzero values in $\mathcal{J}_1$ and $\mathcal{J}_2$ indicate the pixels which are nearby to the sub-pixel matches of $\mathcal{M}$. That is, the closest pixel to a given coordinate vector $\mathbf{x}$ is $\mathbf{p} = \text{round}(\mathbf{x})$, and a nonzero value $\mathcal{J}_k(\mathbf{p})$ is an index to the corresponding item in $\mathcal{M}$. The value $\mathcal{J}_k(\mathbf{p}) = 0$ indicates that pixel $\mathbf{p}$ is not matched.

3.2. *Three-view matching*

The three-view matching approach [29] builds on the ideas of the two-view method [24] but there are some additions and modifications which improve the robustness of matching for view triplets when the trifocal tensor is known. Thus, although the two-view method can be used without knowing the fundamental matrix, the three-view method always requires the trifocal tensor as input. Another difference to the two-view method is that the ordering of seeds in the priority queue $\mathcal{Q}$ is based on a total score s which combines two pairwise similarity scores, s_{ab} and s_{ac}, between the reference view a and the other two views. Hence, at each propagation step, the seed with the best total score is grown.

The three-view method is shown in pseudo-code in Algorithm 2, which is quite similar to Algorithm 1. However, there are certain additional steps (i.e. lines 3, 4, 5, 21, 22, 25, and 26), which are detailed in the following. The data structure for three-view matches is given in Definition 2.

First, since the input to Algorithm 2 is a set of two-view seeds, as in the two-view method, the seeds have to be transformed to three-view seeds by using trifocal transfer [20]. That is, we define a function called `transfer`, which transforms a match $(\mathbf{x}_a, \mathbf{x}_b)$ to the third view, indexed by c, and also computes the local affine transformation $\mathbf{A}_{ac}$ between the reference view a and the view c (line 3 in Algorithm 2). The function is implemented so that it uses $\mathbf{x}_a$, $\mathbf{x}_b$ and $\mathbf{A}_{ab}$ to define three corresponding points in the views a and b and then transforms this local affine basis to the third view by trifocal transfer [20]. Thereafter, $\mathbf{A}_{ac}$ may be solved from the three point correspondences between a and c. Finally, given $\mathbf{x}_c$ and $\mathbf{A}_{ac}$, one may also evaluate the local similarity s_{ac} between views a and c.

Given a three-view seed $\mathbf{s}$, the total score $\mathbf{s}.s$, on which the ordering in $\mathcal{Q}$ is based, combines the pairwise similarities between a and the other two views. This allows the three-view method to perform better than a combination of pairwise

---

Definition 2: Data structure for three-view seed matches

```
struct threeviewseed{ int a, b, c;
                      double xₐ, x_b, x_c, Aₐ_b, Aₐ_c, sₐ_b, sₐ_c, s, v; };
```

$\texttt{struct threeviewseed}\{$ int a, b, c;

double $\mathbf{x}_a$, $\mathbf{x}_b$, $\mathbf{x}_c$, $\mathbf{A}_{ab}$, $\mathbf{A}_{ac}$, s_{ab}, s_{ac}, s, v; $\}$;

Algorithm 2: Three-view match propagation

Input: images $\mathcal{I}_1$, $\mathcal{I}_2$, $\mathcal{I}_3$, two-view seeds $\mathcal{S}_{12}$, $\mathcal{S}_{13}$, $\mathcal{S}_{23}$, thresholds ϵ_d, ϵ_e, t, t_u, z, z_u, $\tilde{z}$,
and trifocal tensor $\mathcal{T}_1^{23}$

Output: list of matches $\mathcal{M}$, matching tables $\mathcal{J}_1$, $\mathcal{J}_2$, $\mathcal{J}_3$

1   Initialize $n=0$, $\mathcal{M}=\emptyset$, $\mathcal{J}_k(\mathbf{p})=0$ for all k, $\mathbf{p}$

2   Compute pairwise similarity scores $\mathbf{s}.s_{ab}$ and uniformity
    scores $\mathbf{s}.v$ for all seeds $\mathbf{s}$, $[\mathbf{s}.s_{ab}, \mathbf{s}.v] = \texttt{sim}(\mathbf{s}, \mathcal{I}_{\mathbf{s}.a}, \mathcal{I}_{\mathbf{s}.b})$

3   Extend the two-view seeds to three views by trifocal
    transfer: $[\mathbf{s}.\mathbf{x}_c, \mathbf{s}.\mathbf{A}_{ac}] = \texttt{transfer}(\mathbf{s}, \mathcal{T}_1^{23})$

4   Compute pairwise similarity scores $\mathbf{s}.s_{ac}$ for all seeds $\mathbf{s}$

5   Combine similarity scores, $\mathbf{s}.s = \texttt{score}(\mathbf{s}.s_{ab}, \mathbf{s}.s_{ac}, z)$

6   Sort the seeds according to the scores $\mathbf{s}.s$

7   Initialize priority queue $\mathcal{Q}$ with sorted seeds

8   while $\mathcal{Q}$ not empty

9   Draw the seed $\hat{\mathbf{q}} \in \mathcal{Q}$ with the best score $\hat{\mathbf{q}}.s$

10   Set $a = \hat{\mathbf{q}}.a$ and $b = \hat{\mathbf{q}}.b$
     (In the following, $\mathcal{I}_a$ and $\mathcal{I}_b$ define new seeds)

11   for each new match $\mathbf{q}^i$ nearby $\hat{\mathbf{q}}$ which satisfies the disparity gradient limit ϵ_d
     and the epipolar constraint ϵ_e

12   Set $\mathbf{q}^i.s_{ab} = -\infty$

13   if $\mathcal{J}_a(\texttt{round}(\mathbf{q}^i.\mathbf{x}_a))=0$ & $\mathcal{J}_b(\texttt{round}(\mathbf{q}^i.\mathbf{x}_b))=0$

14       $[\mathbf{q}^i.s_{ab}, \mathbf{q}^i.v] = \texttt{sim}(\mathbf{q}^i, \mathcal{I}_a, \mathcal{I}_b)$

15   end for

16   Sort matches $\mathbf{q}^i$ according to the scores $\mathbf{q}^i.s_{ab}$

17   for each $\mathbf{q}^i$ satisfying $\mathbf{q}^i.s_{ab} \geq z$ and $\mathbf{q}^i.v \geq t$

18   Set $n = n+1$

19   if $\mathbf{q}^i.s_{ab} \geq z_u$ and $\mathbf{q}^i.v \geq t_u$

20   Update $\mathbf{q}^i.\mathbf{A}_{ab}$, and thereafter $\mathbf{q}^i.a$ and $\mathbf{q}^i.b$

21   Do trifocal transfer for $\mathbf{q}^i$ and compute $\mathbf{q}^i.s_{ac}$, $\mathbf{q}^i.s$

22   if $\mathbf{q}^i.s_{ac} \geq \tilde{z}$ {

23   Set $\mathcal{Q} = \mathcal{Q} \cup \{\mathbf{q}^i\}$ and $\mathcal{M} = \mathcal{M} \cup \{\mathbf{q}^i\}$

24   Set $\mathcal{J}_a(\texttt{round}(\mathbf{q}^i.\mathbf{x}_a))=n$, $\mathcal{J}_b(\texttt{round}(\mathbf{q}^i.\mathbf{x}_b))=n$ }

25   if $\mathcal{J}_c(\texttt{round}(\mathbf{q}^i.\mathbf{x}_c))=0$ & $\mathbf{q}^i.s_{ac} \geq z$

26   Set $\mathcal{J}_c(\texttt{round}(\mathbf{q}^i.\mathbf{x}_c))=n$

27   end for

28 end while

propagations. The scoring function is defined by

$$\texttt{score}(s_{ab}, s_{ac}, z) = \sum_{j \in \{b,c\}} \max \left(0, 1 - \frac{(s_{aj} - 1)^2}{(z-1)^2} \right), \qquad (1)$$

which is a positive function on $[-1, 1]^3$. It is nonzero if either s_{ab} or s_{ac} exceed $z \leq 1$, and it obtains the largest values when they both exceed z and are close to 1.

Most of the computation time of Algorithm 2 is spent in the while loop. The first for loop is the same as in Algorithm 1. Thus, most of the bad candidate matches are rejected already on the basis of the pairwise score s_{ab} and the trifocal transfer and the evaluation of the total score are not necessary for them. In fact, only a fraction of the candidates survive to the second for loop, and hence, the three-view method is only slightly slower than the two-view method.

Finally, the last modifications are related to the acceptance of candidate matches and to the update of the matching tables (lines 22, 25, and 26 in Algorithm 2). Depending on the parameter settings, one may require that an accepted match must be visible in at least one or two pairs of views. That is, by setting $\tilde{z} = -1$, a candidate match is always accepted if its pairwise similarity score s_{ab} exceeds a threshold z and, on the other hand, setting $\tilde{z} = z$ implies that a match is accepted only if both s_{ab} and s_{ac} exceed z (line 22). However, in both cases, a new match is added to the matching table of the third view only if $s_{ac} \geq z$ and the corresponding pixel is not already reserved (lines 25 and 26).

3.3. *General multi-view matching*

As described in Algorithm 2, our current implementation of quasi-dense wide base-line matching can process at most three views simultaneously. However, this does not exclude applying the proposed approach to multi-view datasets with a large number of images, because there are several methods for combining multiple depth maps extracted from different view triplets [9; 37; 32]. Hence, Algorithm 2 could be used in generic multi-view stereo problems by first dividing large image sets into overlapping subsets of three images, then matching these image triplets and converting the obtained three-view matches to depth maps or point clouds, which are finally combined to a single model. In fact, decomposition into manageable subsets is indispensable for very large image sets but it may be advantageous also for smaller datasets in order to improve efficiency [42; 23; 13]. If the image subsets are processed in parallel and the resulting partial reconstructions are merged efficiently, it is possible to implement systems that reconstruct city-scale scenes from thousands or even millions of images within the span of a day on a single PC [13; 12].

Nevertheless, although there are multi-view reconstruction systems that build on two-view matching [51], in some cases it might be useful to be able to perform quasi-dense matching directly for a larger number of views than just two or three. This is the case with large unstructured photo collections since it may be practically not feasible to process all possible view triplets but it is still not obvious how the optimal division into triplets should be done [13]. Hence, analogously to [14], it could be advantageous to extend Algorithm 2 for more than three views. In fact, given camera matrices, this may be relatively straightforward to do since it is easy to transfer the two-view matches to arbitrary number of views by the trifocal transfer (lines 3 and 21 in Algorithm 2) and define the score (1) as the sum of several pairwise terms. This would allow generalization to multiple views. However, our current implementation covers only view triplets and extension to more general settings could be seen as an interesting topic for future research.

4. Examples and experiments

We illustrate the performance of quasi-dense wide baseline matching in experiments with real images. The experiments in Sections 4.1 and 4.2 focus on two-view matching, Section 4.3 compares two-view and three-view approaches, and Sections 4.4 and 4.5 further illustrate three-view matching results.

4.1. *A simple example*

In our first experiment we demonstrate the two-view approach of Algorithm 1 by matching two views without using any additional prior knowledge, i.e., the fundamental matrix is not known a priori. However, in order to get quantitative accuracy estimates, we use sample views which are related by a homography and which were used in [38]. The example image pair is shown in Figure 1 and there is a significant change of scale between the views due to optical zooming. In addition, a seed match, i.e. a pair of elliptical regions, is also illustrated in Figure 1. The seed regions were extracted with the Hessian-Affine region detector [38] and automatically matched using the SIFT descriptor [35].

The match propagation was started from the single seed shown in Figure 1. The growing process resulted in 35589 matches and took 12 seconds on a 1GHz processor. Here we used only one seed in order to illustrate the fact that often already a one or a few correct matches are sufficient, but usually several tentative seed matches can be used since the algorithm is robust to outliers.

Given the grown matches, we fitted a homography to them by using a RANSAC-based estimation procedure. The resulting homography estimate was used to register the images on top of each other and the corresponding difference image is shown in Figure 1(b). Also, the grown quasi-dense matches are illustrated in Figure 1(e), where they are colored according to their homographic transfer error in the second image [20]. The majority of point correspondences have a displacement less than a pixel which indicates that the matches fit well to the estimated homography.

Interestingly, it also seems that our registration result is better than the homography estimate provided by the authors of [38], which is visualized in the last column of Figure 1. Indeed, the difference image in Figure 1(c) shows larger

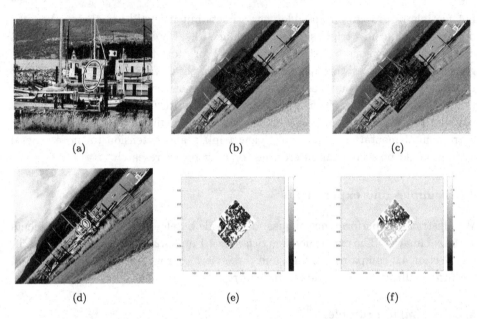

Fig. 1. Matching a pair of views which are related by a homography. (a) and (d): An elliptical region (a seed match) is shown on the original views, which are acquired by rotating and zooming a camera fixed on a tripod [38]. (b) and (c): The difference images obtained by aligning the views with homographies estimated by us and the authors of [38], respectively. (e) and (f): The quasi-dense matches grown from the single seed match are colored according to their distance from the location predicted by the two homographies, ours and theirs [38], respectively. (Image regions that are not visible in the first view have grayvalue 6 and the unmatched regions are white.)

intensity discrepancies than Figure 1(b). In addition, Figure 1(f) shows that the obtained quasi-dense matches are not consistent with the provided homography estimate in the lower left corner of the first image. This suggests that the homography estimate of [38] is not accurate there. Hence, we may note that even such a simple task as homography estimation is error-prone when registration landmarks are sparsely distributed and inaccurately localized. Moreover, our result shows that the quasi-dense approach can clearly improve image registration accuracy in such cases.

4.2. *An experiment with two views of a pair of planes*

The second experiment illustrates the performance of our two-view matching method under different parameter settings. The image pair used in this experiment is the same as in [24] and it is illustrated in Figures 2(a) and 2(d), which show two views of a scene containing two planes: a paper map on a table and a calibration plane orthogonal to the plane of the map. This pair of images is particularly suitable for algorithm evaluation since the calibration plane allows accurate estimation of the homographies which describe the mappings of the planes between

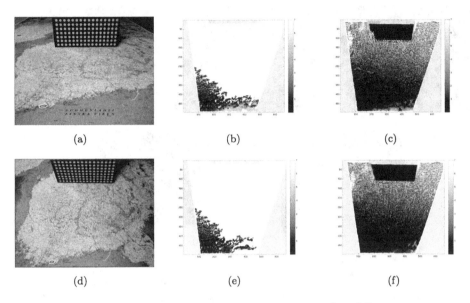

(a)   (b)   (c)

(d)   (e)   (f)

Fig. 2. Matching results for a pair of views of two planes using four different propagation settings. The left column shows the original images and one seed match (small ellipses in the lower left corner of the images). The last two columns illustrate the grown matches obtained by four parameter settings: (b) non-adaptive, (e) non-adaptive with epipolar constraint, (c) adaptive, and (f) adaptive with epipolar constraint. In each case, the matching pixels are colored according to their distance from the true location determined by the known homography of the respective plane. The values over 5 are suppressed to 5, and the non-common image area has grayvalue 6.

the two views, and hence, the obtained quasi-dence matches can be verified with the homographies after propagation. In addition, it can be seen that there is a clear perspective distortion between the views so that a global affine transformation is not a good approximation for the homography of either plane. This allows to demonstrate the local affine adaptation capabilities of Algorithm 1.

The results obtained by Algorithm 1 using four different parameter settings are illustrated in Figure 2. In each case the match propagation was started from a single seed match, which is also shown in Figure 2 and which was extracted using the Hessian-Affine detector [38]. The pictures in the middle column of Figure 2 illustrate the grown quasi-dense matches obtained with the non-adaptive setting, i.e., without updating the affine transformation during propagation. The matches do not grow very far from the seed match because the initial affine transformation estimate is not accurate there due to perspective distortion. By comparing Figures 2(b) and 2(e), it can also be seen that imposing the epipolar constraint reduces the number of erroneous matches.

The adaptive match propagation results, obtained with and without epipolar constraint [24; 25], are shown in the last column of Figure 2, where the grown matches cover almost all the pixels in areas that are visible in both images. Again,

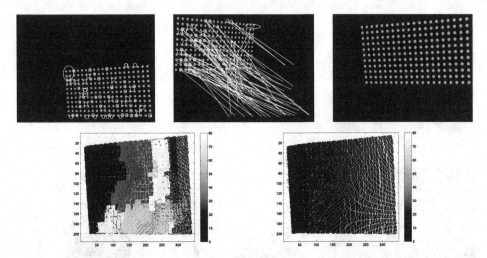

Fig. 3.   Two-view and three-view matching of a plane with repetitive texture. Top: Three views and the elliptical seed regions extracted from the first pair of views. The flow vectors in the second view indicate the tentative seed region correspondences. There are only four correct seeds (cyan colored). Bottom: Disparity maps for the two-view (left) and three-view methods (right). Matched pixels are colored according to their distance to the correct corresponding location.

the epipolar constraint reduces false matches, as expected. The good coverage of matches shows that the update of affine transformation parameters on the basis of local texture properties has been successful and allows the single seed match to propagate into regions where the local geometric transformation between the views differs from the initial one. It is also interesting to note that the matching has expanded from the horizontal plane of the map to the vertical plane of the calibration object although there is a discontinuity in surface orientation. Further, the accuracy of grown matches does not decrease during propagation, i.e., majority of matched points have an error less than a pixel on both planes which indicates that errors do not accumulate and adaptation is stable.

4.3.   *Matching three views of a plane with repetitive texture*

In the third experiment, we compare our two-view and three-view matching approaches by using several view triplets of a planar calibration pattern. As the sample views in Figure 3 show, the calibration pattern has a repetitive texture, which consists of a regular grid of white dots on a black background. Matching such a repetitive pattern is error-prone because the epipolar constraint alone is not sufficient to resolve the ambiguities in local matching. Thus, it is expected that three-view approach performs better than using just two views and our experiment aims to verify this hypothesis. In fact, because we know the true structure of the scene, it is easy to recover the correct homographies from manually picked correspondences and use them to assess the accuracy of obtained matches afterwards.

In detail, the experiment was carried out as follows. We selected a triplet of views, extracted seed matches from two of them [38], and performed both the two-view growing and the three-view growing with the same seeds. This was repeated for 100 triplets. Usually there were many incorrect seeds because the epipolar constraint was not effective in removing them due to repetitive texture. This is illustrated in Figure 3 where only four seeds are correct. However, also in this case, the three-view method (Algorithm 2) produced almost errorless matching whereas the two-view method (Algorithm 1) made many errors. This shows that the trifocal constraint allows to greatly improve the robustness of matching for repetitive textures. The results for all 100 triplets are shown in Table 1, where the total number of matches produced is approximately the same for both methods (i.e. 5.7 million). However, as the quartiles of the error distribution indicate, the matches produced by Algorithm 1 have a larger error. Thus, the two-view method failed in many cases whereas the three-view method was able to produce accurate matches for most of the cases. In this experiment the running time of Algorithm 2 was about 1.2 times the time of Algorithm 1. Hence, the three-view result is usually better than the result of two pairwise propagations and can be computed faster.

4.4. *Reconstruction of a piecewise planar scene*

Our fourth experiment compares the three-view matching method of Algorithm 2 with the multi-view stereo method of [14], which uses patch-based surface representation and relies on repeated patch expansion and filtering stages. The original implementation of [14] is available in source format and has produced good reconstruction results on publicly available benchmark datasets [46; 50].

As our current implementation can process at most three views simultaneously, we concentrated on comparing the basic correspondence growing processes on a dataset of three views. That is, we used our own images of a piecewise planar scene, whose structure can be accurately recovered from the images utilizing prior knowledge about the planes. However, this prior knowledge was only used to obtain the ground truth, i.e. a triangular mesh model of the scene, which was then used for accuracy evaluation of the point cloud reconstructions obtained by the two methods.

The images used in the experiment are illustrated in Figure 4 (top), and they show three planes, i.e., a calibration object consisting of two orthogonal planes on a wooden floor. There are regular patterns of circular dots on the planes which allowed to recover the pose of each plane as well as the camera matrices associated

Table 1.   Experiment with repetitive texture. Quartiles of the error distribution are computed from 100 view triplets of the same kind as the triplet in Figure 3.

Method	No. points	1st quartile	Median	3rd quartile
Two views	5664285	0.33	0.74	11
Three views	5676160	0.079	0.27	0.65

Fig. 4. Top: Three views of a piecewise planar scene. Left: Point cloud obtained by our three-view algorithm. Right: Point cloud computed by Furukawa's approach [14]. Accuracy of these two point clouds is illustrated by the curves with symbols × and ○ in Figure 5, respectively.

to the three views. Given the views and the camera matrices, we reconstructed a point cloud using both Furukawa's approach and our approach (Algorithm 2). In the latter case, the seed matches were automatically extracted and matched using the Hessian-Affine detector [38] and SIFT descriptor [35], and after the correspondence growing by Algorithm 2 each grown match was triangulated from its two reference views (indexed by a and b in Algorithm 2).

The obtained point clouds are illustrated in Figure 4 (bottom). Our reconstruction consists of 155107 points and the match propagation stage took 228 seconds. Furukawa's point cloud has 119932 points, i.e. the centroids of reconstructed patches, and the processing time was 616 seconds, which also includes the time used for initial feature matching. However, as the running time of initial matching is negligible compared to that of patch expansion and filtering, it can be deduced that our correspondence growing is faster than Furukawa's. This is expected as we use only one expansion stage and we do not refine the alignment of matched patches via non-linear optimization during growing as in [14].

The accuracy of reconstructed points is evaluated by computing their distances to the ground truth surface, which is a triangular mesh representing the imaged planes. Further, each matched pixel in the three images corresponds to a point in the point cloud and the associated error is the distance of this point from the ground truth surface. Thus, each pixel is assigned with an error value, and we may assess reconstructions by comparing the error distributions of matched pixels. That is, we discretize the error values into 13 bins and compute the corresponding cumulative error histograms [50], which are visualized in Figure 5. By comparing the curves with symbols × and ○, we may see that the accuracies of points in our

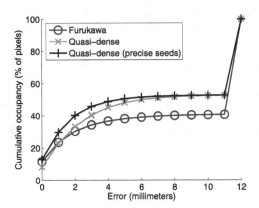

Fig. 5. Evaluation of accuracy for the point cloud reconstructions in Figure 4. Each matched pixel corresponds to a point in the point cloud and the associated error is the distance of this point from the ground truth surface. The curves illustrate the proportion of pixels whose error is less than a threshold. The curves are plotted using discrete error histograms of 13 bins where the last bin contains all pixels with an error > 11.5 mm, also the unmatched ones.

point cloud and Furukawa's point cloud are quite similar but our point cloud is denser. However, this is not very significant difference as the overall coverage of Furukawa's reconstruction is also good. The proportion of matched pixels is much less than 100 % in all cases because both methods used such parameter settings that required the reconstructed patches to be visible in all the three images. Hence, only those scene regions that are unoccluded in all three views could be reconstructed.

In addition, there is also a third curve in Figure 5, denoted by + symbol, and it illustrates the result obtained by our approach when all the circular dots of calibration patterns were used as seed regions (initialized with the correct pairwise affine transformations obtained from the ground truth planes). Although this is not a realistic application scenario for unknown scenes, the result indicates that it might be possible to further improve the accuracy of our approach by iteratively refining the seeds before growing them. In fact, this could be one direction for future developments as such a refinement of seeds would probably not decrease the overall computational efficiency very much since the number of seeds is usually small compared to the final number of grown matches.

4.5. *Examples with benchmark datasets*

In the last experiment, we compare our three-view matching method [29] and Furukawa's method [14] using benchmark datasets from [46]. The datasets contain images of two objects, called Dino and Temple. Three sample images of both objects are shown in Figures 6 and 7. There is also an online evaluation service [46] which allows to compare submitted reconstructions (i.e. mesh models) to ground-truth models of the objects obtained via a laser scanning process. However, since

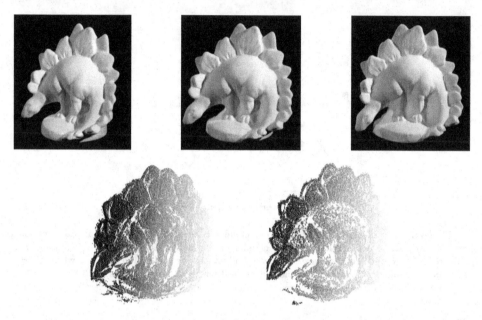

Fig. 6. Top: Three views from Dino dataset [46]. Bottom: Depth maps produced by our approach (left) and Furukawa's approach [14] (right). Note that the points in the depth maps are shown from the viewpoint of the second camera but they are colored according to their distance from the image plane of the first camera.

this chapter mainly presents a basic algorithm for match expansion in two-view and three-view cases and we do not discuss depth map fusion or mesh generation which would be both needed for generating complete object models, we confine ourselves to comparing three-view matching results qualitatively via visualized depth maps.

The experiment was carried out in a similar manner as the previous one in Section 4.4. The results are illustrated with depth maps in Figures 6 and 7 and they basically confirm the findings of the previous section. That is, our approach and Furukawa's approach provide comparable results but our approach is faster. For example, in the Dino example in Figure 6, the number of reconstructed points is 123633 in our point cloud and 90013 in Furukawa's point cloud. The correspondence growing stage of our approach took 442 seconds whereas Furukawa's approach took 621 seconds alltogether.

On the other hand, match expansion is implemented in different ways in the two methods [29; 14] and this may sometimes imply differences to the resulting point clouds. Thus, although the results are generally comparable there may be differences in some cases. For instance, as some parts of the object are not visible in all three views in Figure 7, we ran our algorithm using such parameters that the matching patches were not required to be visible in all views (i.e. we set in $\tilde{z} = -1$ in Algorithm 2). This resulted in slightly improved coverage of matched pixels but without significantly increasing the number of false matches, as shown in Figure 7.

Fig. 7.   Top: Three views from Temple dataset [46]. (The ellipses denote seed matches.) Bottom: Depth maps computed by our approach [29] (left) and Furukawa's approach [14] (right).

However, in order to ensure the robustness of Furukawa's approach it was important to require that all reconstructed patches were visible in all three views. That is, we tried Furukawa's method also so that it accepted patches visible in only two images but this increased outliers and the result was worse than the one shown in Figure 7. This might be due to the fact that the best-first expansion strategy is not used in [14].

In summary, the results with Dino and Temple datasets suggest that accurate correspondence growing is possible without expensive iterative optimization for the pose of matched patches. Hence, although iterative surface refinement is probably necessary at the final stage for the best reconstruction results [14], it might be unnecessary at the correspondence growing stage.

5. Discussion

Besides describing the basic match propagation algorithms for quasi-dense wide baseline matching, this chapter has mainly concentrated on multi-view stereo reconstruction applications, where the algorithms are used for guided matching of calibrated images. That is, the camera matrices are assumed known, as is typical in multi-view stereo problems, and the fundamental matrix or trifocal tensor is used to guide image matching, i.e., to reject matches that do not satisfy multi-view constraints. The experiments of Section 4 demonstrate that our algorithms are computationally efficient and provide results that are comparable to the state of the art [14]. In addition, as our current implementation can process only two or three views simultaneously, we outlined possibilities to extend the algorithms to more generic settings so that they would be directly applicable to arbitrary number of views. As discussed in Section 3.3, this seems to be a promising topic for future research.

However, besides multi-view reconstruction, the basic matching algorithms of Section 3 could be utilized in other applications as well. For example, unguided matching for image registration and multi-view geometry estimation, as illustrated in Section 4.1, is one evident application. In addition, quasi-dense matching can be used for registration of non-rigid deformations and for recognition and retrieval of particular objects [25; 4; 7]. Also, dense pixel-wise motion segmentation from multiple views may benefit from correspondence growing techniques [26].

6. Conclusion

In this chapter, we have described algorithms for quasi-dense wide baseline matching of view pairs and view triplets. The basic idea of the algorithms is to expand a sparse set of seed matches into a quasi-dense set of corresponding points between the views. The expansion is based on the best-first match propagation strategy. The algorithms can be used for unguided matching of view pairs of rigid and non-rigid scenes and for guided matching of view pairs and view triplets of rigid scenes. In the unguided case the quasi-dense approach allows to improve the robustness and accuracy of multi-view geometry estimation. On the other hand, multi-view surface reconstruction is the main application of guided matching, which refers to the case where matching is guided by the fundamental matrix or trifocal tensor. In this case, the algorithms can be used for depth map estimation from view pairs and triplets, and the obtained results are comparable to the state of the art within correspondence growing methods. Importantly, a good matching result is usually achieved with a single growth stage and without iterative refinement for the matched patches during growing. This is promising from the viewpoint of computational efficiency and encourages further studies on the topic.

References

1. Agarwal, S. *et al.* (2009). Building Rome in a day, in *ICCV*.
2. Boykov, Y., Veksler, O. and Zabih, R. (2001). Fast approximate energy minimization via graph cuts, *IEEE Trans. Pattern Anal. Mach. Intell.* **23**, 11, pp. 1222–1239.
3. Campbell, N. D. F., Vogiatzis, G., Hernández, C. and Cipolla, R. (2008). Using multiple hypotheses to improve depth-maps for multi-view stereo, in *ECCV*.
4. Cech, J., Matas, J. and Perdoch, M. (2010). Efficient sequential correspondence selection by cosegmentation, *IEEE Trans. Pattern Anal. Mach. Intell.* **32**, 9, pp. 1568–1581.

5. Cech, J. and Sára, R. (2007). Efficient sampling of disparity space for fast and accurate matching, in *CVPR*.

6. Chauve, A.-L., Labatut, P. and Pons, J.-P. (2010). Robust piecewise-planar 3d reconstruction and completion from large-scale unstructured point data, in *CVPR*.

7. Cho, M., Shin, Y. M. and Lee, K. M. (2010). Unsupervised detection and segmentation of identical objects, in *CVPR*.

8. Collins, R. T. (1996). A space-sweep approach to true multi-image matching, in *CVPR*.

9. Curless, B. and Levoy, M. (1996). A volumetric method for building complex models from range images, in *SIGGRAPH*.

10. Delaunoy, A. *et al.* (2008). Minimizing the multi-view stereo reprojection error for triangular surface meshes, in *BMVC*.

11. Faugeras, O. D. and Keriven, R. (1998). Variational principles, surface evolution, pdes, level set methods, and the stereo problem, *IEEE Transactions on Image Processing* **7**, 3, pp. 336–344.

12. Frahm, J.-M. *et al.* (2010). Building Rome on a cloudless day, in *ECCV*.

13. Furukawa, Y., Curless, B., Seitz, S. M. and Szeliski, R. (2010). Towards Internet-scale multi-view stereo, in *CVPR*.

14. Furukawa, Y. and Ponce, J. (2010). Accurate, dense, and robust multiview stereopsis, *IEEE Trans. Pattern Anal. Mach. Intell.* **32**, 8, pp. 1362–1376.

15. Gallup, D. *et al.* (2007). Real-time plane-sweeping stereo with multiple sweeping directions, in *CVPR*.

16. Gargallo, P. (2008). *Contributions to the Bayesian approach to multi-view stereo*, Ph.D. thesis, Institut National Polytechnique de Grenoble.

17. Goesele, M. *et al.* (2007). Multi-view stereo for community photo collections, in *ICCV*.

18. Grün, A. (1985). Adaptive least squares correlation: a powerful image matching technique, *S. Afr. J. of Photogrammetry, Remote Sensing and Cartography* **14**, 3.

19. Habbecke, M. and Kobbelt, L. (2007). A surface-growing approach to multi-view stereo reconstruction, in *CVPR*.

20. Hartley, R. and Zisserman, A. (2000). *Multiple View Geometry in Computer Vision* (Cambridge).

21. Hiep, V. H., Keriven, R., Labatut, P. and Pons, J.-P. (2009). Towards high-resolution large-scale multi-view stereo, in *CVPR*.

22. Hirschmüller, H. (2008). Stereo processing by semiglobal matching and mutual information, *IEEE Trans. Pattern Anal. Mach. Intell.* **30**, 2, pp. 328–341.

23. Jancosek, M., Shekhovtsov, A. and Pajdla, T. (2009). Scalable multi-view stereo, in *3DIM*.

24. Kannala, J. and Brandt, S. S. (2007). Quasi-dense wide baseline matching using match propagation, in *CVPR*.

25. Kannala, J., Rahtu, E., Brandt, S. S. and Heikkilä, J. (2008). Object recognition and segmentation by non-rigid quasi-dense matching, in *CVPR*.

26. Kannala, J., Rahtu, E., Brandt, S. S. and Heikkilä, J. (2009). Dense and deformable motion segmentation for wide baseline images, in *SCIA*.

27. Kobbelt, L. and Botsch, M. (2004). A survey of point-based techniques in computer graphics, *Computers & Graphics* **28**, 6, pp. 801–814.

28. Kolev, K., Klodt, M., Brox, T. and Cremers, D. (2009). Continuous global optimization in multiview 3d reconstruction, *Int. J. Comput. Vis.* **84**, 1, pp. 80–96.

29. Koskenkorva, P., Kannala, J. and Brandt, S. S. (2010). Quasi-dense wide baseline matching for three views, in *ICPR*.

30. Kutulakos, K. N. and Seitz, S. M. (2000). A theory of shape by space carving, *Int. J. Comput. Vis.* **38**, 3, pp. 199–218.

31. Labatut, P., Keriven, R. and Pons, J.-P. (2007). Efficient multi-view reconstruction of large-scale scenes using interest points, Delaunay triangulation and graph cuts, in *ICCV*.

32. Labatut, P., Pons, J.-P. and Keriven, R. (2009). Robust and efficient surface reconstruction from range data, *Comput. Graph. Forum* **28**, 8, pp. 2275–2290.

33. Lhuillier, M. and Quan, L. (2002). Match propagation for image-based modeling and rendering, *IEEE Trans. Pattern Anal. Mach. Intell.* **24**, 8.

34. Liu, Y., Dai, Q. and Xu, W. (2010). A point-cloud-based multiview stereo algorithm for free-viewpoint video, *IEEE Trans. Vis. Comput. Graph.* **16**, 3, pp. 407–418.

35. Lowe, D. (2004). Distinctive image features from scale invariant keypoints, *Int. J. Comput. Vis.* **60**, pp. 91–110.

36. Martinec, D. (2008). *Robust multiview reconstruction*, Ph.D. thesis, Czech Technical University.

37. Merrell, P. *et al.* (2007). Real-time visibility-based fusion of depth maps, in *ICCV*.

38. Mikolajczyk, K. *et al.* (2005). A comparison of affine region detectors, *Int. J. Comput. Vis.* **65**, pp. 43–72.

39. Mouragnon, E. *et al.* (2009). Generic and real-time structure from motion using local bundle adjustment, *Image and Vision Computing* **27**, pp. 1178–1193.

40. Newcombe, R. A. and Davison, A. J. (2010). Live dense reconstruction with a single moving camera, in *CVPR*.

41. Otto, G. P. and Chau, T. K. W. (1989). "Region-growing" algorithm for matching of terrain images, *Image Vision Comput.* **7**, 2, pp. 83–94.

42. Pollefeys, M. *et al.* (2008). Detailed real-time urban 3D reconstruction from video, *Int. J. Comput. Vis.* **78**, 2-3, pp. 143–167.

43. Scharstein, D. and Szeliski, R. (2002). A taxonomy and evaluation of dense two-frame stereo correspondence algorithms, *Int. J. Comput. Vis.* **47**, 1-3, pp. 7–42.

44. Seitz, S. M. and Dyer, C. R. (1999). Photorealistic scene reconstruction by voxel coloring, *Int. J. Comput. Vis.* **35**, 2, pp. 151–173.

45. Seitz, S. M. *et al.* (2006a). A comparison and evaluation of multi-view stereo reconstruction algorithms, in *CVPR*.

46. Seitz, S. M. *et al.* (2006b). Datasets for evaluation of multi-view stereo reconstruction algorithms, *http://vision.middlebury.edu/mview/* .

47. Sinha, S. N., Mordohai, P. and Pollefeys, M. (2007). Multi-view stereo via graph cuts on the dual of an adaptive tetrahedral mesh, in *ICCV*.

48. Snavely, N., Seitz, S. M. and Szeliski, R. (2008). Modeling the world from internet photo collections, *Int. J. Comput. Vis. (IJCV)* **80**, 2, pp. 189–210.

49. Strecha, C., Tuytelaars, T. and Gool, L. J. V. (2003). Dense matching of multiple wide-baseline views, in *ICCV*.

50. Strecha, C. *et al.* (2008). On benchmarking camera calibration and multi-view stereo for high resolution imagery, in *CVPR*.

51. Tylecek, R. and Sara, R. (2010). Refinement of surface mesh for accurate multi-view reconstruction, *International Journal of Virtual Reality* **9**, 1, pp. 45–54.

52. Xiao, J., Chen, J., Yeung, D.-Y. and Quan, L. (2008). Learning two-view stereo matching, in *ECCV*.

CHAPTER 3.5

COMBINING TRAFFIC SIGN DETECTION WITH 3D TRACKING TOWARDS BETTER DRIVER ASSISTANCE

Radu Timofte[1,*], Victor Adrian Prisacariu[2,*], Luc Van Gool[1] and Ian Reid[2]

[1] *ESAT-PSI-VISICS / IBBT, Katholieke Universiteit Leuven, Belgium*

[2] *Department of Engineering Science, University of Oxford, United Kingdom*

We briefly review the advances in driver assistance systems and present a real-time version that integrates single view detection with region-based 3D tracking of traffic signs. The system has a typical pipeline: detection and recognition of traffic signs in independent frames, followed by tracking for temporal integration. The detection process finds an optimal set of candidates and is accelerated using AdaBoost cascades. A hierarchy of SVMs handles the recognition of traffic sign types. The 2D detections are then employed in simultaneous 2D segmentation and 3D pose tracking, using the known 3D model of the recognized traffic sign. Thus, we achieve not only 2D tracking of the recognized traffic signs, but we also obtain 3D pose information, which we use to establish the relevance of the traffic sign to the driver. The performance of the system is demonstrated by tracking multiple road signs in real-world scenarios.

1. Introduction

Traffic signs play a pivotal role in rendering traffic more efficient and safer. Unfortunately, still many accidents happen because drivers have overlooked a sign. Therefore, increasingly cars are being equipped with vision systems that detect and recognize such signs, in order to assist the driver. Moreover, public authorities are carrying out large-scale campaigns to replace older, hand-drawn maps and manually generated inventories with digital maps and GIS systems. The presence and positions of traffic signs again figure high on the list of data to be surveyed. As signs regularly change (total yearly changes in the traffic sign set are estimated to amount to 7%!), it is important to automate their detection. Otherwise, regular updates are too expensive. In such applications, it is not only important to detect the presence of traffic signs, but their actual position and orientation are crucial as well. Such factors determine whether signs are sufficiently visible and whether they are relevant to traffic participants approaching from specific directions. Thus, traffic sign analysis is not only about finding the signs in images, but also about their 3D positions and orientations, especially in an urban context.

---

*R. Timofte and V. A. Prisacariu contributed equally to this work.

Fig. 1. Importance of determining the traffic sign orientation. The no right turn sign does not point towards the car

Most current work involves combining a detector with a Kalman filter, as in [1, 2], or with a particle filter, as in [3, 4]. These methods rely on a predictable car motion model or reliable feature detectors. For example, [1] and [5] assume the car moves in a straight line and with constant velocity. As feature descriptors, trackers usually use edges or other kinds of information extracted from the traffic sign shape. In [1] the authors explicitly model geometric properties of the traffic sign shape (i.e. a triangle-shaped sign has to be equilateral). This leads to a lack of robustness when subjected to occlusions, deformations or motion blur. In [3] the authors track circular signs and assume these have a colored border on a white interior and that clear edges can be extracted. This approach would not extend to differently shaped signs and is again vulnerable to motion blur. A solution to some of these problems is region based tracking. Regions are more stable than edges so tracking is more robust, and they are less affected by occlusions or motion blur.

To our knowledge, with notable exceptions like [3] and [6], most previous road sign work has been *2D*. That is, the position of the traffic sign was tracked in the *image* rather than in *3D space*. In [3] the authors use inertial sensors mounted on the car to help them obtain an approximation of the 3D pose of the signs with respect to the car. Unfortunately this approach fails when the traffic sign does not point towards the car, as in the case shown in figure 1. Here, the no right turn sign does not have the same orientation as the inertial sensor mounted on the car.

An alternative method is presented in [6], where multiple views, 3D reconstruction, and a Minimum Description Length based method are used. Although a 3D pose is recovered, it is an offline method for 3D mobile mapping purposes.

Our approach integrates a single view detection and recognition step with a multi-object, model-based, 3D tracker. This has several advantages:

(i) we are able to obtain the full 3D pose of the traffic signs in the image, accounting for the case in figure 1,

(ii) the tracking is region based, making it robust to motion blur and occlusions,

(iii) because our tracker processes only a small region in and around the detection we are able to achieve real time performance.

The remainder of this chapter is structured as follows: we review the state-of-the-art in Section 2 and present an overview of our system in Section 3. In Section 4 we detail the single view detection and recognition parts and in Section 5 the energy function used both in the non-maximum supression and the 3D tracking stages, themselves presented in Sections 6 and 7, respectively. Section 8 shows results on a large database with images and videos and we conclude in Section 9.

2. State-of-the-art

2.1. *Single View Detection and Recognition*

A traffic sign is designed to be easily recognizable using color, shape and appearance. Current state-of-the-art work considers these properties either individually or in combination. The standard pipeline starts with a detection phase for localizing candidates in the image space, followed by a recognition phase for setting the labels. The detection often employs a fast segmentation/candidate extraction step, usually done through simple methods, followed by pruning methods to reduce the false detections. The recognition phase aims at fast and robust classification of traffic signs while space/temporal integration (i.e. tracking) can help by exploiting the information from already processed frames to improve performance in the current frame.

Color is important in spotting the traffic sign in cluttered environments. Many approaches use simple crafted thresholds for each color (i.e. red, blue, yellow, black, white) in a suitable color space such as RGB, Hue-Saturation-Intensity/Value (HSI,HSV), or Luv. In [7], for each color, pairs of thresholds for Hue and Saturation channels are picked from the color distribution during training. Figure 3 depicts the extraction of candidates based on color thresholding. Hue is invariant to lighting. This property is exploited in [8] where a shadow-highlight invariant method is proposed. [9] also uses HSV and a dynamic pixel aggregation technique which employs dynamic thresholding to address the hue dependence on external brightness variations. The saturation and intensity values of the images are used to adapt these thresholds. [10] uses a region growing method, which merges neighboring pixels with similar colors. Other color-based approaches train classifiers or probabilistic color models to determine the likelihood of a pixel to belong to a reference color [11].

The regular shapes shared among different traffic sign types render general shape detectors useful. A form of generalized Hough transform is used in [12] for regular polygonal shapes, while [13] uses distance transform matching and a template hierarchy to capture the variety of object shapes. AdaBoost [6, 14] or part-based models [15] are highly accurate in detection and are used in cascades for specific shape classes. In [6, 14, 16] as well as in this paper, color, shape and appearance cues are combined for detection.

Also, for recognition, there is a large variety of methods. Most machine learning techniques have also been used for classifying traffic signs. A recent classification challenge [17] on German traffic signs showed that convolutional neural networks, SVM with an intersection kernel, and Sparse Representation-based classifiers [18] are currently the top performers. The features used by the top contenders were intensity, Histogram of Oriented Gradients (HOGs) [18, 19] and their projections. Color is not a critical feature for traffic sign classification, but for discriminating between traffic signs and background.

The results of traffic sign detection and recognition thus far testify to the great difficulty of the task. In [20] the reported performance is 26% false negatives (missed traffic signs) at a level of 3 false positives (false alarms) per image. [7] applies image thresholding followed by SVM classification. Every traffic sign is detected at least twice in a total of 5000 video frames, with 22 false alarms. Detection rates per view are not given. In both works the thresholds are manually selected. In [21] the search is constrained to road borders and an overhanging strip, by using inertial sensors and estimating 3D geometry. This significantly reduces the false positives, while the false negative rate is 3.8%. These three systems were tested on highways.

Several systems have also been demonstrated off the highways. By restricting the detection to speed, stop and give-way signs, [22] reports a performance of 10^{-4} – 10^{-5} false positive rates for 1% false negatives, but the number of sub-windows per image is not mentioned. [23] reports no false positives in a 150 minute long video, but misses 11% of all traffic signs. In [2, 24] image color thresholding and shape detection are combined, achieving 6.2% false negatives. The number of false positives is not mentioned. [25] proposed a system similar to the one in [7], where the SVM is replaced by a neural network. No quantitative results are presented.

While the majority of the previous contributions work with and report the performance on a rather small subset of sign types, our system (using settings from [6, 16]) handles 62 different types of signs. Moreover, authors usually focus on highway images, whereas our database mainly contains images from smaller roads and streets. This is a more challenging problem as signs tend to be smaller, have more often been smeared with graffiti or stickers, suffer more from occlusions, are often older, and are visible in fewer images. Also, several sign types never appear along highways.

2.2. *3D Region-Based Segmentation and Tracking*

Fast and accurate image segmentation and pose tracking are fundamental tasks in computer vision. Most current studies treat these two tasks independently and often use edge based segmentation, leading to an inability to deal with occlusions or motion blur. A common solution to the edge problem is the use of regions, whose shape and evolution can be effectively modelled using level-set functions [26]. The first paper to use regions in 3D pose recovery was [27]. There the authors represent the contour by a zero level-set of a 3D embedding function and evolve it in a single iterative step, in two stages: first an infinite dimensional level set energy function (with an added shape term) is minimized with respect to the segmentation, in the expectation that the contour will then match the projection of the occluding contour of the 3D object. Second, each point on the contour is back-projected to a ray (represented in Plücker coordinates), and the pose sought that best satisfies the tangency constraints that exist between the 3D object and these rays. The obvious pitfall of this method is that the evolution of the contour is not limited to the space of possible segmentations, which can make the resulting segmentations still inaccurate. The method is improved in [28] where the unconstrained contour evolution stage is removed, and the minimization takes place by evolving the contour (approximately) directly from the 3D pose parameters. The direction of contour evolution is determined by the relative foreground/background membership probabilities of each point, while the amount of evolution is apparently a "tunable" parameter.

In this work we use the variational approach of [29], where the Bibby-Reid energy function of [30] is differentiated with respect to the 6 DoF 3D pose parameters, simultaneously yielding both segmentation and pose. This method has the benefit of running in real time and of being resilient both to occlusions and to motion blur.

3. Overview of the System

An outline of our algorithm is shown in figure 2. It consists of 5 steps. First the signs are detected using a single view detection step. Next the sign in each detection is

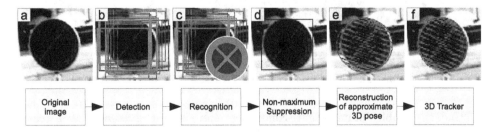

Fig. 2.   Algorithm overview

recognized and the best detection for each object is selected. Next, for each sign in the image, an approximate 3D pose is computed, by combining a predicted 3D pose (from the previous frame, with a constant velocity motion model) with a 3D pose obtained from the detection bounding box. The detection bounding box is then converted to a 3D pose using a 4 point planar pose recovery algorithm. Finally, the 3D tracker is iterated from the approximated pose, which results in a refined 3D pose, for each object in the image. We use the object detection and recognition algorithm from [6] followed by the 3D tracking from [16].

4. Single-View Detection and Recognition

The single view processing starts with a fast candidate extraction process, followed by a pruning/detection process based on AdaBoost cascades, and finally after the further filtering by a background vs. traffic sign SVM classifier, the detections are assigned to the class with the highest estimated probability in the corresponding one-against-all SVM output. The candidate extraction and detection parts follow the work from [6], used also in [16]. The classification part follows settings from [18].

The simplest and most often used extraction method for traffic sign detection is extraction of connected components from a *thresholded image* [7, 25] (see figure 3). The thresholded image is obtained from a color image (RGB in [6, 16] and herein), with color channels (I_R, I_G, I_B), by application of a color threshold $T = (t, a, b, c)^\top$:

$$I(T) = \begin{cases} 1 & a \cdot I_R + b \cdot I_G + c \cdot I_B \geq t \\ 0 & \text{otherwise} \end{cases} \tag{1}$$

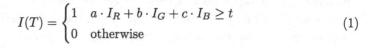

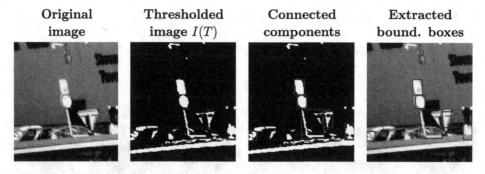

Fig. 3.   **Color-based extraction** method for threshold $T = (0.5, 0.2, -0.4, 1.0)^\top$ [6]

Under variable illumination conditions and in the presence of a cluttered/complex background an extraction method based on only a few manually selected thresholds is insufficient for extracting the traffic signs. It is necessary to

Original image	Extracted region	Bounding box	Rescaled bound. box

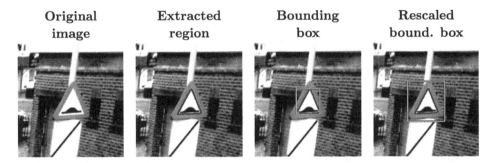

Fig. 4. **Extended threshold** addresses the problem of signs not well locally separable from the background. The bricks have a color similar to the red boundary, the inner white part is extracted and the resulting bounding box is rescaled $\overline{T} = (0.1, -0.433, -0.250, 0.866, 1.6, 1.6)^{\top}$ [6]

combine regions selected by several thresholds $\mathcal{T} = \{T_1, T_2, \dots\}$, in the sense of adding regions (OR-ing operation). The regions that pass are the input for the next stage, i.e. detection. Using more thresholds can lead to overfitting. It lowers the number of false negatives (FN), but adds computational cost and increases the number of false positives (FP).

We search for the optimal subset $\mathcal{T}$ of such thresholds starting from thousands of possible color thresholds. We formulate our search as an Integer Linear Programming (ILP) problem. ILP yields a viable solution within minutes, due to the sparsity of the constraints. Searching for a trade-off between FP and FN is the most straightforward criterion to guide this optimization. To avoid overfitting and to keep the method sufficiently fast, we additionally constrain the number of selected thresholds, the cardinality, card($\mathcal{T}$). The *accuracy*, defined as the average overlap between ground truth (annotation) bounding boxes with the extracted bounding boxes, is important for the subsequent steps in the system. We penalize inaccurate extractions:

$$\mathcal{T}^* = \arg\min_{\mathcal{T}} \left(\text{FP}(\mathcal{T}) + \kappa_1 \cdot \text{FN}(\mathcal{T}) + \kappa_2 \cdot \text{card}(\mathcal{T}) - \kappa_3 \cdot \text{accuracy}(\mathcal{T})\right) \quad (2)$$

where $\text{FP}(\mathcal{T})$ is the number of false positives and $\text{FN}(\mathcal{T})$ is the number of false negatives measured on a training set. The weighting scalars κ_1, κ_2 and κ_3 are learned parameters estimated by cross-validation. The reformulation of problem 2 into an ILP form is described in [6, 14].

The contour of the traffic sign often cannot be separated from the background due to color similarity. In such cases we still have the inner contours that can be extracted by color thresholding. Such thresholding is followed by rescaling the obtained bounding box to include the whole traffic sign. See for example figure 4. The inner part can often define the traffic sign's outline with sufficient accuracy and we therefore introduce the extended threshold — in the sequel simply referred to as threshold:

$$\overline{T} = (\underbrace{t, a, b, c}_{T}, s_r, s_c)^\top \tag{3}$$

which consists of the original threshold T and vertical and horizontal scaling factors (s_r, s_c).

One could try to adapt the set of thresholds to the illumination conditions. To add robustness to the thresholding method itself we adjust the threshold to be *locally stable* in the sense of Maximally Stable Extremal Regions (MSER) [31]. Instead of directly using the bounding box as extracted by the learned threshold (t, a, b, c, s_r, s_c), we use bounding boxes from MSERs detected within the range $[(t-\epsilon, a, b, c, s_r, s_c); (t+\epsilon, a, b, c, s_r, s_c)]$, where ϵ is a parameter of the method. This 'TMSER' method is parametrized by two parameters (ϵ, Δ), as the MSER adds a stability parameter Δ.

Dirty, peeled, partially occluded traffic signs also should pass the color test. Therefore, we also need to employ shape information in order to remain sufficiently selective. We learn fuzzy templates to incorporate small affine transformations and shape variations and we only explicitly determine the position and scale in a 3D Hough accumulator. For more details please refer to [6, 14].

The candidates extracted are verified further by a binary classifier which filters out remaining background regions. It is based on the Viola and Jones Discrete AdaBoost classifier [32]. Detection is performed by cascades of AdaBoost classifiers, followed by an SVM operating on normalized RGB channels, pyramids of HOGs [33] and AdaBoost-selected Haar-like features to prune the background further.

Finally, the detections are further classified as their specific traffic sign type. For each detection we compute pyramids of HOG features as in [34], LDA-project them, and train one-against-all Linear SVM classifiers. The assigned traffic sign type is the one with the highest score estimated from the SVM classifier output.

5. Pixel-Wise Posteriors Energy Function

Before describing our non-maximum suppression and 3D tracking stages we introduce the Bibby-Reid [29, 30] pixel-wise posteriors energy function, which is the basis of both stages.

For any image I with an image domain Ω and known color statistics (foreground and background membership probabilities), the pixel-wise posteriors (PWP) energy function is a function of the contour that separates the image domain into a foreground and a background domain. The PWP energy function has a maximum when the contour best separates the two regions i.e. when all the pixels in the foreground region have a higher probability of being foreground than of being background (and viceversa). Figure 5 shows such an example. Here the contour is denoted with $\mathbf{C}$, the foreground region with Ω_f and the background region with Ω_b. The implicit

Fig. 5. Representation of the object showing the contour of the projection $\mathbf{C}$, the foreground region Ω_f and the background region Ω_b

representation of the contour is used, by representing it as the zero level of a level set function [26].

Of course, the PWP energy function is not the only energy function to measure segmentation quality. It is, however, one of the best behaved ones, as demonstrated in [29, 30], having a large basin of convergence and few local minima. Furthermore, it can be used both for non-maximum suppression and for 3D tracking, as shown in the following sections.

The PWP energy function is:

$$E(\Phi) = -\sum_{x \in \Omega} \log\left(H_e(\Phi)P_f + \left(1 - H_e(\Phi)P_b\right)\right) \qquad (4)$$

Here H_e is the smooth Heaviside function, x is the pixel in the image, Φ the level set embedding function and:

$$P_f = \frac{P(y_i|M_f)}{\eta_f P(y_i|M_f) + \eta_b P(y_i|M_b)}, \quad P_b = \frac{P(y_i|M_b)}{\eta_f P(y_i|M_f) + \eta_b P(y_i|M_b)} \qquad (5)$$

with η_f the number of foreground pixels, η_b the number of background pixels, y_i the color of the i-th pixel, $P(y_i|M_f)$ the foreground model over pixel values y and $P(y_i|M_b)$ the background model. We use RGB images and our models are histograms with 32 bins for each channel, which are updated online, allowing for variations in illumination.

6. Non-Maximum Suppression

At this stage several detections might be available for each object, as shown in figure 6. We need to select only a single, best segmentation, process known as non-maximum suppression. There are several methods for doing this. For example [19] uses mean-shift to search in a 3D scale–location space.

We, on the other hand, have the advantage of knowing color models for the foreground and for the background. This means that, for all the detected

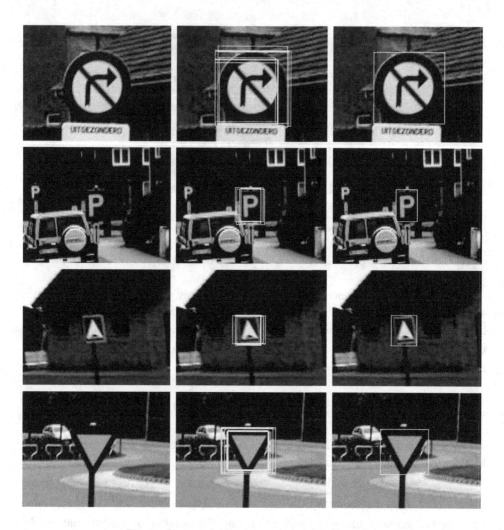

Fig. 6. Non-maximum suppression example results

bounding boxes, we can select the one which best segments the traffic sign from the background. To do this we evaluate the Bibby-Reid energy function described above for all detected bounding boxes and select the one with the highest color matching score.

With b as the bounding box we write:

$$P(b) = \prod_{x \in \Omega} \Big(H_e(a(x)) P_f + \big(1 - H_e(a(x))\big) P_b \Big) \tag{6}$$

where we replace the level set function Φ with a function $a(x)$ indicating membership of pixel x to the inside or outside of the bounding box:

$$a(x) = \begin{cases} 1 & x \text{ inside the bounding box} \\ 0 & \text{otherwise} \end{cases} \tag{7}$$

Obviously, when an object is detected for the first time, we have no color model yet. In this case we choose the detection with the highest SVM score, and initialize the two color models $P(y_i|M_f)$ and $P(y_i|M_b)$.

7. Approximate 3D Pose Reconstruction and 3D Tracking

At this stage of the algorithm we have a single, correct, bounding box for every traffic sign in the image and we need to track their 3D position (rotation and translation). The core of our tracking procedure is the PWP3D algorithm [29]. It assumes a known 3D model and a calibrated camera and it maximizes the above-mentioned pixel-wise posteriors energy function with respect to the pose of the known model. This optimization is done by computing the derivatives of the energy function with respect to the pose parameters and using gradient descent. Note that convergence is not guaranteed within the permitted number of iterations, but in our testing we noticed that an average of 15 iterations is enough for a satisfactory result.

Following [29], we can write:

$$\frac{\partial E}{\partial \lambda_i} = \sum_{x \in \Omega} \frac{P_f - P_b}{H_e(\Phi)P_f + (1 - H_e(\Phi))P_b} \frac{\partial H_e(\Phi)}{\partial \lambda_i} \tag{8}$$

where λ_i are the pose parameters (7 in total). In this work, similar to [29], we use quaternions to describe rotation.

$$\frac{\partial H_e(\Phi(x,y))}{\partial \lambda_i} = \frac{\partial H_e}{\partial \Phi}\left(\frac{\partial \Phi}{\partial x}\frac{\partial x}{\partial \lambda_i} + \frac{\partial \Phi}{\partial y}\frac{\partial y}{\partial \lambda_i}\right) = \delta_e(\Phi)\begin{bmatrix} \frac{\partial \Phi}{\partial x} & \frac{\partial \Phi}{\partial y} \end{bmatrix}\begin{bmatrix} \frac{\partial x}{\partial \lambda_i} \\ \frac{\partial y}{\partial \lambda_i} \end{bmatrix} \tag{9}$$

where δ_e is the Dirac delta function.

At each iteration of the algorithm Φ is recomputed as the signed-distance function of the projected contour. The partial differentials of the level set function with respect to pixel position, $\frac{\partial \Phi}{\partial x}$ and $\frac{\partial \Phi}{\partial y}$, are computed using centered finite differences.

Every 2D point on the contour of the projection of the 3D model has at least one corresponding 3D point $\mathbf{X}$, for which:

$$\begin{bmatrix} x \\ y \end{bmatrix} = \begin{bmatrix} -f_u\frac{X}{Z} - u_0 \\ -f_v\frac{Y}{Z} - v_0 \end{bmatrix} \tag{10}$$

Therefore:

$$\frac{\partial x}{\partial \lambda_i} = -f_u \frac{\partial}{\partial \lambda_i} \frac{X}{Z} = -f_u \frac{1}{Z^2} \left(Z \frac{\partial X}{\partial \lambda_i} - X \frac{\partial Z}{\partial \lambda_i} \right) \tag{11}$$

Similar for y.

Continuing, each point $\mathbf{X}$ in camera coordinates, has a corresponding 3D point $\mathbf{X_0}$ in object coordinates, so:

$$\begin{bmatrix} X \\ Y \\ Z \\ 1 \end{bmatrix} = \mathbf{R} \begin{bmatrix} X_0 \\ Y_0 \\ Z_0 \\ 1 \end{bmatrix} + \begin{bmatrix} t_x \\ t_y \\ t_z \\ 0 \end{bmatrix} \tag{12}$$

where $\mathbf{R}$ is the rotation matrix and is a function of the rotation quaternion and t_x, t_y and t_y are the translation parameters. $\frac{\partial X}{\partial \lambda_i}, \frac{\partial Y}{\partial \lambda_i}$ and $\frac{\partial Z}{\partial \lambda_i}$ follow trivially. For more details the reader is refereed to [29].

There are two issues with the above formulation which need to be addressed in order for us to apply this tracker to our problem. The PWP3D tracker needs an initial 3D pose and values for the foreground/background membership probabilities. Also, the PWP3D tracker can lose tracking, so at each new frame, the 2D detections need to be converted to (approximate) 3D poses to be combined with the 3D tracker. These two problems are addressed in the following sections.

7.1. *Initialization*

Traffic signs can be approximated as planar objects, which means we can use any one of several planar pose recovery algorithms currently available to convert the 2D bounding box surrounding the detection to a 3D pose. We use the currently state-of-the-art algorithm introduced in [35].

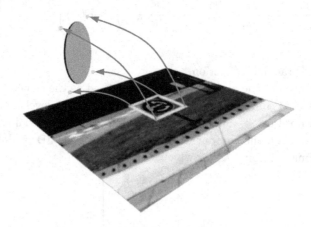

Fig. 7.   3D–2D point correspondences

This algorithm requires (at least) 4 3D–2D point correspondences and works iteratively, minimizing a sum of squared object-space collinearity errors. We use the 4 2D corners of the 2D detection bounding box and relate them to the 4 corners of the bounding box enclosing the 3D model of object, as depicted in figure 7.

An example result is depicted in figure 2e.

7.2. *Detector–Tracker Integration*

The PWP3D tracker will often lose its track if there is no overlap between the positions of the tracked object at consecutive frames. We therefore use the 3D poses recovered from the detection bounding boxes (as presented in the previous subsection), combined with a constant velocity motion model, to re-localize the 3D tracker.

One common solution is to use a Kalman filter for a purely statistical fusion of the tracking data with the approximate poses from the detector. A Kalman filter represents all measurements, system state and noise as multivariate Gaussian distributions. In contrast, in this work we first compute an approximate 3D pose for each 3D sign at the current frame, by combining the detection with a motion model. This serves as an initialization for the PWP3D pose optimization described above. By iterating the tracker rather than doing a purely statistical data fusion we make no pretense on the type of these probability distributions.

We define a constant velocity motion model:

$$v_{t_k}^i = t_{k-1} - t_{k-2}, \quad v_{r_k}^i = r_{k-1}r_{k-2}^{-1} \tag{13}$$

where k is the current frame, $k-1$ is the previous frame, t is the translation and r is the rotation quaternion from the tracker.

There is also a velocity given by the 2D detections:

$$v_{t_k}^{ii} = u_k - u_{k-1}, \quad v_{r_k}^{ii} = p_k p_{k-1}^{-1} \tag{14}$$

where u is the translation and p is the rotation quaternion, obtained from the detector by using the 4 point planar pose recovery algorithm.

When the object is initially detected it is most likely far from the camera, so its rotational motion is less predictable, making the chance of lack of overlap higher. In this case detections should be trusted more. When the object is close to the camera, rotational motion is more predictable so the chance for lack of overlap is smaller. In this case the motion model should be trusted more.

The predicted pose for the current frame becomes:

$$t_k = t_{k-1} + \alpha v_{t_k}^i + \beta v_{t_k}^{ii}, \quad r_k = r_{k-1}q_\alpha v_{r_k}^i q_\beta v_{r_k}^{ii} \tag{15}$$

where k is the current frame, $k-1$ is the previous frame, t is the translation from the tracker and u is the translation from the detector. The variables α, β, q_α and q_β are dependent on the distance between the object and the camera. α and q_α are inversely proportional to this distance while β and q_β are proportional to it.

Finally the predicted pose (t_k, r_k) is refined using the tracker.

Table 1.   Summary of achieved results in single-view detection.

	FN-TS		FN-BB		FP per
	[%]	#/859	[%]	#/2571	2MP img
Extr1 (color)	0.7%	6	1.1%	29	3 281.8
Extr2 (color+TMSER)	0.7%	6	1.1%	28	3 741.7
Extr3 (color+shape)	0.5%	4	0.7%	17	5 206.2
Extr4 (color+TMSER+shape)	0.5%	4	0.7%	17	5 822.0
Det + Extr1	2.3%	20	4.0%	103	2
Det + Extr4	1.9%	16	3.2%	82	2

Extr marks candidate extraction setting. **color** means extended color thresholds, **TMSER** stands for $TMSER(\epsilon, \Delta) = TMSER(0.1, 0.2)$, **shape** is for fuzzy template methods. **FN-BB** means false negative with respect to bounding boxes, **FN-TS** means false negative with respect to traffic signs.

8. Experiments

We report results for different parts of our system: detection and recognition, and 3D tracking. We are using the BelgiumTS[a] database as in our previous work [6, 14, 16]. It contains 13444 traffic sign (TS) annotations in 9006 still images, corresponding to 4565 physically distinct traffic signs visible at less than 50 meters from the camera. BelgiumTS has 3 main subparts: "Training", "2D Testing", and "3D Testing", as well as a classification subset, BelgiumTSC [18].

8.1. *Detection Results*

The Training part of BelgiumTS is used for learning the suitable candidate extraction methods as well as for training the AdaBoost cascades and the SVM classifiers, while the 2D Testing part is used to assess the performance. The method has only been trained for 62 traffic signs classes. In 2D Testing the number of used annotations is 2571, corresponding to 859 physically distinct traffic signs. The output of the cascades is processed further by an SVM classifier that uses Haar-like features, pyramids of HOGs and pixels in RGB space.

The detection and extraction errors (Table 1) are evaluated according to two criteria: either demanding detection every time a sign appears (FN-BB), or only demanding it is detected at least once (FN-TS). On average, a sign is visible in about 3 views. *Coverage* is defined as the ratio between intersection and the union of two areas. A detection is successful if *coverage* ≥ 0.65 when compared to the annotation.

Table 1 shows results of both the candidate extraction (still with an appreciable number of FP, see first four rows) and the final detection (i.e. candidate extraction followed by AdaBoost detector and SVM, see last two rows). The shape extraction significantly increases the number of false positives (see for example the 4th row in the table). The reason is that we keep both the original color bounding boxes

---

[a]BelgiumTS is available at: `http://homes.esat.kuleuven.be/~rtimofte/`

and add all bounding boxes that reflect a good shape match. Combined extraction lowers FN, however. Here we use the **Det+Extr1** setting.

We compare the pipeline outlined so far with a sliding window approach and a part-based model detector. For the sliding window, we train Discrete AdaBoost cascades directly on sampled subwindows from the Training data, instead of the extracted candidates as for **Det+Extr1**. For details please refer to [14]. The number of processed windows per 1628×1236 image is higher than 12 million. Figure 8 shows the complementarity of the sliding window approach to our **Det+Extr1** pipeline. While the individual pipelines have a comparable performance, the combination improves over both. Figure 9 shows cases that could be detected by one pipeline but not the other. In our single-core / single-thread implementations, the **Det+Extr1** pipeline is about 50 times faster than the sliding window pipeline.

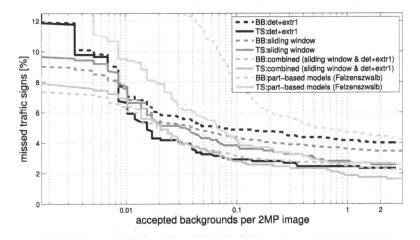

Fig. 8. **Comparison with state-of-the-art methods.** Detection plots for every time a sign appears (**BB**-bounding box level), or demanding it is detected at least once (**TS**-traffic sign level)

We also compare with the state-of-the-art generic object class detector of Felzenszwalb et al. [15], the top performer in the PASCAL VOC Challenge 2009 [36]. We use the scripts from the authors and we train on the Training part of BelgiumTS a model with 5 components which correspond to the basic shapes of the traffic signs. The poorer performance of the part-based detector when compared with our specialized systems (see figure 8) is believed to be caused by the fact that this approach is a generic one and only works on HOG features. However, the part-based model detector is also still far from realtime performance.

8.2. *Recognition Results*

We evaluate several settings for traffic sign recognition on BelgiumTSC [18], the classification subset of BelgiumTS database which contains 62 different sign types,

a) Missed detections:

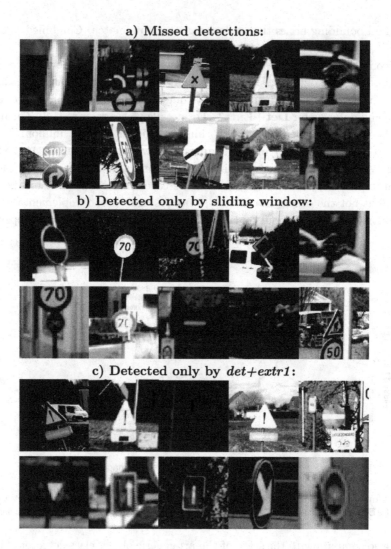

b) Detected only by sliding window:

c) Detected only by *det+extr1*:

Fig. 9. **Complementarity of sliding window and the proposed approach.** Shown are samples where one method fails but the other one is successful (**a,b**), at the same threshold level

4591 training samples and 2534 testing samples. We use raw gray scale pixel values — intensity (I) and pyramids of histograms of oriented gradients (PHOG) as were used in [18], where it was shown that for classification purposes the color information is not essential when the samples are known to be traffic signs. The most discriminative features of the traffic sign are the inner pattern and the shape. This is of course an explicit design goal of traffic signs. We investigate the supervised dimensionality reduction based on Linear Discriminant Analysis (LDA) with regularization [37]. The features are l_2-norm normalized to sum to 1. The LDA regularization parameter is 0.1.

Table 2 depicts the performance achieved using the Sparse Representation-based Classifier (SRC), Linear kernel SVM (LSVM), Intersection Kernel SVM (IKSVM), Polynomial Kernel SVM (POLYSVM) and Radial Basis Function SVM (RBFSVM). The classifiers and their settings are as in [18]. PHOG features are more discriminative than the raw intensity values, but this comes with an increase in time for computing the features and training the classifiers. While provided here, SRC is not applicable for our task as it is 10 to 100 times slower than the other SVM classifiers when run for test. The PHOG features with IKSVM is the winning setting, achieving 97.79% on the BelgiumTSC data. Working on LDA projected features speeds up the training and the testing. SRC benefits greatly from the discriminatively projected features. We use here LSVM+LDA+PHOG because of the high speed and performance.

Table 2.   Recognition rates on the BelgiumTSC database.

Features	I	PHOG	LDA+I	LDA+PHOG
SRC [%]	85.04	92.34	**94.32**	**97.36**
LSVM [%]	90.69	96.68	91.28	96.96
IKSVM [%]	**91.36**	**97.79**	91.67	96.80
POLYSVM [%]	89.94	96.61	91.87	96.80
RBFSVM [%]	90.41	96.57	91.99	97.16

8.3. *3D Tracking Results*

We begin by showing qualitative examples. In figure 10 we show our system tracking a pedestrian crossing sign and obtaining an accurate pose even when the object is far from the camera. Similarly, in figure 11 we show our algorithm successfully tracking a give way sign. An application which comes to mind is to calculate the distance between each sign and the car, to then automatically adjust the speed of the car. Figure 12 shows our system tracking multiple round traffic signs. Currently we estimate each pose independently, though of course relative to the camera each sign undergoes the same rigid transformation. This coupling would results in improved performance.

Fig. 10.   Filmstrip showing frames from a video tracking a pedestrian crossing sign

Fig. 11.   Filmstrip showing frames from a video tracking a triangular sign

Fig. 12.   Filmstrip showing frames from a video tracking multiple round signs

Fig. 13.   Filmstrip showing frames from a video tracking a parking sign, in spite of it being occluded by a tree

As mentioned several times in this paper, our system is able to deal with occlusions. Figure 13 shows our system successfully tracking a parking sign, in spite of it being occluded by a tree.

We compared the performance of our system by tracking a single sign over a distance of 60m (or 60 frames at 36km/s), with and without the 3D tracker. The car was moving along a straight line, with constant speed. Translation should therefore change linearly, while rotation should remain constant. Figure 14 shows the results for a circular sign, figure 15 for a rectangular sign and figure 16 for a triangular sign. At higher distances the object shrinks to only a few tens of pixels, so detection and tracking are prone to errors. Still, the 3D tracker is able to provide smoother values for the translation. Though there may be little difference between using the 3D tracker and using just the 4 point planar pose recovery algorithm with regard to the translation, the tracker must be used to reliably obtain the rotation.

A video processed by our system is available at http://homes.esat.kuleuven.be/~rtimofte. It shows our system tracking multiple objects of different types, orientations and colors, with or without motion blur and occlusions. Our CPU implementation of the detection phase runs at around 20fps on 640×480 images while the GPU based tracker needs up to 20ms per object (on a 640×480 image).

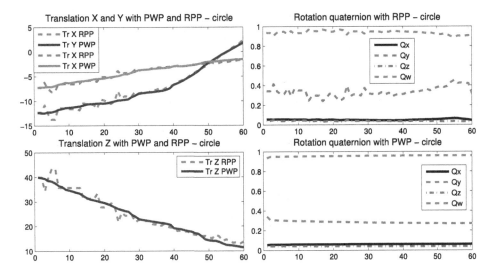

Fig. 14. System performance while tracking a circle shaped sign over 60m, with just the 4 point pose recovery (RPP) and with the tracker (PWP)

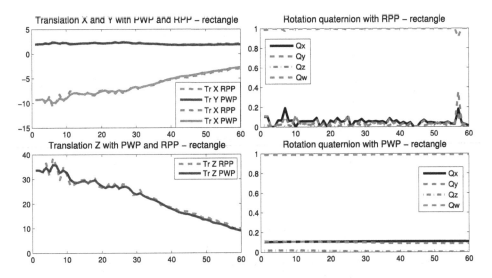

Fig. 15. System performance while tracking a rectangle shaped sign over 60m, with just the 4 point pose recovery (RPP) and with the tracker (PWP)

9. Conclusions

In this chapter we reviewed the advances in driver assistance with respect to traffic sign recognition and proposed a system that can track multiple traffic signs in 3D, from a single view. By integrating accurate detections with 3D region based tracking

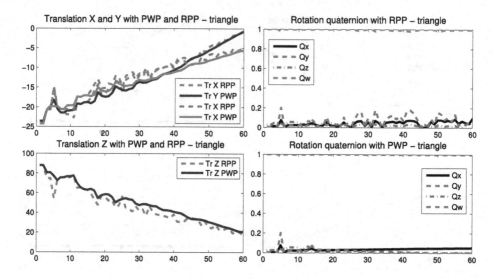

Fig. 16. System performance while tracking a triangle shaped sign over 60m, with just the 4 point pose recovery (RPP) and with the tracker (PWP)

our system is robust to motion blur and occlusions, while still running in real time. Such a system would require more, e.g. person and car 3D detection and tracking, etc.

Acknowledgments

This work was supported by EC FP7-EUROPA and Flemish IBBT-ISBO projects, in continuation of research started under IBBT-URBAN project, and by EPSRC through a DTA grant. The authors thank GeoAutomation for providing the images.

References

[1] C.-Y. Fang, S.-W. Chen, and C.-S. Fuh, Road-sign detection and tracking, *Vehicular Technology, IEEE Transactions on.* **52**, 1329–1341, (2003).

[2] A. Ruta, Y. Li, and X. Liu, Real-time traffic sign recognition from video by class-specific discriminative features, *Pattern Recognition.* **43**, 416–430, (2010).

[3] M. Meuter, A. Kummert, and S. Muller-Schneiders. 3d traffic sign tracking using a particle filter. In *Intelligent Transportation Systems*, pp. 168–173.

[4] L. D. Lopez and O. Fuentes. Color-based road sign detection and tracking. In *ICIAR*, pp. 1138–1147, (2007).

[5] G. Piccioli, E. De Micheli, P. Parodi, and M. Campani. Robust road sign detection and recognition from image sequences. In *Intelligent Vehicles Symposium*, (1994).

[6] R. Timofte, K. Zimmermann, and L. Van Gool. Multi-view traffic sign detection, recognition, and 3d localisation. In *WACV*, Snowbird, Utah, USA, (2009).

[7] S. Maldonado, S. Lafuente, P. Gil, H. Gómez, and F. López, Road-sign detection and recognition based on support vector machines, *IEEE Trans. Intelligent Transportation Systems, pp 264–278.* **8**(2), (2007).

[8] H. Fleyeh, S. Gilani, and M. Dougherty. Road sign detection and recognition using a fuzzy ARTMAP: A case study swedish speed-limit signs. In *10th IASTED International Conference on Artificial Intelligence and Soft Computing*, (2006).

[9] S. Vitabile, G. Pollaccia, G. Pilato, and F. Sorbello. Road signs recognition using a dynamic pixel aggregation technique in the hsv color space. In *ICIAP*, (2001).

[10] M. Lalonde and Y. Li. Road sign recognition - survey of the state of art. Technical Report CRIM-IIT-95/09-35, (1995).

[11] D. Kellmeyer and H. Zwahlen. Detection of highway warning signs in natural video images using color image processing and neural networks. In *ICNN*, (1994).

[12] G. Loy and N. Barnes. Fast shape-based road sign detection for a driver assistance system. In *IROS*, vol. 1, pp. 70–75, (2004).

[13] D. Gavrila and V. Philomin. Real-time object detection for "smart" vehicles. In *ICCV*, pp. 87–93, (1999).

[14] R. Timofte, K. Zimmermann, and L. Van Gool, Multi-view traffic sign detection, recognition, and 3d localisation, *submitted to Machine Vision and Applications.* (2011).

[15] P. F. Felzenszwalb, R. B. Girshick, and D. McAllester. Cascade object detection with deformable part models. In *IEEE Conference on Computer Vision and Pattern Recognition (CVPR)*, (2010).

[16] V. A. Prisacariu, R. Timofte, K. Zimmermann, I. Reid, and L. V. Gool, Integrating object detection with 3d tracking towards a better driver assistance system, *ICPR.* pp. 3344–3347, (2010). doi: http://doi.ieeecomputersociety.org/10.1109/ICPR.2010.816.

[17] J. Stallkamp, M. Schlipsing, J. Salmen, and C. Igel. The German Traffic Sign Recognition Benchmark: A multi-class classification competition. In *submitted to International Joint Conference on Neural Networks*, (2011).

[18] R. Timofte and L. Van Gool. Fast approaches to large-scale classification. In *submitted to International Joint Conference on Neural Networks*, (2011).

[19] N. Dalal and B. Triggs. Histograms of oriented gradients for human detection. In *CVPR 2005*, vol. 1, pp. 886–893, (2005).

[20] S. Lafuente, P. Gil, R. Maldonado, F. López, and S. Maldonado. Traffic sign shape classification evaluation i: Svm using distance to borders. In *IEEE Intelligent Vehicles Symposium, pp 654–658*, (2005).

[21] C. Nunn, A. Kummert, and S. Muller-Schneiders. A novel region of interest selection approach for traffic sign recognition based on 3d modelling. In *IEEE Intelligent Vehicles Symposium, pp 654–658*, (2008).

[22] N. Pettersson, L. Petersson, and L. Andersson. The histogram feature - a resource-efficient weak classifier. In *IEEE Intelligent Vehicles Symposium, pp 678 - 683*, (2008).

[23] F. Moutarde, A. Bargeton, A. Herbin, and L. Chanussot. Robust on-vehicle real-time visual detection of american and european speed limit signs, with a modular traffic signs recognition system. In *IEEE Intelligent Vehicles Symposium*, (2007).

[24] A. Ruta, Y. Li, and X. Liu. Towards real-time traffic sign recognition by class-specific discriminative features. In *British Machine Vision Conference*, (2007).

[25] A. Broggi, P. Cerri, P. Medici, P. Porta, and G. Ghisio. Real time road signs recognition. In *Intelligent Vehicles Symposium, 2007 IEEE*, pp. 981–986 (June, 2007).

[26] S. J. Osher and R. P. Fedkiw, *Level Set Methods and Dynamic Implicit Surfaces.* (Springer, October 2002). ISBN 0387954821.

[27] B. Rosenhahn, T. Brox, and J. Weickert, Three-dimensional shape knowledge for joint image segmentation and pose tracking, *International Journal of Computer Vision.* **73** (3), 243–262, (2007).

[28] C. Schmaltz, B. Rosenhahn, T. Brox, D. Cremers, J. Weickert, L. Wietzke, and G. Sommer. Region-based pose tracking. In *Proc. 3rd Iberian Conference on Pattern Recognition and Image Analysis* (June, 2007).

[29] V. Prisacariu and I. Reid. Pwp3d: Real-time segmentation and tracking of 3d objects. In *Proceedings of the 20th British Machine Vision Conference* (September, 2009).

[30] C. Bibby and I. Reid. Robust real-time visual tracking using pixel-wise posteriors. In *ECCV 2008*, pp. 831–844, (2008).

[31] J. Matas, O. Chum, M. Urban, and T. Pajdla. Robust wide baseline stereo from maximally stable extremal regions. In *BMVC*, pp. 384–393, (2002).

[32] P. Viola and M. Jones. Robust real-time face detection. In *IEEE International Conference on Computer Vision*, vol. 2, pp. 747–757, (2001).

[33] A. Bosch, A. Zisserman, and X. Munoz. Representing shape with a spatial pyramid kernel. In *Proceedings of the 6th ACM international conference on Image and video retrieval*, pp. 401–408, (2007).

[34] S. Maji and J. Malik. Fast and accurate digit classification. Technical Report UCB/EECS-2009-159, EECS Department, University of California, Berkeley, (2009).

[35] C. ping Lu, G. D. Hager, I. C. Society, and E. Mjolsness, Fast and globally convergent pose estimation from video images, *T-PAMI.* **22**, 610–622, (2000).

[36] M. Everingham, L. Van Gool, C. K. I. Williams, J. Winn, and A. Zisserman. The PASCAL Visual Object Classes Challenge 2009 (VOC2009) Results. http://www.pascal-network.org/challenges/VOC/voc2009/workshop/index.html.

[37] D. Cai, X. He, and J. Han, Srda: An efficient algorithm for large-scale discriminant analysis, *IEEE Trans. on Knowl. and Data Eng.* **20**(1), 1–12, (2008).

CHAPTER 3.6

SELF-CALIBRATION OF CAMERA NETWORKS: ACTIVE AND PASSIVE METHODS

Marcel Brückner, Ferid Bajramovic and Joachim Denzler

Chair for Computer Vision, Friedrich Schiller University of Jena,
Ernst-Abbe-Platz 2, 07743 Jena, Germany
{*marcel.brueckner, ferid.bajramovic, joachim.denzler*}*@uni-jena.de*

This chapter discusses the problem of extrinsic self-calibration of camera networks. Self-calibration in this context means that no artificial calibration object or any kind of user interaction are necessary. The cameras estimate their position and orientation in a common world coordinate system only from the images they record. Since many real life multi-camera systems are mounted on pan-tilt units, we will also focus on the active calibration of such a system. After explaining the basic approaches for the relative pose estimation of a camera pair, several uncertainty measures for relative poses are presented. These are used as edge weights in a graph-based approach to calibrate the multi-camera system with a minimal set of relative poses. Furthermore we will show how active camera control can be used to simplify the problem of extracting correct point correspondences and to actively estimate consistent translation scales in a camera triangle. In several experiments it is shown that each step of our active approach improves the calibration accuracy and reduces the number of outliers. The final calibration reaches a high accuracy.

1. Introduction

Nowadays camera and sensor networks play an important role in various applications. Examples of such multi-camera systems cover security applications for surveillance of large public areas, processing of sport events, smart homes, or even other sciences beyond technical disciplines, like multi-camera fly tracking.[1] One important prerequisite of such applications is the calibration of the cameras involved. Without such a calibration, quantitative analysis is hardly possible.

Camera calibration is a well understood problem in computer vision. A multi-camera calibration consists of several parts: the intrinsic calibration of each single camera and the extrinsic calibration of the complete multi-camera system. In this chapter we will focus on the latter. Provided that it is possible to place a so called calibration pattern in the scene, very effective and accurate methods exist in the literature to compute the relations between each individual camera in the camera network with high accuracy. If the calibration pattern is not visible in all cameras,

Fig. 1.   A multi-camera system (left) consisting of six pan-tilt zoom cameras (white circles). The cameras are mounted near the intersection of the pan and tilt axes (right).

calibration becomes more difficult. In the applications mentioned in the beginning, it is rather difficult if not impossible to use a special pattern placed in the scene to calibrate the system. The reason is that either the size of the scene covered by the cameras is too big, or the environment makes it impossible to place a pattern that would be visible in most of the cameras. As a consequence automatic methods are necessary that allow calibration of camera networks without artificially modifying the scene and without user interaction. We call such a procedure camera network self-calibration, i. e. the estimation of the position and orientation of each camera in a common world coordinate system.

During the past decades a shift has been made from passive scene analysis, where a system has no influence on the way image data is acquired, to active approaches, where a system can actively control its sensor acquitsion parameters, like focal length or view point of the cameras. The latter principle has been intensively discussed in the area of active vision during the 80's.[2] In general state estimation, as which many computer vision can be interpreted, can benefit from active sensor data acquisition.[3] Examples from object tracking,[4] object recognition[5] or 3d reconstruction[6] show improvements in quality and robustness of such systems compared to passive ones. The results presented in this chapter are a continuation of our research in the area of active vision, investigated for multi-camera self-calibration. We first present a method where self-calibration must deal with the configuration of the whole camera network and the viewing directions of each individual camera. Such a situation is typical in environments where the cameras are mounted statically without any further degree of freedom. Calibration accuracy is based on the reliability of the estimated point correspondences. In cases, where the images contain ambiguities with respect to point correspondence estimation, the overall quality is limited. In the second part we introduce a method for active self-calibration of a multi-camera system consisting of pan-tilt unit mounted cameras (Fig. 1), i. e. the system is able to optimize image acquisition in order to improve the following self-calibration step. Active camera control aims at maximizing the chance for and number of correct point correspondences. Furthermore we use active camera control to support the global scale computation step in such a network. We compare the two strategies and show that active camera control can be successfully integrated in the area of camera network calibration leading to high accuracy in the estimation of the extrinsic camera parameters.

2. Related Work

Existing methods for extrinsic multi-camera calibration can be very roughly classified by the type of input or scene knowledge they require: 1. Pattern based: a classical (planar) calibration pattern either has to be visible in all images[7,8] or the poses of multiple calibration objects have to be known.[9] 2. LED based: some easily detectable, moving single feature, like an LED in a dark room, is recorded over time.[10–14] 3. Self-calibration (in the broader sense): images are taken from an unknown scene, typically with some (unknown) 3D structure and texture. Clearly, the third class is most appealing from a practical point of view, as it is the most flexible one. From a pure multiple view geometry point of view, multi-camera calibration can be interpreted as a structure from motion (sub)problem.[15]

Snavely et al.[16] and Läbe and Förstner[17] calibrate multi-camera systems by establishing multi-camera point correspondences between two and more cameras. During the calibration, cameras are added incrementally and after each step a bundle adjustment[18] is performed. In case of self-calibration of multiple physical cameras, however, point correspondences have to be extracted from images in a medium or wide baseline situation, which typically leads to a very high portion of outliers. The estimation of point correspondences between three and more cameras is extremely challenging for wide-baseline configurations.

As the relative pose of two cameras can be estimated even in presence of very many outliers,[19,20] many approaches use known relative poses between some camera pairs as input. The quality of the results, of course, highly depends on the relative pose estimates. This problem has been briefly mentioned by Chen, David and Slusallek.[11] It has been more recently addressed by Martinec and Pajdla.[21,22] They weight relative poses by a measure based on the number of inliers found by RANSAC and also on the importance of a relative pose. While their measure is plausible, a theoretical justification is missing.

A further method has been proposed by Vergés-Llahí, Moldovan and Wada.[23] They use an uncertainty measure for relative poses, which consists of a residual and a constraint violation term. Note, however, that constraint violations can be avoided by using the five point algorithm.[19,24] Selection of relative poses is performed by finding triangle connected shortest paths in a graph connecting cameras according to known relative poses with edge weights set to these uncertainty values. This graph-based approach was adopted by Bajramovic and Denzler.[25] They present a much more efficient algorithm for the relative pose selection by building an auxiliary graph based on camera triangles.

However, all of these methods are designed for static cameras and do not use the benefits of active camera control. Sinha and Pollefeys[26] suggest a method where

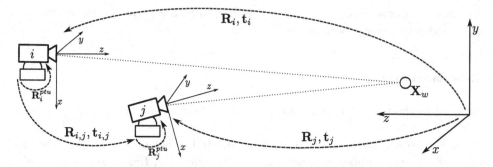

Fig. 2. Each camera has its own coordinate system. The absolute camera pose describes the transformation between the world coordinate system and the camera coordinate system. The transformation between two camera coordinate systems is described by the relative pose.

each pan-tilt zoom camera builds a high resolution panorama image. These images are used for relative pose estimation. However, these huge images can contain many ambiguities which affect the extraction of correct point correspondences. The calibration method of Chippendale and Tobia[27] defines an observer camera which searches for the continuously moving other cameras. If the observer spots some other camera the relative pose between the two cameras is extracted by detecting the circle shape of the camera lens and tracking some special predefined camera movements. The applicability and accuracy of this method highly depends on the distance between the cameras. Brückner and Denzler[28] use active camera control and an image based probability measure for the relative poses to optimize relative poses and to estimate the translation scales and one of the relative poses in each camera triangle.

3. Basics and Notation

This section starts with the basic notations for extrinsic multi-camera calibration. We will then shortly describe the basic ideas for extracting point correspondences and how the relative pose can be estimated based on these.

3.1. *Camera Model and Relative Pose between Cameras*

A world point $\mathbf{X}_w$ is projected to the homogeneous image point

$$\mathbf{x}_i \stackrel{\text{def}}{\cong} \mathbf{K}_i \mathbf{R}_i^{ptu} \left(\mathbf{R}_i \mathbf{X}_w + \mathbf{t}_i \right) \tag{1}$$

of camera i, where $\mathbf{R}_i, \mathbf{t}_i$ is the absolute camera pose or the extrinsic camera parameters (rotation and translation), $\mathbf{K}_i$ is the pinhole matrix[15] and $\mathbf{R}_i^{ptu}$ is the rotation of the pan-tilt unit. We assume the pan and tilt axes to be identical to the Y and X axes of the camera coordinate system, respectively. From this point on, when talking about the camera orientation and position we actually mean $\mathbf{R}_i, \mathbf{t}_i$ with

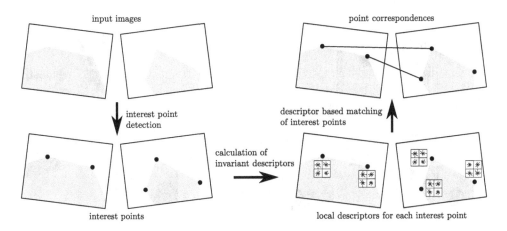

input images

point correspondences

interest point detection

descriptor based matching of interest points

calculation of invariant descriptors

interest points

local descriptors for each interest point

Fig. 3. The three steps of point correspondence extraction.

no pan-tilt rotation $\mathbf{R}^{ptu} = \mathbf{I}$. The relative pose between two cameras i and j is defined as $\mathbf{R}_{i,j} \stackrel{\text{def}}{=} \mathbf{R}_j \mathbf{R}_i^{-1}$ and $\mathbf{t}_{i,j} \stackrel{\text{def}}{=} \mathbf{t}_j - \mathbf{R}_j \mathbf{R}_i^{-1} \mathbf{t}_i$. It is defined such that it maps a 3D point $\mathbf{X}_i$ from the i-th camera coordinate system to the j-th one as follows: $\mathbf{X}_j = \mathbf{R}_{i,j} \mathbf{X}_i + \mathbf{t}_{i,j}$. Note that relative poses have a double subscript, while absolute poses have a single one. The described relationships for multi-camera calibration are illustrated in Fig. 2. When estimating relative poses from images only, the scale of $\mathbf{t}_{i,j}$ cannot be determined. We emphasize unknown scale by an asterisk: $\mathbf{t}_{i,j}^*$. Relative pose estimation will be treated in Section 3.3. Calibrating a multi-camera system with known relative poses up to scale $\mathbf{R}_{i,j}, \mathbf{t}_{i,j}^*$ means computing absolute poses $\mathbf{R}_k, \mathbf{t}_k$ expressed in an arbitrary (but common) world coordinate system up to only one common, unknown scale factor—in other words, we want to calibrate the multi camera system up to a 3D similarity transformation.

3.2. *Extraction of Point Correspondences*

In this section we want to impart the basic idea of point correspondence extraction. With (1) in mind a world point $\mathbf{X}_w$ is projected to the image points $\mathbf{x}_i$ and $\mathbf{x}_j$ of two different cameras i and j. The two image points $\mathbf{x}_i$ and $\mathbf{x}_j$ correspond. The problem of extracting such point correspondences from camera images is an ongoing topic in research. The biggest challenge are so-called wide-baseline camera setups where the distance between the cameras in relation to the camera's distance to the scene and the difference between the viewing directions between the cameras is big. Problematic are also different camera sensor types and qualities. The optimal approach for point correspondence extraction should be invariant to all of these influences. Most approaches consist of the three steps outlined in Fig. 3:

(1) detection of interest points $\mathcal{C}_i = \{\mathbf{x}_1, \ldots, \mathbf{x}_n\}$ and $\mathcal{C}_j = \{\mathbf{x}_1, \ldots, \mathbf{x}_m\}$,
(2) calculating (invariant) local descriptors $\mathbf{des}(\mathbf{x})$ for these interest points,

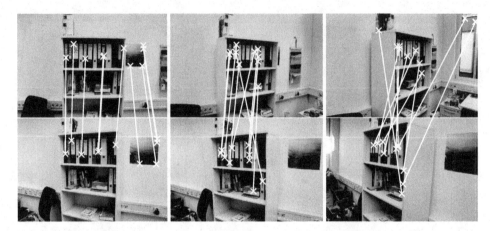

Fig. 4. The ten best point correspondence matches for increasing baselines. Only two out of the ten point correspondences of the wide-baseline image pair on the right are correct. A further increase of the baseline will result in not a single correct point correspondence.

(3) and matching the interest points based on their descriptors which result in point correspondences $\mathbf{d} \in \mathcal{D}$.

An example for an approach that is capable to extract point correspondences from wide-baseline camera setup is the work of Lowe[29] where interest points are detected using a difference of gaussian (DoG) detector, for each interest point a scale-invariant feature transform (SIFT) descriptor is calculated and matching is carried out using a two neareast neighbors matching (2NN). The SIFT descriptor is (to a certain extent) invariant with respect to scaling, 2D rotation, and illumination changes. Some other examples for descriptors are the maximally stable extremal regions (MSER)[30] and the histogram of oriented gradients (HoG).[31]

However, all of these approaches only work to a certain extent. If the cameras have very different viewpoints on a scene, projective influences and occlusions complicate or make it impossible to establish correct point correspondences. Fig. 4 show the results using the approach of Lowe[29] on increasing baselines. Almost all estimated point correspondences of the image pair on the left are correct. But only two out of the ten point correspondences of the wide-baseline image pair on the right are correct. Increasing the baseline any further will result in not a single correct point correspondence.

In Section 5 we will show how active cameras can use rotation and zoom to reduce these projective influences and simplify the extraction of correct point correspondences. Another approach is to avoid point correspondences as far as possible. Instead of matching points of the two interest point sets, the matching probability between two point of the sets is calculated. In Section 4.1 we show how to use these probabilistic point correspondences for common field of view detection and for the evaluation of relative poses. We then use these methods in Section 5.3 for estimating the translation scales and the third relative pose in a camera triangle without any point correspondences and given only two of the relative poses.

3.3. *Relative Pose Estimation*

The relative pose between two cameras cannot be estimated directly. Instead the essential matrix

$$\mathbf{E}_{i,j} \stackrel{\text{def}}{=} \left[\mathbf{t}_{i,j}^*\right]_\times \mathbf{R}_{i,j} \tag{2}$$

needs to be estimated, where $\left[\mathbf{t}_{i,j}^*\right]_\times$ is the skew-symmetric cross-product matrix of the translation vector $\mathbf{t}_{i,j}^*$. The most important property of the essential matrix is the epipolar constraint

$$\tilde{\mathbf{x}}_j^{\mathrm{T}} \mathbf{E}_{i,j} \tilde{\mathbf{x}}_i = 0 \tag{3}$$

for camera normalized image points $\tilde{\mathbf{x}}_i \stackrel{\text{def}}{=} \mathbf{K}_i^{-1}\mathbf{x}_i$. The state-of-the-art approach for estimating the essential matrix of two cameras is the five-point algorithm.[19,24] It needs at least five camera normalized point correspondences. Note that in most cases the five-point algorithm does not have a unique solution, it rather has up to ten solutions. Furthermore, from each of these essential matrices four relative poses can be extracted.[15] However, using the constraint that observed 3D points need to lie in front of both cameras and using additional point correspondences for verification enables to select the correct hypothesis. As mentioned in Section 3.1, the scale of the estimated translation is not known. For the calibration of a multi-camera system these scales need to be estimated consistently.

Since most point correspondence sets contain outliers, the relative pose estimation should be embedded into a robust estimation scheme. These schemes randomly sample a minimal number of point correspondences which are used to estimate a hypothesis for a relative pose. Based on this hypothesis the number of inlier point correspondences (RANdom SAmple Consensus – RANSAC[32]) or a Likelihood (Maximum Likelihood Estimation SAmple Consensus – MLESAC[33]) is calculated. The final estimate is the relative pose with the highest number of inliers or Likelihood, respectively.

A popular distance measure for the RANSAC scheme is the Sampson epipolar distance[15] which approximates the squared reprojection error. Another suitable distance measure is the closest distance between two viewing rays

$$d_e^{i,j}\left(\tilde{\mathbf{x}}_i, \tilde{\mathbf{x}}_j\right) \stackrel{\text{def}}{=} \min_{\lambda_i, \lambda_j} \left\| \left(\lambda_i \mathbf{R}_{i,j} \tilde{\mathbf{x}}_i + \frac{\mathbf{t}_{i,j}}{\|\mathbf{t}_{i,j}\|_2}\right) - \lambda_j \tilde{\mathbf{x}}_j \right\|_2 \quad \text{with} \quad \lambda_i, \lambda_j > 0 \ . \tag{4}$$

Since the translation is normalized to unit length, it is possible to define the inlier threshold relative to the camera distance. The scale factors λ_i and λ_j need to be positive which affects the direction of the viewing rays and is similar to the constraint of 3D points to lie in front of both cameras.

4. Uncertainty-Based Multi-Camera Calibration

This section will present several uncertainty-based methods for multi-camera calibration. these methods cover the problem of common field of view detection, the uncertainty of relative poses and the calibration of a complete multi-camera system based on these uncertainties. These methods form also an important basis for the active approaches described in Section 5.

4.1. *Common Field of View Detection*

Common field of view detection consists of deciding which image pairs show a common part of the world. It is beneficial for active and passive calibration approaches. During a passive calibration it can be used to separate image pairs which cannot be used to estimate a relative pose because the images do not show a common part of the world.[34] It can also be used to orientate two pan-tilt units in a way that they can estimate their relative pose.[35]

In Ref. 35 we presented and compared different geometric and probabilistic image similarity measures for common field of view detection. We will briefly describe the probabilistic method which gave the best results in our experiments. The main idea consists of using the normalized joint entropy of point correspondence probability distributions as a measure. A low entropy indicates peaked distributions due to unambiguously matchable points. High entropy values, however, result from more or less uniform correspondence distributions, which indicate unrelated images.

Given the interest points and descriptors of two camera images, a conditional correspondence probability distribution

$$p\left(\mathbf{x}_j \mid \mathbf{x}_i\right) \propto \exp\left(-\frac{d_d^{i,j} - d_N(\mathbf{x}_i)}{\lambda_d \, d_N(\mathbf{x}_i)}\right), \tag{5}$$

for each $\mathbf{x}_i$ is constructed, where λ_d is the inverse scale parameter of the exponential distribution, $d_d^{i,j} = \text{dist}(\mathbf{des}(\mathbf{x}_i), \mathbf{des}(\mathbf{x}_j))$ is the Euclidean distance between the descriptors of the points $\mathbf{x}_i$ and $\mathbf{x}_j$, and $d_N(\mathbf{x}_i) = \min_j(d_d^{i,j})$ denotes the distance of the nearest neighbor of the point $\mathbf{x}_i$. Each of the resulting conditional probability distributions $p(\mathbf{x}_j \mid \mathbf{x}_i)$ has to be normalized such that $\sum_{\mathbf{x}_j \in \mathcal{C}_j} p(\mathbf{x}_j \mid \mathbf{x}_i) = 1$ holds.

The conditional probability distributions are used to calculate the normalized joint entropy which is defined as

$$H(\mathcal{C}_i, \mathcal{C}_j) \stackrel{\text{def}}{=} -\frac{1}{\eta} \sum_{\mathbf{x}_i \in \mathcal{C}_i} \sum_{\mathbf{x}_j \in \mathcal{C}_j} p(\mathbf{x}_i) p\left(\mathbf{x}_j \mid \mathbf{x}_i\right) \log\left(p(\mathbf{x}_i) p\left(\mathbf{x}_j \mid \mathbf{x}_i\right)\right), \tag{6}$$

where $\eta = \log(nm)$ is the maximum joint entropy and $p(\mathbf{x}_i)$ is a uniform distribution if no prior information about the interest points is available. The joint entropy is maximized if all conditional probability distributions $p\left(\mathbf{x}_j \mid \mathbf{x}_i\right)$ are uniform. It is minimized if every interest point in the first image has a unique corresponding point

camera dependency graph $\mathcal{G}_R$ triangle graph $\mathcal{G}_T$ extended triangle graph $\mathcal{G}_E$

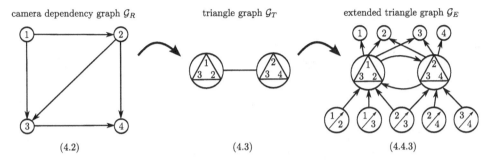

(4.2)   (4.3)   (4.4.3)

Fig. 5. From left to right: a camera dependency graph with four cameras and five known relative poses, the associated triangle graph, and the extended triangle graph. Figure similar to Ref. 25.

with an identical descriptor in the second image. For further details the reader is referred to Ref. 35.

In Ref. 28 we presented the probability of relative poses

$$p\left(\mathbf{R}_{i,j}, \mathbf{t}_{i,j}^*\right) \propto \sum_{\tilde{\mathbf{x}}_i \in \mathcal{C}_i} \sum_{\tilde{\mathbf{x}}_j \in \mathcal{C}_j} p\left(\mathbf{R}_{i,j}, \mathbf{t}_{i,j}^* \mid \tilde{\mathbf{x}}_j, \tilde{\mathbf{x}}_i\right) p\left(\tilde{\mathbf{x}}_j \mid \tilde{\mathbf{x}}_i\right) p\left(\tilde{\mathbf{x}}_i\right), \tag{7}$$

which incorporates image and geometric information. Here,

$$p\left(\mathbf{R}_{i,j}, \mathbf{t}_{i,j}^* \mid \tilde{\mathbf{x}}_j, \tilde{\mathbf{x}}_i\right) \overset{\text{def}}{=} \exp\left(-d_e^{i,j}\left(\tilde{\mathbf{x}}_i, \tilde{\mathbf{x}}_j\right) / \lambda_e\right) \tag{8}$$

is an exponential distribution using the distance measure of (4) and the inverse scale parameter λ_e, $p\left(\tilde{\mathbf{x}}_j \mid \tilde{\mathbf{x}}_i\right)$ is the conditional correspondence probability of (5) and $p\left(\tilde{\mathbf{x}}_i\right)$ is a uniform distribution if no prior information about the interest points is available. The probability distribution (7) is a very reliable measure for the quality of a relative pose. We will use it during our active calibration approach in Section 5 as well as part of an uncertainty measure for relative poses in the following sections.

4.2. *Camera Dependency Graph*

A multi-camera system together with the set of all known relative poses can be represented by the *camera dependency graph*[23] $\mathcal{G}_\mathrm{R} = (\mathcal{V}, \mathcal{E})$: each camera is a vertex $i \in \mathcal{V} = \{1, \ldots, c\}$ and camera i is connected to camera j iff the relative pose $\mathbf{R}_{i,j}, \mathbf{t}_{i,j}^*$ is known. An example is given in Figure 5 (left). Note that edges are directed to distinguish $\mathbf{R}_{i,j}, \mathbf{t}_{i,j}^*$ from $\mathbf{R}_{j,i}, \mathbf{t}_{j,i}^*$. As relative poses can be easily inverted, however, all edges will be treated as bidirectional.

4.3. *Calibration via Triangulation*

A *triangle* in $\mathcal{G}_\mathrm{R}$ consists of three cameras and three known relative poses. If one of the scales $\|\mathbf{t}_{i,j}\|$ is known (or fixed), the remaining two scales can be computed

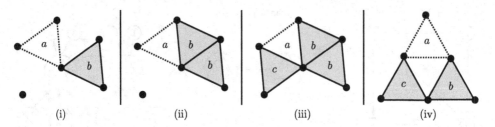

Fig. 6. Situations that can occur when inserting a scaled camera triangle into a camera dependency graph. The inserted triangle *a* has doted lines. Existing triangles are gray and share the same letter if they are in a triangle connected subgraph.

by triangulation.[15] By treating one triangle after the other such that all consecutive triangles have a common edge, all scales can be computed up to one common unknown scale parameter. The calibration algorithm can be formulated nicely by using the *triangle graph* $\mathcal{G}_{\mathrm{T}} = (\mathcal{T}, \mathcal{E}_{\mathrm{T}})$, which is defined as follows (Figure 5, center): each triangle in $\mathcal{G}_{\mathrm{R}}$ becomes a vertex in $\mathcal{G}_{\mathrm{T}}$, two such vertices are connected iff the two triangles have a common edge in $\mathcal{G}_{\mathrm{R}}$. Once the scales have been estimated, the absolute poses can be extracted from the relative poses based on the equations in Section 3.1. These steps can be integrated into the following *relative pose calibration algorithm*: First choose a starting triangle $(i, j, k) \in \mathcal{T}$ and set $\|\mathbf{t}_{i,j}\| = 1$, $\mathbf{R}_i = \mathbf{I}$, and $\mathbf{t}_i = 0$. Then traverse $\mathcal{G}_{\mathrm{T}}$, e.g. via breadth first search (BFS). When visiting a triangle, estimate the missing scale factors via triangulation and extract the missing absolute poses.

In the case that camera triangles are not scaled in this subsequent order (e. g. like in the active calibration approach of Section 5), a camera dependency graph can be used to ensure a consistent scaling. Before the first triangle is inserted, this graph consists of one vertex for each camera and has no edges. If the relative poses and scales of a camera triangle are known, it is inserted into this new graph.

The problem is that the scaling is only consistent within each triangle. Hence, four different situations need to be distinguished when inserting a new relative pose triangle into the camera dependency graph. These situations are illustrated in Fig. 6. The first situation is the trivial case of inserting a single triangle without conflicting edges. In the second case the inserted triangle shares a common edge with a triangle connected subgraph. This situation requires a rescaling of the triangle. The scale factor is defined by the relation between the translation lengths of the two common edges (the translation direction of these is identical). The third case creates a triangle connection between two prior simple connected parts of the graph. This requires rescaling the triangle and one of the two graph parts. The triangle in the fourth case cannot be inserted because the correct scaling between the three subgraphs is undefined. After inserting a camera triangle into the graph we need to check if two edges of some camera triangle are in the same triangle connected sub graph. In this case the relative pose of the third edge results from the poses of these two edges.

4.4. Selection of Relative Poses

In the relative pose calibration algorithm, the traversal order is unspecified. Furthermore, in case of a dense camera dependency graph, only a rather small subset of triangles has to be visited in order to calibrate all cameras. Thus, the traversal order corresponds to a selection of triangles and thus relative poses. We calibrate along shortest triangle connected paths in the camera dependency graph. Each relative pose edge is weighted by an uncertainty measure.

An uncertainty measure should not only capture the local precision of a relative pose estimate, but also the global uncertainty caused by ambiguities. Furthermore, it should be well suited for a theoretically sound formulation of relative pose selection as an optimization problem. Engels and Nistér[36] proposed a sampling based approach to estimate the global uncertainty of a relative pose estimate. Based on that, we will define three uncertainty measures.

4.4.1. Uncertainty Measures

First, we will briefly summarize their approach in a slightly different formulation (focussing on the five point variant). Let $\mathcal{D}$ denote the set of all point correspondences. The posterior probability density function of a relative pose $\mathbf{R}, \mathbf{t}^*$ is denoted by $p(\mathbf{R}, \mathbf{t}^* \mid \mathcal{D}) \propto p(\mathcal{D} \mid \mathbf{R}, \mathbf{t}^*)p(\mathbf{R}, \mathbf{t}^*)$. We assume a flat prior $p(\mathbf{R}, \mathbf{t}^*)$ if nothing is known about the relative pose of a camera pair. Very similar to Engels and Nistér,[36] we define the data likelihood based on the Cauchy distribution as follows:

$$
p(\mathcal{D} \mid \mathbf{R}, \mathbf{t}^*) \propto \left(\prod_{\mathbf{d} \in \mathcal{D}} \frac{\alpha}{\alpha^2 + s(\mathbf{R}, \mathbf{t}^*, \mathbf{d})} \right)^{|\mathcal{D}|^{-\phi}}, \tag{9}
$$

where α is a scale parameter, which is typically set to 1, ϕ widens peaks of the distribution ($\phi = 0.5$ is recommended, $\phi = 0$ corresponds to assuming independent point correspondences), and $s(\mathbf{R}, \mathbf{t}^*, \mathbf{d})$ denotes the Sampson epipolar error,[15,36] which approximates the squared reprojection error of the point correspondence $\mathbf{d}$ and the relative pose $\mathbf{R}, \mathbf{t}^*$ together with the known intrinsic calibration.

The density over translation directions is given by $p(\mathbf{t}^* \mid \mathcal{D}) = \int p(\mathbf{R}, \mathbf{t}^* \mid \mathcal{D})d\mathbf{R}$. Engels and Nistér[36] approximate the integral using Laplace's method. Instead, we use the even simpler approximation $p(\mathbf{t}^* \mid \mathcal{D}) \propto \max_{\mathbf{R}} p(\mathbf{R}, \mathbf{t}^* \mid \mathcal{D})$, which equals the Laplace approximation if $\det(\nabla^2_{\mathbf{R}} \log p(\mathbf{R}, \mathbf{t}^* \mid \mathcal{D}))$ is independent of $\mathbf{t}^*$.[37]

The density $p(\mathbf{t}^* \mid \mathcal{D})$ is represented discretely over the range of $\mathbf{t}^*$ by a $c \times c$ matrix $\mathbf{A}$. The idea consists of generating relative pose hypotheses by sampling, computing the according indices (a, b) into $\mathbf{A}$, and keeping the maximum for each entry of $\mathbf{A}$. As a byproduct, an estimate for $\mathbf{t}^*$ can be computed, which we denote by $\hat{\mathbf{t}}^*$ in this section: $\hat{\mathbf{t}}^* \approx \operatorname{argmax}_{\mathbf{t}^*} p(\mathbf{t}^* \mid \mathcal{D})$. The following algorithm contains further details:

(1) Initialize $\mathbf{A} = 0$ and $\gamma = 0$ (storing the maximum occuring value of $p(\mathbf{R}, \mathbf{t}^* \mid \mathcal{D})$).

(2) Repeat m times:

    (a) Draw a sample from $\mathcal{D}$ and estimate the relative pose $\mathbf{R}, \mathbf{t}^*$.

    (b) Normalize $\mathbf{t}^*$ to length 1, choose the sign such that $\mathbf{t}^*$ is in the upper unit-hemisphere, and map the first two coordinates to discrete matrix indices (a, b).

    (c) Set $\mathbf{A}(a, b) = \max(\mathbf{A}(a, b), p(\mathbf{R}, \mathbf{t}^* \mid \mathcal{D}))$.

    (d) If $\gamma > p(\mathbf{R}, \mathbf{t}^* \mid \mathcal{D})$, set $\gamma = p(\mathbf{R}, \mathbf{t}^* \mid \mathcal{D})$ and $\hat{\mathbf{t}}^* = \mathbf{t}^*$.

(3) Normalize $\mathbf{A}$ such that $\sum_{a,b} \mathbf{A}(a, b) = 1$.

We choose a rather low resolution of $c = 100$ for $\mathbf{A}$ and only perform $m = 10000$ sampling iterations. Nevertheless, this still gives a reasonable approximation to $p(\mathbf{t}^* \mid \mathcal{D})$.

In experiments we have shown[25] that equation (9) does not work well in case of a large amount of outliers. As an alternative, we propose the following data likelihood based on the Blake-Zisserman distribution[a]:[15]

$$p(\mathcal{D} \mid \mathbf{R}, \mathbf{t}^*) \propto \left(\prod_{\mathbf{d} \in \mathcal{D}} \left(\exp \left(-\frac{s(\mathbf{R}, \mathbf{t}^*, \mathbf{d})}{\sigma^2} \right) + \epsilon \right) \right)^{|\mathcal{D}|^{-\phi}}, \qquad (10)$$

where σ^2 is the variance of the Gaussian component (we set $\sigma = 1$) and ϵ defines the relative weight of the uniform component (we set $\epsilon = 0.0002$).

We investigate the following uncertainty measures $\omega(\hat{\mathbf{t}}^*)$ based on $p(\mathbf{t}^* \mid \mathcal{D})$, which is approximated by $\mathbf{A}$:

(1) The local *information* measure $\omega_I(\hat{\mathbf{t}}^*) = -\log p(\hat{\mathbf{t}}^* \mid \mathcal{D})$.

(2) The global *entropy* measure $\omega_E(\hat{\mathbf{t}}^*) = -\int p(\mathbf{t}^* \mid \mathcal{D}) \log p(\mathbf{t}^* \mid \mathcal{D}) d\mathbf{t}^*$.

(3) The *smoothed information* measure, defined as the information of the smoothed density: $\omega_S(\hat{\mathbf{t}}^*) = -\log \int \mathcal{N}(\mathbf{t}^*; \hat{\mathbf{t}}^*, \mathbf{\Sigma}) p(\mathbf{t}^* \mid \mathcal{D}) d\mathbf{t}^*$, where $\mathcal{N}(\mathbf{t}^*; \hat{\mathbf{t}}^*, \mathbf{\Sigma})$ denotes the normal distribution with mean $\hat{\mathbf{t}}^*$ and covariance matrix $\mathbf{\Sigma}$ (we set $\mathbf{\Sigma} = \sqrt{5}\mathbf{I}$). Note that the integral can be interpreted as a variant of the confidence measure used by Engels and Nistér[36] (using a Gaussian kernel instead of a confidence area).

In our experiments[25] the best results were obtained by using the global entropy measure.

As mentioned before the quality of the relative pose estimates depends on the quality of the point correspondences. Hence, the active calibration approach of Section 5 uses a purely image based uncertainty measure: the information of the relative pose

$$\omega\left(\mathbf{R}, \mathbf{t}^*\right) \overset{\text{def}}{=} -\log\left(p\left(\mathbf{R}, \mathbf{t}^*\right)\right), \qquad (11)$$

---

[a]If we make the reasonable assumption that outlier points only occur within the image area, the uniform part of the distribution can be limited to finite support and we actually get a valid probability density function.

where $p(\mathbf{R}, \mathbf{t}^*)$ is the probability of (7). This also avoids the discretization effects of the other weights.

4.4.2. *Criterion for the Selection of Relative Poses*

Calibrating the pose of a camera j relative to a reference camera i involves a set $\mathcal{P} = \{(i_1, j_1), \ldots, (i_l, j_l)\}$ of relative poses such that $\mathcal{P}$ is a triangle connected path from i to j. As pointed out before, there are often several alternative sets $\mathcal{P}$. Based on the uncertainty measures defined above, we assign an uncertainty to each set $\mathcal{P}$ and choose the set with minimum uncertainty. Assuming independence, we have $p(\mathcal{P} \mid \mathcal{D}) = \prod_k p(\mathbf{t}^*_{i_k, j_k} \mid \mathcal{D})$. This leads to the following expression for all three uncertainty measures: $\omega(\mathcal{P}) = \sum_k \omega(\mathbf{t}^*_{i_k, j_k})$. As an example, we present the derivation for the information measure: $\omega_{\mathrm{I}}(\mathcal{P}) = -\log p(\mathcal{P} \mid \mathcal{D}) = -\log \prod_k p(\mathbf{t}^*_{i_k, j_k} \mid \mathcal{D}) = \sum_k -\log p(\mathbf{t}^*_{i_k, j_k} \mid \mathcal{D}) = \sum_k \omega_{\mathrm{I}}(\mathbf{t}^*_{i_k, j_k})$.

Choosing a suitable set $\mathcal{P}$ with minimum uncertainty is equivalent to finding a shortest triangle connected path in the camera dependency graph, where the weight of an edge (k, l) is defined as the uncertainty $\omega(\mathbf{t}^*_{k,l})$ or $\omega(\mathbf{R}_{k,l}, \mathbf{t}^*_{k,l})$.

Using this approach to calibrate a whole multi camera system involves shortest triangle connected paths from i to all other cameras. Note, however, that the union $\mathcal{U}$ of these paths also has to be triangle connected. While the common reference camera only guarantees that $\mathcal{U}$ is connected, using a common reference edge instead actually guarantees that $\mathcal{U}$ is *triangle* connected. Hence, we use shortest paths from a common relative pose to all cameras. We choose the reference edge $e \in \mathcal{E}$ with minimum total uncertainty $\omega(\mathcal{U})$. In other words, we use the *shortest* triangle connected *shortest* paths subgraph.

4.4.3. *Computing Shortest Triangle Connected Paths*

In this section, we present an efficient algorithm for our selection criterion. First, we construct the directed *extended triangle graph* $\mathcal{G}_{\mathrm{E}}$ as follows (Figure 5, right): Each triangle becomes a vertex. Additionally, we add an *entry vertex* for each relative pose and an *exit vertex* for each camera. Each entry vertex is connected to each triangle containing the respective relative pose. The weight of such an edge is the summed uncertainty of all relative poses in the triangle. Each triangle is also connected to the exit vertex of each camera it contains. These edges have zero weight. Each pair of triangles, which is connected in $\mathcal{G}_{\mathrm{T}}$, is connected in both directions. The weight of each edge is the summed uncertainty of the *two* new relative poses in the target triangle (i.e. the common relative pose is ignored).

The graph is defined such that a shortest path from an entry vertex to an exit vertex corresponds to a shortest triangle connected path in $\mathcal{G}_{\mathrm{R}}$ with fixed first edge. We can thus use a standard shortest paths algorithm (e.g. Dijkstra) to compute the shortest paths from an entry vertex e to all exit vertices (*not* to *all* vertices in the graph!), and also the according shortest paths tree. This tree corresponds

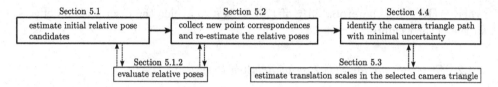

Fig. 7. The three steps of our multi-camera calibration method. Each step uses active camera control to a different extent and is described in the indicated section.

to a triangle connected subgraph $\mathcal{U}_e$ of the camera dependency graph, which contains triangle connected shortest paths from e to all cameras. We apply Dijkstra to all entry vertices e and choose the resulting subgraph $\mathcal{U}_e$ with minimum total uncertainty $\omega(\mathcal{U}_e)$.

5. Active Self-Calibration of Multi-Camera Systems

In Ref. 35 we presented an approach for calibration of an active multi-camera system consisting of c pan-tilt zoom cameras. For each camera the intrinsic parameters for different zoom steps are assumed to be known. Our calibration method consists of three steps which are illustrated in Fig. 7: an initial relative pose estimation with an evaluation of the relative poses, an optimization of these relative poses and a final estimation of the translation scale factors. Each step uses active camera control in a different way and to a different extent. Throughout this section we use image points which are normalized with respect to the camera and pan-tilt rotation $\tilde{\mathbf{x}} \overset{\text{def}}{=} \mathbf{R}^{ptu-1}\mathbf{K}^{-1}\mathbf{x}$.

5.1. *Initial Relative Pose Estimation and Evaluation*

5.1.1. *Initial Relative Pose Estimation*

Given the intrinsic parameters, each camera records as many images as necessary to cover its complete environment. Now each camera pair searches for image pairs sharing a common field of view (Section 4.1). This search can be viewed as a prematching of point correspondences which considers the local environment of each interest point. Hence, it decreases the chance of ambiguities disturbing the point matching process. Between each of these image pairs point correspondences are extracted (Section 3.2). Based on all extracted point correspondences of a camera pair we estimate the relative pose (Section 3.3).

Instead of selecting a single best pose, we select the m_p best poses based on the number of inliers. Since most of these poses are quite similar, we additionally constrain the selection to take only relative poses that satisfy a minimum rotation difference θ_R and translation difference θ_t to the already selected relative poses.

| backmost camera (without zoom) | backmost camera (with adjusted zoom) | front camera |

Fig. 8. Two cameras which are rotated in a way that their optical axes are (almost) aligned. After adjusting the zoom of the backmost camera, the camera images are quite similar. This simplifies the estimation of point correspondences.

5.1.2. *Evaluation of Relative Poses*

Now, each camera pair i, j performs the following procedure for each of its m_p relative pose candidates. First, the two cameras are rotated in a way that they look into the same direction and their optical axes are aligned (or a setup as close as possible to this). Camera i has to look in the direction $-\mathbf{R}_{i,j}\mathbf{t}_{i,j}$ and camera j looks at $-\mathbf{t}_{i,j}$. Additionally we search for the zoom step z of the backmost camera i with the highest image similarity by

$$\underset{z}{\operatorname{argmin}} H\left(\mathcal{C}_i\left(z\right), \mathcal{C}_j\right) , \tag{12}$$

where $\mathcal{C}_i\left(z\right)$ is the interest point set of camera i at zoom step z and $H\left(\mathcal{C}_i, \mathcal{C}_j\right)$ is the normalized joint entropy (6). From each of the resulting camera images (see Fig. 8 for an example) interest points are extracted. Now, the cameras repeat the first step, but in the opposite direction. The result of this procedure is a set of interest points $\mathcal{C}_i$ and $\mathcal{C}_j$ for each of the two cameras i and j. Given these interest point sets we want to evaluate the relative pose candidate. Therefore we calculate the probability of the relative poses (7). For each camera pair the relative pose candidate with the highest probability is selected.

5.2. *Actively Optimizing the Relative Poses*

Given the initial relative poses $\mathbf{R}_{i,j}, \mathbf{t}_{i,j}$ we optimize these poses by steering each camera pair in a way that it can easily establish new point correspondences.

As mentioned in Section 3.2, the biggest problem in finding correct point correspondences are projective influences. These influences depend on the relation between camera distance and scene distance and the difference in the viewing directions between the cameras. To reduce these influences we first rotate the two cameras in a way that their optical axes are aligned and adjust the zoom as described in Section 5.1. Fig. 8 gives an impression of the resulting camera images. Again, this procedure is repeated for the opposite direction and yields in an interest point set for each camera. Similar to the initial calibration we extract point correspondences and use these to estimate the relative pose. Since we expect the descriptors of two corresponding points to be very similar due to the high similarity of the camera images, we choose a stricter rejection threshold for the two nearest neighbors matching than in the initial calibration. The estimated relative poses are evaluated as described in Section 5.1. For each camera pair the reestimated relative pose will only be used if it has a higher probability than the initial relative pose.

5.3. *Active Estimation of the Translation Scale Factors*

At this point of our calibration we have for each camera pair i, j a relative pose $\mathbf{R}_{i,j}, \mathbf{t}_{i,j}^*$ and a probability of this pose $p\left(\mathbf{R}_{i,j}, \mathbf{t}_{i,j}^*\right)$. As mentioned in Section 3.1 we need to estimate consistent scales for all translations.

For the final calibration we use the uncertainty-based selection of relative poses described in Section 4, using the information of the relative pose (11) as edge weight. For each triangle of the triangle path the translation scales need to be estimated. Camera triangles with a low uncertainty are scaled first. We note that an inserted triangle does not need to be triangle connected to some prior triangle connected subgraph. Thus several single camera triangles can be inserted before some of the build a triangle connected subgraph which reduces the propagation of errors.

In order to estimate the scale factors in a camera triangle, traditional methods use either all three relative poses in the triangle (Section 4.3) or they try to establish point correspondences between all three camera images.[16,17] As described in Section 3.2, the problem of extracting correct point correspondences is very error-prone due to ambiguities and projective influences.

Our proposed method does not need any point correspondences at all and uses only two of the three relative poses. Instead we use active camera control and the image based probability distribution of (7). This reduces the number of required relative poses and totally avoids the chance of outlier point correspondences.

We want to scale the camera triangle (i, j, k), where $\mathbf{R}_{i,k}, \mathbf{t}_{i,k}$ has the highest uncertainty in the triangle and the relative pose graph has no edge between the vertices i and k. Now we simultaneously estimate the third relative pose $\mathbf{R}_{i,k}, \mathbf{t}_{i,k}$ and all translation scale factors of the triangle.

This is done by rotating camera i and k simultaneously around the plane normal of the camera triangle. In the beginning both cameras look into the direction defined by the translation $\mathbf{t}_{i,j}$. Now, we search for the rotation angle α that

$$\max_{\alpha} p\left(\mathbf{R}_{i,k}, \mathbf{t}_{i,k}\left(\alpha\right)\right) \text{ with } \mathbf{R}_{i,k} \stackrel{\text{def}}{=} \mathbf{R}_{j,k}\mathbf{R}_{i,j} \text{ and } \mathbf{t}_{i,k}\left(\alpha\right) \stackrel{\text{def}}{=} \mathbf{R}_{j,k}\mathbf{t}_{i,j} + \lambda_s \mathbf{t}_{j,k} , \quad (13)$$

where we assume $\|\mathbf{t}_{i,j}\|_2 = \|\mathbf{t}_{j,k}\|_2$, the scale factor $\lambda_s \stackrel{\text{def}}{=} \sin\left(\alpha\right) / \sin\left(\Pi - \alpha - \beta\right)$ arises from the law of sines and $\beta \stackrel{\text{def}}{=} \arccos\left(\left(\mathbf{t}_{i,j}^T \mathbf{t}_{k,j}\right) / \left(\|\mathbf{t}_{i,j}\|_2 \|\mathbf{t}_{k,j}\|_2\right)\right)$ is the angle between the translation vectors $\mathbf{t}_{i,j}$ and $\mathbf{t}_{k,j}$. The probability $p\left(\mathbf{R}_{i,k}, \mathbf{t}_{i,k}\left(\alpha\right)\right)$ is the probability of (7). There will be only one rotation angle α where the two

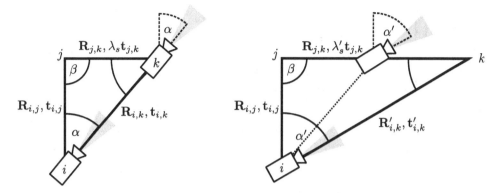

Fig. 9. A camera triangle (i, j, k). Cameras i and k rotate with angle α around the plane normal of the camera triangle. The scale λ_s depends on the angle α. There will be only one angle α where the optical axes of both cameras are aligned (left). At this point the triangle is correctly scaled. In all other cases the cameras will not look into the same direction and the scaling between the relative poses is incorrect (right).

cameras i and k look exactly into the same direction. For a clearer understanding the described relations are visualized in Fig. 9. The procedure is repeated in the opposite direction which results in estimating the inverse relative pose $\mathbf{R}_{k,i}, \mathbf{t}_{k,i}$. Again, we decide between these two poses based on their probability.

6. Experiments

6.1. *Experimental Setup*

In our experiments we use a multi-camera system consisting of five to six Sony DFW-VL500 cameras with a resolution of 640×480 pixels. Each camera is mounted on a Directed Perception PTU-46-17.5 pan-tilt unit. We use a slightly modified version of this pan-tilt unit which allows to mount the camera quite close to the intersection of the pan and tilt axes (Fig. 1, right). We test our method on a total of 37 calibrations with 6 different setups of the multi-camera system. An example setup can be found in Fig. 1 (left).

In order to generate ground truth we use the calibration software of the University Kiel[38] which uses a pattern based calibration method[7] with non-linear refinement. The intrinsic parameters of five zoom steps of each camera are estimated using the self-calibration method of.[39] The radial distortion of the images is corrected using the two parameter radial distortion model of.[40] We also estimate the hand-eye calibration[41] between camera and pan-tilt unit. All parameters of the calibration approaches are set to the values suggested by the authors.[25,28,35]

As explained in Section 5.3, we can only calibrate up to a common scale factor. In order to compare our calibration with the ground truth, we scale our calibration

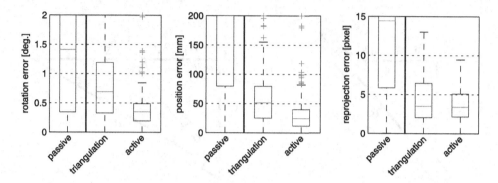

Fig. 10.   The rotation, position and reprojection errors using different extents of active calibration. We present the results of the completely **passive** approach,[25] a partially active approach which uses the active optimization and the image based probability measure but **triangulation** for the scale estimation, and the completely **active** approach. Some of the results are truncated.

result by the median of the factors $\|\mathbf{t}_{i,j}^{\text{GT}}\|_2/\|\mathbf{t}_{i,j}\|_2$ of all camera pairs i, j, where $\mathbf{t}_{i,j}^{\text{GT}}$ is the ground truth translation.

6.2. *Results*

We present the results of our experiments in Fig. 10. Three different calibration approaches are compared:

(1) The **passive** uncertainty based calibration method described in Section 4. We manually rotate the cameras to ensure that they share a common field of view for this passive method.
(2) An approach which uses the relative poses estimated by the active optimization of Section 5.2, the image based probability (7) for weighting a camera dependency graph, and estimates the scales using **triangulation** (Section 4.3).
(3) The completely **active** approach of Section 5.

For each of the calibration approaches the rotation error in degree, the position error in millimeters and the reprojection error in pixel is presented using box plots (the box depicts the 0.25 and 0.75 quantiles, the line in the middle is the median and crosses are outliers, for further details please refer to Ref. 42).

The results show that the (partially) active approaches outperform the **passive** approach. The difference between the approaches **triangulation** and **active** is only the scale estimation. While the reprojection error is quite similar, the **active** scaling approach achieves a clearly lower rotation and position error. For the **active** approach we achieve a median rotation error of 0.35 degree, a median position error of 24.8 millimeters and a median reprojection error of 3.42 pixels. This accuracy is sufficient to exploit the epipolar constraint to collect multi-camera point correspondences. These multi point correspondences can then be used for a final bundle adjustment[18] which would further increase the calibration accuracy.

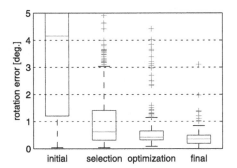 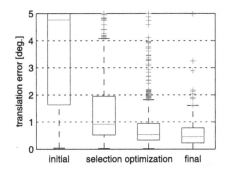

Fig. 11. The rotation and translation errors of the relative poses for the different steps of the active calibration approach. Some of the results are truncated.

Fig. 11 shows the rotation and the translation error in degree of the relative poses during the different steps of the **active** calibration approach. Note that it is not possible to present a position error in each step as long as the translation scales are unknown. The results show that each step refines the calibration accuracy and the number of outlier relative poses is reduced.

We also investigate the repeatability of the camera zoom by switching between the zoom steps and calibrating the intrinsic parameters several times. The calculated coefficients of variation for the intrinsic parameters lie in a magnitude of 10^{-3} which indicates good repeatability.

7. Conclusions

In this chapter we discussed the problem of extrinsic self-calibration of a multi-camera system. During such a self-calibration the cameras estimate their position and orientation in a common global coordinate system only from the images they record. No artificial calibration object or user interaction are needed. Many real life multi-camera systems consist of pan-tilt mounted cameras that are able to change their point of view by rotating. Hence, we also focused on the problems that arise with such active cameras and how active camera control can be used to improve the accuracy of extrinsic multi-camera calibration.

We started with a short explanation of the basic steps for relative pose estimation between two camera. These steps included the extraction of point correspondences between the cameras and the robust estimation of the relative pose based on these correspondences. Since the translation of the relative pose of a camera pair can only be estimated up to an unknown scale factor, consistent translation scale factors need to be estimated in order to completely calibrate a multi-camera system from pairwise relative poses. During this calibration only a small subset of the relative poses between the camera pairs is necessary. Thus, we presented an approach to select a set of camera pairs based on their uncertainty. This approach uses a graph where each edge represents the relative pose between two cameras. The edges are weighted by the uncertainty of the relative pose and we presented several uncertainty measure for relative poses. The algorithms finds minimal uncertainty triangle connected paths to each camera in the multi-camera system starting from a camera pair. Along these paths we estimated the missing translation scales by triangulation of the relative poses.

A totally different approach for scaling a camera triangle is enabled when using actively controlled pan-tilt cameras. We described our approach which is able to estimate the third relative pose and consistent scales for all poses in a camera triangle. In this context we also presented an image-based measure for the probability of a relative pose. This measure was also used to weight the relative pose graph. Furthermore we showed how to simplify the extraction of point correspondences by controlling the camera orientation and the camera's zoom in a way that reduces projective influences. In order to adjust the zoom of two cameras we also presented an image similarity measure which can also be used to detect the common field of views between two pan-tilt cameras.

In our experiments we compared our active approach with a purely passive approach. We showed that each step of our active calibration approach improves the calibration and reduces the number of outliers. For the final calibration we achieve a high accuracy with a median of 0.35 degree rotation error and 24.8 mm position error. Based on the median reprojection error of 3.42 pixels, multi-image point correspondences can easily be extracted which enables a further improvement by using bundle adjustment.

Acknowledgements

This research was partially funded by the Carl Zeiss Foundation (Carl-Zeiss-Stiftung).

References

1. A. D. Straw, K. Branson, T. R. Neumann, and M. H. Dickinson, Multi-camera real-time three-dimensional tracking of multiple flying animals, *Journal of the Royal Society: Interface.* **8**, 395–409, (2011).
2. J. Denzler. Active Vision. In *Principles of 3D Image Analysis and Synthesis*, pp. 131–140. Kluwer Academic Publishers, (2000).
3. J. Denzler and C. Brown, Information Theoretic Sensor Data Selection for Active Object Recognition and State Estimation, *IEEE Transactions on Pattern Analysis and Machine Intelligence (PAMI).* **24**(2), 145–157, (2002).
4. J. Denzler, M. Zobel, and H. Niemann. Information Theoretic Focal Length Selection for Real-Time Active 3-D Object Tracking. In *Proceedings of the International Conference on Computer Vision (ICCV)*, pp. 400–407, (2003).
5. F. Deinzer, C. Derichs, J. Denzler, and H. Niemann, A framework for actively selecting viewpoints in object recognition, *International Journal of Pattern Recognition and Artificial Intelligence.* **23**(4), 765–799, (2009).

6. M. Trummer, C. Munkelt, and J. Denzler. Online Next-Best-View Planning for Accuracy Optimization Using an Extended E-Criterion. In *Proceedings of IAPR 20th International Conference on Pattern Recognition (ICPR)*, pp. 1642–1645, (2010).

7. Z. Zhang. Flexible Camera Calibration by Viewing a Plane from Unknown Orientations. In *Proceedings of the International Conference on Computer Vision (ICCV)*, pp. 666–673, (1999).

8. Z. Zhang, A Flexible New Technique for Camera Calibration, *IEEE Transactions on Pattern Analysis and Machine Intelligence (PAMI)*. **22**(11), 1330–1334, (2000).

9. I. Kitahara, H. Saito, S. Akimichi, T. Ono, Y. Ohta, and T. Kanade. Large-scale Virtualized Reality. In *Proceedings of the IEEE Conference on Computer Vision and Pattern Recognition (CVPR), Technical Sketches*, (2001).

10. P. Baker and Y. Aloimonos. Complete Calibration of a Multi-camera Network. In *Proceedings of the IEEE Workshop on Omnidirectional Vision (OMNIVIS)*, pp. 134–144, (2000).

11. X. Chen, J. Davis, and P. Slusallek. Wide area camera calibration using virtual calibration objects. In *Proceedings of the IEEE Conference on Computer Vision and Pattern Recognition (CVPR)*, vol. 2, pp. 520–527, (2000).

12. T. Svoboda, H. Hug, and L. Van Gool. ViRoom–Low Cost Synchronized Multicamera System and Its Self-calibration. In *Proceedings of the Annual Symposium of the German Association for Pattern Recognition (DAGM)*, pp. 515–522, (2002).

13. J. P. Barreto and K. Daniilidis. Wide Area Multiple Camera Calibration and Estimation of Radial Distortion. In *Proceedings of the IEEE Workshop on Omnidirectional Vision (OMNIVIS)*, (2004).

14. T. Svoboda, D. Martinec, and T. Pajdla, A Convenient Multi-Camera Self-Calibration for Virtual Environments, *PRESENCE: Teleoperators and Virtual Environments*. **14** (4), 407–422, (2005).

15. R. Hartley and A. Zisserman, *Multiple View Geometry in Computer Vision*. (Cambridge University Press, 2003).

16. N. Snavely, S. Seitz, and R. Szeliski. Photo tourism: exploring photo collections in 3D. pp. 835–846, (2006).

17. T. Läbe and W. Förstner. Automatic relative orientation of images. In *Proceedings of the 5th Turkish-German Joint Geodetic Days*, (2006).

18. B. Triggs, P. McLauchlan, R. Hartley, and A. Fitzgibbon. Bundle adjustmenta modern synthesis. In *Proceedings of the International Workshop on Vision Algorithms*, vol. 1883, pp. 153–177. Springer, (1999).

19. H. Stewénius, C. Engels, and D. Nistér, Recent Developments on Direct Relative Orientation, *ISPRS Journal of Photogrammetry and Remote Sensing*. **60**(4), 284–294, (2006).

20. M. Brückner, F. Bajramovic, and J. Denzler. Experimental Evaluation of Relative Pose Estimation Algorithms. In *Proceedings of the International Conference on Computer Vision Theory and Applications (VISAPP)*, vol. 2, pp. 431–438, (2008).

21. D. Martinec and T. Pajdla. Robust Rotation and Translation Estimation in Multiview Reconstruction. In *Proceedings of the IEEE Conference on Computer Vision and Pattern Recognition (CVPR)*, pp. 1–8, (2007).

22. D. Martinec and T. Pajdla. 3D Reconstruction by Gluing Pair-Wise Euclidean Reconstructions, or "How to Achieve a Good Reconstruction from Bad Images". In *3DPVT '06: Proceedings of the Third International Symposium on 3D Data Processing, Visualization, and Transmission (3DPVT'06)*, pp. 25–32, (2006).

23. J. Vergés-Llahí, D. Moldovan, and T. Wada. A new reliability measure for essential matrices suitable in multiple view calibration. In *Proceedings of the International*

Conference on Computer Vision Theory and Applications (VISAPP), vol. 1, pp. 114–121, (2008).

24. D. Nistér, An efficient solution to the five-point relative pose problem, *IEEE Transactions on Pattern Analysis and Machine Intelligence (PAMI)*. **26**, 756–770, (2004).

25. F. Bajramovic and J. Denzler. Global uncertainty-based selection of relative poses for multi cameracalibration. In *Proceedings of the British Machine Vision Conference (BMVC)*, vol. 2, pp. 745–754, (2008).

26. S. Sinha and M. Pollefeys. Towards calibrating a pan-tilt-zoom cameras network. In *Proceedings of the IEEE Workshop on Omnidirectional Vision (OMNIVIS)*, (2004).

27. P. Chippendale and F. Tobia. Collective calibration of active camera groups. In *IEEE Conference on Advanced Video and Signal Based Surveillance*, pp. 456–461, (2005).

28. M. Brückner and J. Denzler. Active self-calibration of multi-camera systems. In *Proceedings of the Annual Symposium of the German Association for Pattern Recognition (DAGM)*, pp. 31–40. Springer (September, 2010).

29. D. G. Lowe, Distinctive Image Features from Scale-Invariant Keypoints, *International Journal on Computer Vision (IJCV)*. **60**(2), 91–110, (2004).

30. J. Matas, O. Chum, M. Urban, and T. Pajdla. Robust wide baseline stereo from maximally stable extremal regions. In *Proceedings of the British Machine Vision Conference (BMVC)*, pp. 384–393, (2002).

31. N. Dalal and B. Triggs. Histograms of oriented gradients for human detection. In *Proceedings of the IEEE Conference on Computer Vision and Pattern Recognition (CVPR)*, vol. 1, pp. 886–893, (2005).

32. M. A. Fischler and R. C. Bolles, Random Sample Consensus: a Paradigm for Model Fitting with Applications to Image Analysis and Automated Cartography, *Communications of the ACM*. **24**(6), 381–395, (1981).

33. P. Torr and A. Zisserman, MLESAC: A New Robust Estimator with Application to Estimating Image Geometry, *Computer Vision and Image Understanding*. **78**(1), 138–156, (2000). ISSN 1077-3142.

34. F. Bajramovic, M. Brückner, and J. Denzler. Using common field of view detection for multi camera calibration. In *Proceedings of the Vision Modeling and Visualization Conference (VMV)*, pp. 113–120, (2009).

35. M. Brückner, F. Bajramovic, and J. Denzler. Geometric and probabilistic image dissimilarity measures for common field of view detection. In *Proceedings of the IEEE Conference on Computer Vision and Pattern Recognition (CVPR)*, pp. 2052–2057, (2009).

36. C. Engels and D. Nistér. Global uncertainty in epipolar geometry via fully and partially data-driven sampling. In *ISPRS Workshop BenCOS: Towards Benchmarking Automated Calibration, Orientation and Surface Reconstruction from Images*, pp. 17–22, (2005).

37. D. MacKay, *Information Theory, Inference, and Learning Algorithms*. (Cambridge University Press, 2003).

38. I. Schiller. MIP - MultiCameraCalibration, (2010). http://mip.informatik.uni-kiel.de/tiki-index.php?page=Calibration, last visited on 16-11-2010.

39. F. Bajramovic and J. Denzler. Self-calibration with Partially Known Rotations. In *Proceedings of the Annual Symposium of the German Association for Pattern Recognition (DAGM)*, pp. 1–10, (2007).

40. J. Heikkila and O. Silvén. A Four-step Camera Calibration Procedure with Implicit Image Correction. In *Proceedings of the IEEE Conference on Computer Vision and Pattern Recognition (CVPR)*, pp. 1106–1112, (1997).

41. R. Y. Tsai and R. K. Lenz, A new technique for fully autonomous and efficient 3d robotics hand/eye calibration, *IEEE Transactions on Robotics and Automation.* **5**(3), 345–357, (1989).

42. R. McGill, J. Tukey, and W. A. Larsen, Variations of Boxplots, *The American Statistician.* **32**, 12–16, (1978).

CHAPTER 3.7

A PERFORMANCE EVALUATION OF ROBOT LOCALIZATION METHODS IN OUTDOOR TERRAINS

Ehsan Fazl-Ersi and John K. Tsotsos

Department of Computer Science and Engineering, York University,
Toronto, M3J 1P3, Ontario, Canada
{efazl, tsotsos}@cse.yorku.ca

This chapter presents an overview of two common processes involved in most visual robot localization techniques, namely data association and robust motion estimation, and compares the performance of the available state-of-the-art solutions for each of them, in the context of outdoor robot localization, where the robot is subject to 6-DOF motions. For data association, where the task is to match different sets of observations over time, we compare the performance of different image feature detectors and descriptors to determine which is the most appropriate to use in the context of 6-DOF outdoor localization. For motion estimation, where the goal is to compute the transformation that would bring each observation into the best alignment with its match, we perform a comparative study on common fitting methods to find out which is best for approximating the robot pose that is consistent with as many matched landmarks as possible. Our experiments with different combinations of data association and motion estimation techniques show the superiority of the Hessian-Affine feature detector and the SIFT feature descriptor for data association, and the Hough Transform for robust motion estimation.

1. Introduction

One of the fundamental requirements for an autonomous mobile robot is localization, i.e., the capability of knowing where it is located within the environment. Robots should be able to localize themselves in order to navigate in the environment, compute a path to a target destination, and recognize that the target destination has already been reached. A recent focus of localization research is the use of vision sensors (i.e., camera) due to their advantages over laser, ultrasonic and sonar range finders. Cameras are information rich, relatively inexpensive and easily available. Additionally, 3D information is simple to obtain, either immediately with stereo vision, or in a delayed manner by triangulating over several frames in monocular configurations. However, visual localization is more challenging particularly in outdoor environments where the robot is subject to six degree-of-freedom (DOF) motion.

There are two types of localization: local and global. Local localization, also known as pose maintenance, aims at determining the current pose of the robot relative to its previous pose(s) by estimating the local motion of the robot. Some successful examples of local localization techniques are the works of Se et al. [1] and Klippenstein et al. [2]. Global localization on the other hand, aims at estimating the pose of the robot relative to a global metric map, when no prior knowledge about its pose and motion is available. The works of Se et al. [3] and Siagian et al. [4] are examples of global localization techniques.

Although different techniques are used for local and global localization, they usually share two common steps, *data association* and *robust motion estimation*.

Data association involves matching the current set of observations with the past ones. In visual localization, observations are usually natural 3D landmarks in the environment, identified by feature extraction methods from the images taken by the robot. Given that there are many feature detection and description techniques available to identify and characterize features in images, the main questions are which detector and descriptor are the most appropriate to be used in an outdoor localization system, and how much impact the choice among these methods can have on the accuracy of localization. Although several comparative studies have already been performed on feature detectors and descriptors (e.g., [5] and [6]), they are mainly in the context of object recognition or use simple planar test scenes for evaluation which are not comparable to the dynamic complex scenes encountered in outdoor robotic experiments.

Motion estimation involves computing the transformation that would bring each observation into the best alignment with its match. In local localization, the estimated motion is relative to the previous pose of the robot, while in global localization, it is relative to the origin of the world coordinate (map). Each three pairs of matched 3D landmarks can be used to estimate robot's pose. However, the poses estimated from different pairs of matches may not be the same (or even close). This is mainly due to the presence of outliers in the result of data association. Therefore, a robust fitting method should be applied to approximate the pose that is consistent with as many matched landmarks as possible. Although almost all visual localization systems rely on a robust fitting approach for consistent pose estimation, we know of no comparative study on these techniques in the context of 6-DOF outdoor localization.

The goal of this chapter is to present an overview of data association and motion estimation and compare different solutions available for each. More specifically, in this study, comparisons are carried out for three feature detectors (i.e., Hessian-Affine [5], DoG [7] and MSER [8]), three feature descriptors (i.e., SIFT [7], Cross-Correlation and Moments [9]), and three robust motion estimation solutions, based on RANSAC [10], Hough Transform [11], and Least Squares [12] fitting approaches, in the context of outdoor 6-DOF localization.

The remainder of this chapter is organized as follows. In Sections 2 and 3, we overview the problems of data association and robust motion estimation and briefly describe the techniques used in our comparative study. Section 4 presents our experimental setup, evaluation criteria, and the result of our comparisons. Finally, in Section 5, we conclude the chapter and discuss directions for future work.

2. Data Association

As mentioned in the introduction, one of the most challenging tasks in visual localization is the data association process, which involves correctly identifying which features in images are the projections of a common scene point. Successful data association requires the extracted features from the image frames to be distinctive, and robust against viewpoint variations and changes in lighting. Traditional approaches to data association, such as those based on the widely used Harris detector [13] and KLT tracker [14], are sensitive to scale and orientation of the images and therefore are not suited for localization in rough outdoor terrains in which the robot (and therefore the camera) is subject to six degree-of-freedom motion. Therefore, it is necessary to use techniques that are invariant to scale, rotation and even affine changes (to compensate for view-point variation). In this study, we evaluate several invariant feature detectors and descriptors that have been successfully used for several vision-based applications (e.g., object recognition). In the following, the local detectors and descriptors used in our comparisons are briefly described.

2.1. *Detectors*

2.1.1. *Difference-of-Gaussian (DoG)*

Lowe [7] has proposed a method to identify locations in image scale space that are invariant with respect to image scaling and rotation, and are minimally affected by noise and small distortions. A scale space is built for the image, $I(x)$, using a difference-of-Gaussian (DoG) function, $D(x, \sigma)$. It can be efficiently obtained from the difference of two Gaussians at adjacent scales, separated by a factor of k:

$$D(x, \sigma) = \big(G(x, k\sigma) - G(x, \sigma)\big) * I(x) \qquad (1)$$

Features are localized at locations x that are extrema both in the image plane and along the scale coordinate of $D(x, \sigma)$ function. Such points are detected by comparing the DoG of each pixel with its 26 neighbours in 3×3 regions at the current and adjacent scales. Since the scale coordinate is only sampled on discrete levels, the responses at neighbouring scales are interpolated in order to improve the accuracy of scale selection. In practice, this is done by fitting a second order polynomial to each candidate point and its two closest neighbours.

DoG detector, which mainly reacts to blob-like structures, has been successfully used in many vision applications including several localization systems (e.g., [1] and [3]). Fig. 1.b shows the result of applying this detector on a sample image from our database.

2.1.2. *Maximally Stable Extremal Region (MSER)*

Maximally stable extremal region detector was developed by Matas et al. [8] and reported in an extensive comparative study [5] as the best detector in the context of matching. This detector identifies regions in the image where local binarization is stable over a large range of thresholds. MSER detection is closely related to thresholding, since each extremal region is a connected component of a thresholded image. The algorithm evaluates the stability of extremal regions by testing all thresholds. Those regions corresponding to thresholds for which the rate of change of area as a function of change of threshold is at a local minimum, are selected as maximally stable extremal regions (MSERs).

The detected regions are robust against rotation, scale, and certain amount of view-point changes. The result of applying this detector on a sample image from our database is shown in Fig. 1.c.

2.1.3. *Hessian-Affine*

Hessian-Affine detector was developed by Mikolajczyk et al. [5] and was reported as the second best detector in the context of matching, in an extensive comparative study [5]. This detector localizes points x in space at the local maxima of the Hessian determinant, and in scale at the local maxima of the

Fig. 1. Result of applying the DoG (b), the MSER (c), and the Hessian-Affine (d) detectors on a sample image (a). For the DoG detector, the length and the direction of each arrow show the scale and orientation of the feature, respectively. For the other two detectors, each ellipse shows the shape of the feature.

Laplacian-of-Gaussian. Hessian matrix H is defined as:

$$H(x, \sigma_D) = \begin{bmatrix} I_x^2(x, \sigma_D) & I_{xy}(x, \sigma_D) \\ I_{xy}(x, \sigma_D) & I_y^2(x, \sigma_D) \end{bmatrix} \tag{2}$$

The second derivatives used in this matrix are computed with Gaussian kernels of scale σ_D (differentiation scale) and give strong responses to blob-like structures for which the signal variations lie on the blob boundaries [5]. Laplacian-of-Gaussian provides higher localization accuracy in scale-space than DOG method.

Given the set of initial points extracted at their characteristic scales, the iterative estimation of elliptical affine region [15] is applied to determine the affine neighborhood of each point. Similar to MSER detector, Hessian-Affine approach is also robust against common image variations. The result of applying this detector on a sample image from our database is shown in Fig. 1.d.

2.2. *Descriptors*

For each detected feature, a support region is created using the scale, orientation and shape (elliptical or circular) of the feature, to determine the pixels which contribute to the descriptor computation. All support regions are then mapped to a circular region of constant radius (i.e., 20 pixels in our experiments, similar to [6]) to obtain scale and affine invariance.

2.2.1. *Cross Correlation*

This technique is the simplest way of describing image regions, where a vector consisting of pixel intensities is created for each region (with a length equal to the number of pixels within the region), normalized with its mean and standard deviation. Cross-correlation can then be used to compute a similarity score between two description vectors. This descriptor has been widely used in many vision-based applications, including several localization methods (e.g., [2]).

2.2.2. *Scale Invariant Feature Transform (SIFT)*

SIFT is a gradient distribution based descriptor developed by Lowe [7]. Gradient distribution based descriptors have been widely used for many vision applications including localization and mapping. In many comparative studies in the context of matching and recognition (e.g., [6]) this type of descriptor is reported as the most accurate one. SIFT is represented by a 3D (two locations and one orientation) histogram of gradient locations and orientations. The contribution of each pixel to the location and orientation bins is weighted by its gradient magnitude. The quantization of gradient locations and orientations makes the descriptor robust to small geometric distortions and small errors in the region detection. The descriptor produces a 128-dimensional vector for each feature. Euclidean distance is used to compare SIFT description vectors. Fig. 2 shows the result of feature matching when SIFT descriptors are used with different feature detection methods.

2.2.3. *Moment Invariants*

Generalized moment invariants have been introduced by Van Gool et al. [9] to describe the multispectral nature of the image data. In an extensive comparative study in the context of recognition [6], Moments were reported as the best low-dimensional descriptor. For discrete cases such as a digitized image $I(x, y)$, the Moments of order $(p + q)$ are defined as:

$$m_{pq} = \sum_x \sum_y x^p y^q I(x, y) \tag{3}$$

for $p, q \geq 0$. The *central moments* are then defined as:

$$\mu_{pq} = \sum_x \sum_y \left(x - \frac{m_{10}}{m_{00}}\right)^p \left(y - \frac{m_{10}}{m_{00}}\right)^q I(x, y) \tag{4}$$

and the normalized central moments are obtained by:

$$\eta_{pq} = \frac{\mu_{pq}}{\mu_{00}} \tag{5}$$

Normalized central moments are computed up to the second order, resulting in 20-dimensional description vectors. Mahalanobis distance is used to compare Moment description vectors.

3. Motion Estimation

Given a set of current 3D landmarks P_t and their correspondences from past observations, P_{t-1}, a 6-DOF motion, defined by the 3×3 rotation matrix R and the 3×1 translation vector T, maps each landmark P onto its correspondence P':

$$P' = RP + T \tag{6}$$

$T = (\Delta X, \Delta Y, \Delta Z)$ indicates the translations in X, Y and Z axes and R corresponds to rotations around each of these axes: *yaw*, $\Delta \alpha$, rotation around X, *pitch*, $\Delta \theta$, rotation around Y, and *roll*, $\Delta \beta$, rotation around Z. Given R and T, the current location (X_c, Y_c, Z_c) and orientation $(\alpha_c, \theta_c, \beta_c)$ of the robot in world coordinate can be determined by the following equations respectively:

$$\begin{bmatrix} X_c \\ Y_c \\ Z_c \end{bmatrix} = -R^{-1}T + \begin{bmatrix} X_p \\ Y_p \\ Z_p \end{bmatrix} \tag{7}$$

(a)

(b)

(c)

Fig. 2. (a), (b) and (c) show the feature matching results on two consecutive image frames, when the sift descriptor is used with the Hessian-affine, DoG and MSER detectors, respectively. For clarity purposes, only 50% of the matches (with the highest similarity scores) are shown for each pair. In each pair of images, yellow circles show the location of the features and the blue lines connect the corresponding features.

$$\begin{bmatrix} a_c \\ \theta_c \\ \beta_c \end{bmatrix} = \begin{bmatrix} a_p \\ \theta_p \\ \beta_p \end{bmatrix} + \begin{bmatrix} \Delta\alpha \\ \Delta\theta \\ \Delta\beta \end{bmatrix} \tag{8}$$

In the above equations, $(X_p, Y_p, Z_p, \alpha_p, \theta_p, \beta_p)$ refers to the previous known pose of the robot. In global localization, it is usually equal to the origin of the world coordinate $(0,0,0,0,0,0)$.

It is well known that R and T can be estimated from three non-collinear matched landmarks [16]. However, in most cases there exist many more matched landmarks than three. Furthermore, poses estimated form different triplets of matched landmarks may not be the same due to the presence of outliers in the result of data association. Therefore, a robust fitting method should be applied to approximate the pose that is consistent with as many matched landmarks as possible. In this chapter, we evaluate robust fitting methods that have been widely used for vision-based applications (including robot localization). A localization algorithm based on each of these techniques is described in the following, and the comparative results are presented in Section 4.

3.1. *Least Squares*

Least Squares is one of the most commonly used fitting strategies employed in many localization methods (e.g., [1] and [3]). In a system of equations of the form $Ax = b$, where A is an $m \times n$ matrix (m for the number of equations and n for the number of unknowns), if a solution does not exist, then in many cases a vector x can be found that is closest to providing a solution to the system. In other words, we search for an x, such that $\|Ax - b\|$ is minimized. Such an x is known as the least squares solution to the over-determined system [12].

In the context of localization, given the set of matched landmarks, rather than solving directly for the 6-DOF robot's motion, $s = (\Delta X, \Delta Y, \Delta Z, \Delta\alpha, \Delta\theta, \Delta\beta)$, a vector of correction x is computed to be subtracted from an initial motion estimate s_0:

$$s = s_0 - x \tag{9}$$

s_0 in local localization is usually computed from odometry information (if available) and in global localization is either a null vector or an initial estimate computed by other techniques (e.g., Hough Transform, RANSAC). In our experiments with Least Squares as the (sole) robust motion estimator, we set s_0 to a null vector.

x can be obtained by minimizing the vector of error measurements ε, i.e., the re-projection error of 3D landmarks. For a known camera calibration matrix K, ε is defined as:

$$\varepsilon = \begin{bmatrix} \varepsilon_0^t \\ \varepsilon_1^t \\ \vdots \\ \varepsilon_k^t \end{bmatrix} = \begin{bmatrix} p_t^0 - K(RP_{t-1}^0 + T) \\ p_t^1 - K(RP_{t-1}^1 + T) \\ \vdots \\ p_t^k - K(RP_{t-1}^k + T) \end{bmatrix} \tag{10}$$

where p_t^k is the position of the feature point in the reference frame (right or left) corresponding to 3D landmark P_t^k.

3.2. Hough Transform

One way to cluster the matched landmarks that are in agreement with the same estimated motion is to record the estimations from all the possible triples of landmarks, and then look for the motion that obtains the maximum votes. This quite general technique is known as the Hough Transform [11]. Considering that there might be many correspondences (sometimes hundreds) between the current set of observations and the previous ones, evaluating all possible triplets computationally would be very expensive. Therefore, we only use the N best matches (with highest matching scores) for this approach, where N can be chosen according to the available computational resources.

A six dimensional search space is used in our Hough Transform localization approach, three for location and three for orientation. We use relatively broad bin sizes of 5 degrees for yaw, pitch and roll orientations, and $5cm$ for X, Y, and Z directions. Also, in order to save memory and speed up the search process we use an implementation similar to [7] based on hash table.

For each possible triplet of non-collinear matched landmarks, a motion is estimated and a vote is placed in the corresponding Hough bins. Votes are also placed in the neighboring bins to avoid the problem of boundary effects in bin assignments. Our Hough-based approach is different from the one proposed by Se et al. [3] which computes all the possible motions from each *single* pair of matched landmarks.

The bin with the maximum number of votes is selected as the initial estimate for the robot's motion and used by Least Squares technique (Eq. 10) to refine the motion and remove outliers (matched landmarks with high re-projection errors). Least Squares is re-applied on the inlier matches to obtain a more accurate motion.

Applying Least Squares on the result of Hough Transform helps to compensate for the use of broad bin sizes used in our implementation.

3.3. *RANSAC*

An alternative to the above technique is the RANSAC approach, which searches for a random subset of matched landmarks that leads to a motion on which many of the corresponding landmarks agree. RANSAC has been a commonly used approach for robust fitting for many years, and has been applied to many vision applications, including robot localization. The algorithm starts by randomly selecting three pairs of non-collinear matched landmarks, and estimating a hypothesis pose. All matched landmarks are evaluated based on their re-projection errors (Eq. 10) and those with errors less than a predefined threshold (e.g., 3 pixels in our experiments) are counted as supports for the hypothesis motion. The random selection, motion estimation, and support seeking steps are repeated for m_{ransac} times or until a hypothesis is found which is consistent with n_{ransac} percent of all matched landmarks. The estimated motion with maximum support is then selected and used by a Least Squares technique for the inliers to obtain a better estimate for the motion. In our experiments, we set m_{ransac} and n_{ransac} to 1000 and 70, respectively.

4. Experiments

For our experiments, we use a publicly available database of stereo images [17] taken from a $9Km$ trajectory in Little Bit, Pennsylvania. Each pair of stereo images is tagged with a robot pose measured by an RTK GPS, which is accurate to several cm in XY and to $10cm$ in Z [17]. All images are gray-scale, with 384×512 pixels.

The database is very large, containing around 46,000 pairs of stereo images. We selected a set of 50 test samples from the database, each consisting of two ordered stereo pairs. The first stereo pair and the GPS pose tagged to it are used to estimate the robot's pose for the second stereo pair, and then the difference between the estimated pose and the GPS pose tagged to the second stereo pair forms the localization error.

Test samples are selected so that they cover most common localization challenges. Several samples contain less textured images, several with large motion between the first and the second stereo pairs, and some with moving objects (e.g., vehicle) in the scene.

4.1. Results and Discussions

For detection and description of invariant feature points, we used the binary codes available at [18], with the default parameters proposed by the authors of each technique. We have evaluated all possible combinations of detection, description and robust motion estimation techniques, resulting in 27 localization algorithms.

The first criterion we used to evaluate localization algorithms is the number of times they fail to provide a response, which could be due to two main reasons[a]: (i) the detector fails to extract an adequate number of features from images, which is usually the case for less textured images or when the images contain feature types different from those the detector responds to, (ii) the data association approach (detector + descriptor) cannot find enough matched landmarks (at least 3) between the two stereo images, which might happen when there is relatively large motion.

In our experiments, the highest number of failures was produced by the (MSER, Moment) combination, which failed to provide a response for 27% of the test samples. All combinations of DoG and Hessian-Affine detectors provided a response to all test cases. Among descriptors, the highest number of failures were obtained by Moments, while the other two descriptors provided a response to all test cases, when used in combination with DoG and Hessian-Affine detectors, and similarly failed for 21% of the cases when used with MSER detector.

Although MSER is a very powerful detector and achieved the highest accuracy in several comparative studies (e.g., [5]) in the context of matching and recognition, it mainly responds to structured scenes. Therefore it did not work well when applied to the outdoor unstructured scenes used in our experiment.

In order to evaluate the localization accuracy of each of the algorithms, two error measurements were used: (i) position error, as the Euclidean distance in 3D (X, Y, Z) between the ground truth robot's position and the estimated one, and (ii) heading error, as the average of differences between the ground truth yaw, pitch, and roll of the robot (obtained from IMU reading) and the estimated ones.

Fig. 3 shows the performance of different detectors, descriptors and robust motion estimation techniques in our experiments. Note that only those test samples were used in error calculation for which all algorithms successfully provided a response.

---

[a] In our experiments, we found only the detectors and the descriptors responsible in failing or succeeding of the localization algorithms to provide a response, no matter which robust motion estimation approach being used.

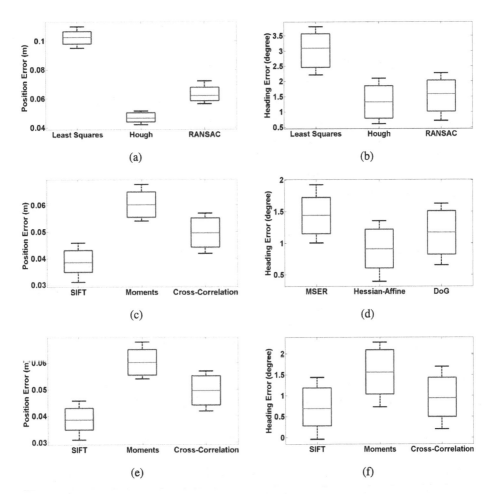

Fig. 3. Average position and heading errors for different motion estimation, feature detection, and feature description techniques. The errors for each motion estimation technique are computed by averaging the errors over test samples, detectors, and descriptors. As can be seen in (a) and (b), algorithms based on Hough Transform achieved lower average errors in comparison to the algorithms based on Least Squares and RANSAC. (c) and (d) show the average errors for different feature detection technique. Errors are computed by averaging the errors of Hough-based algorithms over test samples, and different descriptors. Similarly, for descriptors, (e) and (f), errors are computed by averaging the errors of Hough-based algorithms over test samples, and different detectors.

With regards to robust motion estimation, localization algorithms based on the Hough Transform achieved the best average performance both in terms of position and heading estimation, followed by those based on RANSAC and Least Squares, respectively. The main reason for the success of the Hough Transform could be the fact that it carries out a complete and exhaustive search over all possible triplets of the *available*[b] matched landmarks.

On the other hand, the low accuracy of the algorithms based on Least Squares is due to the fact that Least Squares [12] is extremely sensitive to outliers and noise, therefore if the number of outliers is relatively large (which might be the case for many test samples), it fails to find a good model.

Overall, the lowest errors were achieved by the (Hessian-Affine, SIFT, Hough) combination, with a position error of $3.1cm$ and a heading error of 0.72 degrees (averaged over all test samples). Fig. 4 shows the performance of this algorithm for several test cases.

Using the algorithm with lowest localization error, we performed a local localization experiment on a small segment of the database [17], consisting of 1000 successive stereo images with a trajectory length of $\sim310m$. The ground truth pose tagged to the first stereo pair was used as the initial pose. Starting from the second stereo pair, the algorithm estimated the robot's motion between each stereo pair and its previous one and cumulatively updated the robot's pose. Fig. 5 shows the result. The position and heading errors between the pose estimated for the last stereo pair and its ground truth pose were $\sim6m$ and ~2 degrees, respectively. For a $310m$ trajectory with several turnings, the results are promising as the localization algorithm lacked two main components, *error modeling* [1] and *bundle adjustment* [16]. Furthermore, unlike many existing local localization approaches (e.g., [1] and [2]), no odometry, IMU (Inertial Measurement Unit) or GPS data were used to provide initial estimates for the robot's local motions.

The computational time of the localization algorithms depends on the computational complexity of the algorithms used for detection, description and robust motion estimation. Among the detectors, the fastest one is the MSER detector with a time complexity of $O(n \log \log n)$ [5], where n is the number of pixels. Although Hessian-Affine and DoG are very similar in terms of time complexity for detecting features (i.e., both of $O(n)$), Hessian-Affine performs slower as it involves a shape adoption process to make the regions robust against affine changes.

For description computation, it is obvious that the Cross-Correlation based descriptor is the fastest, followed by the Moments and SIFT. The computational time for feature matching directly depends on the dimensionality of the descriptors, fastest for the Moments and slowest for the Cross-Correlation based descriptions.

---

[b] Note that in most cases we only use a subset of the matched landmarks not all of them.

PE: 1.37*cm* - HE: 0.41°   PE: 0.83*cm* - HE: 0.23°

PE: 2.10*cm* - HE: 0.55°   PE: 3.01*cm* - HE: 0.93°

PE: 0.75*cm* - HE: 0.18°   PE: 2.76*cm* - HE: 0.76°

Fig. 4. Localization result - Position Error (PE) and Heading Error (HE) - for some of the test samples. Each image shows the matching features that are used to estimate the motion. The red circles show the location of the image features in the current frame (the image shown) and the blue lines show how the location of the features moved from the previous frame.

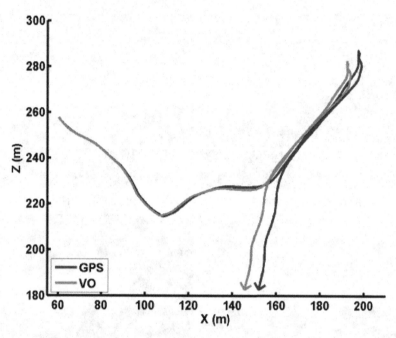

Fig. 5. Localization result on a rough terrain from a publicly available database [17]. In this experiment, the length of the trajectory is $310.23m$ and the robot has yaw, roll and pitch movements of $[-135°, +95°]$, $[-5°, +2°]$ and $[-12°, +12°]$, respectively. The figure shows the trajectory of the robot, computed by accumulating the separate motions estimated by the (Hessian-Affine, SIFT, Hough) method between consecutive frames, in comparison to the ground truth data (GPS and IMU). Although it is well understood that trajectory estimation from motion accumulation results in significant positional and angular drifts after certain amount of movement, the small error of the system (~$6m$ positional drift and ~$2°$ angular drift after traveling for $306m$ on a rough terrain) reveals the robustness of the method in estimating large 6-DOF motions.

For robust motion fitting approaches, given the parameters we used in our experiments, the computational time for the Least Squares technique was significantly lower than the other approaches. Between RANSAC and Hough Transform, RANSAC usually performed much faster depending on the number of matched landmarks available. This was expected, since usually more motion estimation operations are performed in Hough Transform as it carries out a complete and exhaustive search over all possible triplet of the matched landmarks.

In our local localization experiment with (Hessian-Affine, SIFT, Hough) algorithm on a standard PC with a 2.4 GHz CPU, the computational time was $1.1s$ in average.

5. Conclusions

In this chapter, we have reviewed several techniques for data association and robust motion estimation for visual localization. A comparative study on 27 localization algorithms, developed by combining different detectors, descriptors, and robust motion estimators, showed the superiority of the Hessian-Affine feature detector and the SIFT feature descriptor for data association, and the Hough Transform for robust motion estimation.

A potential direction for future work is to extend the comparative study and evaluate more detectors, descriptors and robust fitting techniques in the context of localization. Furthermore, considering that different detectors respond to different types of features in the scene, we would like to explore the impact of using complementary detectors for global localization in structured environments, where usually different types of features are available in the environment.

References

1. S. Se et al. Mobile Robot Localization and Mapping with Uncertainty Using Scale-Invariant Visual Landmarks. *International Journal of Robotics Research*, vol. 21, No. 8, pp. 735-758 (2002).
2. J. Klippenstein et al. Feature Initialization for Bearing Only Visual SLAM Using Triangulation and the Unscented Transform. *Proceedings of IEEE International Conference on Mechatronics and Automation*, pp. 157-164 (2007).
3. S. Se et al. Vision-based Global Localization and Mapping for Mobile Robots. *IEEE Transactions on Robotics*, Vol. 21, No. 3, pp. 364-375 (2005).
4. C. Siagian and L. Itti. Rapid Biologically-Inspired Scene Classification Using Features Shared with Visual Attention. *IEEE Transactions on Pattern Analysis and Machine Intelligence*, Vol. 29, No. 2, pp. 300-312 (2007).
5. K. Mikolajczyk et al. A Comparison of Affine Region Detectors. *International Journal of Computer Vision*, Vol. 65, No. 1, pp. 43-72 (2006).
6. K. Mikolajczyk and C. Schmid. A Performance Evaluation of Local Descriptors. *Proceedings of International Conference on Computer Vision and Pattern Recognition*, pp. 257-263 (2003).
7. D. Lowe. Distinctive Image Features from Scale-Invariant Keypoints. *International Journal of Computer Vision*, Vol. 60, No. 2, pp. 91-110 (2004).
8. J. Matas et al. Robust Wide Baseline Stereo from Maximally Stable Extremal Regions. *Proceedings of British Machine Vision Conference*, pp. 384-396 (2002).
9. L. Van Gool et al. Affine/Photometric Invariants for Planar Intensity Patterns. *Proceedings of European Conference on Computer Vision*, pp. 642-651 (1996).
10. N. Fischler and R.C. Bolles. Random Sample Consensus: A Paradigm for Model Fitting with Application to Image Analysis and Automated Cartography. *Communication of the ACM*, Vol. 24, No. 6, pp. 381-395 (1981).

11. P.V.C. Hough. Machine Analysis of Bubble Chamber Pictures. Proceedings of International Conference on High Energy Accelerators and Instrumentation, pp. 554-556 (1959).

12. P.J. Rousseeuw and A.M. Leroy. Robust Regression and Outlier Detection, John Wiley and Sons, New York (1987).

13. C. Harris and M. Stephens. A Combined Corner and Edge Detector. *Proceedings of the 4th Alvey Vision Conference*, pp. 147-151 (1988).

14. B.D. Lucas and T. Kanade. An Iterative Image Registration Technique with an Application to Stereo Vision. *Proceedings of International Joint Conference on Artificial Intelligence*, pp. 674-679 (1981).

15. T. Lindeberg and J. Gårding. Shape-Adapted Smoothing in Estimation of 3D Depth Cues from Affine Distortions of Local 2D Brightness Structure. *Proceedings of European Conference on Computer Vision*, pp. 389-400 (1994).

16. R. Hartley and A. Zisserman. Multiple View Geometry in Computer Vision. Cambridge University Press (2003).

17. K. Konolige et al. Large-Scale Visual Odometry for Rough Terrain. *Proceedings of International Symposium on Research in Robotics* (2007).

18. http://www.robots.ox.ac.uk/~vgg/research/affine/detectors.html

INDEX

Call for Book Proposals

Direct your submissions to *cchen@umassd.edu*

Series in

Computer
Vision

Series Editor

C.H. CHEN

Professor Emeritus
Department of Electrical and Computer Engineering
University of Massachusetts Dartmouth, USA

In recent years, there has been significant progress in computer vision in theory and methodology and enormous advancement on the application front, accompanied by rapid progress in vision systems and technology.

It is hoped that this book series in computer vision can capture most of the important and recent progress and results in computer vision. The target audiences cover researchers, engineers, scientists and professionals in many disciplines including computer science and engineering, mathematics, physics, biology, and medical areas, etc.

Topics include (but not limited to)

- Computer vision theory and methodology (algorithms)

- Computer vision applications in biometrics, biomedicine, etc biometrics, biomedicine, etc

- Robotic vision

- New vision sensors, software and hardware systems and technology

World Scientific
www.worldscientific.com

Imperial College Press
www.icpress.co.uk

Preferred Publisher of Leading Thinkers